SCHAUM'S

PRINCIPLES AND PROBLEMS OF

ELEMENTARY ALGEBRA

by BARNETT RICH, Ph.D.

Chairman, Department of Mathematics

BROOKLYN TECHNICAL HIGH SCHOOL

NEW YORK CITY

SCHAUM'S OUTLINE SERIES

McGRAW-HILL PUBLISHING COMPANY

New York St. Louis San Francisco Auckland Bogotá Caracas Hamburg Lisbon London Madrid Mexico Milan Montreal New Delhi Oklahoma City Paris San Juan São Paulo Singapore Sydney Tokyo Toronto

ISBN 07-052244-8

35 36 37 38 39 40 41 42 43 44 45 SH SH 89

Preface

The chief objective and central purpose of this book is *TO PROVIDE MAXIMUM HELP FOR STUDENTS AND MAXIMUM SERVICE FOR TEACHERS.*

PROVIDING HELP FOR STUDENTS:

The learning of algebra is of crucial importance. Without a solid foundation in algebra, all subsequent work in mathematics is seriously impaired. This book has been designed to improve the learning of algebra far beyond that of the typical and traditional book in the subject. Those who need help in algebra will find this text most useful for these reasons:

(1) *Learning Each Rule, Formula and Principle.*
Each important rule, formula and principle is stated in simple language, is made to stand out in distinctive bold type and is applied to one or more sets of solved problems.

(2) *Learning Each Procedure.*
Each algebraic procedure is developed step-by-step, with each step immediately applied to problems alongside of the procedure.

(3) *Learning Each Set of Solved Problems.*
Each set of solved problems is used to clarify and illustrate a rule or principle. The particular character of each such set is indicated by a title. There are more than 2700 *CAREFULLY SELECTED AND FULLY SOLVED PROBLEMS!*

(4) *Learning Each Set of Supplementary Problems.*
Each set of supplementary problems provides further application of a rule or principle. A guide number with each such set refers a student needing help to the related set of solved problems. There are over 3300 *ADDITIONAL RELATED SUPPLEMENTARY PROBLEMS.* Each of these contains its required answer and, where needed, further aids to solution.

(5) *Understanding Each Chapter.*
Understanding is emphasized throughout the book. Introductory units in each chapter are used for this purpose. These units clarify the underlying ideas of the chapter which are essential for such understanding.

(6) *Enriching the Learning of Algebra.*
The work of students can be considerably enriched by the challenging materials and problems contained in this text. These enrichment materials look ahead to the 10th and 11th years of mathematics and substantially enlarge the scope of the standard traditional course of study in algebra.

(7) *Accelerating the Learning of Algebra.*
The readiness and willingness of an able student will enable him to accelerate the algebra course and, through the use of this text, accomplish the work of the standard course of study in much less time.

PROVIDING SERVICE FOR TEACHERS:

Teachers of algebra will find this text most useful for these reasons:

(1) *Teaching Each Chapter.*

Each chapter has a central unifying theme. Each chapter is divided into two to twelve major units which support the central theme of the chapter. In turn, these units or chapter divisions are arranged in a graded sequence for greater teaching effectiveness.

(2) *Teaching Each Unit.*

Each of the chapter divisions is a master lesson. Each such unit contains the materials needed for the full teaching development of those principles and procedures which are related to the unit.

(3) *Teaching Sets of Solved Problems.*

Using this text, the teacher and the class can completely cover many more solved problems. In so doing, the class will gain greater understanding of the way in which each principle applies in many varied situations. In order that lesson development may proceed from the simple to the more difficult, the solved problems have been carefully graded from one set to the other and also within sets.

Furthermore, the text seeks to present the best solution or solutions of each problem. Where more than one method of solution is equally valuable, each such method is included. These methods have been selected on the basis of the author's extensive practical classroom experience.

(4) *Preparing Homework Assignments.*

The preparation of homework assignments is facilitated because the supplementary problems are related to the sets of solved problems. It is suggested that in class the underlying principle and the major steps used in solving problems be given greatest attention. At home, students can complete the solved problems and then proceed to do those supplementary problems which are related to the solved ones.

OTHERS WHO WILL FIND THIS TEXT MOST ADVANTAGEOUS:

The text can be used most profitably by many others besides students and teachers. In this group, we include (1) the parents of algebra students who wish to refer their children to the wealth of self-study materials available in the text, or who may wish to refresh their own memory of algebra in order to help their children directly; (2) the supervisor who seeks to accelerate select groups or to provide enrichment materials to supplement an algebra course of study; (3) the person who seeks to review algebra quickly and effectively; and (4) the person who seeks to learn algebra through independent study.

The author acknowledges his deep gratitude to Mr. Henry Hayden for the artistry, arrangement and typographical design of each page of the book. These have added immeasurably to the effective presentation of its content.

BARNETT RICH

Brooklyn Technical High School

April, 1960

Contents

Chapter 1. *FROM ARITHMETIC TO ALGEBRA* PAGE

1. Representing Numbers by Letters 1
2. Interchanging Numbers in Addition 2
3. Interchanging Numbers in Multiplication 3
4. Symbolizing the Operations in Algebra 3
5. Expressing Addition and Subtraction Algebraically 4
6. Expressing Multiplication and Division Algebraically 5
7. Expressing Two or More Operations Algebraically 5
8. Order In Which Fundamental Operations Are Performed 6
9. The Uses of Parentheses: Changing the Order of Operations 7
10. Multiplying Factors in Terms: Numerical and Literal Coefficient 8
11. Repeated Multiplying of a Factor: Base, Exponent and Power 9
12. Combining Like and Unlike Terms 11

SUPPLEMENTARY PROBLEMS 12

Chapter 2. *SIMPLE EQUATIONS AND THEIR SOLUTIONS*

1. Kinds of Equalities: Equations and Identities 17
2. Translating Verbal Statements into Equations 18
3. Solving Simple Equations Using Inverse Operations 19
4. Rules of Equality for Solving Equations 20
5. Using Division to Solve an Equation 21
6. Using Multiplication to Solve an Equation 23
7. Using Subtraction to Solve an Equation 25
8. Using Addition to Solve an Equation 26
9. Using Two or More Operations to Solve an Equation 27

SUPPLEMENTARY PROBLEMS 31

Chapter 3. *SIGNED NUMBERS*

1. Understanding Signed Numbers: Positive and Negative Numbers 36
2. Using Number Scales for Signed Numbers 37
3. Adding Signed Numbers 40
4. Simplifying the Addition of Signed Numbers 41
5. Subtracting Signed Numbers 42
6. Multiplying Signed Numbers 44
7. Finding Powers of Signed Numbers 46
8. Dividing Signed Numbers 47
9. Evaluating Expressions Having Signed Numbers 48

SUPPLEMENTARY PROBLEMS 49

Chapter 4. *MONOMIALS AND POLYNOMIALS*

PAGE

1. Understanding Monomials and Polynomials 53
2. Adding Monomials 53
3. Arranging and Adding Polynomials 54
4. Subtracting Monomials 55
5. Subtracting Polynomials 56
6. Using Parentheses and Other Grouping Symbols to Add or Subtract Polynomials 57
7. Multiplying Monomials and Powers of the Same Base 59
8. Multiplying a Polynomial by a Monomial 60
9. Multiplying Polynomials 61
10. Dividing Powers and Monomials 62
11. Dividing a Polynomial by a Monomial 63
12. Dividing a Polynomial by a Polynomial 64

SUPPLEMENTARY PROBLEMS 65

Chapter 5. *EQUATIONS OF THE FIRST DEGREE IN ONE UNKNOWN*

1. Reviewing the Solution of First Degree Equations Having Positive Roots 71
2. Solving First Degree Equations Having Negative Roots 73
3. Solving Equations by Transposing 74
4. Solving Equations Containing Parentheses 75
5. Solving Equations Containing One Fraction or Fractions Having the Same Denominator 76
6. Solving Equations Containing Fractions Having Different Denominators: Lowest Common Denominator (L.C.D.) 77
7. Solving Equations Containing Decimals 78
8. Solving Literal Equations 79

SUPPLEMENTARY PROBLEMS 81

Chapter 6. *FORMULAS*

1. Understanding Polygons, Circles and Solids 86
2. Formulas for Perimeters and Circumferences: Linear Measure 89
3. Formulas for Areas: Square Measure 92
4. Formulas for Volumes: Cubic Measure 95
5. Deriving Formulas 98
6. Transforming Formulas 100
7. Finding the Value of an Unknown in a Formula 101

SUPPLEMENTARY PROBLEMS 103

Chapter 7. *GRAPHS OF LINEAR EQUATIONS*

1. Understanding Graphs 109
2. Graphing Linear Equations 112
3. Solving a Pair of Linear Equations Graphically 116
4. Deriving a Linear Equation from a Table of Values 118

SUPPLEMENTARY PROBLEMS 120

Chapter 8. *EQUATIONS OF FIRST DEGREE IN TWO UNKNOWNS*

1. Solving a Pair of Equations by Addition or Subtraction 124
2. Solving a Pair of Equations by Substitution 126

SUPPLEMENTARY PROBLEMS 128

PAGE

Chapter 9. *PROBLEM-SOLVING*

1. Number Problems Having One Unknown 130
2. Number Problems Having Two Unknowns 131
3. Consecutive Integer Problems 133
4. Age Problems 135
5. Ratio Problems 136
6. Angle Problems 139
7. Perimeter Problems 141
8. Coin or Stamp Problems 143
9. Cost and Mixture Problems 144
10. Investment or Interest Problems 146
11. Motion Problems 148
12. Combination Problems 153

SUPPLEMENTARY PROBLEMS 154

Chapter 10. *SPECIAL PRODUCTS AND FACTORING*

1. Understanding Factors and Products 167
2. Factoring a Polynomial Having a Common Monomial Factor 168
3. Squaring a Monomial 169
4. Finding the Square Root of a Monomial 170
5. Finding the Product of the Sum and Difference of Two Numbers 171
6. Factoring the Difference of Two Squares 172
7. Finding the Product of Two Binomials with Like Terms 172
8. Factoring Trinomials in Form of $x^2 + bx + c$ 174
9. Factoring a Trinomial in Form of $ax^2 + bx + c$ 174
10. Squaring a Binomial 175
11. Factoring a Perfect Square Trinomial 176
12. Completely Factoring Polynomials 177

SUPPLEMENTARY PROBLEMS 178

Chapter 11. *FRACTIONS*

1. Understanding Fractions 184
2. Changing Fractions to Equivalent Fractions 185
3. Reciprocals and Their Uses 186
4. Reducing Fractions to Lowest Terms 188
5. Multiplying Fractions 190
6. Dividing Fractions 191
7. Adding or Subtracting Fractions Having the Same Denominator 191
8. Adding or Subtracting Fractions Having Different Denominators 192
9. Simplifying Complex Fractions 194

SUPPLEMENTARY PROBLEMS 195

Chapter 12. *ROOTS AND RADICALS*

1. Understanding Roots and Radicals 201
2. Understanding Rational and Irrational Numbers 202
3. Finding the Square Root of a Number by Using a Graph 204
4. Finding the Square Root of a Number by Using a Table 205
5. Computing the Square Root of a Number 206
6. Simplifying the Square Root of a Product 208
7. Simplifying the Square Root of a Fraction 209
8. Adding and Subtracting Square Roots of Numbers 210
9. Multiplying Square Roots of Numbers 211
10. Dividing Square Roots of Numbers 212
11. Rationalizing the Denominator of a Fraction 213
12. Solving Radical Equations 214

SUPPLEMENTARY PROBLEMS 216

Chapter 13. *QUADRATIC EQUATIONS IN ONE UNKNOWN* PAGE

1. Understanding Quadratic Equations in One Unknown 221
2. Solving Quadratic Equations by Factoring 222
3. Solving Incomplete Quadratic Equations 223
4. Solving a Quadratic Equation by Completing the Square 225
5. Solving a Quadratic Equation by Quadratic Formula 227
6. Solving Quadratic Equations Graphically 228

SUPPLEMENTARY PROBLEMS 230

Chapter 14. *THE VARIABLE: DIRECT, INVERSE, JOINT AND POWER VARIATION*

1. Understanding the Variable 234
2. Understanding Direct Variation 235
3. Understanding Inverse Variation 238
4. Understanding Joint Variation 241
5. Understanding Power Variation 242

SUPPLEMENTARY PROBLEMS 244

Chapter 15. *INDIRECT MEASUREMENT*

1. Indirect Measurement: Using Triangles Drawn to Scale 248
2. $D = SN$: A Formula of Indirect Measurement for Figures Drawn to Scale on Maps, Graphs, Models and Blueprints 249

SUPPLEMENTARY PROBLEMS 250

Chapter 16. *LAW OF PYTHAGORAS, PROPORTIONS AND SIMILAR TRIANGLES*

1. Law of Pythagoras 251
2. Proportions: Equal Ratios 254
3. Similar Triangles 255

SUPPLEMENTARY PROBLEMS 257

Chapter 17. *TRIGONOMETRY*

1. Understanding Trigonometric Ratios 260
2. Solving Trigonometry Problems 263
3. Angles of Elevation and Depression 265

SUPPLEMENTARY PROBLEMS 267

Chapter 18. *SUPPLEMENTARY TOPICS*

1. Solving Problems Graphically 270
2. Understanding Slope of a Line 271
3. Understanding Descriptive Statistics 273
4. Understanding Congruent Triangles 275
5. Understanding Symmetry 277
6. Using a Symbol to Simplify Variation 278

SUPPLEMENTARY PROBLEMS 280

Chapter 19. *REVIEWING ARITHMETIC*

1. Reviewing Whole Numbers 282
2. Reviewing Fractions 284
3. Reviewing Decimals 286
4. Reviewing Per Cents and Percentage 288
5. Reviewing Statistical Graphs 289

Table of Approximate Square Roots 290

Table of Natural Trigonometric Functions 292

Index 293

Chapter 1 From ARITHMETIC To ALGEBRA

1. REPRESENTING NUMBERS BY LETTERS

In this chapter, we are going to lead you from arithmetic to algebra. Underlying algebra as well as arithmetic are the four fundamental operations:

1. ADDITION (sum)
2. SUBTRACTION (difference)
3. MULTIPLICATION (product)
4. DIVISION (quotient)

The answer for each operation is the word enclosed in parentheses. The answer in subtraction may also be called the remainder.

In algebra, **letters are used to represent numbers.** By using letters and mathematical symbols, short algebraic statements replace lengthy verbal statements. Note how this is done in the following examples:

Verbal Statements	Algebraic Statements
1. Seven times a number reduced by the same number equals six times the number.	*1.* $7n - n = 6n$
2. The sum of twice a number and three times the same number equals five times that number.	*2.* $2n + 3n = 5n$
3. The perimeter of a square equals four times the length of one of its sides.	*3.* $p = 4s$

In the first example, "$7n$" is used instead of "seven times a number". When multiplying a number by a letter, the multiplication sign may be omitted. Multiplication may also be indicated by using a multiplication sign, a raised dot or parentheses.

Thus, seven times a number may be shown by $7\times n$, $7\cdot n$, $7(n)$ or $7n$.

Omitting the multiplication sign, as in $7n$, is the preferred method of indicating multiplication. However, the multiplication sign may not be omitted when multiplying two numbers. "Seven times four" may be written as 7×4, $7\cdot 4$ or $7(4)$ but never as 74.

1.1. Stating Products Without Multiplication Signs

a) $7\times y$	*Ans.* $7y$	*e*) $b\times c\times d$	*Ans.* bcd
b) $3\times 5\times a$	*Ans.* $15a$	*f*) $7\times 11\times h\times k$	*Ans.* $77hk$
c) $l\times w$	*Ans.* lw	*g*) $\frac{1}{2}\times 8\times n$	*Ans.* $4n$
d) $10\times r\times s$	*Ans.* $10rs$	*h*) $.07\times p\times q\times t$	*Ans.* $.07pqt$

1.2. Changing Verbal Statements To Algebraic Equations

Using letters and symbols, replace each verbal statement by an algebraic equation:

a) If six times a number is reduced by the same number, the result must be five times the number.
Ans. $6n - n = 5n$

b) The sum of twice a number, three times the same number and four times the same number is equivalent to nine times the number. *Ans.* $2n + 3n + 4n = 9n$

c) Increasing a number by itself and 20 is the same as doubling the number and adding 20.
Ans. $n + n + 20 = 2n + 20$

d) The area of a rectangle is equal to the product of its length and width. *Ans.* $A = lw$

2. INTERCHANGING NUMBERS IN ADDITION

Addends are numbers being added. Their sum is the answer obtained.

Thus, in $5+3=8$, the addends are 5 and 3. Their sum is 8.

Numerical addends are numbers used as addends.

Thus, in $3+4+6=13$; 3, 4 and 6 are numerical addends.

Literal addends are letters used to represent numbers being added.

Thus, in $a+b=8$, a and b are literal addends.

RULE: Interchanging addends does not change their sum.

Thus, $2+3=3+2$ and $3+4+6=4+6+3$.

In general, $a+b=b+a$ and $b+c+a=a+b+c$.

Interchanging addends may be used

(*1*) **to simplify addition.**

Thus, $25+82+75$ by interchanging becomes $25+75+82$.
The sum is $100+82=182$.

(*2*) **to check addition.**

Thus, numbers may be added downwards and checked upwards:

add down ↓	check up ↑
148	148
357	357
762	762
1267	1267

(*3*) **to rearrange addends in a preferred order.**

Thus, $b+c+a$ becomes $a+b+c$ if the literal addends are to be arranged alphabetically. Also, $3+x$ becomes $x+3$ if the literal addend is to precede the numerical addend.

2.1. Interchanging Addends To Simplify Addition

Simplify each addition by interchanging addends:

a) $20+73+280$	*c*) $\frac{3}{4}+2\frac{1}{2}+1\frac{1}{4}$	*e*) $1.95+2.65+.05+.35$
b) $42+113+58$	*d*) $1\frac{1}{2}+2\frac{2}{7}+\frac{1}{2}+\frac{1}{7}$	*f*) $9.4+18.7+1.3+.6$

Ans.

a) $20+280+73$; $300+73=373$	*c*) $\frac{3}{4}+1\frac{1}{4}+2\frac{1}{2}$; $2+2\frac{1}{2}=4\frac{1}{2}$	*e*) $1.95+.05+2.65+.35$; $2+3=5$
b) $42+58+113$; $100+113=213$	*d*) $1\frac{1}{2}+\frac{1}{2}+2\frac{2}{7}+\frac{1}{7}$; $2+2\frac{3}{7}=4\frac{3}{7}$	*f*) $9.4+.6+18.7+1.3$; $10+20=30$

2.2. Rearranging Addends

Rearrange the addends so that literal addends are arranged alphabetically and precede numerical addends:

a) $3+b$	*c*) $d+10+e$	*e*) $15+x+10$	*g*) $w+y+x$
b) $c+a$	*d*) $c+12+b$	*f*) $20+s+r$	*h*) $b+8+c+a$

Ans.

a) $b+3$	*c*) $d+e+10$	*e*) $x+25$	*g*) $w+x+y$
b) $a+c$	*d*) $b+c+12$	*f*) $r+s+20$	*h*) $a+b+c+8$

3. INTERCHANGING NUMBERS IN MULTIPLICATION

Factors are numbers being multiplied. Their product is the answer obtained.

Thus, in $5 \times 3 = 15$, the factors are 5 and 3. Their product is 15.

Numerical factors are numbers used as factors.

Thus, in $2 \times 3 \times 5 = 30$; 2, 3 and 5 are numerical factors.

Literal factors are letters used to represent numbers being multiplied.

Thus, in $ab = 20$, a and b are literal factors.

RULE: Interchanging factors does not change their product.

Thus, $2 \times 5 = 5 \times 2$ and $2 \times 4 \times 5 = 2 \times 5 \times 4$.

In general, $ab = ba$ and $cba = abc$.

Interchanging factors may be used

(*1*) **to simplify multiplication.**

Thus, $4 \times 13 \times 25$ by interchanging becomes $4 \times 25 \times 13$.

The product is $100 \times 13 = 1300$.

(*2*) **to check multiplication.**

Thus, since $24 \times 75 = 75 \times 24$,

24	check: 75
×75	×24
120	300
168	150
1800	1800

(*3*) **to rearrange factors in a preferred order.**

Thus, bca becomes abc if the literal factors are arranged alphabetically. Also, $x3$ becomes $3x$ if the numerical factor is to precede the literal factor.

3.1. Simplifying Multiplication

Simplify each multiplication by interchanging factors:

a) $2 \times 17 \times 5$ | *c*) $7\frac{1}{2} \times 7 \times 4$ | *e*) $1.25 \times 4.4 \times 4 \times 5$

b) $25 \times 19 \times 4 \times 2$ | *d*) $33\frac{1}{3} \times 23 \times 3$ | *f*) $.33 \times 225 \times 3\frac{1}{3} \times 4$

Ans. *a*) $2 \times 5 \times 17$
$10 \times 17 = 170$

b) $25 \times 4 \times 19 \times 2$
$100 \times 38 = 3800$

c) $7\frac{1}{2} \times 4 \times 7$
$30 \times 7 = 210$

d) $33\frac{1}{3} \times 3 \times 23$
$100 \times 23 = 2300$

e) $1.25 \times 4 \times 4.4 \times 5$
$5 \times 22 = 110$

f) $.33 \times 3\frac{1}{3} \times 225 \times 4$
$1.1 \times 900 = 990$

3.2. Rearranging Factors

Rearrange the factors so that literal factors are arranged alphabetically and follow numerical factors:

a) $b\,3$ *b*) $c\,a$ *c*) $d\,10\,e$ *d*) $c\,12\,b$ *e*) $15\,x\,10$ *f*) $20\,s\,r$ *g*) $w\,y\,x$ *h*) $b\,35\,c\,a$

Ans. *a*) $3b$ *b*) ac *c*) $10de$ *d*) $12bc$ *e*) $150x$ *f*) $20rs$ *g*) wxy *h*) $35abc$

4. SYMBOLIZING THE OPERATIONS IN ALGEBRA

The symbols for the fundamental operations are as follows:

1. ADDITION: +
2. SUBTRACTION: −
3. MULTIPLICATION: ×, ( ), ·, no sign
4. DIVISION: ÷, :, fraction bar

Thus, $n + 4$ means "add n and 4". $n - 4$ means "subtract 4 from n".

$4 \times n$, $4(n)$, $4 \cdot n$, $4n$ mean "multiply n and 4". $n \div 4$, $n:4$, $\frac{n}{4}$ mean "divide n by 4".

RULE: Division by zero is an impossible operation. Hence, $\frac{3}{0}$ is meaningless.

Thus, $4 \div 0$ or $x \div 0$ is impossible.

Hence, $\frac{4}{0}$ or $\frac{x}{0}$ is meaningless.

Also, $\frac{4}{n}$ is meaningless if $n = 0$.

4.1. Symbols In Multiplication

Symbolize each using multiplication signs:

a) 8 times 11 | *c*) b times c | *e*) 5 multiplied by a and the result divided by b

b) 8 times x | *d*) 8 divided by x | *f*) d divided by the product of 7 and e

Ans. *a*) 8×11, $8 \cdot 11$, $8(11)$ or $(8)(11)$ | *c*) $b \cdot c$ or bc (avoid $b \times c$) | *e*) $\frac{5a}{b}$

b) $8 \cdot x$ or $8x$ (avoid $8 \times x$) | *d*) $\frac{8}{x}$ or $8 \div x$ ($\frac{8}{x}$ is preferred) | *f*) $\frac{d}{7e}$

4.2. Division By Zero

When is each division impossible ?

a) $\frac{10}{b}$ *b*) $\frac{a}{3c}$ *c*) $\frac{8}{x-5}$ *d*) $\frac{7}{xy}$

Ans. *a*) if $b = 0$, *b*) if $c = 0$, *c*) if $x = 5$, *d*) if $x = 0$ or $y = 0$

5. EXPRESSING ADDITION AND SUBTRACTION ALGEBRAICALLY

In algebra, changing verbal statements into algebraic expressions is of major importance. The operations of addition and subtraction are denoted by words such as the following:

WORDS DENOTING ADDITION		WORDS DENOTING SUBTRACTION	
sum	more than	difference	less than
plus	greater than	minus	smaller than
gain	larger than	lose	fewer than
increase	enlarge	decrease	shorten
rise	grow	drop	depreciate
expand	augment	lower	diminish

In adding two numbers, the numbers (addends) may be interchanged.

Thus, "the sum of n and 20" may be represented by $n + 20$ or $20 + n$.

But in subtracting one number from another, the numbers may not be interchanged.

Thus, "a number less 20" may be represented by $n - 20$ but **not** by $20 - n$.

Also, "20 minus a number" may be represented by $20 - n$ but **not** by $n - 20$.

5.1. Expressing Addition Algebraically

If n represents a number, express algebraically:

a) the sum of the number and 7 | *c*) the number increased by 9 | *e*) 20 enlarged by the number

b) the number plus 8 | *d*) 15 plus the number | *f*) 25 augmented by the number

Ans. *a*) $n + 7$ or $7 + n$ | *c*) $n + 9$ or $9 + n$ | *e*) $20 + n$ or $n + 20$

b) $n + 8$ or $8 + n$ | *d*) $15 + n$ *or* $n + 15$ | *f*) $25 + n$ or $n + 25$

5.2. Expressing Subtraction Algebraically

If n represents a number, express algebraically:

a) the difference if the number is subtracted from 15 *Ans.* $15 - n$

b) the number diminished by 20 *Ans.* $n - 20$

c) 25 less than the number *Ans.* $n - 25$

d) 25 less the number *Ans.* $25 - n$

e) the difference if 15 is subtracted from the number *Ans.* $n - 15$

f) 50 subtracted from the number *Ans.* $n - 50$

g) the number subtracted from 50 *Ans.* $50 - n$

h) the number reduced by 75 *Ans.* $n - 75$

5.3. Changing Verbal Statements Into Algebraic Expressions

Express algebraically:

	Ans.
a) the no. of lb of a weight that is 10 lb heavier than w lb	*a*) $w+10$
b) the no. of mi in a distance that is 40 mi farther than d mi	*b*) $d+40$
c) the no. of degrees in a temperature 50° hotter than t degrees	*c*) $t+50$
d) the no. of dollars in a price \$60 cheaper than p dollars	*d*) $p-60$
e) the no. of mph in a speed 30 mph (miles per hour) faster than r mph	*e*) $r+30$
f) the no. of ft in a length of l ft expanded 6 ft	*f*) $l+6$
g) the no. of oz in a weight that is 10 oz lighter than w oz	*g*) $w-10$
h) the no. of yd in a distance that is 120 ft shorter than d yd	*h*) $d-40$

6. EXPRESSING MULTIPLICATION AND DIVISION ALGEBRAICALLY

WORDS DENOTING MULTIPLICATION		WORDS DENOTING DIVISION	
multiplied by	double	divided by	ratio
times	triple or treble	quotient	half
product	quadruple		
twice	quintuple		

In multiplying two numbers, the numbers (factors) may be interchanged.

Thus, "the product of n and 10" may be represented by $n10$ or $10n$. The latter is preferred.

In dividing one number by another, the numbers may not be interchanged.

Thus, "a number divided by 20" may be represented by $\frac{n}{20}$ but **not** by $\frac{20}{n}$.

Also, "20 divided by a number" may be represented by $\frac{20}{n}$ but **not** by $\frac{n}{20}$.

6.1. Representing Multiplication Or Division

What statement may be represented by each?

a) $5x$ *b*) $\frac{y}{5}$ *c*) $\frac{5w}{7}$

Ans. *a*) *1.* 5 multiplied by x	*b*) *1.* y divided by 5	*c*) *1.* five-sevenths of w
2. 5 times x	*2.* quotient of y and 5	*2.* $5w$ divided by 7
3. product of 5 and x	*3.* ratio of y to 5	*3.* quotient of $5w$ and 7
	4. one-fifth of y	*4.* ratio of $5w$ to 7

7. EXPRESSING TWO OR MORE OPERATIONS ALGEBRAICALLY

Parentheses ( ) are used to treat an expression as a single number.

Thus, to double the sum of 4 and x, write $2(4+x)$.

7.1. Expressing Two Operations Algebraically

Express algebraically:

	Ans.
a) a increased by twice b	*a*) $a+2b$
b) twice the sum of a and b	*b*) $2(a+b)$
c) 30 decreased by three times c	*c*) $30-3c$
d) three times the difference of 30 and c	*d*) $3(30-c)$
e) 50 minus the product of 10 and p	*e*) $50-10p$
f) the product of 50 and the sum of p and 10	*f*) $50(p+10)$
g) 100 increased by the quotient of x and y	*g*) $100+\frac{x}{y}$
h) the quotient of x and the sum of y and 100	*h*) $\frac{x}{y+100}$
i) the average of s and 20	*i*) $\frac{s+20}{2}$

7.2. More Difficult Expressions

Express algebraically:

	Ans.
a) half of a increased by the product of 25 and b	*a*) $\frac{a}{2} + 25b$
b) four times c decreased by one-fifth of d	*b*) $4c - \frac{d}{5}$
c) half the sum of m and twice n	*c*) $\frac{m+2n}{2}$
d) the average of m, r and 80	*d*) $\frac{m+r+80}{3}$
e) 60 diminished by one-third the product of 7 and x	*e*) $60 - \frac{7x}{3}$
f) twice the sum of e and 30 diminished by 40	*f*) $2(e+30) - 40$
g) two-thirds the sum of n and three-sevenths of p	*g*) $\frac{2}{3}(n + \frac{3p}{7})$
h) the product of a and b decreased by twice the difference of c and d	*h*) $ab - 2(c-d)$
i) the quotient of x and 10 minus four times their sum	*i*) $\frac{x}{10} - 4(x+10)$

7.3. Changing Verbal Statements Into Algebraic Expressions

Express algebraically:

	Ans.
a) a speed in mph that is 30 mph faster than twice another of r mph	*a*) $2r + 30$
b) a weight in lb that is 20 lb less than three times another of w lb	*b*) $3w - 20$
c) a temperature in degrees that is 15° colder than two-thirds another of $t°$	*c*) $\frac{2t}{3} - 15$
d) a price in cents that is 25¢ cheaper than another of D dollars	*d*) $100D - 25$
e) a length in inches that is 8 in. longer than another of f ft.	*e*) $12f + 8$

8. *ORDER IN WHICH FUNDAMENTAL OPERATIONS ARE PERFORMED*

In evaluating or finding the value of an expression containing numbers, the operations involved must be performed in a certain order. Note in the following, how **multiplication and division precede addition and subtraction !**

To Evaluate A Numerical Expression Not Containing Parentheses

Evaluate: *a*) $3 + 4 \times 2$ *b*) $5 \times 4 - 18 \div 6$

Procedure:	**Solution:** (*a*)	(*b*)
1) Do **multiplications and divisions (M & D)** in order from left to right:	1) $3 + 4 \times 2$ $3 + 8$	1) $5 \times 4 - 18 \div 6$ $20 - 3$
2) Do remaining **additions and subtractions (A & S)** in order from left to right:	2) 11 *Ans.*	2) 17 *Ans.*

To Evaluate An Algebraic Expression Not Containing Parentheses

Evaluate $x + 2y - \frac{z}{5}$ when $x = 5$, $y = 3$, $z = 20$.

Procedure:	**Solution:**
1) **Substitute** the value given for each letter:	1) $x + 2y - \frac{z}{5}$ $5 + 2(3) - \frac{20}{5}$
2) Do **multiplications and divisions (M & D)** in order from left to right:	2) $5 + 6 - 4$
3) Do remaining **additions and subtractions (A & S)** in order from left to right:	3) 7 *Ans.*

8.1. Evaluating Numerical Expressions

Evaluate:

Procedure:	a)	b)	c)
	$24 \div 4 + 8$	$24 + 8 \div 4$	$8 \times 6 - 10 \div 5 + 12$
1) Do **M & D**:	$6 + 8$	$24 + 2$	$48 - 2 + 12$
2) Do **A & S**:	14 *Ans.*	26 *Ans.*	58 *Ans.*

8.2. Evaluating Algebraic Expressions

Evaluate if $a=8$, $b=10$, $x=3$:

Procedure:	a)	b)	c)
	$4b - \frac{a}{4}$	$12x + ab$	$\frac{3a}{4} + \frac{4b}{5} - \frac{2x}{3}$
1) Substitute:	$4 \times 10 - \frac{8}{4}$	$12 \times 3 + 8 \times 10$	$\frac{3}{4} \times 8 + \frac{4}{5} \times 10 - \frac{2}{3} \times 3$
2) Do **M & D**:	$40 - 2$	$36 + 80$	$6 + 8 - 2$
3) Do **A & S**:	38 *Ans.*	116 *Ans.*	12 *Ans.*

8.3. Evaluating When A Letter Represents 0

Evaluate if $w=4$, $x=2$, and $y=0$:

	a)	b)	c)	d)	e)	f)
	$wx + y$	$w + xy$	$\frac{w+y}{x}$	$\frac{xy}{w}$	$\frac{x}{w+y}$	$\frac{wx}{y}$
Solutions:	$4 \times 2 + 0$	$4 + 2 \times 0$	$\frac{4+0}{2}$	$\frac{2 \times 0}{4}$	$\frac{2}{4+0}$	$\frac{4 \times 2}{0}$
	$8 + 0$	$4 + 0$	$\frac{4}{2}$	$\frac{0}{4}$	$\frac{2}{4}$	$\frac{8}{0}$
	8 *Ans.*	4 *Ans.*	2 *Ans.*	0 *Ans.*	$\frac{1}{2}$ *Ans.*	meaningless *Ans.*

9. THE USES OF PARENTHESES: CHANGING THE ORDER OF OPERATIONS

Parentheses may be used

(1) to treat an expression as a single number.

Thus, $2(x+y)$ represents twice the sum of x and y.

(2) to replace the multiplication sign.

Thus, $4(5)$ represents the product of 4 and 5.

(3) to change the order of operations in evaluating.

Thus, to evaluate $2(4+3)$, **add** 4 and 3 in the parentheses **before multiplying;** that is, $2(4+3) = 2 \cdot 7 = 14$. Compare this with $2 \cdot 4 + 3 = 8 + 3 = 11$.

To Evaluate An Algebraic Expression Containing Parentheses

Evaluate $2(a+b) + 3a - \frac{b}{2}$ if $a=7$, $b=2$.

Procedure:	Solution:
1) **Substitute** the value given for each letter:	1) $2(7+2) + 3 \cdot 7 - \frac{2}{2}$
2) **Evaluate inside parentheses:**	2) $2 \cdot 9 + 3 \cdot 7 - \frac{2}{2}$
3) Do **multiplications and divisions (M & D)** in order from left to right:	3) $18 + 21 - 1$
4) Do remaining **additions and subtractions (A & S)** in order from left to right:	4) 38 *Ans.*

9.1. Evaluating Numerical Expressions Containing Parentheses

Evaluate:

Procedure:	a) $3(4-2)+12$	b) $7-\frac{1}{2}(14-6)$	c) $8+\frac{1}{3}(4+2)$	d) $20-5(4-1)$
1) Do ( ):	$3\cdot 2+12$	$7-\frac{1}{2}\cdot 8$	$8+\frac{1}{3}\cdot 6$	$20-5\cdot 3$
2) Do **M&D**:	$6+12$	$7-4$	$8+2$	$20-15$
3) Do **A&S**:	18 *Ans.*	3 *Ans.*	10 *Ans.*	5 *Ans.*

9.2. Evaluating Algebraic Expressions Containing Parentheses

Evaluate if $a=10$, $b=2$, and $x=12$:

Procedure:	a) $3(x+2b)-30$	b) $8+2(\frac{a}{b}+x)$	c) $3x-\frac{1}{2}(a+b)$
1) **Substitute:**	$3(12+2\cdot 2)-30$	$8+2(\frac{10}{2}+12)$	$3\cdot 12-\frac{1}{2}(10+2)$
2) Do ( ):	$3\cdot 16-30$	$8+2\cdot 17$	$36-\frac{1}{2}\cdot 12$
3) Do **M & D**:	$48-30$	$8+34$	$36-6$
4) Do **A&S**:	18 *Ans.*	42 *Ans.*	30 *Ans.*

9.3. Evaluating When A Letter Represents 0

Evaluate if $w=1$, $y=4$ and $x=0$:

a) $x(2w+3y)$	b) $y(wx+5)$	c) $\frac{1}{2}(y+\frac{x}{w})+5$	d) $\frac{y}{w+x}-w(y+x)$
$0(2\cdot 1+3\cdot 4)$	$4(1\cdot 0+5)$	$\frac{1}{2}(4+\frac{0}{1})+5$	$\frac{4}{1+0}-1(4+0)$
$0\cdot 14$	$4\cdot 5$	$\frac{1}{2}\cdot 4+5$	$4-4$
0 *Ans.*	20 *Ans.*	7 *Ans.*	0 *Ans.*

10. MULTIPLYING FACTORS IN TERMS: NUMERICAL AND LITERAL COEFFICIENT

A **term** is a number or the product of numbers.

Thus, 5, $8y$, cd, $3wx$ and $\frac{2}{3}rst$ are terms.

Also, the expression $8y+5$ consists of two terms $8y$ and 5.

A **factor of a term** is each of the numbers multiplied to form the term.

Thus, 8 and y are factors of the term $8y$; 3, w and x are factors of the term $3wx$.

Also, 5 and $(a+b)$ are the factors of the term $5(a+b)$.

Any factor or group of factors of a term is a **coefficient** of the product of the remaining factors.

Thus, in $3abc$, 3 is the **numerical coefficient** of abc while abc is the **literal coefficient** of 3. Other examples are the following:

	TERM	NUMERICAL COEFFICIENT	LITERAL COEFFICIENT
(1)	xy	1	xy
(2)	$\frac{3y}{7}$	$\frac{3}{7}$	y
(3)	$\frac{3ab}{5c}$	$\frac{3}{5}$	$\frac{ab}{c}$

An **expression** contains one or more terms connected by plus or minus signs.
Thus, $2x$, $3ab+2$ and $5a-2b-7$ are expressions.

10.1. Expressions Containing Terms

State the number of terms and the terms in each expression.

a) $8abc$	*Ans.*	1 term : $8abc$
b) $8+a+bc$	*Ans.*	3 terms : 8, a and bc
c) $8a+bc$	*Ans.*	2 terms : $8a$ and bc
d) $3b+c+d$	*Ans.*	3 terms : $3b$, c and d
e) $3+bcd$	*Ans.*	2 terms : 3 and bcd
f) $3(b+c)+d$	*Ans.*	2 terms : $3(b+c)$ and d

10.2. Factors Of Terms

State the factors of the following, disregarding 1 and the product itself:

a) 21	*Ans.* 3 and 7	*e*) $\frac{1}{3}m$	*Ans.* $\frac{1}{3}$ and m
b) 121	*Ans.* 11 and 11	*f*) $\frac{n}{5}$	*Ans.* $\frac{1}{5}$ and n
c) rs	*Ans.* r and s	*g*) $\frac{n+3}{5}$	*Ans.* $\frac{1}{5}$ and $(n+3)$
d) $5cd$	*Ans.* 5, c and d	*h*) $3(x+2)$	*Ans.* 3 and $(x+2)$

10.3. Numerical and Literal Coefficients

State each numerical and literal coefficient:

	a) y	*b*) $\frac{4x}{5}$	*c*) $\frac{w}{7}$	*d*) $.7abc$	*e*) $8(a+b)$
Numerical Coefficient:	1	$\frac{4}{5}$	$\frac{1}{7}$	.7	8
Literal Coefficient:	y	x	w	abc	$(a+b)$

11. REPEATED MULTIPLYING OF A FACTOR: BASE, EXPONENT AND POWER

$$\text{BASE}^{\text{EXPONENT}} = \text{POWER}$$

In $2\cdot2\cdot2\cdot2\cdot2$, the factor 2 is being multiplied repeatedly. This may be written in a shorter form as 2^5 where the repeated factor 2 is the **base** while the small 5 written above and to the right of 2 is the **exponent**. The answer 32 is called the fifth **power** of 2.

An **exponent** is a number which indicates how many times another number, the **base**, is being used as a repeated factor. The **power** is the answer thus obtained. Thus, since $3\cdot3\cdot3\cdot3$ or $3^4=81$, 3 is the base, 4 is the exponent and 81 is the fourth power of 3.

TABLE OF POWERS

The table of powers contains the first five powers of the most frequently used numerical bases 1, 2, 3, 4, 5 and 10. It will be very useful to learn these.

BASE \ EXPONENT	1	2	3	4	5
1	1	1	1	1	1
2	2	4	8	16	32
3	3	9	27	81	243
4	4	16	64	256	1024
5	5	25	125	625	3125
10	10	100	1,000	10,000	100,000

Literal Bases: Squares and Cubes

Square

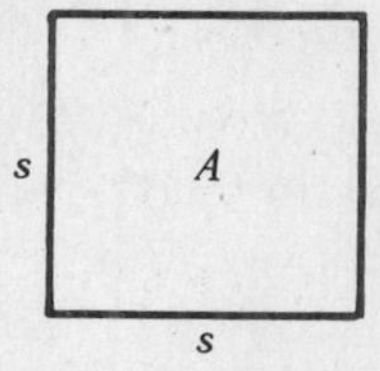

The area of a square with a side s, is found by multiplying s by s. This may be written as $A = s^2$ and read "Area equals s-square". Here, A is the second power of s. See adjoining figure.

The volume of a cube with a side s, is found by multiplying s three times; that is, $s \cdot s \cdot s$. This may be written as $V = s^3$ and read, "Volume equals s-cube". Here, V is the third power of s.

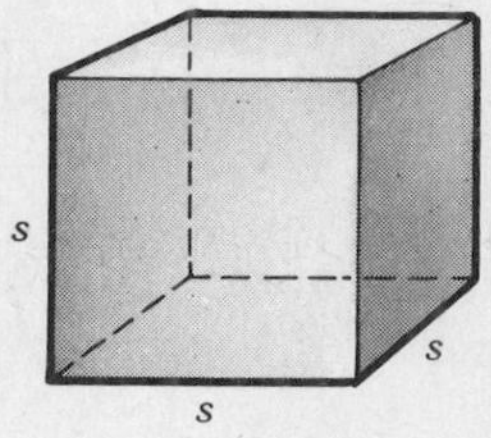

Cube

Reading Powers

"b^2" is read as "b-square", "b to the second power", "b-second" or "b to the second".

"x^3" is read as "x-cube", "x to the third power", "x-third" or "x to the third".

11.1. Writing As Bases And Exponents

Write each, using bases and exponents:

	$a)$ $5\cdot5\cdot5$	$c)$ $2\cdot8\cdot8\cdot8\cdot8$	$e)$ $bbccc$	$g)$ $(3y)(3y)(3y)$	$i)$ $7rr(s-8)$	$k)$ $\frac{7w}{xx}$
	$b)$ $3\cdot3\cdot7\cdot7$	$d)$ $bbbbb$	$f)$ $12bccd$	$h)$ $2(a+b)(a+b)$	$j)$ $\frac{yy}{x}$	$l)$ $\frac{2ttt}{5vvvv}$
Ans.	$a)$ 5^3	$c)$ $2\cdot8^4$	$e)$ b^2c^3	$g)$ $(3y)^3$	$i)$ $7r^2(s-8)$	$k)$ $\frac{7w}{x^2}$
	$b)$ $3^2 7^2$	$d)$ b^5	$f)$ $12bc^2d$	$h)$ $2(a+b)^2$	$j)$ $\frac{y^2}{x}$	$\frac{2t^3}{5v^4}$

11.2. Writing Without Exponents

Write each without exponents:

	$a)$ 2^6	$d)$ x^5	$g)$ $(2x)^3$	$j)$ $\frac{a^5}{b^2}$
	$b)$ $3\cdot4^2$	$e)$ $10y^4z^2$	$h)$ $6(5y)^2$	$k)$ $\frac{2(a+b)^2}{c^5}$
	$c)$ $5\cdot7^3\cdot8$	$f)$ $8rs^2t^3$	$i)$ $4(a-b)^2$	$l)$ $\frac{a^2+b^2}{c^3-d^3}$
Ans.	$a)$ $2\cdot2\cdot2\cdot2\cdot2\cdot2$	$d)$ $xxxxx$	$g)$ $(2x)(2x)(2x)$	$j)$ $\frac{aaaaa}{bb}$
	$b)$ $3\cdot4\cdot4$	$e)$ $10yyyyzz$	$h)$ $6(5y)(5y)$	$k)$ $\frac{2(a+b)(a+b)}{ccccc}$
	$c)$ $5\cdot7\cdot7\cdot7\cdot8$	$f)$ $8rssttt$	$i)$ $4(a-b)(a-b)$	$l)$ $\frac{aa+bb}{ccc-ddd}$

11.3. Evaluating Powers

Evaluate (The table of powers may be used to check values):

	$a)$ 3^5	$d)$ 2^2+3^2	$g)$ $1^2\cdot1^3\cdot1^4$	$j)$ $10+3\cdot2^2$	$m)$ $(3+4^2)(3^3-5^2)$
	$b)$ 5^4	$e)$ 2^3+3^3	$h)$ $2^3 5^2$	$k)$ $8\cdot10^2-3^3$	$n)$ $\frac{4^4}{2^5}$
	$c)$ 10^3	$f)$ 10^4-4^4	$i)$ $\frac{1}{2}\cdot2^4\cdot3^2$	$l)$ $\frac{1}{2}\cdot4^2-\frac{1}{3}\cdot3^2$	$o)$ $\frac{3^3+2^5}{10^2}$
Ans.	$a)$ 243	$d)$ $4+9=13$	$g)$ $1\cdot1\cdot1=1$	$j)$ $10+3\cdot4=22$	$m)$ $19\cdot2=38$
	$b)$ 625	$e)$ $8+27=35$	$h)$ $8\cdot25=200$	$k)$ $8\cdot100-27=773$	$n)$ $\frac{256}{32}=8$
	$c)$ 1,000	$f)$ $10{,}000-256$ $=9{,}744$	$i)$ $\frac{1}{2}\cdot16\cdot9=72$	$l)$ $\frac{1}{2}\cdot16-\frac{1}{3}\cdot9=5$	$o)$ $\frac{27+32}{100}=\frac{59}{100}$

11.4. Evaluating Powers Of Fractions And Decimals

Evaluate:

a) $.2^4$	d) $1000(.2^3)$	g) $(\frac{1}{2})^3$	j) $100(\frac{1}{5})^2$
b) $.5^2$	e) $\frac{1}{3}(.3^2)$	h) $(\frac{2}{3})^2$	k) $32(\frac{3}{2})^3$
c) $.01^3$	f) $200(.4^4)$	i) $(\frac{5}{3})^4$	l) $80(\frac{5}{2})^3$

Ans. a) .0016	d) $1000(.008) = 8$	g) $\frac{1}{2}\cdot\frac{1}{2}\cdot\frac{1}{2} = \frac{1}{8}$	j) $100(\frac{1}{25}) = 4$
b) .25	e) $\frac{1}{3}(.09) = .03$	h) $\frac{2}{3}\cdot\frac{2}{3} = \frac{4}{9}$	k) $32(\frac{27}{8}) = 108$
c) .000001	f) $200(.0256) = 5.12$	i) $\frac{5}{3}\cdot\frac{5}{3}\cdot\frac{5}{3}\cdot\frac{5}{3} = \frac{625}{81}$	l) $80(\frac{125}{8}) = 1250$

11.5. Evaluating Powers Of Literal Bases

Evaluate if $a = 5$, $b = 1$ and $c = 10$:

a) a^3	d) $2a^2$	g) $(\frac{1}{2}c)^2$	j) $a^2 + c^2$	m) $c^2(a+b)$
b) b^4	e) $(2a)^2$	h) $(\frac{b}{3})^2$	k) $(c+3b)^2$	n) $c(a-b^2)$
c) c^2	f) $(a+2)^2$	i) $\frac{4a^2}{c}$	l) $5(a^2-b^2)$	o) $3b(a^3-c^2)$

Ans. a) $5\cdot5\cdot5 = 125$	d) $2\cdot25 = 50$	g) $5^2 = 25$	j) $25 + 100 = 125$	m) $100\cdot6 = 600$
b) $1\cdot1\cdot1\cdot1 = 1$	e) $10^2 = 100$	h) $(\frac{1}{3})^2 = \frac{1}{9}$	k) $13^2 = 169$	n) $10\cdot4 = 40$
c) $10\cdot10 = 100$	f) $7^2 = 49$	i) $\frac{100}{10} = 10$	l) $5\cdot24 = 120$	o) $3\cdot25 = 75$

12. COMBINING LIKE AND UNLIKE TERMS

Like terms or **similar terms** are terms having the same literal factors, each with the same base and same exponent.

Thus:

LIKE TERMS	UNLIKE TERMS
$7x$ and $5x$	$7x$ and $5y$
$8a^2$ and a^2	$8a^2$ and a^3
$5rs^2$ and $2rs^2$	$5rs^2$ and $2r^2s$

NOTE: **Like terms must have a common literal coefficient.**

To Combine Like Terms Being Added Or Subtracted

	Combine: a) $7x + 5x - 3x$	b) $8a^2 - a^2$
Procedure:	**Solutions:**	
1) Add or subtract numerical coefficients:	1) $7 + 5 - 3 = 9$	1) $8 - 1 = 7$
2) Keep common literal coefficient:	2) $9x$ *Ans.*	2) $7a^2$ *Ans.*

12.1. Combine Like Terms

Combine:

a) $8+5-3$	*Ans.* 10	e) $7x-4x-x$	*Ans.* $2x$	i) $12c^2+c^2-7c^2$	*Ans.* $6c^2$
b) $7-4-1$	*Ans.* 2	f) $20r+30r-40r$	*Ans.* $10r$	j) $8ab+7ab$	*Ans.* $15ab$
c) $20+30-45$	*Ans.* 5	g) $2a^2+a^2$	*Ans.* $3c^2$	k) $6r^2s-2r^2s$	*Ans.* $4r^2s$
d) $8b+5b-3b$	*Ans.* $10b$	h) $13y^2-2y^2$	*Ans.* $11y^2$	l) $30x^3y^2-25x^3y^2$	*Ans.* $5x^3y^2$

12.2. Simplifying Expressions By Combining Like Terms

Simplify each expression by combining like terms:

a) $18a + 12a - 10$ *Ans.* $30a - 10$	d) $2x^2 + 3x^2 - y^2$ *Ans.* $5x^2 - y^2$	g) $6b + 20b + 2c - c$ *Ans.* $26b + c$
b) $18a + 12 - 10$ *Ans.* $18a + 2$	e) $2x^2 + 3y^2 - x^2$ *Ans.* $x^2 + 3y^2$	h) $6b + 20c + 2b - c$ *Ans.* $8b + 19c$
c) $18a + 12 - 10a$ *Ans.* $8a + 12$	f) $2x^2 + 3y^2 - y^2$ *Ans.* $2x^2 + 2y^2$	i) $6c + 20b + 2b - c$ *Ans.* $22b + 5c$

12.3. Combining Like Terms With Fractional And Decimal Coefficients

Combine like terms:

a) $8x + 3\frac{1}{4}x$	*e*) $\frac{2}{3}b^2 + \frac{1}{2}b^2$	*i*) $4.8w + 6.5w - 6.3w$
b) $7x - 2\frac{1}{3}x$	*f*) $1\frac{2}{3}c^2 - 1\frac{1}{3}c^2$	*j*) $1\frac{1}{2}y^2 + 3\frac{3}{4}y^2 - 2\frac{3}{4}y^2$
c) $\frac{5}{3}b - b$	*g*) $2.3ab + 11.6ab$	*k*) $3xy + 1.2xy - .6xy$
d) $3c - \frac{2}{3}c$	*h*) $.58cd^2 - .39cd^2$	*l*) $1.7x^2y + .9x^2y - 2.05x^2y$

Ans. *a*) $11\frac{1}{4}x$	*e*) $1\frac{1}{6}b^2$	*i*) $5w$
b) $4\frac{2}{3}x$	*f*) $\frac{1}{3}c^2$	*j*) $2\frac{1}{2}y^2$
c) $\frac{2}{3}b$	*g*) $13.9ab$	*k*) $3.6xy$
d) $2\frac{1}{3}c$	*h*) $.19cd^2$	*l*) $.55x^2y$

12.4. Combining Like Terms Representing Line Segments

Represent the length of each entire line:

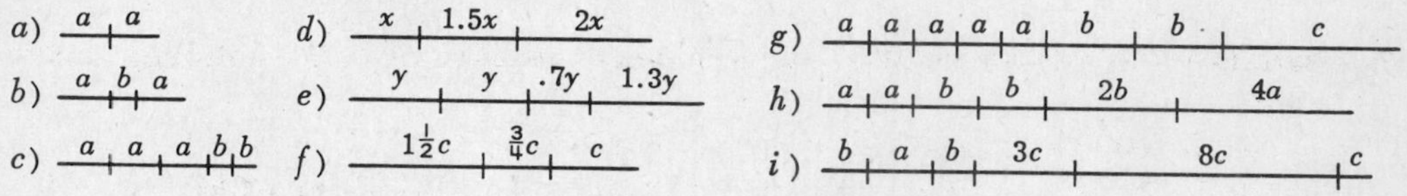

Ans. *a*) $2a$	*d*) $4.5x$	*g*) $5a + 2b + c$
b) $2a + b$	*e*) $4y$	*h*) $6a + 4b$
c) $3a + 2b$	*f*) $3\frac{1}{4}c$	*i*) $a + 2b + 12c$

SUPPLEMENTARY PROBLEMS

The numbers in parentheses indicate where to find the same type of example in this chapter. Refer to these examples for any help you may require.

1. State each product without multiplication signs: **(1.1)**

a) $8 \times w$	*c*) $b \times c \times d$	*e*) $.5 \times y \times z$	*g*) $2\frac{1}{3} \times g \times h \times n$
b) $2 \times 8 \times a$	*d*) $10 \times l \times m$	*f*) $\frac{2}{3} \times 12 \times t$	*h*) $.15 \times 100 \times q \times r$

Ans. *a*) $8w$ *b*) $16a$ *c*) bcd *d*) $10lm$ *e*) $.5yz$ *f*) $8t$ *g*) $2\frac{1}{3}ghn$ *h*) $15qr$

2. Using letters and symbols, replace each verbal statement by an algebraic equation: **(1.2)**

a) Three times a number added to eight times the same number is equivalent to eleven times the number. *Ans.* $3n + 8n = 11n$

b) The difference between ten times a number and one-half of the same number is exactly the same as nine and one-half times the number. *Ans.* $10n - \frac{1}{2}n = 9\frac{1}{2}n$

c) The perimeter of an equilateral triangle is equal to three times the length of one of the sides. *Ans.* $p = 3s$

d) The area of a square is found by multiplying the length of a side by itself. *Ans.* $A = ss$

3. Simplify each addition by interchanging addends: **(2.1)**

a) $64 + 138 + 36$	*c*) $1\frac{1}{3} + \frac{3}{5} + 6\frac{2}{3}$	*e*) $12\frac{1}{2}\% + 46\% + 87\frac{1}{2}\%$
b) $15 + 78 + 15 + 170$	*d*) $2\frac{1}{8} + \frac{13}{16} + \frac{3}{16} + \frac{3}{8}$	*f*) $5.991 + 1.79 + .21 + .009$

Ans. *a*) $64 + 36 + 138$	*c*) $1\frac{1}{3} + 6\frac{2}{3} + \frac{3}{5}$	*e*) $12\frac{1}{2}\% + 87\frac{1}{2}\% + 46\%$
$100 + 138 = 238$	$8 + \frac{3}{5} = 8\frac{3}{5}$	$100\% + 46\% = 146\%$
b) $15 + 15 + 170 + 78$	*d*) $2\frac{1}{8} + \frac{3}{8} + \frac{13}{16} + \frac{3}{16}$	*f*) $5.991 + .009 + 1.79 + .21$
$200 + 78 = 278$	$2\frac{1}{2} + 1 = 3\frac{1}{2}$	$6 + 2 = 8$

4. Rearrange the addends so that literal addends are arranged alphabetically and precede numerical addends: (2.2)

a) $10 + d$ | *c*) $15 + g + f$ | *e*) $d + b + e + a$ | *g*) $5 + m + 4 + j + 11$
b) $y + x$ | *d*) $17 + r + q + 13$ | *f*) $s + 12 + p + 48$ | *h*) $v + w + 16 + t + 50$

Ans. *a*) $d + 10$ | *c*) $f + g + 15$ | *e*) $a + b + d + e$ | *g*) $j + m + 20$
b) $x + y$ | *d*) $q + r + 30$ | *f*) $p + s + 60$ | *h*) $t + v + w + 66$

5. Simplify each multiplication by interchanging factors: (3.1)

a) $5 \times 26 \times 40$ | *c*) $10\frac{1}{2} \times 7 \times 2$ | *e*) $3.75 \times .15 \times 20 \times 4$
b) $17 \times 12 \times 6$ | *d*) $303 \times 8 \times 1\frac{2}{3}$ | *f*) $66\frac{2}{3}\% \times 50 \times 27$

Ans. *a*) $5 \times 40 \times 26$
$200 \times 26 = 5200$
b) $17 \times 6 \times 12$
$102 \times 12 = 1224$
c) $10\frac{1}{2} \times 2 \times 7$
$21 \times 7 = 147$
d) $303 \times 1\frac{2}{3} \times 8$
$505 \times 8 = 4040$
e) $3.75 \times 4 \times .15 \times 20$
$15 \times 3 = 45$
f) $66\frac{2}{3}\% \times 27 \times 50$
$\frac{2}{3} \times 27 \times 50 = 18 \times 50 = 900$

6. Rearrange the factors so that literal factors are arranged alphabetically and follow numerical factors: (3.2)

a) $r8$ *b*) cab *c*) $qp7$ *d*) $4v\,3t$ *e*) $x\,5y\,7w$ *f*) $def\,13c$ *g*) $2h\,5k\,10$ *h*) $r\,11sm\,4$ *i*) $cd\,13ab$
Ans. *a*) $8r$ *b*) abc *c*) $7pq$ *d*) $12tv$ *e*) $35wxy$ *f*) $13cdef$ *g*) $100hk$ *h*) $44mrs$ *i*) $13abcd$

7. Express each using symbols of operation: (4.1)

a) 10 added to r
b) 10 subtracted from r
c) r subtracted from 20
d) the sum of 7, x and y
e) the product of 8, m and n
f) 5 times p times q
g) two-thirds of c
h) one-half of b multiplied by h
i) f divided by 7
j) 25 divided by x
k) the product of u and v divided by 9.

Ans. *a*) $r + 10$ *c*) $20 - r$ *b*) $r - 10$ *d*) $x + y + 7$ | *e*) $8mn$ *g*) $\frac{2}{3}c$ *f*) $5pq$ *h*) $\frac{1}{2}bh$ | *i*) $\frac{f}{7}$ *k*) $\frac{uv}{9}$ *j*) $\frac{25}{x}$

8. When is each division impossible ? (4.2)

a) $\frac{7}{d}$ *b*) $\frac{r}{t}$ | *c*) $\frac{3}{4x}$ *d*) $\frac{5}{a-8}$ | *e*) $\frac{b}{7-c}$ *f*) $\frac{10}{2x-4}$ | *g*) $\frac{50}{w-y}$ *h*) $\frac{100}{pq}$ | *i*) $\frac{45}{x-2y}$

Ans. *a*) if $d = 0$ *b*) if $t = 0$ | *c*) if $x = 0$ *d*) if $a = 8$ | *e*) if $c = 7$ *f*) if $x = 2$ | *g*) if $w = y$ *h*) if $p = 0$ or $q = 0$ | *i*) if $x = 2y$

9. If n represents a number, express algebraically: (5.1, 5.2)

a) 25 more than the number
b) 30 greater than the number
c) the sum of the number and 35
d) the number increased by 40
e) 45 plus the number
f) 50 added to the number
g) 30 less than the number
h) 35 fewer than the number
i) 40 less the number
j) 45 decreased by the number
k) 50 minus the number
l) 55 subtracted from the number.

Ans. *a*) $n+25$ or $25+n$ *d*) $n+40$ or $40+n$
b) $n+30$ or $30+n$ *e*) $n+45$ or $45+n$
c) $n+35$ or $35+n$ *f*) $n+50$ or $50+n$
g) $n - 30$ *j*) $45 - n$
h) $n - 35$ *k*) $50 - n$
i) $40 - n$ *l*) $n - 55$

10. Express algebraically: (5.3)

a) the no. of tons of a weight that is 15 tons lighter than w tons. *Ans.* $w - 15$
b) the no. of ft in a length that is 50 ft shorter than l ft. $l - 50$
c) the no. of sec in a time interval that is 1 minute less than t sec. $t - 60$
d) the no. of cents in a price that is \$1 more than p cents. $p + 100$
e) the no. of ft per sec (fps) in a speed that is 20 fps slower than r fps. $r - 20$

f) the no. of ft in a distance that is 10 yd farther than d ft. *Ans.* $d+30$
g) the no. of sq ft in an area that is 30 sq ft greater than A sq ft. $A+30$
h) the no. of degrees in a temperature that is 40° colder than $t°$. $t-40$
i) the no. of floors in a building that is 8 floors higher than f floors. $f+8$
j) the no. of yr in an age 5 yr younger than a yr. $a-5$

11. Express algebraically: **(6.1)**

a) x times 3
b) one-eighth of b
c) product of 12 and y
d) three-eighths of r
e) 10 divided by y
f) quotient of y and 10

Ans. *a*) $3x$ *b*) $\frac{b}{8}$ or $\frac{1}{8}b$ *c*) $12y$ *d*) $\frac{3}{8}r$ or $\frac{3r}{8}$ *e*) $\frac{10}{y}$ *f*) $\frac{y}{10}$

12. Express algebraically: **(7.1 and 7.2)**

a) b decreased by one-half c
b) one-third of g decreased by 5
c) four times r divided by 9
d) the average of m and 60
e) three-quarters of x less y
f) twice d less 25
g) 8 more than the product of 5 and x
h) four times the sum of r and 9
i) the average of 60, m, p and q
j) the ratio of b to three times c.

Ans. *a*) $b-\frac{c}{2}$ *b*) $\frac{g}{3}-5$ *c*) $\frac{4r}{9}$ *d*) $\frac{m+60}{2}$ *e*) $\frac{3x}{4}-y$ *f*) $2d-25$ *g*) $5x+8$ *h*) $4(r+9)$ *i*) $\frac{m+p+q+60}{4}$ *j*) $\frac{b}{3c}$

13. Express algebraically: **(7.3)**

a) a distance in yd that is 25 yd shorter than three times another of d yd. *Ans.* $3d-25$
b) a weight in oz that is 5 oz more than twice another of w oz. $2w+5$
c) a temperature in degrees that is 8° warmer than five times another of $T°$. $5T+8$
d) a price in dollars that is \$50 dearer than one-half another of p dollars. $\frac{p}{2}+50$
e) a price in cents that is 50¢ cheaper than one-third another of p cents. $\frac{p}{3}-50$
f) a length in ft that is 2 ft longer than y yd. $3y+2$

14. Evaluate: **(8.1)**

a) $40-2\times5$
b) $3\times8-2\times5$
c) $40\div2+5$
d) $3+8-2\times5$
e) $16\div2-\frac{1}{2}\cdot10$
f) $3+8\times2\times5$
g) $40\times2-40\div2$
h) $3+8\times2-5\div10$

Ans. *a*) 30 *b*) 14 *c*) 25 *d*) 1 *e*) 3 *f*) 83 *g*) 60 *h*) $18\frac{1}{2}$

15. Evaluate if $a=5$, $b=6$ and $c=10$: **(8.2)**

a) $a+b-c$ *Ans.* 1
b) $a+2b$ *Ans.* 17
c) $a+\frac{b}{2}$ *Ans.* 8
d) $\frac{a+b}{2}$ *Ans.* $5\frac{1}{2}$
e) $\frac{3c}{a}$ *Ans.* 6
f) $3+\frac{c}{a}$ *Ans.* 5
g) $\frac{4}{5}c$ or $\frac{4c}{5}$ *Ans.* 8
h) $\frac{2}{3}b+\frac{3}{2}c$ *Ans.* 19
i) $\frac{a+b}{c-9}$ *Ans.* 11
j) $\frac{ab}{c}$ *Ans.* 3
k) $5a+4b-2c$ *Ans.* 29
l) $6c-2ab$ *Ans.* 0
m) $a+\frac{c-b}{2}$ *Ans.* 7
n) $a+c-\frac{b}{2}$ *Ans.* 12
o) $\frac{a+c-b}{3}$ *Ans.* 3

16. Evaluate if $x=3$, $y=2$ and $z=0$: **(8.3)**

a) $x+y+z$ *Ans.* 5
b) $x-y-z$ *Ans.* 1
c) $x(y+z)$ *Ans.* 6
d) $z(x+y)$ *Ans.* 0
e) $y(x+z)$ *Ans.* 6
f) $\frac{z}{x}$ *Ans.* 0
g) $\frac{x}{z}$ *Ans.* meaningless
h) xyz *Ans.* 0
i) $xy+z$ *Ans.* 6
j) $x+yz$ *Ans.* 3
k) $xz+yz$ *Ans.* 0
l) $\frac{z}{x+y}$ *Ans.* 0
m) $\frac{x}{y+z}$ *Ans.* $1\frac{1}{2}$
n) $x+\frac{z}{y}$ *Ans.* 3
o) $\frac{y+z}{x}$ *Ans.* $\frac{2}{3}$

17. Evaluate: **(9.1)**

a) $5(8+2)$ *Ans.* 50
b) $5(8-2)$ *Ans.* 30
c) $8+2(5-3)$ *Ans.* 12
d) $8(2\cdot5-3)$ *Ans.* 56
e) $8\cdot2(5-3)$ *Ans.* 32
f) $3(6+2\cdot5)$ *Ans.* 48
g) $(3\cdot6+2)5$ *Ans.* 100
h) $3(6+2)5$ *Ans.* 120
i) $4(4\cdot4-4)$ *Ans.* 48
j) $(4+4)4-4$ *Ans.* 28
k) $(4+4)(4-4)$ *Ans.* 0
l) $4+4(4-4)$ *Ans.* 4

18. Evaluate if $a=4$, $b=3$ and $c=5$: **(9.2)**

a) $a(b+c)$ *Ans.* 32
b) $b(c-a)$ *Ans.* 3
c) $c(a-b)$ *Ans.* 5
d) $\frac{1}{2}(a+b+c)$ *Ans.* 6
e) $\frac{1}{2}(a+b)+c$ *Ans.* $8\frac{1}{2}$
f) $3(b+2c)$ *Ans.* 39
g) $3(b+2)c$ *Ans.* 75
h) $(3b+2)c$ *Ans.* 55
i) $3a+2(c-b)$ *Ans.* 16
j) $3(a+2c)-b$ *Ans.* 39
k) $3(a+2c-b)$ *Ans.* 33
l) $3(a+2)(c-b)$ *Ans.* 36

19. Evaluate if $x=6$, $y=4$, $w=2$ and $z=0$: **(9.3)**

a) $x(y+w)$ *Ans.* 36
b) $z(w+x)$ *Ans.* 0
c) $w(x-y)$ *Ans.* 4
d) $wx(y+z)$ *Ans.* 48
e) $x+y(w+z)$ *Ans.* 14
f) $x+z(y-w)$ *Ans.* 6
g) $z\div(x+w)$ *Ans.* 0
h) $wy\div(z+x)$ *Ans.* $\frac{4}{3}$
i) $(w+x)(y-z)$ *Ans.* 32

20. State the number of terms and the terms in each expression: **(10.1)**

a) $5xyz$ *Ans.* 1 term: $5xyz$
b) $5+xyz$ *Ans.* 2 terms: 5 and xyz
c) $5+x+y+z$ *Ans.* 4 terms: $5, x, y$ and z
d) $3a+bc$ *Ans.* 2 terms: $3a$ and bc
e) $3ab+c$ *Ans.* 2 terms: $3ab$ and c
f) $3a(b+c)$ *Ans.* 1 term: $3a(b+c)$

21. State the factors of the following, disregarding 1 and the product itself: **(10.2)**

a) 77 *b*) 25 *c*) pq *d*) $\frac{3}{4}x$ *e*) $\frac{w}{10}$ *f*) $8(x-5)$ *g*) $\frac{y-2}{4}$

Ans. *a*) 7 and 11, *b*) 5 and 5, *c*) p and q, *d*) $\frac{3}{4}$ and x, *e*) $\frac{1}{10}$ and w, *f*) 8 and $(x-5)$, *g*) $\frac{1}{4}, (y-2)$

22. State each numerical and literal coefficient: **(10.3)**

a) w *b*) $\frac{1}{8}x$ *c*) $\frac{n}{10}$ *d*) $.03ab$ *e*) $\frac{3y}{10}$ *f*) $\frac{2a}{3b}$ *g*) $\frac{3}{5}(a-b)$

Ans.

	(*a*)	(*b*)	(*c*)	(*d*)	(*e*)	(*f*)	(*g*)
NUMERICAL COEFFICIENT:	1	$\frac{1}{8}$	$\frac{1}{10}$	.03	$\frac{3}{10}$	$\frac{2}{3}$	$\frac{3}{5}$
LITERAL COEFFICIENT:	w	x	n	ab	y	$\frac{a}{b}$	$a-b$

23. Write each, using bases and exponents: **(11.1)**

a) $7\cdot3\cdot3$ *b*) $7xyyy$ *c*) $\frac{7x}{yyy}$ *d*) $(7x)(7x)$ *e*) $(a+5)(a+5)$ *f*) $\frac{2rrw}{5stvv}$

Ans. *a*) $7\cdot3^2$ *b*) $7xy^3$ *c*) $\frac{7x}{y^3}$ *d*) $(7x)^2$ *e*) $(a+5)^2$ *f*) $\frac{2r^2w}{5stv^2}$

24. Write each without exponents: **(11.2)**

a) $4\cdot7^2$ *b*) $\frac{1}{2}y^4$ *c*) $\frac{5a}{b^4}$ *d*) $(ab)^3$ *e*) $(x+2)^2$ *f*) $\frac{a^2-b^3}{c+d^2}$

Ans. *a*) $4\cdot7\cdot7$ *b*) $\frac{1}{2}yyyy$ *c*) $\frac{5a}{bbbb}$ *d*) $(ab)(ab)(ab)$ *e*) $(x+2)(x+2)$ *f*) $\frac{aa-bbb}{c+dd}$

25. Evaluate (The table of powers may be used to check values): **(11.3)**

a) 3^3-2^3 *b*) $5^2\cdot2^5$ *c*) 10^3+5^4 *d*) $10^2\div2$ *e*) $1^2+2^2+3^2$ *f*) $5^3\div5$ *g*) $1^5+1^4+1^3+1^2$ *h*) $2^5-4\cdot2^2$ *i*) $5\cdot1^3-3\cdot1^5$ *j*) $\frac{1}{2}\cdot2^2+\frac{1}{3}\cdot3^3$

Ans. *a*) 19, *b*) 800 *c*) 1625, *d*) 50 *e*) 14, *f*) 25 *g*) 4, *h*) 16 *i*) 2, *j*) 11

26. Evaluate: **(11.4)**

a) $.1^2\cdot9^2$ *b*) $.3\cdot4^2$ *c*) $3^2 4^2$ *d*) $40(\frac{1}{2})^3$ *e*) $(\frac{2}{5})^3$ *f*) $\frac{2^3}{5^2}$ *g*) $\frac{10}{.1^2}$

Ans. *a*) .81 *b*) 4.8 *c*) 144 *d*) 5 *e*) $\frac{8}{125}$ *f*) $\frac{8}{25}$ *g*) 1000

27. Evaluate if $a=3$ and $b=2$: **(11.5)**

a) a^2b *Ans.* 18
b) ab^2 *Ans.* 12
c) $(ab)^2$ *Ans.* 36
d) $a+b^2$ *Ans.* 7
e) $(a+b)^2$ *Ans.* 25
f) a^2+b^2 *Ans.* 13
g) a^3b *Ans.* 54
h) $(ab)^3$ *Ans.* 216
i) a^3-b^3 *Ans.* 19
j) $(a-b)^3$ *Ans.* 1
k) a^2b^3 *Ans.* 72
l) a^3b^2 *Ans.* 108

28. Evaluate if $w=1$, $x=3$, $y=4$: **(11.5)**

a) $2w^2$ *Ans.* 2
b) $(2w)^2$ *Ans.* 4
c) $(x+2)^2$ *Ans.* 25
d) y^2+x^2 *Ans.* 25
e) $(y+x)^2$ *Ans.* 49
f) y^2-x^2 *Ans.* 7
g) $(y-x)^2$ *Ans.* 1
h) $(w+x+y)^2$ *Ans.* 64
i) $w^2+x^2+y^2$ *Ans.* 26
j) y^3x *Ans.* 192
k) yx^3 *Ans.* 108
l) $(yx)^3$ *Ans.* 1728

29. Combine: **(12.1)**

a) $20+10-18$
b) $10-6-1$
c) $6x+5x+x$
d) $6x^2+5x^2+x^2$
e) $13y^2-y^2+10y^2$
f) $27w^5-22w^5$
g) $3pq+11pq-pq$
h) $2abc+abc-3abc$
i) $36a^2b^2c-23a^2b^2c$
j) $8(a+b)-2(a+b)$
k) $11(x^2+y^2)+4(x^2+y^2)$
l) $5(x+y)^2-(x+y)^2$

Ans. a) 12 b) 3 c) $12x$ d) $12x^2$ e) $22y^2$ f) $5w^5$ g) $13pq$ h) 0 i) $13a^2b^2c$ j) $6(a+b)$ k) $15(x^2+y^2)$ l) $4(x+y)^2$

30. Simplify each expression by combining like terms: **(12.2)**

a) $13b+7b-6$ *Ans.* $20b-6$
b) $13b+7-6$ *Ans.* $13b+1$
c) $13b+7b-6b$ *Ans.* $14b$
d) $13b^2+7b^2-6b$ *Ans.* $20b^2-6b$
e) $13b^2+7b-6b^2$ *Ans.* $7b^2+7b$
f) $5y^2+3y^2+10y-2$ *Ans.* $8y^2+10y-2$
g) $5y^2+3y+10y-2y$ *Ans.* $5y^2+11y$
h) $5y^2+3y+10y^2-2y$ *Ans.* $15y^2+y$
i) $5y+3y^2+10y^2-2y^2$ *Ans.* $11y^2+5y$
j) $5+3y+10y^2-2$ *Ans.* $10y^2+3y+3$

31. Combine like terms: **(12.3)**

a) $3y+7\frac{2}{3}y$
b) $20\frac{1}{2}a-11a$
c) $\frac{7}{3}c-\frac{1}{3}c$
d) $5d-\frac{d}{5}$
e) $\frac{x^2}{2}-\frac{x^2}{3}$
f) $\frac{y^2}{2}+\frac{y^2}{6}$
g) $7.1ab+3.9ab-2.7ab$
h) $2.12c^2-1.09c^2-.55c^2$
i) $3\frac{5}{12}xy^2+4\frac{1}{6}xy^2-2xy^2$

Ans. a) $10\frac{2}{3}y$ b) $9\frac{1}{2}a$ c) $2c$ d) $4\frac{4}{5}d$ e) $\frac{x^2}{6}$ f) $\frac{2}{3}y^2$ g) $8.3ab$ h) $.48c^2$ i) $5\frac{7}{12}xy^2$

32. Represent the length of each entire line: **(12.4)**

a) 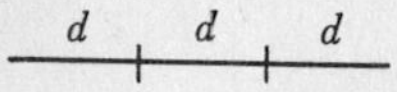

d) 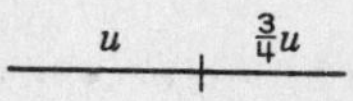

g) 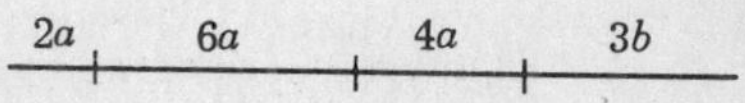

b) 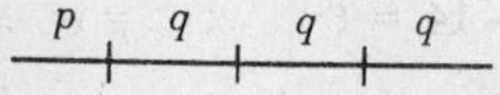

e) 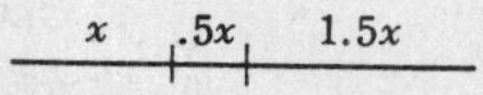

h) 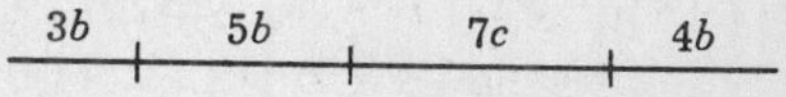

c) 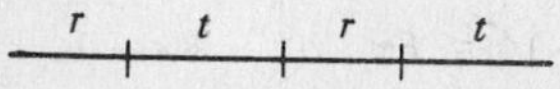

f) 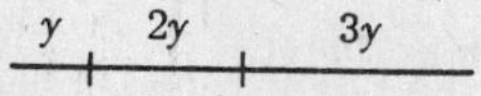

i) 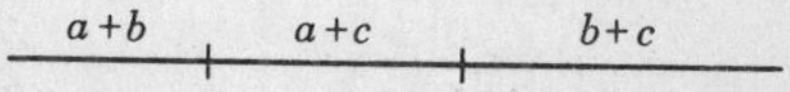

Ans. a) $3d$ b) $p+3q$ c) $2r+2t$ d) $1\frac{3}{4}u$ e) $3x$ f) $6y$ g) $12a+3b$ h) $12b+7c$ i) $2a+2b+2c$

Chapter 2

SIMPLE EQUATIONS and their SOLUTIONS

1. KINDS OF EQUALITIES: EQUATIONS AND IDENTITIES

An **equality** is a mathematical statement that two expressions are equal, or have the same value.

Thus, $2n = 6$ and $2n + 3n = 5n$ are equalities.

In an equality, the expression to the left of the equal sign is called the **left member** or **left side** of the equality; the expression to the right is the **right member** or **right side** of the equality.

Thus, in $6n = 3n - 9$, $6n$ is the left member or left side.
while $3n - 9$ is the right member or right side.

An **equation** is an equality in which the unknown or unknowns may have only a particular value or values. An equation is a conditional equality.

Thus, $2n = 12$ is an equation since n may have only one value, 6.

An **identity** is an equality in which a letter or letters may have any value. An identity is an unconditional equality.

Thus, $2n + 3n = 5n$ and $x + y = y + x$ are identities since there is no restriction on the values that n, x and y may have.

A **root of an equation** is any number which when substituted for the unknown will make both sides of the equation equal. A root is said to **satisfy the equation.**

Thus, 6 is a root of $2n = 12$, while 5 or any other number is not.

Checking an equation is the process of substituting a particular value for an unknown to see if the value will make both sides equal.

Thus, check in $2x + 3 = 11$ for $x = 4$ and $x = 5$ as follows:

$2x + 3 = 11$	$2x + 3 = 11$	**Note.**
$2(4) + 3 \stackrel{?}{=} 11$	$2(5) + 3 \stackrel{?}{=} 11$	(*1*) The symbol $\stackrel{?}{=}$ is read "should equal".
$8 + 3 \stackrel{?}{=} 11$	$10 + 3 \stackrel{?}{=} 11$	(2) The symbol $\neq$ is read "does not equal".
$11 = 11$	$13 \neq 11$	

Hence, 4 is a root of $2x + 3 = 11$ since it satisfies the equation.

1.1. Checking an Equation

By checking, determine which value is a root of each equation:

a) Check $2n + 3n = 25$ for $n = 5$ and $n = 6$

Check: $n = 5$	$n = 6$
$2n + 3n = 25$	$2n + 3n = 25$
$2(5) + 3(5) \stackrel{?}{=} 25$	$2(6) + 3(6) \stackrel{?}{=} 25$
$10 + 15 \stackrel{?}{=} 25$	$12 + 18 \stackrel{?}{=} 25$
$25 = 25$	$30 \neq 25$

Ans. 5 is a root of $2n + 3n = 25$

b) Check $8x - 14 = 6x$ for $x = 6$ and $x = 7$

Check: $x = 6$	$x = 7$
$8x - 14 = 6x$	$8x - 14 = 6x$
$8(6) - 14 \stackrel{?}{=} 6(6)$	$8(7) - 14 \stackrel{?}{=} 6(7)$
$48 - 14 \stackrel{?}{=} 36$	$56 - 14 \stackrel{?}{=} 42$
$34 \neq 36$	$42 = 42$

Ans. 7 is a root of $8x - 14 = 6x$

1.2. Checking an Identity

By checking the identity $4(x+2) = 4x + 8$, show that x may have any of the following values:

a) $x = 10$	*b*) $x = 6$	*c*) $x = 4\frac{1}{2}$	*d*) $x = 3.2$
Check:			
a) $4(x+2) = 4x + 8$	*b*) $4(x+2) = 4x + 8$	*c*) $4(x+2) = 4x + 8$	*d*) $4(x+2) = 4x + 8$
$4(10+2) \stackrel{?}{=} 4(10) + 8$	$4(6+2) \stackrel{?}{=} 4(6) + 8$	$4(4\frac{1}{2}+2) \stackrel{?}{=} 4(4\frac{1}{2}) + 8$	$4(3.2+2) \stackrel{?}{=} 4(3.2) + 8$
$4(12) \stackrel{?}{=} 40 + 8$	$4(8) \stackrel{?}{=} 24 + 8$	$4(6\frac{1}{2}) \stackrel{?}{=} 18 + 8$	$4(5.2) \stackrel{?}{=} 12.8 + 8$
$48 = 48$	$32 = 32$	$26 = 26$	$20.8 = 20.8$

2. TRANSLATING VERBAL STATEMENTS INTO EQUATIONS

In algebra, a verbal problem is solved when the value of its unknown (or unknowns) is found. In the process it is necessary to "translate" verbal statements into equations. The first step is to choose a letter to represent the unknown.

Thus, by letting n represent the unknown number, "Twice what number equals 12" becomes "$2n = 12$".

2.1. Translating Statements into Equations

Translate into an equation, letting n represent the number:

(You need not find the value of the unknown.)

a) 4 less than what number equals 8 ? — *Ans.* *a*) $n - 4 = 8$

b) One-half of what number equals 10 ? — *b*) $\frac{n}{2} = 10$

c) Ten times what number equals 20 ? — *c*) $10n = 20$

d) What number increased by 12 equals 17 ? — *d*) $n + 12 = 17$

e) Twice what number added to 8 is 16 ? — *e*) $2n + 8 = 16$

f) 15 less than three times what number is 27? — *f*) $3n - 15 = 27$

g) The sum of what number and twice the same number is 18 ? — *g*) $n + 2n = 18$

h) What number and 4 more equals five times the number ? — *h*) $n + 4 = 5n$

i) Twice the sum of a certain number and five is 24. What is the number ? — *i*) $2(n+5) = 24$

2.2. Matching Statements and Equations

Match the statements in Column 1 with the equations in Column 2:

Column 1	Column 2
1. The product of 8 and a number is 40.	*a*) $n - 8 = 40$
2. A number increased by 8 is 40.	*b*) $8(n+8) = 40$
3. 8 less than a number equals 40.	*c*) $8n = 40$
4. Eight times a number less 8 is 40.	*d*) $\frac{n}{8} = 40$
5. Eight times the sum of a number and 8 is 40.	*e*) $8n - 8 = 40$
6. One-eighth of a number is 40.	*f*) $n + 8 = 40$

Ans. *1* and *c*, *2* and *f*, *3* and *a*, *4* and *e*, *5* and *b*, *6* and *d*.

2.3. Representing Unknowns

Represent the unknown by a letter and obtain an equation for each problem:

(You need not solve each equation.)

a) A man worked for 5 hours and earned \$8.75. What was his hourly wage ?

Ans. *a*) Let w = his hourly wage in dollars. Then, $5w = 8.75$

b) How old is Henry now, if ten years ago, he was 23 years old ?

b) Let H = Henry's age now. Then, $H - 10 = 23$

c) After gaining 12 lb., Mary weighed 120 lb. What was her previous weight?

d) A baseball team won four times as many games as it lost. How many games did it lose, if it played a total of 100 games?

Ans. *c*) Let M = Mary's previous weight in lb. Then, $M + 12 = 120$

d) Let n = no. of games lost and $4n$ = no. of games won. Then, $n + 4n = 100$

3. SOLVING SIMPLE EQUATIONS USING INVERSE OPERATIONS

For the present, we will study simple equations containing a single unknown having one value. **To solve a simple equation** is to find this value of the unknown. This value is the root of the equation.

Thus, the equation $2n = 12$ is solved when n is found to equal 6.

To solve an equation, think of it as asking a question such as in each of the following equations:

Equation	Question Asked By Equation	Finding Root Of Equation
1. $n + 4 = 12$	What number plus 4 equals 12?	$n = 12 - 4 = 8$
2. $n - 4 = 12$	What number minus 4 equals 12?	$n = 12 + 4 = 16$
3. $4n = 12$	What number multiplied by 4 equals 12?	$n = 12 \div 4 = 3$
4. $\frac{n}{4} = 12$	What number divided by 4 equals 12?	$n = 12 \cdot 4 = 48$

Note the two operations involved in each of the above cases:

1. The equation, $n + 4 = 12$, involving **addition** is solved by **subtracting** 4 from 12.
2. The equation, $n - 4 = 12$, involving **subtraction** is solved by **adding** 4 to 12.
3. The equation, $4n = 12$, involving **multiplication** is solved by **dividing** 4 into 12.
4. The equation, $\frac{n}{4} = 12$, involving **division** is solved by **multiplying** 4 by 12.

Inverse operations are two operations such that if one is involved with the unknown in the equation, then the other is used to solve the equation.

Rule 1. Addition and subtraction are inverse operations.
Thus, in $n+6=10$, n and 6 are **added**. To find n, **subtract** 6 from 10.
In $n-3=9$, 3 is **subtracted** from n. To find n, **add** 3 to 9.

Rule 2. Multiplication and division are inverse operations.
Thus, in $7n = 35$, n and 7 are **multiplied**. To find n, **divide** 7 into 35.
In $\frac{n}{11} = 4$, n is **divided** by 11. To find n, **multiply** 11 by 4.

Note. In **3.1**, **Rule 1** is applied to such equations as $x - 10 = 2$ and $w - 20 = 12$. Later in the chapter, the solution of such equations as $10 - x = 2$ and $20 - w = 12$ will be considered. In **3.2**, **Rule 2** is applied to such equations as $\frac{x}{3} = 12$ and $\frac{y}{5} = 10$. Later in the chapter, the solution of such equations as $\frac{3}{x} = 12$ and $\frac{5}{y} = 10$ will be considered.

3.1. Rule 1: Addition and Subtraction are Inverse Operations

Solve each equation:

Equations Involving Addition Of Unknown	Solutions Requiring Subtraction *Ans.*	Equations Involving Subtraction From Unknown	Solutions Requiring Addition *Ans.*
a) $x + 3 = 8$	*a*) $x = 8 - 3$ or 5	*e*) $x - 10 = 2$	*e*) $x = 2 + 10$ or 12
b) $5 + y = 13$	*b*) $y = 13 - 5$ or 8	*f*) $w - 20 = 12$	*f*) $w = 12 + 20$ or 32
c) $15 = a + 10$	*c*) $a = 15 - 10$ or 5	*g*) $18 = a - 13$	*g*) $a = 18 + 13$ or 31
d) $28 = 20 + b$	*d*) $b = 28 - 20$ or 8	*h*) $21 = b - 2$	*h*) $b = 21 + 2$ or 23

3.2. Rule 2: Multiplication and Division are Inverse Operations

Solve each equation:

Equations Involving Multiplication Of Unknown	Solutions Requiring Division *Ans.*	Equations Involving Division Of Unknown	Solutions Requiring Multiplication *Ans.*
a) $3x = 12$	*a*) $x = \frac{12}{3}$ or 4	*e*) $\frac{x}{3} = 12$	*e*) $x = 12 \cdot 3$ or 36
b) $12y = 3$	*b*) $y = \frac{3}{12}$ or $\frac{1}{4}$	*f*) $\frac{y}{12} = 3$	*f*) $y = 3 \cdot 12$ or 36
c) $35 = 7a$	*c*) $a = \frac{35}{7}$ or 5	*g*) $4 = \frac{a}{7}$	*g*) $a = 4 \cdot 7$ or 28
d) $7 = 35b$	*d*) $b = \frac{7}{35}$ or $\frac{1}{5}$	*h*) $7 = \frac{b}{4}$	*h*) $b = 7 \cdot 4$ or 28

3.3. Solving by Using Inverse Operations

Solve each equation, showing operation used to solve:

	Ans.		*Ans.*
a) $x + 5 = 20$	*a*) $x = 20 - 5$ or 15	*i*) $14 = a - 7$	*i*) $a = 14 + 7$ or 21
b) $x - 5 = 20$	*b*) $x = 20 + 5$ or 25	*j*) $14 = 7a$	*j*) $a = \frac{14}{7}$ or 2
c) $5x = 20$	*c*) $x = \frac{20}{5}$ or 4	*k*) $14 = \frac{a}{7}$	*k*) $a = 14(7)$ or 98
d) $\frac{x}{5} = 20$	*d*) $x = 20(5)$ or 100	*l*) $b - 8 = 2$	*l*) $b = 2 + 8$ or 10
e) $10 + y = 30$	*e*) $y = 30 - 10$ or 20	*m*) $8b = 2$	*m*) $b = \frac{2}{8}$ or $\frac{1}{4}$
f) $10y = 30$	*f*) $y = \frac{30}{10}$ or 3	*n*) $\frac{b}{8} = 2$	*n*) $b = 2(8)$ or 16
g) $\frac{y}{10} = 30$	*g*) $y = 30(10)$ or 300	*o*) $24 = 6 + c$	*o*) $c = 24 - 6$ or 18
h) $14 = a + 7$	*h*) $a = 14 - 7$ or 7	*p*) $6 = 24c$	*p*) $c = \frac{6}{24}$ or $\frac{1}{4}$

4. RULES OF EQUALITY FOR SOLVING EQUATIONS

1. **The Addition Rule of Equality**

   To keep an equality, equal numbers may be **added to** both sides.

2. **The Subtraction Rule of Equality**

   To keep an equality, equal numbers may be **subtracted from** both sides.

3. **The Multiplication Rule of Equality**

   To keep an equality, both sides may be **multiplied by** equal numbers.

4. **The Division Rule of Equality**

   To keep an equality, both sides may be **divided by** equal numbers, except by zero.

These four rules may be summed up in one rule:

The Rule of Equality for All Operations

To keep an equality, the same operation, using equal numbers, may be performed on both sides, except division by zero.

To understand these **Rules of Equality**, think of an **equality** as a **scale in balance**.

If only one side of a balanced scale is changed, the scale becomes unbalanced. To balance the scale, exactly the same change must be made on the other side. Similarly, if only one side of an equality is changed, the two sides are no longer equal. To keep an equality, exactly the same change must be made on both sides.

Balanced Scales

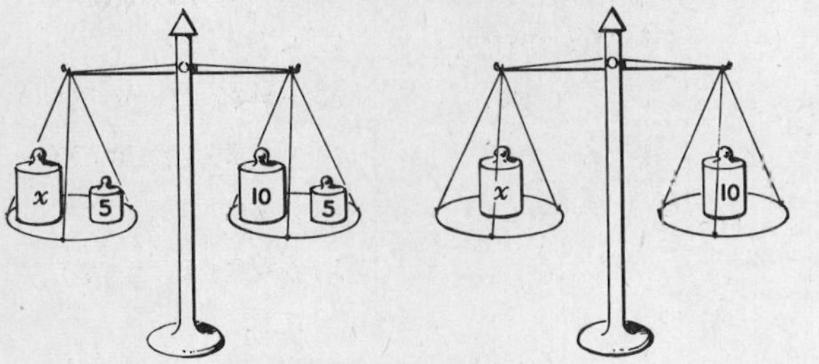

Thus, if 5 is subtracted from both sides of a balanced scale, the scale is still in balance.

EQUALITIES

$$\begin{array}{rcl} x + 5 &=& 15 \\ -5 &=& -5 \\ \hline x &=& 10 \end{array}$$

If 5 is subtracted from both sides of an equality, an equality remains.

4.1. Using Rules of Equality

State the equality rule used to solve each equation:

a) $\begin{array}{rcl} x+15 &=& 21 \\ -15 &=& -15 \\ \hline x &=& 6 \end{array}$ *b*) $\begin{array}{rcl} 40 &=& r-8 \\ +8 &=& +8 \\ \hline 48 &=& r \end{array}$ *c*) $\begin{array}{rcl} 25 &=& 5m \\ \frac{25}{5} &=& \frac{5m}{5} \\ 5 &=& m \end{array}$ *d*) $\begin{array}{rcl} \frac{n}{8} &=& 3 \\ 8\cdot\frac{n}{8} &=& 8\cdot 3 \\ n &=& 24 \end{array}$ *e*) $\begin{array}{rcl} 24x &=& 8 \\ \frac{24x}{24} &=& \frac{8}{24} \\ x &=& \frac{1}{3} \end{array}$

Ans. a) subtraction rule, *b*) addition rule, *c*) division rule, *d*) multiplication rule, *e*) division rule.

5. USING DIVISION TO SOLVE AN EQUATION

Division Rule of Equality

To keep an equality, both sides may be divided by equal numbers, except by zero.

To Solve an Equation Using the Division Rule of Equality

Procedure:	Solve: *a*) $2n = 16$	*b*) $16n = 2$
	Solutions:	
1) **Divide both sides of the equation by the coefficient or multiplier of the unknown:**	$D_2 \quad \frac{2n}{2} = \frac{16}{2}$	$D_{16} \quad \frac{16n}{16} = \frac{2}{16}$
	Ans. $n = 8$	*Ans.* $n = \frac{1}{8}$
2) **Check** the original equation:	**Check:** $2n = 16$	$16n = 2$
	$2(8) \stackrel{?}{=} 16$	$16(\frac{1}{8}) \stackrel{?}{=} 2$
	$16 = 16$	$2 = 2$

Note 1. "D" is a convenient symbol for "dividing both sides".
"D_2" means "divide both sides by 2".

Note 2. A common factor may be eliminated in $\frac{\overset{1}{\cancel{2}}n}{\cancel{2}}$ and $\frac{\overset{1}{\cancel{16}}n}{\cancel{16}}$.

5.1. Solving Equations with Integral Coefficients

Solve each equation:

a) $7x = 35$

$\mathbf{D}_7$ $\dfrac{7x}{7} = \dfrac{35}{7}$

Ans. $x = 5$

Check:

$7x = 35$

$7(5) \stackrel{?}{=} 35$

$35 = 35$

b) $35y = 7$

$\mathbf{D}_{35}$ $\dfrac{35y}{35} = \dfrac{7}{35}$

Ans. $y = \frac{1}{5}$

Check:

$35y = 7$

$35(\frac{1}{5}) \stackrel{?}{=} 7$

$7 = 7$

c) $33 = 11z$

$\mathbf{D}_{11}$ $\dfrac{33}{11} = \dfrac{11z}{11}$

Ans. $3 = z$

Check:

$33 = 11z$

$33 \stackrel{?}{=} 11(3)$

$33 = 33$

d) $11 = 33w$

$\mathbf{D}_{33}$ $\dfrac{11}{33} = \dfrac{33w}{33}$

Ans. $\frac{1}{3} = w$

Check:

$11 = 33w$

$11 \stackrel{?}{=} 33(\frac{1}{3})$

$11 = 11$

5.2. Division in Equations with Decimal Coefficients

Solve each equation:

a) $.3a = 9$

$\mathbf{D}_{.3}$ $\dfrac{.3a}{.3} = \dfrac{9}{.3}$

Ans. $a = 30$

Check:

$.3a = 9$

$.3(30) \stackrel{?}{=} 9$

$9 = 9$

b) $1.2b = 48$

$\mathbf{D}_{1.2}$ $\dfrac{1.2b}{1.2} = \dfrac{48}{1.2}$

Ans. $b = 40$

Check:

$1.2b = 48$

$1.2(40) \stackrel{?}{=} 48$

$48 = 48$

c) $15 = .05c$

$\mathbf{D}_{.05}$ $\dfrac{15}{.05} = \dfrac{.05c}{.05}$

Ans. $300 = c$

Check:

$15 = .05c$

$15 \stackrel{?}{=} .05(300)$

$15 = 15$

5.3. Solving Equations with Percents as Coefficients

Solve each equation (*Hint: First, replace each percent by a decimal.*):

a) $22\% s = 88$

Since $22\% = .22$,

$\mathbf{D}_{.22}$ $\dfrac{.22s}{.22} = \dfrac{88}{.22}$

Ans. $s = 400$

Check:

$22\% s = 88$

$(.22)(400) \stackrel{?}{=} 88$

$88 = 88$

b) $75\% t = 18$

Since $75\% = .75$,

$\mathbf{D}_{.75}$ $\dfrac{.75t}{.75} = \dfrac{18}{.75}$

Ans. $t = 24$

Check:

$75\% t = 18$

$(.75)(24) \stackrel{?}{=} 18$

$18 = 18$

c) $72 = 2\% n$

Since $2\% = .02$,

$\mathbf{D}_{.02}$ $\dfrac{72}{.02} = \dfrac{.02n}{.02}$

Ans. $3600 = n$

Check:

$72 = 2\% n$

$72 \stackrel{?}{=} (.02)(3600)$

$72 = 72$

5.4. Solving Equations with Like Terms on One Side

Solve each equation (*Hint: First, collect like terms.*):

a) $60 = 7x - x$

$60 = 6x$

$\mathbf{D}_6$ $\dfrac{60}{6} = \dfrac{6x}{6}$

Ans. $10 = x$

Check:

$60 = 7x - x$

$60 \stackrel{?}{=} 70 - 10$

$60 = 60$

b) $3x + 5x = 48$

$8x = 48$

$\mathbf{D}_8$ $\dfrac{8x}{8} = \dfrac{48}{8}$

Ans. $x = 6$

Check:

$3x + 5x = 48$

$18 + 30 \stackrel{?}{=} 48$

$48 = 48$

c) $7x - 2x = 55$

$5x = 55$

$\mathbf{D}_5$ $\dfrac{5x}{5} = \dfrac{55}{5}$

Ans. $x = 11$

Check:

$7x - 2x = 55$

$77 - 22 \stackrel{?}{=} 55$

$55 = 55$

5.5. Division Rule in a Wage Problem

John worked 7 hr. and earned $8.54. What was his hourly wage?

Solution: Let h = his hourly wage in $.

Then, $7h = 8.54$

D_7 $\quad \frac{7h}{7} = \frac{8.54}{7}$

$h = 1.22$

Ans. John's hourly wage was $1.22.

Check (the problem):

In 7 hr., John should earn $8.54. Hence,

$7(\$1.22) \stackrel{?}{=} \$8.54$

$\$8.54 = \$8.54$

5.6. Division Rule in a Commission Problem

Mr. Black's commission rate was 3%. If he earned $66 in commission, how much did he sell?

Solution: Let s = Mr. Black's sales in $.

Then, $3\%s$ or $.03s = 66$

$D_{.03}$ $\quad \frac{.03s}{.03} = \frac{66}{.03}$

$s = 2200$

Ans. Mr. Black's sales were $2200.

Check (the problem):

At 3%, Mr. Black's commission should be $66. Hence,

$3\% \text{ of } \$2200 \stackrel{?}{=} \$66$

$(.03)(\$2200) \stackrel{?}{=} \$66$

$\$66 = \$66$

6. USING MULTIPLICATION TO SOLVE AN EQUATION

Multiplication Rule of Equality

To keep an equality, both sides may be multiplied by equal numbers.

To Solve an Equation Using the Multiplication Rule of Equality

Procedure:	Solve: *a*) $\frac{w}{3} = 5$	*b*) $10 = \frac{x}{7}$
	Solutions:	
1. Multiply both sides of the equation by the divisor of the unknown:	M_3 $\quad 3\cdot\frac{w}{3} = 5\cdot 3$	M_7 $\quad 7\cdot 10 = \frac{x}{7}\cdot 7$
	Ans. $\quad w = 15$	*Ans.* $\quad 70 = x$
2. Check the original equation:	**Check:** $\frac{w}{3} = 5$	**Check:** $10 = \frac{x}{7}$
	$\frac{15}{3} \stackrel{?}{=} 5$	$10 \stackrel{?}{=} \frac{70}{7}$
	$5 = 5$	$10 = 10$

Note 1. "M" is a convenient symbol for "multiplying both sides". "M_3" means "multiply both sides by 3".

Note 2. A common factor may be eliminated in $\overset{1}{\cancel{3}}\cdot\frac{w}{\cancel{3}}$ and $\frac{x}{\cancel{7}}\cdot\overset{1}{\cancel{7}}$

Dividing by a Fraction

To divide by a fraction, invert the fraction and multiply. Thus, $8 \div \frac{2}{3} = 8 \times \frac{3}{2}$ or 12. Hence, multiplying by $\frac{3}{2}$ is equivalent to dividing by $\frac{2}{3}$.

To Solve an Equation Whose Unknown has a Fractional Coefficient

Procedure:	Solve: *a*) $\frac{2}{3}x = 8$	*b*) $\frac{5}{3}y = 25$
	Solutions:	
1. Multiply both sides of the equation by the fractional coefficient inverted: *(instead of dividing by the fractional coefficient)*	$M_{3/2}$ $\quad \frac{3}{2}\cdot\frac{2}{3}x = 8\cdot\frac{3}{2}$	$M_{3/5}$ $\quad \frac{3}{5}\cdot\frac{5}{3}y = 25\cdot\frac{3}{5}$
	Ans. $\quad x = 12$	*Ans.* $\quad y = 15$
2. Check the original equation:	**Check:** $\frac{2}{3}x = 8$	**Check:** $\frac{5}{3}y = 25$
	$\frac{2}{3}\cdot 12 \stackrel{?}{=} 8$	$\frac{5}{3}\cdot 15 \stackrel{?}{=} 25$
	$8 = 8$	$25 = 25$

6.1. Solving Equations with Integral Divisors

Solve each equation:

a)	*b)*	*c)*	*d)*
$\frac{x}{8} = 4$	$\frac{1}{3}y = 12$	$20 = \frac{z}{10}$	$.2 = \frac{w}{40}$
M_8 $8 \cdot \frac{x}{8} = 4 \cdot 8$	M_3 $3 \cdot \frac{1}{3}y = 12 \cdot 3$	M_{10} $10 \cdot 20 = \frac{z}{10} \cdot 10$	M_{40} $40(.2) = \frac{w}{40} \cdot 40$
Ans. $x = 32$	*Ans.* $y = 36$	*Ans.* $200 = z$	*Ans.* $8 = w$
Check:	**Check:**	**Check:**	**Check:**
$\frac{x}{8} = 4$	$\frac{1}{3}y = 12$	$20 = \frac{z}{10}$	$.2 = \frac{w}{40}$
$\frac{32}{8} \stackrel{?}{=} 4$	$\frac{1}{3}(36) \stackrel{?}{=} 12$	$20 \stackrel{?}{=} \frac{200}{10}$	$.2 \stackrel{?}{=} \frac{8}{40}$
$4 = 4$	$12 = 12$	$20 = 20$	$.2 = .2$

6.2. Solving Equations with Decimal Divisors

Solve each equation:

a)	*b)*	*c)*
$\frac{a}{.5} = 4$	$\frac{b}{.08} = 400$	$1.5 = \frac{c}{1.2}$
$M_{.5}$ $.5(\frac{a}{.5}) = 4(.5)$	$M_{.08}$ $.08(\frac{b}{.08}) = 400(.08)$	$M_{1.2}$ $1.2(1.5) = (\frac{c}{1.2})1.2$
Ans. $a = 2$	*Ans.* $b = 32$	*Ans.* $1.8 = c$

(Check your answers.)

6.3. Solving Equations with Fractional Coefficients

Solve each equation:
Hint: Multiply by the fractional coefficient inverted.

a)	*b)*	*c)*
$\frac{2}{5}x = 10$	$1\frac{1}{3}w = 30$	$c - \frac{1}{4}c = 24$
Solutions:	$\frac{4}{3}w = 30$	$\frac{3}{4}c = 24$
$M_{5/2}$ $\frac{5}{2} \cdot \frac{2}{5}x = 10(\frac{5}{2})$	$M_{3/4}$ $\frac{3}{4} \cdot \frac{4}{3}w = 30(\frac{3}{4})$	$M_{4/3}$ $\frac{4}{3} \cdot \frac{3}{4}c = 24(\frac{4}{3})$
Ans. $x = 25$	*Ans.* $w = 22\frac{1}{2}$	*Ans.* $c = 32$

(Check your answers.)

6.4. Solving Equations with Percents as Coefficients

Solve each equation:
Hint. Replace a percent by a fraction if the percent equals an easy fraction.

a)	*b)*	*c)*
$66\frac{2}{3}\% s = 22$	$87\frac{1}{2}\% t = 35$	$120\% w = 72$
Solutions:		
$\frac{2}{3}s = 22$	$\frac{7}{8}t = 35$	$\frac{6}{5}w = 72$
$M_{3/2}$ $\frac{3}{2} \cdot \frac{2}{3}s = 22(\frac{3}{2})$	$M_{8/7}$ $\frac{8}{7} \cdot \frac{7}{8}t = 35(\frac{8}{7})$	$M_{5/6}$ $\frac{5}{6} \cdot \frac{6}{5}w = 72(\frac{5}{6})$
Ans. $s = 33$	*Ans.* $t = 40$	*Ans.* $w = 60$

(Check your answers.)

6.5. Multipiication Rule in Distance Problem

After traveling 84 mi., Henry found that he had gone three-fourths of the entire distance to his home. What is the total distance to his home ?

Solution:

Let d = the total distance in mi.

then, $\frac{3}{4}d = 84$

$M_{4/3}$ $\frac{4}{3}\cdot\frac{3}{4}d = 84\left(\frac{4}{3}\right)$

$d = 112$

Ans. The total distance is 112 miles.

Check (the problem):

The 84 miles traveled should be $\frac{3}{4}$ of the entire distance.

Hence,

$84 \text{ mi.} \overset{?}{=} \frac{3}{4}(112 \text{ mi.})$

$84 \text{ mi.} = 84 \text{ mi.}$

6.6. Multiplication Rule in Investment Problem

Mr. White receives 5% on a stock investment. If his interest at the end of one year was $140, how large was his investment?

Solution:

Let s = the sum invested in $.

then, $5\% s$ or $\frac{s}{20} = 140$

M_{20} $20\cdot\frac{s}{20} = 140(20)$

$s = 2800$

Ans. The investment was $2800.

Check (the problem):

5% of the investment should be $140.

Hence,

$5\%(\$2800) \overset{?}{=} \$140$

$\$140 = \$140$

7. USING SUBTRACTION TO SOLVE AN EQUATION

Subtraction Rule of Equality

To keep an equality, equal numbers may be subtracted from both sides.

To Solve an Equation Using the Subtraction Rule of Equality

Procedure:		Solve: a) $w + 12 = 19$	b) $28 = 11 + x$
1. **Subtract from both sides the number added to the unknown:**	**Solutions:**	$w + 12 = 19$	$28 = 11 + x$
	S_{12}	$-12 = -12$	S_{11} $-11 = -11$
	Ans.	$w = 7$	*Ans.* $17 = x$
2. **Check** the original equation:	**Check:**	$w + 12 = 19$	$28 = 11 + x$
		$7 + 12 \overset{?}{=} 19$	$28 \overset{?}{=} 11 + 17$
		$19 = 19$	$28 = 28$

Note. "S" is a convenient symbol for "subtracting from both sides".
"S_{11}" means "subtract 11 from both sides".

7.1. Subtraction Rule in Equations Containing Integers

Solve each equation:

a) $r + 8 = 13$
S_8 $-8 = -8$
Ans. $r = 5$

Check:

$r + 8 = 13$
$5 + 8 \overset{?}{=} 13$
$13 = 13$

b) $15 + t = 60$
S_{15} $-15 = -15$
Ans. $t = 45$

Check:

$15 + t = 60$
$15 + 45 \overset{?}{=} 60$
$60 = 60$

c) $110 = s + 20$
S_{20} $-20 = -20$
Ans. $90 = s$

Check:

$110 = s + 20$
$110 \overset{?}{=} 90 + 20$
$110 = 110$

7.2. Subtraction Rule in Equations Containing Fractions or Decimals

Solve each equation:

a)

$$b + \frac{1}{3} = 3\frac{2}{3}$$
$$S_{\frac{1}{3}} \quad -\frac{1}{3} = -\frac{1}{3}$$

Ans. $b = 3\frac{1}{3}$

Check:

$$b + \frac{1}{3} = 3\frac{2}{3}$$
$$3\frac{1}{3} + \frac{1}{3} \stackrel{?}{=} 3\frac{2}{3}$$
$$3\frac{2}{3} = 3\frac{2}{3}$$

b)

$$2\frac{3}{4} + c = 8\frac{1}{2}$$
$$S_{2\frac{3}{4}} \quad -2\frac{3}{4} = -2\frac{3}{4}$$

Ans. $c = 5\frac{3}{4}$

Check:

$$2\frac{3}{4} + c = 8\frac{1}{2}$$
$$2\frac{3}{4} + 5\frac{3}{4} \stackrel{?}{=} 8\frac{1}{2}$$
$$8\frac{1}{2} = 8\frac{1}{2}$$

c)

$$20.8 = d + 6.9$$
$$S_{6.9} \quad -6.9 = -6.9$$

Ans. $13.9 = d$

Check:

$$20.8 = d + 6.9$$
$$20.8 \stackrel{?}{=} 13.9 + 6.9$$
$$20.8 = 20.8$$

7.3. Subtraction Rule in Problem Solving

After an increase of 22¢, the price of grade A eggs rose to 81¢. What was the original price?

Solution:

Let p = original price in ¢.

Then, $p + 22 = 81$

$$S_{22} \quad -22 = -22$$
$$p = 59$$

Ans. The original price was 59¢.

Check (the problem):

After increasing 22¢, the new price should be 81¢.

Hence,

$$59¢ + 22¢ \stackrel{?}{=} 81¢$$
$$81¢ = 81¢$$

7.4. Subtraction Rule in Problem Solving

Harold's height is 5 ft. 3 in. If he is 9 in. taller than John, how tall is John?

Solution:

Let J = John's height in ft.

Then, $J + \frac{3}{4} = 5\frac{1}{4}$ (9 in. = $\frac{3}{4}$ ft.)

$$S_{3/4} \quad -\frac{3}{4} = -\frac{3}{4}$$
$$J = 4\frac{1}{2}$$

Ans. John is $4\frac{1}{2}$ ft. or 4 ft. 6 in. tall.

Check (the problem):

9 in. more than John's height should equal 5 ft. 3 in.

$$4\tfrac{1}{2}\text{ ft.} + \tfrac{3}{4}\text{ ft.} \stackrel{?}{=} 5\tfrac{1}{4}\text{ ft.}$$
$$5\tfrac{1}{4}\text{ ft.} = 5\tfrac{1}{4}\text{ ft.}$$

8. USING ADDITION TO SOLVE AN EQUATION

Addition Rule of Equality

To keep an equality, equal numbers may be added to both sides.

To Solve an Equation Using the Addition Rule of Equality

Procedure:	Solve:	*a)* $n - 19 = 21$	*b)*	$17 = m - 8$
	Solutions:			
		$n - 19 = 21$		$17 = m - 8$
1. Add to both sides the number subtracted from the unknown:	A_{19}	$+19 = +19$	A_8	$+8 = +8$
	Ans.	$n = 40$	*Ans.*	$25 = m$
2. **Check** the original equation:	**Check:**	$n - 19 = 21$	**Check:**	$17 = m - 8$
		$40 - 19 \stackrel{?}{=} 21$		$17 \stackrel{?}{=} 25 - 8$
		$21 = 21$		$17 = 17$

Note. "**A**" is a convenient symbol for "adding to both sides".
"A_{19}" means "add 19 to both sides".

8.1. Addition Rule in Equations Containing Integers

Solve each equation:

a) $w - 10 = 19$
$\mathbf{A}_{10}$ $\quad +10 = +10$
Ans. $w = 29$

Check:

$w - 10 = 19$
$29 - 10 \stackrel{?}{=} 19$
$19 = 19$

b) $x - 19 = 10$
$\mathbf{A}_{19}$ $\quad +19 = +19$
Ans. $x = 29$

Check:

$x - 19 = 10$
$29 - 19 \stackrel{?}{=} 10$
$10 = 10$

c) $7 = y - 82$
$\mathbf{A}_{82}$ $\quad +82 = \ +82$
Ans. $89 = y$

Check:

$7 = y - 82$
$7 \stackrel{?}{=} 89 - 82$
$7 = 7$

d) $82 = z - 7$
$\mathbf{A}_{7}$ $\quad +7 = \ +7$
Ans. $89 = z$

Check:

$82 = z - 7$
$82 \stackrel{?}{=} 89 - 7$
$82 = 82$

8.2. Addition Rule in Equations Containing Fractions or Decimals

Solve each equation:

a) $h - \frac{3}{8} = 5\frac{1}{4}$
$\mathbf{A}_{3/8}$ $\quad +\frac{3}{8} = \frac{3}{8}$
Ans. $h = 5\frac{5}{8}$

b) $j - 20\frac{7}{12} = 1\frac{1}{12}$
$\mathbf{A}_{20\frac{7}{12}}$ $\quad +20\frac{7}{12} = 20\frac{7}{12}$
Ans. $j = 21\frac{2}{3}$

c) $12.5 = m - 2.9$
$\mathbf{A}_{2.9}$ $\quad +2.9 = \ +2.9$
Ans. $15.4 = m$

(Check your answers.)

8.3. Addition Rule in Problem Solving

A drop of 8° brought the temperature to 64°. What was the original temperature?

Solution:

Let t = original temperature in °.

Then, $t - 8 = 64$
$\mathbf{A}_{8}$ $\quad +8 = +8$
$t = 72$

Ans. The original temperature was 72°.

Check (the problem):

The original temperature, dropped 8°, should become 64°.

Hence,

$72° - 8° \stackrel{?}{=} 64°$
$64° = 64°$

8.4. Addition Rule in Problem Solving

After giving 15 marbles to Sam, Joe has 43 left. How many did Joe have originally?

Solution:

Let m = the original no. of marbles.

Then, $m - 15 = 43$
$\mathbf{A}_{15}$ $\quad +15 = +15$
$m = 58$

Ans. Sam had 58 marbles at first.

Check (the problem):

The original number of marbles, less 15, should be 43.

Hence,

58 marbles − 15 marbles $\stackrel{?}{=}$ 43 marbles
43 marbles = 43 marbles

9. USING TWO OR MORE OPERATIONS TO SOLVE AN EQUATION

In equations where two operations are performed upon the unknown, two inverse operations are needed to solve the equation.

Thus, in $2x + 7 = 19$, the two operations upon the unknown are **multiplication and addition.** To solve, use **division and subtraction**, performing subtraction first.

Also, in $\frac{x}{3} - 5 = 2$, the two operations upon the unknown are **division and subtraction.** To solve, use **multiplication and addition**, performing addition first.

To Solve Equations Using Two Inverse Operations

Procedure:	Solve: *a*) $2x + 7 = 19$	*b*) $\frac{x}{3} - 5 = 2$
	Solutions: *a*) $2x + 7 = 19$	*b*) $\frac{x}{3} - 5 = 2$
1. Perform **addition** to undo subtraction, or **subtraction** to undo addition:	S_7 $\quad -7 = -7$ $2x = 12$	A_5 $\quad +5 = +5$ $\frac{x}{3} = 7$
2. Perform **multiplication** to undo division, or **division** to undo **multiplication**:	D_2 $\quad \frac{2x}{2} = \frac{12}{2}$ *Ans.* $x = 6$	M_3 $\quad 3 \cdot \frac{x}{3} = 3 \cdot 7$ *Ans.* $x = 21$
3. **Check** in the original equation:	**Check:** $2x + 7 = 19$ $2(6) + 7 \stackrel{?}{=} 19$ $19 = 19$	**Check:** $\frac{x}{3} - 5 = 2$ $\frac{21}{3} - 5 \stackrel{?}{=} 2$ $2 = 2$

9.1. Using Two Inverse Operations to Solve an Equation

Solve each equation:

a) $2x + 7 = 11$	*b*) $3x - 5 = 7$	*c*) $\frac{x}{3} + 5 = 7$	*d*) $\frac{x}{5} - 3 = 7$
S_7 $\quad -7 = -7$	A_5 $\quad +5 = +5$	S_5 $\quad -5 = -5$	A_3 $\quad +3 = 3$
$2x = 4$	$3x = 12$	$\frac{x}{3} = 2$	$\frac{x}{5} = 10$
D_2 $\quad \frac{2x}{2} = \frac{4}{2}$	D_3 $\quad \frac{3x}{3} = \frac{12}{3}$	M_3 $\quad 3 \cdot \frac{x}{3} = 3 \cdot 2$	M_5 $\quad 5 \cdot \frac{x}{5} = 5 \cdot 10$
Ans. $x = 2$	*Ans.* $x = 4$	*Ans.* $x = 6$	*Ans.* $x = 50$
Check:	**Check:**	**Check:**	**Check:**
$2x + 7 = 11$	$3x - 5 = 7$	$\frac{x}{3} + 5 = 7$	$\frac{x}{5} - 3 = 7$
$2(2) + 7 \stackrel{?}{=} 11$	$3(4) - 5 \stackrel{?}{=} 7$	$\frac{6}{3} + 5 \stackrel{?}{=} 7$	$\frac{50}{5} - 3 \stackrel{?}{=} 7$
$4 + 7 \stackrel{?}{=} 11$	$12 - 5 \stackrel{?}{=} 7$	$2 + 5 \stackrel{?}{=} 7$	$10 - 3 \stackrel{?}{=} 7$
$11 = 11$	$7 = 7$	$7 = 7$	$7 = 7$

9.2. Solving Equations with Like Terms on the Same Side

Solve each equation (*Hint: Combine like terms first.*):

a) $8n + 4n - 3 = 9$	*b*) $13n + 4 + n = 39$	*c*) $10 = 7 + n - \frac{n}{2}$
$12n - 3 = 9$	$14n + 4 = 39$	$10 = 7 + \frac{n}{2}$
A_3 $\quad +3 = 3$	S_4 $\quad -4 = -4$	S_7 $\quad -7 = -7$
$12n = 12$	$14n = 35$	$3 = \frac{n}{2}$
D_{12} $\quad \frac{12n}{12} = \frac{12}{12}$	D_{14} $\quad \frac{14n}{14} = \frac{35}{14}$	M_2 $\quad 2 \cdot 3 = (\frac{n}{2})2$
Ans. $n = 1$	*Ans.* $n = 2\frac{1}{2}$	*Ans.* $6 = n$

(*Check your answers.*)

9.3. Solving Equations with Like Terms on Both Sides

Solve each equation (*Hint: First, add or subtract to collect like terms on the same side.*):

a) $5n = 40 - 3n$	*b*) $4u + 5 = 5u - 30$	*c*) $3r + 10 = 2r + 20$
A_{3n} $\quad +3n = +3n$	A_{30} $\quad +30 = +30$	S_{10} $\quad -10 = -10$
$8n = 40$	$4u + 35 = 5u$	$3r = 2r + 10$
	S_{4u} $\quad -4u = -4u$	S_{2r} $\quad -2r = -2r$
D_8 $\quad \frac{8n}{8} = \frac{40}{8}$	*Ans.* $35 = u$	*Ans.* $r = 10$
Ans. $n = 5$		

(*Check your answers.*)

9.4. Solving Equations in which the Unknown is a Divisor

Solve each equation (*Hint: First, multiply both sides by the unknown.*):

a) $\frac{8}{x} = 2$
- M_x: $x\left(\frac{8}{x}\right) = 2x$
- $8 = 2x$
- D_2: $\frac{8}{2} = \frac{2x}{2}$
- *Ans.* $4 = x$

b) $12 = \frac{3}{y}$
- M_y: $12y = \left(\frac{3}{y}\right)y$
- $12y = 3$
- D_{12}: $\frac{12y}{12} = \frac{3}{12}$
- *Ans.* $y = \frac{1}{4}$

c) $\frac{7}{x} = \frac{1}{5}$
- M_x: $x \cdot \frac{7}{x} = \frac{1}{5}x$
- $7 = \frac{x}{5}$
- M_5: $5 \cdot 7 = \left(\frac{x}{5}\right)5$
- *Ans.* $35 = x$

d) $\frac{1}{7} = \frac{3}{x}$
- M_x: $\frac{1}{7}x = \frac{3}{x} \cdot x$
- $\frac{x}{7} = 3$
- M_7: $7 \cdot \frac{x}{7} = 3(7)$
- *Ans.* $x = 21$

(*Check your answers.*)

9.5. Solving Equations in which the Unknown is being Subtracted

Solve each equation (*Hint: First, add the unknown to both sides.*):

a) $10 - w = 3$
- A_w: $+w = w$
- $10 = w + 3$
- S_3: $-3 = -3$
- *Ans.* $7 = w$

b) $10 = 50 - n$
- A_n: $+n = +n$
- $n + 10 = 50$
- S_{10}: $-10 = -10$
- *Ans.* $n = 40$

c) $25 - 3n = 13$
- A_{3n}: $+3n = +3n$
- $25 = 3n + 13$
- S_{13}: $-13 = -13$
- $12 = 3n$
- D_3: $\frac{12}{3} = \frac{3n}{3}$
- *Ans.* $4 = n$

d) $8 = 50 - 7n$
- A_{7n}: $+7n = +7n$
- $7n + 8 = 50$
- S_8: $-8 = -8$
- $7n = 42$
- D_7: $\frac{7n}{7} = \frac{42}{7}$
- *Ans.* $n = 6$

(*Check your answers.*)

9.6. Solving Equations whose Unknown has a Fractional Coefficient

a) Solve: $\frac{3}{8}x = 9$

Solutions:

Using One Operation

a) $\frac{3}{8}x = 9$
- $M_{8/3}$: $\frac{8}{3} \cdot \frac{3}{8}x = \frac{8}{3} \cdot 9$
- *Ans.* $x = 24$

Using Two Operations

a) $\frac{3}{8}x = 9$
- M_8: $8 \cdot \frac{3}{8}x = 8 \cdot 9$
- $3x = 72$
- D_3: $\frac{3x}{3} = \frac{72}{3}$
- *Ans.* $x = 24$

b) Solve: $25 = \frac{5}{4}x$

Solutions:

Using One Operation

b) $25 = \frac{5}{4}x$
- $M_{4/5}$: $\frac{4}{5} \cdot 25 = \frac{4}{5} \cdot \frac{5}{4}x$
- *Ans.* $20 = x$

Using Two Operations

b) $25 = \frac{5}{4}x$
- M_4: $4(25) = 4 \cdot \frac{5}{4}x$
- $100 = 5x$
- D_5: $\frac{100}{5} = \frac{5x}{5}$
- *Ans.* $20 = x$

(*Check your answers.*)

9.7. Solving Equations whose Unknown has a Fractional Coefficient

a) $\frac{3}{4}y - 5 = 7$
- A_5: $+5 = +5$
- $\frac{3}{4}y = 12$
- $M_{4/3}$: $\frac{4}{3} \cdot \frac{3}{4}y = \frac{4}{3} \cdot 12$
- *Ans.* $y = 16$

b) $8 + \frac{2}{7}b = 20$
- S_8: $-8 = -8$
- $\frac{2}{7}b = 12$
- $M_{7/2}$: $\frac{7}{2} \cdot \frac{2}{7}b = \frac{7}{2} \cdot 12$
- *Ans.* $b = 42$

c) $48 - \frac{5}{3}w = 23$
- $A_{\frac{5}{3}w}$: $+\frac{5}{3}w = +\frac{5}{3}w$
- $48 = \frac{5}{3}w + 23$
- S_{23}: $-23 = -23$
- $25 = \frac{5}{3}w$
- $M_{3/5}$: $\frac{3}{5} \cdot 25 = \frac{3}{5} \cdot \frac{5}{3}w$
- *Ans.* $15 = w$

(*Check your answers.*)

9.8. Using Two Operations in Problem Solving

a) How many boys are there in a class of 36 pupils if the number of girls is 6 more?

Solution:

a) Let b = the number of boys

Then $b + 6$ = the number of girls

$$b + b + 6 = 36$$

S_6 $$2b + 6 = 36$$

D_2 $$2b = 30$$

$$b = 15$$

Ans. There are 15 boys.

b) How many boys are there in a class of 36 pupils if the number of girls is three times as many?

Solution:

b) Let b = the number of boys

Then $3b$ = the number of girls

$$b + 3b = 36$$

D_4 $$4b = 36$$

$$b = 9$$

Ans. There are 9 boys.

(*Check your answers.*)

9.9. Using Two Operations in Problem Solving

Paul has $2.60 in his bank. By adding equal deposits each week for 20 weeks, he hopes to have $7.80. How much should each weekly deposit be?

Solution:

Let d = no. of cents in each deposit

Then $$20d + 260 = 780$$

S_{260} $$-260 = -260$$

D_{20} $$20d = 520$$

$$d = 26$$

Ans. He must deposit 26¢ a week.

Check (the problem):

20 deposits and $2.60 should equal the total of $7.80. Hence,

$$20(\$.26) + \$2.60 \stackrel{?}{=} \$7.80$$
$$\$5.20 + \$2.60 \stackrel{?}{=} \$7.80$$
$$\$7.80 = \$7.80$$

9.10. Using Two Operations in Problem Solving

Mr. Richards sold his house for $9000. His loss amounted to two-fifths of his cost. What did the house cost him?

Solution:

Let c = the cost in $

Then $$9000 = c - \frac{2}{5}c$$

$$9000 = \frac{3}{5}c$$

$M_{5/3}$ $$\frac{5}{3} \cdot 9000 = \frac{5}{3} \cdot \frac{3}{5}c$$

$$15{,}000 = c$$

Ans. The cost was $15,000.

Check (the problem):

If $15,000 is the cost, the loss is $\frac{2}{5} \cdot \$15{,}000$ or $6000. The selling price of $9000 should be the cost minus the loss. Hence,

$$\$9000 \stackrel{?}{=} \$15{,}000 - \$6000$$
$$\$9000 = \$9000$$

SUPPLEMENTARY PROBLEMS

1. By checking, determine which value is a root of the equation: (1.1)
   a) $3x + 4x = 42$ for $x = 4, 6$ and 8 — *Ans.* $x = 6$
   b) $3n + 14 = 47$ for $n = 9, 10$ and 11 — *Ans.* $n = 11$
   c) $6y - 48 = 2y$ for $y = 8, 10$ and 12 — *Ans.* $y = 12$

2. By checking, show that x may have any of the following values in the identity $2(x-3) = 2x - 6$:
   a) $x = 10$, b) $x = 6$, c) $x = 4\frac{1}{2}$, d) $x = 3.1$ (1.2)

3. Translate into an equation, letting n represent the number: (2.1)
   (You need not find the value of the unknown.)
   a) What number diminished by 8 equals 13? — *Ans.* $n - 8 = 13$
   b) Two-thirds of what number equals 10? — *Ans.* $\frac{2}{3}n = 10$
   c) Three times the sum of a number and six is 33. What is the number? — *Ans.* $3(n + 6) = 33$
   d) What number increased by 20 equals three times the same number? — *Ans.* $n + 20 = 3n$
   e) What number increased by 5 equals twice the same number decreased by 4? — *Ans.* $n + 5 = 2n - 4$

4. Match the statement in *Column 1* with the equations in *Column 2*: (2.2)

Column 1	*Column 2*
1. The sum of 8 and twice a number is 18.	a) $\frac{n}{8} + 2 = 18$
2. Twice a number less 8 is 18.	b) $8(n - 2) = 18$
3. Twice the sum of a number and 8 is 18.	c) $\frac{1}{2}(8 - n) = 18$
4. Eight times the difference of a number and 2 is 18.	d) $2(n + 8) = 18$
5. One-half the difference of 8 and a number is 18.	e) $2n + 8 = 18$
6. 2 more than one-eighth of a number is 18.	f) $\frac{n}{2} - 8 = 18$
7. 8 less than half a number is 18.	g) $2n - 8 = 18$

   Ans. *1* and *e*, *2* and *g*, *3* and *d*, *4* and *b*, *5* and *c*, *6* and *a*, *7* and *f*.

5. Letting n represent the number of games lost, obtain an equation for each problem: (2.3)
   (You need not solve each equation.)
   a) A team won three times as many games as it lost.
   It played a total of 52 games. — *Ans.* a) $n + 3n = 52$
   b) A team won 20 games more than it lost.
   It played a total of 84 games. — *Ans.* b) $n + n + 20 = 84$
   c) A team won 15 games less than twice the number lost.
   It played a total of 78 games. — *Ans.* c) $n + 2n - 15 = 78$

6. Solve each equation: (3.1)

a) $a+5 = 9$	*Ans.* $a = 4$	e) $x+11 = 21+8$	*Ans.* $x = 18$	i) $45 = m - 13$	*Ans.* $m = 58$
b) $7+b = 15$	*Ans.* $b = 8$	f) $27+13 = 18+y$	*Ans.* $y = 22$	j) $22 = n - 50$	*Ans.* $n = 72$
c) $20 = c+12$	*Ans.* $c = 8$	g) $h - 6 = 14$	*Ans.* $h = 20$	k) $x - 42 = 80 - 75$	*Ans.* $x = 47$
d) $75 = 55+d$	*Ans.* $d = 20$	h) $k - 14 = 6$	*Ans.* $k = 20$	l) $100 - 31 = y - 84$	*Ans.* $y = 153$

7. Solve each equation: (3.2)

a) $4p = 48$	d) $4n = 2$	g) $\frac{t}{5} = 6$	j) $\frac{y}{12} = \frac{3}{2}$
b) $10r = 160$	e) $12w = 4$	h) $\frac{u}{65} = 1$	k) $\frac{a}{10} = \frac{2}{5}$
c) $25s = 35$	f) $24x = 21$	i) $\frac{x}{15} = 4$	l) $\frac{1}{3}b = \frac{5}{6}$

Ans.

a) $p = 12$	d) $n = \frac{1}{2}$	g) $t = 30$	j) $y = 18$
b) $r = 16$	e) $w = \frac{1}{3}$	h) $u = 65$	k) $a = 4$
c) $s = \frac{7}{5}$ or $1\frac{2}{5}$	f) $x = \frac{7}{8}$	i) $x = 60$	l) $b = \frac{5}{2}$ or $2\frac{1}{2}$

8. Solve each equation: (3.3)

a) $n+8 = 24$ *e*) $3+y = 15$ *i*) $16 = y-20$ *m*) $x+\frac{1}{3} = 9$

b) $n-8 = 24$ *f*) $15 = y-3$ *j*) $16 = \frac{y}{20}$ *n*) $x-\frac{1}{3} = 9$

c) $8n = 24$ *g*) $15 = 3y$ *k*) $\frac{y}{20} = 16$ *o*) $\frac{1}{3}x = 9$

d) $\frac{n}{8} = 24$ *h*) $15 = \frac{y}{3}$ *l*) $16+y = 20$ *p*) $\frac{x}{9} = \frac{1}{3}$

Ans. *a*) $n = 16$ *c*) $n = 3$ *e*) $y = 12$ *g*) $y = 5$ *i*) $y = 36$ *k*) $y = 320$ *m*) $x = 8\frac{2}{3}$ *o*) $x = 27$
b) $n = 32$ *d*) $n = 192$ *f*) $y = 18$ *h*) $y = 45$ *j*) $y = 320$ *l*) $y = 4$ *n*) $x = 9\frac{1}{3}$ *p*) $x = 3$

9. Solve each equation: (3.3)

a) $x+11 = 14$ *f*) $h-3 = 7\frac{1}{2}$ *k*) $11r = 55$ *p*) $6\frac{1}{2} = \frac{l}{2}$

b) $11+y = 24$ *g*) $35 = m-20\frac{1}{3}$ *l*) $44s = 44$ *q*) $1.7 = \frac{n}{3}$

c) $22 = 13+a$ *h*) $17\frac{3}{4} = n-2\frac{1}{4}$ *m*) $10t = 5$ *r*) $100 = \frac{h}{.7}$

d) $45 = b+33$ *i*) $x+1.2 = 5.7$ *n*) $8x = 3$ *s*) $24 = \frac{t}{.5}$

e) $z-9 = 3$ *j*) $10.8 = y-3.2$ *o*) $3y = 0$ *t*) $.009 = \frac{x}{1000}$

Ans. *a*) $x = 3$ *d*) $b = 12$ *g*) $m = 55\frac{1}{3}$ *j*) $y = 14$ *m*) $t = \frac{1}{2}$ *p*) $l = 13$ *s*) $t = 12$
b) $y = 13$ *e*) $z = 12$ *h*) $n = 20$ *k*) $r = 5$ *n*) $x = \frac{3}{8}$ *q*) $n = 5.1$ *t*) $x = 9$
c) $a = 9$ *f*) $h = 10\frac{1}{2}$ *i*) $x = 4.5$ *l*) $s = 1$ *o*) $y = 0$ *r*) $h = 70$

10. State the equality rule used in each: (4.1)

a)
$$6r = 30$$
$$\frac{6r}{6} = \frac{30}{6}$$
$$r = 5$$

c)
$$30 = \frac{r}{6}$$
$$6(30) = 6\cdot\frac{r}{6}$$
$$180 = r$$

e)
$$100x = 5$$
$$\frac{100x}{100} = \frac{5}{100}$$
$$x = \frac{1}{20}$$

b)
$$30 = r-6$$
$$+6 = +6$$
$$36 = r$$

d)
$$30 = 6+r$$
$$-6 = -6$$
$$24 = r$$

f)
$$100 = \frac{y}{5}$$
$$5(100) = 5(\tfrac{y}{5})$$
$$500 = y$$

Ans. Addition rule in (*b*), Subtraction rule in (*d*), Multiplication rule in (*c*) and (*f*), Division rule in (*a*) and (*e*).

11. Solve each equation: (5.1)

a) $12x = 60$ *b*) $60y = 12$ *c*) $24 = 2z$ *d*) $2 = 24w$ *e*) $6r = 9$ *f*) $9s = 6$ *g*) $10 = 4t$ *h*) $4 = 10u$

Ans. *a*) $x = 5$ *b*) $y = \frac{1}{5}$ *c*) $12 = z$ *d*) $\frac{1}{12} = w$ *e*) $r = \frac{3}{2}$ *f*) $s = \frac{2}{3}$ *g*) $\frac{5}{2} = t$ *h*) $\frac{2}{5} = u$

12. Solve each equation: (5.2)

a) $.7a = 21$ *c*) $24 = .06c$ *e*) $.1h = 100$ *g*) $25.2 = .12k$
b) $1.1b = 55$ *d*) $18 = .009d$ *f*) $.6j = .96$ *h*) $7.5 = .015m$

Ans. *a*) $a = 30$ *c*) $400 = c$ *e*) $h = 1000$ *g*) $210 = k$
b) $b = 50$ *d*) $2000 = d$ *f*) $j = 1.6$ *h*) $500 = m$

13. Solve each equation: (5.3)

a) $10\% s = 7$ *c*) $18 = 3\% n$ *e*) $5\% m = 13$ *g*) $.23 = 1\% y$
b) $25\% t = 3$ *d*) $14 = 70\% w$ *f*) $17\% x = 6.8$ *h*) $3.69 = 90\% z$

Ans. *a*) $s = 70$ *c*) $600 = n$ *e*) $m = 260$ *g*) $23 = y$
b) $t = 12$ *d*) $20 = w$ *f*) $x = 40$ *h*) $4.1 = z$

14. Solve each equation: (5.4)

a) $14 = 3x - x$ *b*) $7y + 3y = 50$ *c*) $8z - 3z = 45$ *d*) $132 = 10w + 3w - w$ *e*) $24 = 4\frac{1}{2}x - \frac{x}{2}$ *f*) $4y + 15y = 57$ *g*) $7\frac{1}{2}z - 7z = 28$ *h*) $15w - 3w - 2w = 85$

Ans. *a*) $7 = x$ *b*) $y = 5$ *c*) $z = 9$ *d*) $11 = w$ *e*) $6 = x$ *f*) $y = 3$ *g*) $z = 56$ *h*) $w = 8\frac{1}{2}$

15. Harry earned \$9.63. What was his hourly wage if he worked *a*) 3 hr., *b*) 9 hr., *c*) $\frac{1}{2}$ hr.? (5.5)
Ans. *a*) \$3.21, *b*) \$1.07, *c*) \$19.26

16. Mr. Brown's commission rate is 5%. How much did he sell if his commissions were *a*) \$85, *b*) \$750, *c*) \$6.20 ? *Ans.* *a*) \$1700, *b*) \$15,000, *c*) \$124 (5.6)

17. Solve each equation: (6.1)

a) $\frac{x}{3} = 2$ *b*) $\frac{1}{7}y = 12$ *c*) $16 = \frac{z}{5}$ *d*) $3 = \frac{1}{50}w$ *e*) $\frac{a}{2} = 3$ *f*) $\frac{1}{30}b = 20$ *g*) $.6 = \frac{c}{10}$

Ans. *a*) $x = 6$ *b*) $y = 84$ *d*) $80 = z$ *d*) $150 = w$ *e*) $a = 6$ *f*) $b = 600$ *g*) $6 = c$

18. Solve each equation: (6.2)

a) $\frac{a}{.7} = 10$ *b*) $\frac{b}{.02} = 600$ *c*) $30 = \frac{c}{2.4}$ *d*) $11 = \frac{d}{.05}$ *e*) $\frac{m}{.4} = 220$ *f*) $\frac{n}{.01} = 3$

Ans. *a*) $a = 7$ *b*) $b = 12$ *c*) $72 = c$ *d*) $.55 = d$ *e*) $m = 88$ *f*) $n = .03$

19. Solve each equation: (6.3)

a) $\frac{3}{4}x = 21$ *b*) $\frac{4}{3}y = 32$ *c*) $\frac{3x}{2} = 9$ *d*) $45 = \frac{5}{9}y$ *e*) $2\frac{1}{5}z = 55$ *f*) $2c + \frac{1}{2}c = 10$

Ans. *a*) $x = 28$ *b*) $y = 24$ *c*) $x = 6$ *d*) $y = 81$ *e*) $z = 25$ *f*) $c = 4$

20. Solve each equation: (6.4)

a) $37\frac{1}{2}\%s = 15$ *b*) $60\%t = 60$ *c*) $16\frac{2}{3}\%n = 14$ *d*) $150\%r = 15$ *e*) $83\frac{1}{3}\%w = 35$

Hint: $37\frac{1}{2}\% = \frac{3}{8}$ $60\% = \frac{3}{5}$ $16\frac{2}{3}\% = \frac{1}{6}$ $150\% = 1\frac{1}{2}$ or $\frac{3}{2}$ $83\frac{1}{3}\% = \frac{5}{6}$

Ans. *a*) $s = 40$ *b*) $t = 100$ *c*) $n = 84$ *d*) $r = 10$ *e*) $w = 42$

21. On a trip, John covered a distance of 35 mi. What was the total distance of the trip if the distance traveled was *a*) $\frac{5}{6}$ of the total distance, *b*) 70% of the total distance? (6.5)
Ans. *a*) 42 mi., *b*) 50 mi.

22. Mr. Reynolds receives 7% per year on a stock investment. How large is his investment if, at the end of one year, his interest is *a*) \$28, *b*) \$350, *c*) \$4.27? *Ans.* *a*) \$400, *b*) \$5000, *c*) \$61 (6.6)

23. Solve each equation: (7.1)

a) $r + 25 = 70$ *b*) $31 + t = 140$ *c*) $18 = s + 3$ *d*) $842 = 720 + u$ *e*) $x + 130 = 754$ *f*) $116 + y = 807$ *g*) $259 = s + 237$ *h*) $901 = 857 + w$

Ans. *a*) $r = 45$ *b*) $t = 109$ *c*) $15 = s$ *d*) $122 = u$ *e*) $x = 624$ *f*) $y = 691$ *g*) $22 = s$ *h*) $44 = w$

24. Solve each equation: (7.2)

a) $b + \frac{2}{3} = 7\frac{2}{3}$ *b*) $1\frac{1}{2} + c = 8\frac{3}{4}$ *c*) $35.4 = d + 23.2$ *d*) $87.4 = 80.6 + e$ *e*) $f + \frac{5}{8} = 3\frac{1}{2}$ *f*) $8\frac{1}{6} + g = 10\frac{5}{6}$ *g*) $7.28 = m + .79$ *h*) $15.87 = 6.41 + n$

Ans. *a*) $b = 7$ *b*) $c = 7\frac{1}{4}$ *c*) $12.2 = d$ *d*) $6.8 = e$ *e*) $f = 2\frac{7}{8}$ *f*) $c = 2\frac{2}{3}$ *g*) $6.49 = m$ *h*) $9.46 = n$

25. The price of eggs rose 29¢. What was the original price if the new price is *a*) 70¢, *b*) $1.05 ?
Ans. *a*) 41¢, *b*) 76¢ (7.3)

26. Will is 8 in. taller than George. How tall is George if Will's height is *a*) 5 ft. 2 in., *b*) 4 ft. 3 in. ?
Ans. *a*) 4 ft. 6 in., *b*) 3 ft. 7 in. (7.4)

27. Solve each equation: (8.1)

a) $w - 8 = 22$ *c*) $40 = y - 3$ *e*) $m - 140 = 25$ *g*) $158 = p - 317$
b) $x - 22 = 8$ *d*) $3 = z - 40$ *f*) $n - 200 = 41$ *h*) $256 = r - 781$

Ans. *a*) $w = 30$ *c*) $43 = y$ *e*) $m = 165$ *g*) $475 = p$
b) $x = 30$ *d*) $43 = z$ *f*) $n = 241$ *h*) $1037 = r$

28. Solve each equation: (8.2)

a) $h - \frac{7}{8} = 8\frac{3}{4}$ *c*) $28.4 = m - 13.9$ *e*) $p - 1\frac{5}{12} = 1\frac{7}{12}$ *g*) $.03 = s - 2.07$
b) $j - 34\frac{1}{2} = 65$ *d*) $.37 = n - 8.96$ *f*) $r - 14\frac{2}{3} = 5\frac{1}{3}$ *h*) $5.84 = t - 3.06$

Ans. *a*) $h = 9\frac{5}{8}$ *c*) $42.3 = m$ *e*) $p = 3$ *g*) $2.10 = s$
b) $j = 99\frac{1}{2}$ *d*) $9.33 = n$ *f*) $r = 20$ *h*) $8.90 = t$

29. What was the original temperature if a drop of 12° brought the temperature to (8.3)
a) 75°, *b*) $14\frac{1}{2}$°, *c*) $6\frac{1}{4}$° ? *Ans.* *a*) 87°, *b*) $26\frac{1}{2}$°, *c*) $18\frac{1}{4}$°

30. Solve each equation: (9.1)

a) $2x + 5 = 9$ *e*) $2x - 5 = 9$ *i*) $\frac{x}{4} + 3 = 7$ *m*) $\frac{x}{4} - 3 = 7$
b) $4x + 11 = 21$ *f*) $4x - 11 = 21$ *j*) $\frac{x}{5} + 2 = 10$ *n*) $\frac{x}{5} - 2 = 10$
c) $20 = 3x + 8$ *g*) $60 = 10x - 20$ *k*) $17 = \frac{x}{2} + 15$ *o*) $3 = \frac{x}{12} - 7\frac{1}{4}$
d) $13 = 6 + 7x$ *h*) $11 = 6x - 16$ *l*) $25 = \frac{x}{10} + 2$ *p*) $5\frac{1}{2} = \frac{x}{8} - 4$

Ans. *a*) $x = 2$ *e*) $x = 7$ *i*) $x = 16$ *m*) $x = 40$
b) $x = 2\frac{1}{2}$ *f*) $x = 8$ *j*) $x = 40$ *n*) $x = 60$
c) $x = 4$ *g*) $x = 8$ *k*) $x = 4$ *o*) $x = 123$
d) $x = 1$ *h*) $x = 4\frac{1}{2}$ *l*) $x = 230$ *p*) $x = 76$

31. Solve each equation: (9.2)

a) $10n + 5n - 6 = 9$ *c*) $25 = 19 + 20n - 18n$ *e*) $19n - 10 + n = 80$ *g*) $40 = 25t + 22 - 13t$
b) $7m + 10 - 2m = 45$ *d*) $35 = 6p + 8 + 3p$ *f*) $3\frac{1}{2}r + r + 2 = 20$ *h*) $145 = 10 + 7.6s - 3.1s$

Ans. *a*) $n = 1$ *b*) $m = 7$ *c*) $n = 3$ *d*) $p = 3$ *e*) $n = 4\frac{1}{2}$ *f*) $r = 4$ *g*) $t = \frac{3}{2}$ or $1\frac{1}{2}$ *h*) $s = 30$

32. Solve each equation: (9.3)

a) $5r = 2r + 27$ *c*) $10r - 11 = 8r$ *e*) $13b = 15 + 3b$ *g*) $9u = 16u - 105$
b) $2r = 90 - 7r$ *d*) $18 - 5a = a$ *f*) $100 + 3\frac{1}{2}t = 23\frac{1}{2}t$ *h*) $5x + 3 - 2x = x + 8$

Ans. *a*) $r = 9$ *b*) $r = 10$ *c*) $r = 5\frac{1}{2}$ *d*) $a = 3$ *e*) $b = 1\frac{1}{2}$ *f*) $t = 5$ *g*) $u = 15$ *h*) $x = 2\frac{1}{2}$

33. Solve each equation: (9.4)

a) $\frac{40}{x} = 5$ *c*) $14 = \frac{28}{y}$ *e*) $\frac{32}{n} = 8$ *g*) $4 = \frac{15}{w}$
b) $\frac{5}{x} = 40$ *d*) $28 = \frac{14}{y}$ *f*) $\frac{3}{n} = 2$ *h*) $15 = \frac{90}{w}$

Ans. *a*) $x = 8$ *b*) $x = \frac{1}{8}$ *c*) $y = 2$ *d*) $y = \frac{1}{2}$ *e*) $n = 4$ *f*) $n = 1\frac{1}{2}$ *g*) $w = 3\frac{3}{4}$ *h*) $w = 6$

34. Solve each equation: (9.5)

a) $12-w = 2$ *c*) $84-r = 70$ *e*) $8\frac{1}{2}-t = 5\frac{1}{2}$ *g*) $8.7-u = 7.8$

b) $24 = 27-n$ *d*) $90 = 105-s$ *f*) $\frac{1}{2} = \frac{3}{4}-s$ *h*) $3.25 = 5.37-v$

Ans. *a*) $w=10$ *b*) $n=3$ *c*) $r=14$ *d*) $s=15$ *e*) $t=3$ *f*) $s=\frac{1}{4}$ *g*) $u=.9$ *h*) $v=2.12$

35. Solve each equation: (9.6,9.7)

a) $\frac{7}{8}x = 21$ *d*) $1\frac{1}{2}w = 15$ *g*) $\frac{4}{5}n+6 = 22$ *j*) $10 = \frac{2}{9}r+8$

b) $\frac{3}{4}x = 39$ *e*) $2\frac{1}{3}b = 35$ *h*) $10+\frac{6}{5}m = 52$ *k*) $6 = 16-\frac{5t}{3}$

c) $\frac{5}{4}y = 15$ *f*) $2c+2\frac{1}{2}c = 54$ *i*) $30-\frac{3}{2}p = 24$ *l*) $3s+\frac{s}{3}-7 = 5$

Ans. *a*) $x=24$ *d*) $w=10$ *g*) $n=20$ *j*) $r=9$

b) $x=52$ *e*) $b=15$ *h*) $m=35$ *k*) $t=6$

c) $y=12$ *f*) $c=12$ *i*) $p=4$ *l*) $s=3.6$

36. Solve the equations: (9.1 to 9.7)

a) $20 = 3x-10$ *Ans.* *a*) $x=10$

b) $20 = \frac{x}{3}-10$ *b*) $x=90$

c) $15 = \frac{3}{4}y$ *c*) $y=20$

d) $17 = 24-z$ *d*) $z=7$

e) $3 = \frac{39}{x}$ *e*) $x=13$

f) $\frac{15}{x} = \frac{5}{4}$ *f*) $x=12$

g) $5c = 2c+4.5$ *g*) $c=1.5$

h) $.30-g = .13$ *h*) $g=.17$

i) $\frac{3}{4}n+11\frac{1}{2} = 20\frac{1}{2}$ *i*) $n=12$

j) $6w+5w-8 = 8w$ *j*) $w=2\frac{2}{3}$

k) $\frac{x}{2}+27 = 30$ *Ans.* *k*) $x=6$

l) $8x+3 = 43$ *l*) $x=5$

m) $21 = \frac{7}{5}w$ *m*) $w=15$

n) $60 = 66-12w$ *n*) $w=\frac{1}{2}$

o) $10b-3b = 49$ *o*) $b=7$

p) $12b-5 = .28+b$ *p*) $b=.48$

q) $6d-.8 = 2d$ *q*) $d=.2$

r) $40-.5h = 5$ *r*) $h=70$

s) $40-\frac{3}{5}m = 37$ *s*) $m=5$

t) $12t-2t+10 = 9t+12$ *t*) $t=2$

37. How many girls are there in a class of 30 pupils if *a*) the number of boys is 10 less, *b*) the number of boys is four times as many? *Ans.* *a*) 20 girls *b*) 6 girls (9.8)

38. Charles has $3.70 in his bank and hopes to increase this to $10 by making equal deposits each week. How much should he deposit if he deposits money for *a*) 14 wk., *b*) 5 wk.? (9.9)
Ans. *a*) 45¢ *b*) $1.26

39. Mr. Barr sold his house for $12,000. How much did the house cost him if his loss was *a*) $\frac{1}{3}$ of the cost, *b*) 20% of the cost? *Ans.* *a*) $18,000 *b*) $15,000 (9.10)

Chapter 3

SIGNED NUMBERS

1. UNDERSTANDING SIGNED NUMBERS: POSITIVE AND NEGATIVE NUMBERS

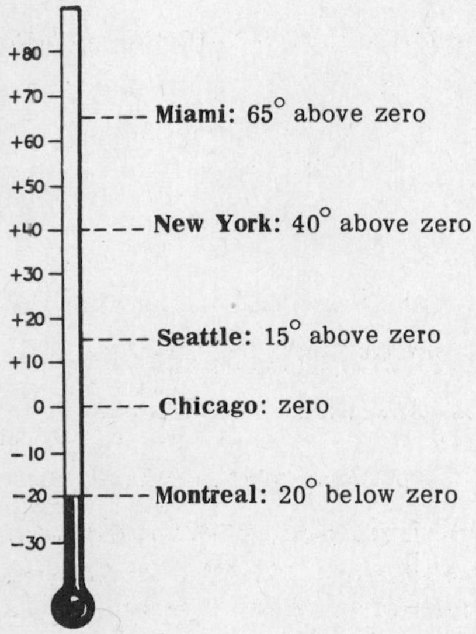

The temperatures listed, in the adjoining figure, are the Fahrenheit temperatures for five cities on a winter's day. Note how these temperatures have been shown on a **number scale.** On this scale, plus and minus signs are used to distinguish between temperatures above zero and those below zero. Such numbers are called **signed numbers.**

Signed numbers are positive or negative numbers used to represent quantities that are opposites of each other.

Thus, if $+25$ represents 25° above zero, -25 represents 25° below zero.

While a minus sign (−) is used to indicate a negative number, a positive number may be shown by a plus sign (+) or by no sign at all.

Thus, 35° above zero may be indicated by $+35$ or 35.

The following table illustrates pairs of opposites which may be represented by $+25$ and -25.

+ 25	− 25
$25 deposited	$25 withdrawn
25 MPH faster	25 MPH slower
25 lb. gained	25 lb. lost
25 mi. to the north	25 mi. to the south

The **absolute or numerical value of a signed number** is the number which remains when the sign is removed.

Thus, 25 is the absolute value of $+25$ or -25.

1.1. Words Opposite in Meaning

State the words that are opposite in meaning to the following:

a) gain	*e*) north	*i*) deposit	*m*) A.D.
b) rise	*f*) east	*j*) asset	*n*) expand
c) above	*g*) right	*k*) earnings	*o*) accelerate
d) up	*h*) forward	*l*) receipt	*p*) clockwise

Ans.

a) loss	*e*) south	*i*) withdrawal	*m*) B.C.
b) fall	*f*) west	*j*) liability	*n*) contract
c) below	*g*) left	*k*) spendings	*o*) decelerate
d) down	*h*) backward	*l*) payment	*p*) counterclockwise

1.2. Expressing Quantities as Signed Numbers

State the quantity represented by each signed number:

a) By -10, if $+10$ means 10 yd. gained
b) By -5, if $+5$ means \$5 earned
c) By $+15$, if -15 means 15 mi. south
d) By $+8$, if -8 means 8 hr. earlier
e) By -100, if $+100$ means 100 ft. east
f) By -3, if $+3$ means 3 steps right
g) By 20, if -20 means 20 oz. underweight
h) By 5, if -5 means 5 flights down.

Ans. *a*) 10 yd. lost *b*) \$5 spent *c*) 15 mi. north *d*) 8 hr. later *e*) 100 ft. west *f*) 3 steps left *g*) 20 oz. overweight *h*) 5 flights up

1.3. Absolute or Numerical Value of a Signed Number

State (1) the absolute values of each pair of numbers and (2) the difference of these absolute values: *a*) $+25$ and -20, *b*) -3.5 and $+2.4$, *c*) $18\frac{3}{4}$, $-16\frac{3}{4}$.

Ans. *a*) (1) 25, 20 (2) 5 *b*) (1) 3.5, 2.4 (2) 1.1 *c*) (1) $18\frac{3}{4}$, $16\frac{3}{4}$ (2) 2

2. *USING NUMBER SCALES FOR SIGNED NUMBERS*

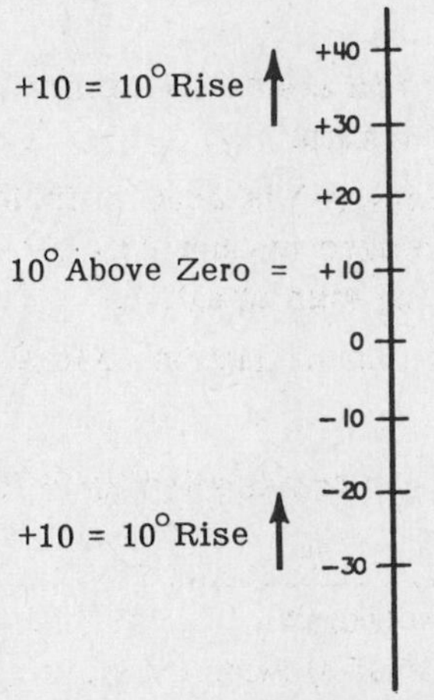

Vertical Number Scale

Constructing Number Scales

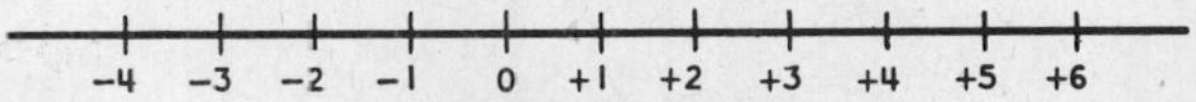

Horizontal Number Scale

Examine both horizontal and vertical number scales. Notice how positive and negative numbers are placed on opposite sides of 0. Since 0 is the starting point, it is called the origin.

In marking off signed numbers, equal lengths on a scale must have equal values. On the horizontal scale shown, each interval has a value of 1 while the intervals on the vertical scale have a value of 10. In a problem, other values may be chosen for each interval, depending on the numbers in the problem.

Number scales may be used for the following purposes:

(1) To understand the meanings of signed numbers.
(2) To show which of two signed numbers is the greater.
(3) To solve problems with signed numbers.
(4) To understand operations on signed numbers.
(5) To understand graphs in which horizontal and vertical number scales are combined. (*Such graphs will be studied in a later chapter.*)

(1) Using number scales to understand the meanings of signed numbers:

A signed number may refer to (*1*) **a position** or (*2*) **a change in position.**

Thus, $+10$ may refer to the 10° above zero position on a temperature scale. However, $+10$ may also show a 10° rise in temperature.

Note on the vertical scale, the use of $+10$ as a rise of 10°. The rise of 10° from any temperature is shown by arrows having the same size and direction. Note the two such arrows from $+30$ to $+40$ and from -30 to -20. These arrows, in mathematics, are called **vectors.** Since signed numbers involve direction, they are also known as **directed numbers.**

(2) Using number scales to show which of two signed numbers is the greater:

Horizontal Number Scale

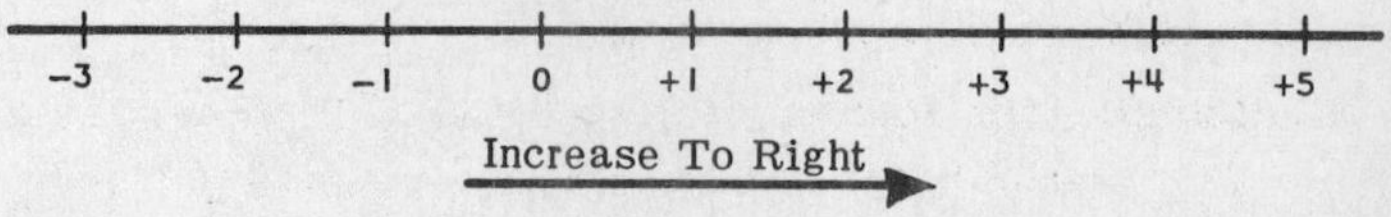

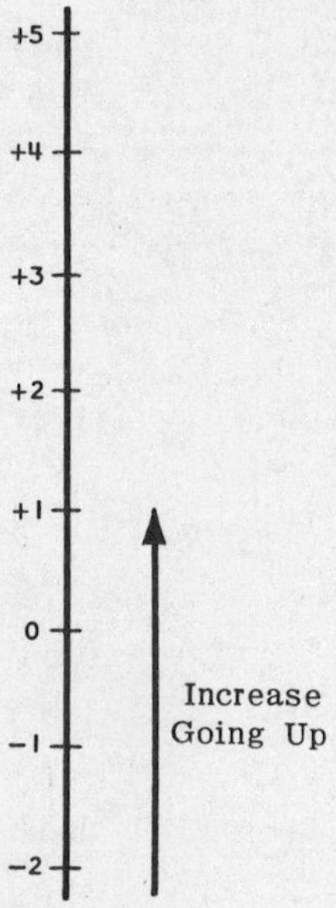

Vertical Number Scale

On a horizontal scale, signed numbers increase to the right; on a vertical scale, the increase is upwards. From this, we obtain the following **rules for comparing signed numbers:**

Rule 1. Any positive number is greater than 0.

Rule 2. Any negative number is less than 0.

Rule 3. Any positive number is greater than any negative number. Thus +1 is greater than −1000.

Rule 4. The greater of two positive numbers has the greater absolute value. Thus, +20 is greater than +10.

Rule 5. The greater of two negative numbers has the smaller absolute value. Thus, −10 is greater than −20. It is better to owe $10 than to owe $20.

(3) Using number scales to solve problems:

Such solutions are shown in examples **2.3** to **2.6**.

(4) Using number scales to understand operations on signed numbers:

This is done throughout the remainder of this chapter.

(5) Using number scales to understand graphs:

In a later chapter, the horizontal number scale and vertical number scale are combined into the graph invented by Descartes in 1637. This graph is of the greatest importance in mathematics and science.

2.1. Two Meanings of Signed Numbers

On a temperature scale, state two meanings of *a*) +25, *b*) − 25, *c*) 0.

Ans. *a*) +25 means (*1*) 25° above zero or (*2*) a rise of 25° from any temperature.
b) −25 means (*1*) 25° below zero or (*2*) a drop of 25° from any temperature.
c) 0 means (*1*) zero degrees or (*2*) no change in temperature.

Note. As a number in arithmetic, 0 means "nothing". Do not confuse this meaning of zero with the other two.

2.2. Comparing Signed Numbers

Which is greater ?

a) $+\frac{1}{4}$ or 0   *b*) +50 or +30   *c*) − 30 or 0   *d*) − 30 or −10   *e*) +10 or −100

(Refer to the rules for comparing signed numbers.)

Ans. *a*) $+\frac{1}{4}$ **(Rule 1)**   *b*) +50 **(Rule 4)**   *c*) 0 **(Rule 2)**   *d*) −10 **(Rule 5)**   *e*) +10 **(Rule 3)**

2.3 to 2.6. Use a Number Scale to Solve Problems

2.3. Beginning with the main floor, an elevator went up 4 floors, then up two more and then down 8. Find its location after making these changes.

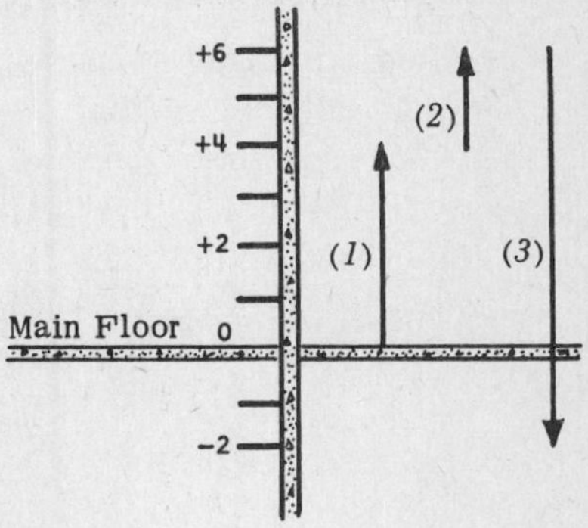

Solution: Each interval is one floor. The origin, 0, indicates the main floor. The three changes are shown by arrows:
(*1*) 4 floors up, (*2*) 2 floors up, (*3*) 8 floors down
The final location is -2.

Ans. 2 floors below the main floor.

2.4. Beginning with 3° below zero, the temperature changed by rising 9°, then dropping 12° and finally rising 6°. Find the final temperature.

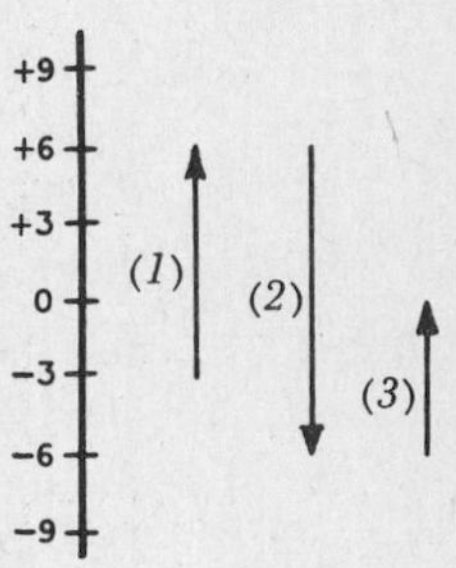

Solution: Each interval is 3°. The temperature began at 3° below zero, -3 on the scale. Arrows show the temperature changes:
(*1*) 9° rise, (*2*) 12° drop, (*3*) 6° rise
The final temperature reading is 0.

Ans. zero degrees

2.5. A football player gained 50 yd. and reached his opponent's 10 yd. line. Where did he begin his gain?

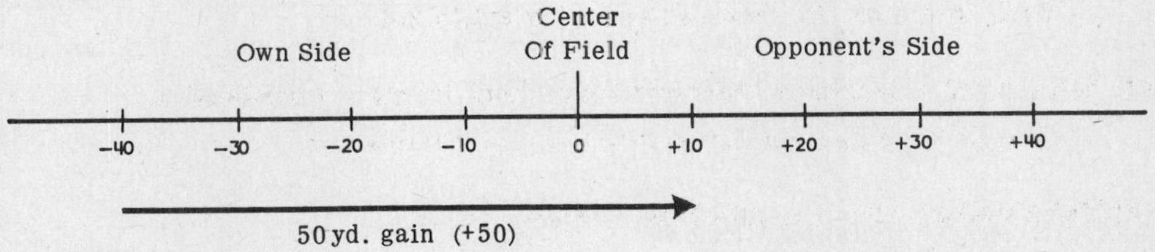

Solution: Each interval is 10 yd. The 50 yd. gain is shown by an arrow. The origin, 0, indicates the center of the field. Since the arrow ends at +10 and has 50 yd. length, it began at -40.

Ans. He began his gain from his 40 yd. line.

2.6. A plane started from a point 150 mi. west of its base and flew directly east, reaching a point 150 mi. east of the base. How far did it travel?

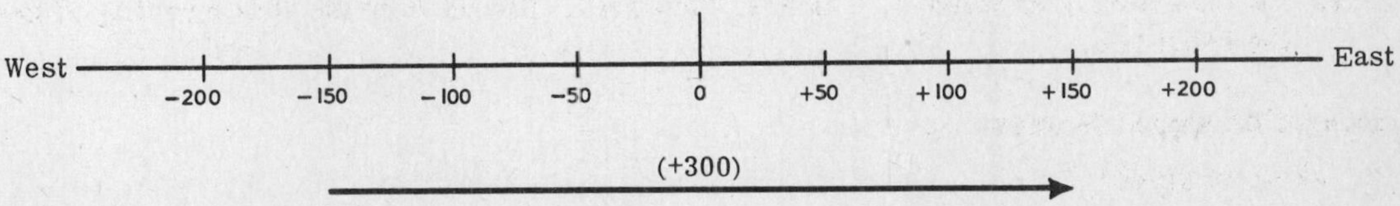

Solution: Each interval is 50 mi. The origin indicates the base. The arrow begins at -150 and ends at $+150$. It has a length of 300 and points eastward.

Ans. 300 mi. eastward

3. ADDING SIGNED NUMBERS

"Add", in algebra, means "combine".

In algebra, **adding signed numbers** means **combining them** to obtain a single number which represents the total or combined effect.

Thus, if Mary first gains 10 lb. and then loses 15 lb., the total effect of the two changes is a loss of 5 lb. This is shown by adding signed numbers: $(+10) + (-15) = -5$.

Uses of the symbol "+": The symbol "+", when used in adding two signed numbers, has two meanings: (*1*) "+" may mean "add" or
(*2*) "+" may mean "positive number".

Thus $(-8) + (+15)$ means **add positive** 15 to negative 8.

Rules For Adding Signed Numbers

Rule 1. To add two signed numbers with like signs, add their absolute values. To this result, prefix the common sign.

Thus, to add (+7) and (+3) or to add (−7) and (−3), add the absolute values 7 and 3. To the result 10, prefix the common sign.

Hence, $(+7) + (+3) = +10$ and $(-7) + (-3) = -10$.

Rule 2. To add two signed numbers with unlike signs, subtract the smaller absolute value from the other. To this result, prefix the sign of the number having the larger absolute value.

Thus, to add (+7) and (−3) or (−7) and (+3), subtract the absolute value 3 from the absolute value 7. To the result 4, prefix the sign of the number having the larger absolute value. Hence, $(+7) + (-3) = +4$ and $(-7) + (+3) = -4$.

T H I N K ! IN ADDING TWO SIGNED NUMBERS:
Add Absolute Values for Like Signs, Subtract for Unlike.

Rule 3. Zero is the sum of two signed numbers with unlike signs and the same absolute value. Such signed numbers are **opposites** of each other.

Thus, $(+27) + (-27) = 0$. The numbers +27 and −27 are opposites of each other.

3.1. Combining by Means of Signed Numbers

Using signed numbers, find each sum:

a) 10 yd. gained plus 5 yd. lost.
b) 8 steps right plus 10 steps left.
c) \$5 earned plus \$5 spent.
d) 80 mi. south plus 100 mi. south.

Solutions:

a) 10 yd. gained plus 5 yd. lost
$(+10) + (-5)$
Ans. +5 or 5 yd. gained

b) 8 steps right plus 10 steps left
$(+8) + (-10)$
Ans. −2 or 2 steps left

c) \$5 earned plus \$5 spent
$(+5) + (-5)$
Ans. 0 or no change

d) 80 mi. south plus 100 mi. south
$(-80) + (-100)$
Ans. −180 or 180 mi. south

3.2. Rule 1. Adding Signed Numbers with Like Signs

Add: *a*) (+8), (+2) *b*) (−8), (−2) *c*) (−30), $(-14\frac{1}{2})$

Procedure:	**Solutions:** *a*) $(+8) + (+2)$	*b*) $(-8) + (-2)$	*c*) $(-30) + (-14\frac{1}{2})$
1. Add absolute values:	$8 + 2 = 10$	$8 + 2 = 10$	$30 + 14\frac{1}{2} = 44\frac{1}{2}$
2. Prefix common sign:	**Ans.** +10	**Ans.** −10	**Ans.** $-44\frac{1}{2}$

3.3. Rule 2. Adding Signed Numbers with Unlike Signs

Procedure:	Add: a) (+7), (−5)	b) (−17), (+10)
	Solutions:	
	a) (+7) + (−5)	b) (−17) + (+10)
1. Subtract absolute values:	7 − 5 = 2	17 − 10 = 7
2. Prefix sign of number having larger absolute value:	Sign of +7 is +.	Sign of −17 is −.
	Ans. +2	**Ans.** −7

3.4. Rule 3. Adding Signed Numbers which are Opposites of Each Other

Procedure:	Add: a) (−18), (+18)	b) $(+30\frac{1}{2})$, $(-30\frac{1}{2})$	c) (−1.75), (+1.75)
	Solutions:		
	a) (−18) + (+18)	b) $(+30\frac{1}{2}) + (-30\frac{1}{2})$	c) (−1.75) + (+1.75)
Sum is always zero:	**Ans.** 0	**Ans.** 0	**Ans.** 0

3.5. Rules 1, 2 and 3. Adding Signed Numbers

Add +25 to *a*) +30, *b*) −30, *c*) −25.

Ans. *a*) (+30) + (+25) = +55 **(Rule 1)**
b) (−30) + (+25) = − 5 **(Rule 2)**
c) (−25) + (+25) = 0 **(Rule 3)**

To −20, add *d*) −30, *e*) +10, *f*) +20

Ans. *d*) (−20) + (−30) = −50 **(Rule 1)**
e) (−20) + (+10) = −10 **(Rule 2)**
f) (−20) + (+20) = 0 **(Rule 3)**

4. SIMPLIFYING THE ADDITION OF SIGNED NUMBERS

To simplify the writing used in adding signed numbers:

(1) Parentheses may be omitted.
(2) The symbol "+" may be omitted when it means "add".
(3) If the first signed number is positive, its + sign may be omitted.
Thus, 8 + 9 − 10 may be written instead of (+8) + (+9) + (−10).

To Simplify Adding Signed Numbers — Add: (+23), (−12), (−8), (+10)

Procedure:	**Solution:**
1. Add all the positive numbers: (*Their sum is positive.*)	1. Add +'s: 23 + 10 = 33
2. Add all the negative numbers: (*Their sum is negative.*)	2. Add −'s: −12 − 8 = −20
3. Add the resulting sums:	3. Add sums: 33 − 20
	Ans. 13

4.1. Simplifying the Addition of Signed Numbers

Express in simplified form, horizontally and vertically; then add:

a) (+27) + (−15) + (+3) + (−5), *b*) (−8) + (+13) + (−20) + (+9), *c*) (+11.2) + (+13.5) + (−6.7) + (+20.9)

Simplified Horizontal Forms:

a) 27 − 15 + 3 − 5 *b*) −8 + 13 − 20 + 9 *c*) 11.2 + 13.5 − 6.7 + 20.9

Simplified Vertical Forms:

a)	b)	c)
27	− 8	11.2
−15	13	13.5
3	−20	− 6.7
− 5	9	20.9
Ans. 10	*Ans.* − 6	*Ans.* 38.9

4.2. Adding Positives and Negatives Separately

Add: *a*) (+27), (−15), (+3), (−5) *b*) (−8), (+13), (−20), (+9) *c*) (+11.2), (+13.5), (−6.7), (+20.9), (−3.1)

Solutions:

a) Simplify:
27 − 15 + 3 − 5

Add +'s	Add −'s
27	−15
3	−5
30	−20

Add results: 30 − 20

Ans. 10

b) Simplify:
−8 + 13 − 20 + 9

Add +'s	Add −'s
13	−8
9	−20
22	−28

Add results: 22 − 28

Ans. −6

c) Simplify:
11.2 + 13.5 − 6.7 + 20.9 − 3.1

Add +'s	Add −'s
11.2	−6.7
13.5	−3.1
20.9	−9.8
45.6	

Add results: 45.6 − 9.8

Ans. 35.8

4.3. Using Signed Numbers and Number Scales to Solve Problems

In May, Tom added deposits of $30 and $20. Later, he withdrew $40 and $30. Find the change in his balance due to these changes using *a*) signed numbers and *b*) number scale:

a) **Signed Number Solution**

Add	Add +'s	Add −'s
+30	30	−40
+20	20	−30
−40	50	−70
−30		

50 − 70 = −20

Ans. $20 less

b) **Number Scale Solution**

The last of the 4 arrows ends at −20. This means $20 less than the original balance.

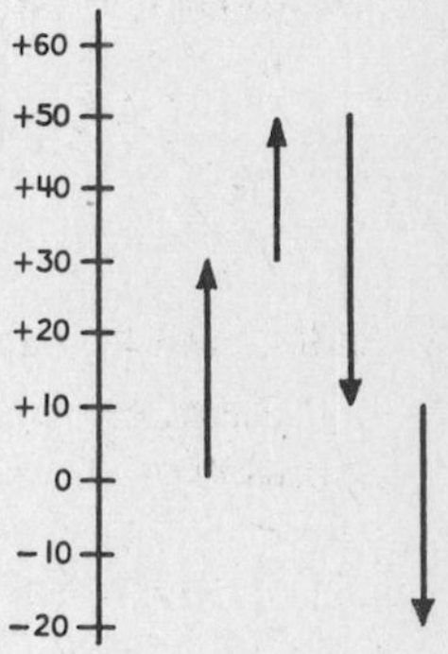

5. SUBTRACTING SIGNED NUMBERS

The symbol "−", used in subtracting signed numbers, has two meanings:

(1) "−" may mean "subtract" or

(2) "−" may mean "negative number".

Thus, (+8) − (−15) means **subtract negative** 15 from positive 8.

Using Subtraction to Find the Change From One Position to Another

Subtraction may be used to find the change from one position to another. (See 5.6 to 5.8.)

Thus, to find a temperature change from 10° below zero to 20° above zero, subtract (−10) from (+20). The result is +30, meaning a rise of 30°.

Rule for Subtracting Signed Numbers

Rule 1. To subtract a positive number, **add** its opposite negative.

Thus, to subtract (+10), add (−10). For example, (+18) − (**+10**)
= (+18) + (**−10**). *Ans.* +8

Rule 2. To subtract a negative number, **add** its opposite positive.

Thus, to subtract (−10), add (+10). For example, (+30) − (**−10**)
= (+30) + (**+10**). *Ans.* +40

THINK! **To Subtract a Signed Number, Add Its Opposite.**

5.1. Rule 1. Subtracting a Positive Number

Procedure:	Subtract: *a*) (+8) from (+29)	*b*) (+80) from (+80)	*c*) (+18) from (−40)
	Solutions:		
To **subtract** a positive, **add** its opposite negative:	*a*) (+29) − (+8) (+29) + (−8) **Ans.** 21	*b*) (+80) − (+80) (+80) + (−80) **Ans.** 0	*c*) (−40) − (+18) (−40) + (−18) **Ans.** −58

5.2. Rule 2. Subtracting a Negative Number

Procedure:	Subtract: *a*) (−7) from (+20)	*b*) (−67) from (−67)	*c*) (−27) from (−87)
	Solutions:		
To **subtract** a negative, **add** its opposite positive:	*a*) (+20) − (−7) (+20) + (+7) **Ans.** 27	*b*) (−67) − (−67) (−67) + (+67) **Ans.** 0	*c*) (−87) − (−27) (−87) + (+27) **Ans.** −60

5.3. Subtracting Vertically

Subtract the lower number from the upper:

a)	*b*)	*c*)
+40 +10	+40 +55	−40 −75

Solutions: (To subtract a signed number, add its opposite.)

a)	*b*)	*c*)
+40 → +40 −(+10) → +(−10) *Ans.* +30	+40 → +40 −(+55) → +(−55) *Ans.* −15	−40 → −40 −(−75) → +(+75) *Ans.* +35

5.4. Rule 1 and 2. Subtracting Signed Numbers

Subtract +3 from *a*) +11, *b*) −15.

From −8, subtract *c*) −15, *d*) −5.

Solutions:

a)	*b*)	*c*)	*d*)
(+11) − (+3) (+11) + (−3) 11 − 3 *Ans.* 8	(−15) − (+3) (−15) + (−3) −15 − 3 *Ans.* −18	(−8) − (−15) (−8) + (+15) −8 + 15 *Ans.* +7	(−8) − (−5) (−8) + (+5) −8 + 5 *Ans.* −3

5.5. Combining Addition and Subtraction of Signed Numbers

Combine:

a)	*b*)	*c*)
(+10) + (+6) − (−2) (+10) + (+6) + (+2) 10 + 6 + 2 *Ans.* 18	(+8) − (−12) − (+5) + (−3) (+8) + (+12) + (−5) + (−3) 8 + 12 − 5 − 3 *Ans.* 12	(−11) − (+5) + (−7) (−11) + (−5) + (−7) −11 − 5 − 7 *Ans.* −23

5.6. Finding the Change Between Two Signed Numbers

Find the change from +20 to −60, using *a*) number scale and *b*) signed numbers:

a) **Number Scale Solution**

From

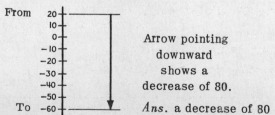

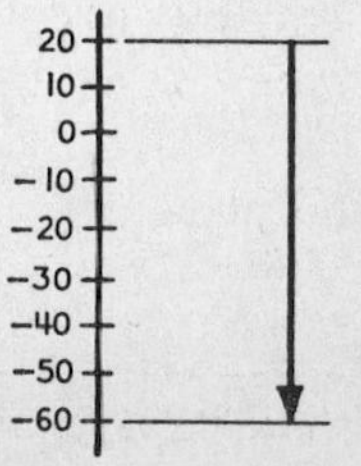

To

Arrow pointing downward shows a decrease of 80.

Ans. a decrease of 80

b) **Signed Number Solution**

Subtract 20 from −60,

−60 − (+20)

−60 + (−20)

−80

5.7. Finding the Distance Between Two Levels

Using (*1*) a number scale and (*2*) signed numbers, find the distance from 300 ft. below sea level to *a*) 800 ft. below sea level, *b*) sea level, *c*) 100 ft. above sea level.

Signed Number Solution

a) $-800-(-300)$	*b*) $0-(-300)$	*c*) $100-(-300)$
$-800+(+300)$	$0+(+300)$	$100+(+300)$
-500	$+300$	$+400$
Ans. 500 ft. down	*Ans.* 300 ft. up	*Ans.* 400 ft. up

(*Show how each arrow indicates an answer.*)

Number Scale Solution

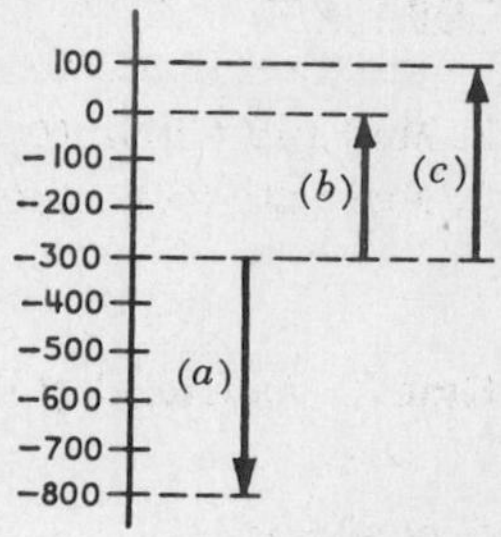

5.8. Finding a Temperature Change

On Monday, the temperature changed from $-10°$ to $20°$. On Tuesday, the change was from $20°$ to $-20°$. Find each change, using *a*) number scale and *b*) signed numbers.

Signed Number Solution

MONDAY CHANGE	TUESDAY CHANGE
$20-(-10)$	$-20-(+20)$
$20+(+10)$	$-20+(-20)$
$+30$	-40
Ans. Rise of 30°	*Ans.* Drop of 40°

(*Show how each arrow indicates an answer.*)

Number Scale Solution

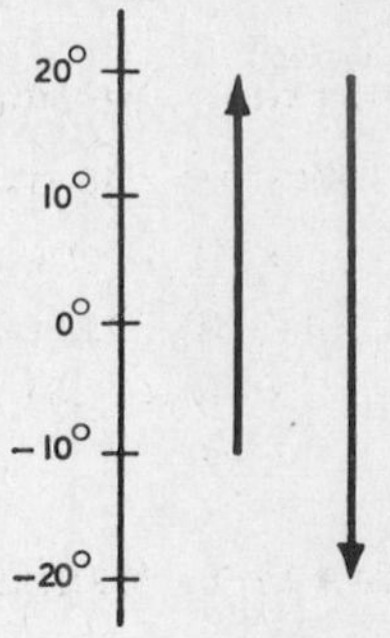

6. MULTIPLYING SIGNED NUMBERS

Rules for Multiplying Two Signed Numbers

Rule 1. To multiply two signed numbers with like signs, multiply their absolute values and make the product positive.

Thus, $(+5)(+4) = +20$ and $(-5)(-4) = +20$.

Rule 2. To multiply two signed numbers with unlike signs, multiply their absolute values and make the product negative.

Thus, $(-7)(+2) = -14$ and $(+7)(-2) = -14$.

THINK!

IN MULTIPLYING TWO SIGNED NUMBERS:
Like Signs Result in Positive Product, Unlike Signs in Negative.

Rule 3. Zero times any signed number equals zero.

Thus, $0(-8\frac{1}{2}) = 0$ and $(44.7)(0) = 0$.

Rules for Multiplying More Than Two Signed Numbers

Rule 4. Make the product positive if all the signed numbers are positive or if there is an even number of negatives.

Thus, $(+10)(+4)(-3)(-5) = +600$.

Rule 5. Make the product negative if there is an odd number of negatives.

Thus, $(+10)(-4)(-3)(-5) = -600$.

Rule 6. Make the product zero if any number is zero.

Thus, $(-5)(+82)(0)(316\frac{1}{2}) = 0$.

6.1. Rule 1. Multiplying Signed Numbers with Like Signs

Multiply:	a) (+5)(+9)	b) (−5)(−11)	c) (−3)(−2.4)
Procedure:	**Solutions:**		
	a) (+5)(+9)	b) (−5)(−11)	c) (−3)(−2.4)
1. Multiply absolute values:	5(9) = 45	5(11) = 55	3(2.4) = 7.2
2. Make product positive:	**Ans.** +45	**Ans.** +55	**Ans.** +7.2

6.2. Rule 2. Multiplying Signed Numbers with Unlike Signs

Multiply:	a) (+8)(−9)	b) (−20)(+5)	c) $(+3\frac{1}{2})(-10)$
Procedure:	**Solutions:**		
	a) (+8)(−9)	b) (−20)(+5)	c) $(+3\frac{1}{2})(-10)$
1. Multiply absolute values:	8(9) = 72	20(5) = 100	$3\frac{1}{2}(10) = 35$
2. Make product negative:	**Ans.** −72	**Ans.** −100	**Ans.** −35

6.3. Rules 1 to 3. Multiplying Signed Numbers

Multiply +10 by *a*) +18, *b*) 0, *c*) −13

Ans. *a*) (+10)(+18) = +180 **(Rule 1)**
b) (+10)(0) = 0 **(Rule 3)**
c) (+10)(−13) = −130 **(Rule 2)**

Multiply −20 by *d*) +2.5, *e*) 0, *f*) $-2\frac{1}{4}$

d) (−20)(+2.5) = −50 **(Rule 2)**
e) (−20) 0 = 0 **(Rule 3)**
f) $(-20)(-2\frac{1}{4}) = +45$ **(Rule 1)**

6.4. Rules 4 to 6. Multiplying More Than Two Signed Numbers

Multiply:

a) (+2)(+3)(+4) *Ans.* +24 **(Rule 4)**
b) (+2)(+3)(−4) *Ans.* −24 **(Rule 5)**
c) (+2)(−3)(−4) *Ans.* +24 **(Rule 4)**
d) (−2)(−3)(−4) *Ans.* −24 **(Rule 5)**
e) (+2)(+3)(−4)(0) *Ans.* 0 **(Rule 6)**
f) (−1)(−2)(−5)(−10) *Ans.* +100 **(Rule 4)**
g) (−1)(−2)(+5)(+10) *Ans.* +100 **(Rule 4)**
h) (−1)(+2)(+5)(+10) *Ans.* −100 **(Rule 5)**
i) (−1)(−2)(−5)(+10) *Ans.* −100 **(Rule 5)**
j) (−1)(−2)(+5)(+10) 0 *Ans.* 0 **(Rule 6)**

6.5. Using Signed Numbers to Solve Problems

Complete each statement, using signed numbers to obtain the answer:

a) If George deposits $5 each week, then after 3 weeks his bank balance will be (   ).

Let +5 = $5 weekly deposit
and +3 = 3 weeks later

Then, (+5)(+3) = +15
Ans. $15 more

b) If Henry has been spending $5 a day, then 4 days ago he had (   ).

Let −5 = $5 daily spending
and −4 = 4 days earlier

Then, (−5)(−4) = +20
Ans. $20 more

c) If the temperature falls 6° each day, then after 3 days it will be (   ).

Let −6 = 6° daily fall
and +3 = 3 days later

Then, (−6)(+3) = −18
Ans. 18° lower

d) If a team has been gaining 10 yd. on each play, then 3 plays ago it was (   ).

Let +10 = 10 yd. gain per play
and −3 = 3 plays earlier

Then, (+10)(−3) = −30
Ans. 30 yd. farther back

e) If a car has been traveling west at 30 MPH, then 3 hr. ago it was (   ).

Let −30 = 30 MPH rate westward
and −3 = 3 hr. earlier

Then, (−30)(−3) = +90
Ans. 90 mi. farther east

7. FINDING POWERS OF SIGNED NUMBERS

Rules for Finding Powers of Signed Numbers

(Keep in mind that the power is the answer obtained.)

Rule 1. For positive base, power is always positive.
Thus, $(+2)^3$ or $2^3 = +8$ since $(+2)^3 = (+2)(+2)(+2)$.

Rule 2. For negative base having even exponent, power is positive.
Thus, $(-2)^4 = +16$ since $(-2)^4 = (-2)(-2)(-2)(-2)$.
Note. -2^4 means "the negative of 2^4". Hence, $-2^4 = -16$.

Rule 3. For negative base having odd exponent, power is negative.
Thus, $(-2)^5 = -32$ since $(-2)^5 = (-2)(-2)(-2)(-2)(-2)$.

THINK!

FOR NEGATIVE BASE:
Power is Positive when Exponent is Even, Negative when Odd.

7.1. Rule 1. Finding Powers when Base is Positive

Find each power:

a) 3^2 *Ans.* 9	*d*) $(+5)^3$ *Ans.* 125	*g*) $(\frac{2}{3})^2$ *Ans.* $\frac{4}{9}$
b) 2^4 *Ans.* 16	*e*) $(+7)^2$ *Ans.* 49	*h*) $(\frac{1}{2})^3$ *Ans.* $\frac{1}{8}$
c) 10^3 *Ans.* 1000	*f*) $(+1)^{10}$ *Ans.* 1	*i*) $(\frac{1}{10})^4$ *Ans.* $\frac{1}{10,000}$

7.2. Rules 2 and 3. Finding Powers when Base is Negative

Find each power:

a) $(-3)^2$ *Ans.* 9	*d*) $(-1)^{105}$ *Ans.* -1	*g*) $(-\frac{1}{2})^2$ *Ans.* $\frac{1}{4}$
b) $(-3)^3$ *Ans.* -27	*e*) $(-.5)^2$ *Ans.* .25	*h*) $(-\frac{1}{3})^3$ *Ans.* $-\frac{1}{27}$
c) $(-1)^{100}$ *Ans.* 1	*f*) $(-.2)^3$ *Ans.* $-.008$	*i*) $(-\frac{2}{5})^3$ *Ans.* $-\frac{8}{125}$

7.3. Rules 1 to 3. Finding Powers of Signed Numbers

Find each power:

a) 3^2 *Ans.* 9	*d*) $(\frac{1}{2})^3$ *Ans.* $\frac{1}{8}$	*g*) $(-1)^{100}$ *Ans.* 1
b) -3^2 *Ans.* -9	*e*) $(-\frac{1}{2})^3$ *Ans.* $-\frac{1}{8}$	*h*) -1^{100} *Ans.* -1
c) $(-3)^2$ *Ans.* 9	*f*) $-(\frac{1}{2})^3$ *Ans.* $-\frac{1}{8}$	*i*) $(-1)^{111}$ *Ans.* -1

7.4. Finding Bases, Exponents or Powers

Complete each:

a) $(-3)^? = 81$ *Ans.* 4	*c*) $(?)^5 = -32$ *Ans.* -2	*e*) $(-\frac{3}{4})^3 = ?$ *Ans.* $-\frac{27}{64}$
b) $2^? = 128$ *Ans.* 7	*d*) $(?)^{121} = -1$ *Ans.* -1	*f*) $-.2^4 = ?$ *Ans.* $-.0016$

8. DIVIDING SIGNED NUMBERS

Rules for Dividing Signed Numbers

Rule 1. To divide two signed numbers with like signs, divide the absolute value of the first by that of the second and make the quotient positive.

Thus, $\frac{+8}{+2} = +4$ and $\frac{-8}{-2} = +4$.

Rule 2. To divide two signed numbers with unlike signs, divide the absolute value of the first by that of the second and make the quotient negative.

Thus, $\frac{+12}{-4} = -3$ and $\frac{-12}{+4} = -3$.

THINK!

IN DIVIDING SIGNED NUMBERS:
Like Signs Result in Positive Quotient, Unlike Signs in Negative.

Rule 3. Zero divided by any signed number is zero.

Thus, $\frac{0}{+17} = 0$ and $\frac{0}{-17} = 0$.

Rule 4. Dividing a signed number by zero is an impossible operation.

Thus, $(+18) \div 0$ or $(-18) \div 0$ are impossible.

$\frac{18}{0}$, $\frac{+18}{0}$ or $\frac{-18}{0}$ are meaningless.

Combining Multiplying and Dividing of Signed Numbers

To multiply and divide signed numbers at the same time:

(1) Isolate their absolute values to obtain the absolute value of the answer.

(2) Find the sign of the answer, as follows:

Rule 5. Make the sign of the answer positive if all the numbers are positive or if there is an even number of negatives.

Thus, $\frac{(+12)(+5)}{(+3)(+2)}$, $\frac{(+12)(+5)}{(-3)(-2)}$, and $\frac{(-12)(-5)}{(-3)(-2)}$ equal $+10$.

Rule 6. Make the sign of the answer negative if there is an odd number of negatives.

Thus, $\frac{(+12)(+5)}{(+3)(-2)}$, $\frac{(+12)(-5)}{(-3)(-2)}$, and $\frac{(-12)(-5)}{(+3)(-2)}$ equal -10.

Rule 7. Make the answer zero if one of the numbers in the dividend is 0.

Thus, $\frac{(+53)(0)}{(-17)(-84)} = 0$.

Rule 8. A zero divisor makes the operation impossible.

Thus, $\frac{(+53)(-17)}{(+84)\,0}$ is meaningless.

8.1. Rule 1. Dividing Signed Numbers with Like Signs

Procedure:	Divide: *a*) (+12) by (+6)	*b*) (−12) by (−4)	*c*) (−24.6) by (−3)
	Solutions:		
	a) $\frac{+12}{+6}$	*b*) $\frac{-12}{-4}$	*c*) $\frac{-24.6}{-3}$
1. Divide absolute values:	$\frac{12}{6} = 2$	$\frac{12}{4} = 3$	$\frac{24.6}{3} = 8.2$
2. Make quotient positive:	**Ans.** +2	**Ans.** +3	**Ans.** +8.2

8.2. Rule 2. Dividing Signed Numbers with Unlike Signs

Divide: *a*) (+20) by (−5) | *b*) −24 by +8 | *c*) +8 by −16

Procedure:

Solutions:

Procedure	*a*)	*b*)	*c*)
	$\frac{+20}{-5}$	$\frac{-24}{+8}$	$\frac{+8}{-16}$
1. Divide absolute values:	$\frac{20}{5} = 4$	$\frac{24}{8} = 3$	$\frac{8}{16} = \frac{1}{2}$
2. Make quotient negative:	**Ans.** -4	**Ans.** -3	**Ans.** $-\frac{1}{2}$

8.3. Rules 1 to 4. Dividing Signed Numbers

Divide +9 by *a*) +3, *b*) 0, *c*) −10

Ans. *a*) $\frac{+9}{+3} = +3$ **(Rule 1)**

b) $\frac{+9}{0}$ is meaningless **(Rule 4)**

c) $\frac{+9}{-10} = -\frac{9}{10}$ **(Rule 2)**

Divide each by −12: *d*) +36, *e*) 0, *f*) −3

d) $\frac{+36}{-12} = -3$ **(Rule 2)**

e) $\frac{0}{-12} = 0$ **(Rule 3)**

f) $\frac{-3}{-12} = +\frac{1}{4}$ **(Rule 1)**

8.4. Rules 5 and 6: Combining Multiplying and Dividing Signed Numbers

Solve: *a*) $\frac{(+12)(+8)}{(-3)(-4)}$ | *b*) $\frac{(-4)(-2)(-3)}{(-5)(-10)}$

Procedure:	*a*)	*b*)
1. Isolate absolute values:	$\frac{(12)(8)}{(3)(4)} = 8$	$\frac{(4)(2)(3)}{(5)(10)} = \frac{12}{25}$
2. Find the sign of the answer:	$\frac{(+)(+)}{(-)(-)} = +$	$\frac{(-)(-)(-)}{(-)(-)} = -$
	Ans. $+8$	**Ans.** $-\frac{12}{25}$

8.5. Rules 7 and 8. Zero in Dividend or Divisor

Solve: *a*) $\frac{(+27)0}{(-3)(+15)}$ *Ans.* 0 **(Rule 7)** *b*) $\frac{(+27)(-3)}{(+15)0}$ *Ans.* meaningless **(Rule 8)**

9. *EVALUATING EXPRESSIONS HAVING SIGNED NUMBERS*

To evaluate an expression having signed numbers:

1) Substitute the values given for the letters, enclosing them in parentheses.
2) Perform the operations in the correct order, doing the power operations first.

9.1. Evaluating Expressions Having One Letter

Evaluate if $y = -2$:

a) $2y + 5$	*b*) $20 - 3y$	*c*) $4y^2$	*d*) $2y^3$	*e*) $20 - y^5$
$2(-2) + 5$	$20 - 3(-2)$	$4(-2)^2$	$2(-2)^3$	$20 - (-2)^5$
$-4 + 5$	$20 + 6$	$4(4)$	$2(-8)$	$20 - (-32)$
Ans. 1	*Ans.* 26	*Ans.* 16	*Ans.* −16	*Ans.* 52

9.2. Evaluating Expressions Having Two Letters

Evaluate if $x = -3$ and $y = +2$:

a)	b)	c)	d)
$2x - 5y$	$20 - 2xy$	$3xy^2$	$2x^2 - y^2$
$2(-3) - 5(+2)$	$20 - 2(-3)(+2)$	$3(-3)(+2)^2$	$2(-3)^2 - (+2)^2$
$-6 - 10$	$20 + 12$	$3(-3)(4)$	$2(+9) - (+4)$
Ans. -16	*Ans.* 32	*Ans.* -36	*Ans.* 14

9.3. Evaluating Expressions Having Three Letters

Evaluate if $x = +1$, $y = -2$, $z = -3$:

a)	b)	c)	d)
$4xy^2 + z$	$x^2 + y^2 - z^2$	$\frac{3x - 5y}{2z}$	$\frac{2y^2}{x - z}$
$4(+1)(-2)^2 + (-3)$	$(+1)^2 + (-2)^2 - (-3)^2$	$\frac{3(+1) - 5(-2)}{2(-3)}$	$\frac{2(-2)^2}{+1 - (-3)}$
$4(1)(4) - 3$	$1 + 4 - 9$	$\frac{3 + 10}{-6}$	$\frac{2 \cdot 4}{1 + 3}$
Ans. 13	*Ans.* -4	*Ans.* $-2\frac{1}{6}$	*Ans.* 2

SUPPLEMENTARY PROBLEMS

1. State the quantity represented by each signed number: (1.2)
   a) By +12, if −12 means 12 yd. lost.
   b) By −7, if +7 means 7 flights up.
   c) By +25, if −25 means 25 mi. westward.
   d) By −32, if +32 means $32 earned.
   e) By +15, if −15 means $15 withdrawn.
   f) By −13, if +13 means 13 steps forward.

   Ans. a) 12 yd. gained b) 7 flights down c) 25 mi. eastward d) $32 spent e) $15 deposited f) 13 steps backward

2. State (*1*) the absolute value of each pair of numbers and (*2*) the differences of these absolute values: a) +40 and −32 b) −3.7 and +1.5 c) $-40\frac{1}{2}$ and $-10\frac{1}{2}$ (1.3)

   Ans. a) (*1*) 40, 32; (*2*) 8 b) (*1*) 3.7, 1.5; (*2*) 2.2 c) (*1*) $40\frac{1}{2}$, $10\frac{1}{2}$; (*2*) 30

3. On a temperature scale, state the meanings of a) −17, b) +15. (2.1)

   Ans. a) −17 means (*1*) 17° below zero or (*2*) a drop of 17° from any temperature.
   b) +15 means (*1*) 15° above zero or (*2*) a rise of 15° from any temperature.

4. Which is greater? a) $-\frac{1}{2}$ or 0, b) $-\frac{1}{2}$ or −30, c) $+\frac{1}{2}$ or $-2\frac{1}{2}$, d) −110 or −120. (2.2)

   Ans. a) 0 b) $-\frac{1}{2}$ c) $+\frac{1}{2}$ d) −110

5. Beginning with 5° above zero, the temperature rose 2°, then dropped 8° and finally rose 4°. Find the change in temperature and the final temperature after the changes. (2.4)

   Ans. Change in temperature was a drop of 2° from the initial temperature of +5°. Final temperature after the changes was 3° above zero.

6. A plane starting from a point 200 mi. east of its base flew directly west to a point 300 mi. west of the base. How far did it travel? (2.6)

   Ans. 500 mi. westward

7. Using signed numbers, find each sum: (3.1)

a) 20 lb. gained plus 7 lb. lost.
b) 17 lb. lost plus 3 lb. gained.
c) \$5 spent plus \$15 spent.
d) \$5 gained plus \$15 spent.
e) 8° rise plus 20° rise.
f) 8° drop plus 8° rise.

Ans. *a*) $(+20) + (-7) = +13$ or 13 lb. gained
b) $(-17) + (+3) = -14$ or 14 lb. lost
c) $(-5) + (-15) = -20$ or \$20 spent
d) $(+5) + (-15) = -10$ or \$10 spent
e) $(+8) + (+20) = +28$ or 28° rise
f) $(-8) + (+8) = 0$ or no change

8. Add: *a*) $(+10\frac{1}{2}), (+7\frac{1}{2})$ *b*) $(-1.4), (-2.5)$ *c*) $(+5), (+3), (+12)$ *d*) $(-1), (-2), (-3)$ (3.2)

Ans. *a*) $+18$ *b*) -3.9 *c*) $+20$ *d*) -6

9. Add: *a*) $(+6\frac{1}{4}), (-3\frac{1}{2})$ *b*) $(-.23), (+.18)$ *c*) $(+23), (-13), (+12)$ *d*) $(+5), (-8), (+17)$ (3.3)

Ans. *a*) $+2\frac{3}{4}$ *b*) $-.05$ *c*) $+22$ *d*) $+14$

10. Add: *a*) $(+25.7), (-25.7)$ *b*) $(+120), (-120), (+19)$ *c*) $(+28), (-16), (-28), (+16)$ (3.4)

Ans. *a*) 0 *b*) $+19$ *c*) 0

11. Add -32 to (3.5)

a) -25 *Ans.* -57 *c*) $+32$ *Ans.* 0
b) $+25$ *Ans.* -7 *d*) $+100$ *Ans.* $+68$

To $+11$, add

e) -29 *Ans.* -18 *g*) $+117$ *Ans.* $+128$
f) -11 *Ans.* 0 *h*) -108 *Ans.* -97

12. Express in simplified vertical form and add: (4.1)

a) $(+11) + (+3) + (-9) + (-8)$ *b*) $(-42) + (+42) + (-85)$ *c*) $(+13.7) + (+2.4) - (8.9)$

Ans. *a*)

$$\begin{array}{r} 11 \\ 3 \\ -9 \\ \underline{-8} \\ \textit{Sum} = -3 \end{array}$$

b)

$$\begin{array}{r} -42 \\ +42 \\ \underline{-85} \\ \textit{Sum} = -85 \end{array}$$

c)

$$\begin{array}{r} 13.7 \\ 2.4 \\ \underline{-8.9} \\ \textit{Sum} = 7.2 \end{array}$$

13. Express in simplified horizontal form and add: (4.1, 4.2)

a) $(-14), (-7), (+22), (-35)$ *b*) $(+3\frac{1}{4}), (+8\frac{1}{2}), (-40\frac{3}{4})$ *c*) $(-1.78), (-3.22), (+16)$

Ans. *a*) $-14 - 7 + 22 - 35$, *Sum* $= -34$ *b*) $3\frac{1}{4} + 8\frac{1}{2} - 40\frac{3}{4}$, *Sum* $= -29$ *c*) $-1.78 - 3.22 + 16$, *Sum* $= 11$

14. In a football game, a team gained 8 yd. on the first play, gained 1 yd. on the second play, lost 12 yd. on the third play and lost 6 yd. on the fourth play. Find the change in position due to these changes, using *a*) signed numbers and *b*) a number scale: (4.3)

a) $(+8) + (+1) + (-12) + (-6)$
$8 + 1 - 12 - 6$
-9

b) 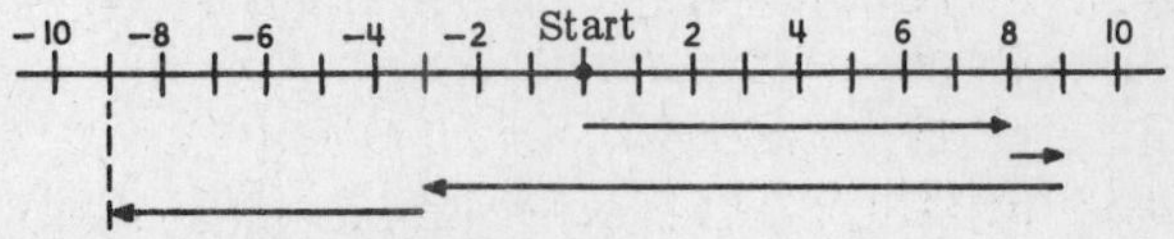

Ans. Total change is 9 yd. loss.

15. Subtract: (5.1)

a) $(+11)$ from $(+16)$
b) $(+47)$ from (-47)
c) $(+3.5)$ from $(+7.2)$
d) $(+7.2)$ from (-3.5)
e) $(+3\frac{1}{3})$ from $(+11\frac{2}{3})$
f) $(+17\frac{2}{3})$ from (-8)

Ans. *a*) $+5$ *b*) -94 *c*) $+3.7$ *d*) -10.7 *e*) $+8\frac{1}{3}$ *f*) $-25\frac{2}{3}$

16. Subtract: (5.2)

a) (-11) from $(+16)$
b) (-47) from $(+47)$
c) $(-.81)$ from $(+.92)$
d) $(-.23)$ from $(-.27)$
e) $(-7\frac{1}{5})$ from $(+2\frac{4}{5})$
f) $(-3\frac{5}{6})$ from (-4)

Ans. *a*) $+27$ *b*) $+94$ *c*) $+1.73$ *d*) $-.04$ *e*) $+10$ *f*) $-\frac{1}{6}$

17. Subtract the lower number from the upper: (5.3)

a) +22	*b*) +17	*c*) −30.7	*d*) −.123	*e*) $-13\frac{2}{3}$	*f*) $-27\frac{1}{2}$
+17	+22	+30.7	+.265	$-10\frac{1}{3}$	$+\ 3\frac{3}{4}$

Ans. *a*) + 5 *b*) − 5 *c*) −61.4 *d*) −.388 *e*) $-\ 3\frac{1}{3}$ *f*) $-31\frac{1}{4}$

18. Subtract +4 from *a*) $+8\frac{1}{2}$, *b*) +4, *c*) +2.3, and *d*) −25. (5.4)
From −10 subtract *e*) −23, *f*) −8, *g*) −3.9, and *h*) $+20\frac{1}{4}$.

Ans. *a*) $+4\frac{1}{2}$, *b*) 0, *c*) −1.7, *d*) −29, *e*) +13, *f*) −2, *g*) −6.1, *h*) $-30\frac{1}{4}$

19. Combine: *a*) (+11) + (+7) − (+24) *d*) (+25) − (−6) − (−22) + (+40) (5.5)
b) (−11) − (−5) − (−14) *e*) (−3.7) + (−2.4) − (+7.8) + (−11.4)
c) (+.13) − (+.07) − (+.32) *f*) $(+2\frac{1}{2}) - (-1\frac{1}{4}) - (+5\frac{3}{4}) - (-7)$

Ans. *a*) −6, *b*) +8, *c*) −.26 *d*) +93, *e*) −25.3, *f*) +5

20. Find the distance from 500 ft. above sea level to *a*) 1200 ft. above sea level, *b*) sea level, *c*) 2000 ft. below sea level. (5.7) *Ans.* *a*) 700 ft. up, *b*) 500 ft. down, *c*) 2500 ft. down

21. On a Monday, the hourly temperatures from 1 P.M. to 6 P.M. were (5.8)

1 P.M.	2 P.M.	3 P.M.	4 P.M.	5 P.M.	6 P.M.
-8°	-5°	0°	4°	-4°	-20°

Find each hourly temperature change.

Ans. *a*) 3° rise from 1 to 2 P.M., *b*) 5° rise from 2 to 3 P.M., *c*) 4° rise from 3 to 4 P.M., *d*) 8° drop from 4 to 5 P.M., *e*) 16° drop from 5 to 6 P.M.

22. Multiply: *a*) (+3)(+22) *c*) (−5)(−4) *e*) $(+8)(+2\frac{1}{2})$ *g*) $(-2\frac{1}{2})(-1\frac{1}{2})$ (6.1)
b) (+.3)(+2.2) *d*) (−.05)(−.04) *f*) $(-35)(-2\frac{1}{5})$ *h*) $(+\frac{5}{4})(+\frac{15}{2})$

Ans. *a*) +66, *b*) +.66, *c*) +20, *d*) +.002, *e*) +20, *f*) +77, *g*) $+3\frac{3}{4}$, *h*) $+9\frac{3}{8}$

23. Multiply: *a*) (−20)(+6) *c*) (−.8)(+.11) *e*) $(+6)(-7\frac{2}{3})$ *g*) $(-\frac{7}{2})(+\frac{11}{4})$
b) (+13)(−8) *d*) (+3.4)(−21) *f*) $(-2\frac{1}{7})(+21)$ *h*) $(+2\frac{1}{5})(-3\frac{2}{5})$

Ans. *a*) −120, *b*) −104, *c*) −.088, *d*) −71.4, *e*) −46, *f*) −45, *g*) $-\frac{77}{8}$, *h*) $-7\frac{12}{25}$

24. Multiply +8 by *a*) +7, *b*) 0, *c*) −4. Multiply −12 by *d*) +.9, *e*) $-1\frac{1}{2}$, *f*) −8.21 (6.3)

Ans. *a*) +56, *b*) 0, *c*) −32 *Ans.* *d*) −10.8, *e*) +18, *f*) +98.52

25. Multiply: *a*) (+3)(+4)(+12) *d*) (−1)(−1)(+1)(−1) *g*) $(+\frac{1}{2})(+\frac{1}{2})(+\frac{1}{2})(-\frac{1}{2})$ (6.4)
b) (+.3)(+.4)(+1.2) *e*) (−2)(−2)(−2)(−2) *h*) $(-\frac{1}{2})(+8)(-\frac{1}{4})(+16)$
c) (+.3)(−4)(−.12) *f*) (−2)(+5)(−5)(+4) *i*) $(-1\frac{1}{2})(+2\frac{1}{2})(+3\frac{1}{2})$

Ans. *a*) +144, *b*) +.144, *c*) +.144, *d*) −1, *e*) +16, *f*) +200, *g*) $-\frac{1}{16}$, *h*) +16, *i*) $-13\frac{1}{8}$

26. Complete each statement. In each case, indicate how signed numbers may be used to obtain the answer. (6.5)

a) If George withdraws $10 each week, then after 5 weeks, his bank balance will be ( ).
b) If Tom has been earning $15 a day, then in 3 days, he will have ( ).
c) If the temperature has risen 8° each day, then five days ago, it was ( ).
d) If a car has been traveling east at 40 MPH, then 3 hours ago, it was ( ).
e) If a school decreases in register 20 pupils per day, then 12 days ago, the register was ( ).

Ans. *a*) $50 less (−10)(+5) = −50 *b*) $45 more (+15)(+3) = +45 *c*) 40° less (+8)(−5) = −40 *d*) 120 mi. farther west (+40)(−3) = −120 *e*) 240 pupils more (−20)(−12) = +240

27. Find each power: *a*) 2^3, *b*) $(+4)^2$, *c*) 10^4, *d*) $.3^2$, *e*) $(+.2)^3$, *f*) $(\frac{1}{4})^2$, *g*) $(\frac{1}{10})^3$, *h*) $(\frac{2}{5})^2$ (7.1)
Ans. *a*) 8, *b*) 16, *c*) 10,000, *d*) .09, *e*) .008, *f*) $\frac{1}{16}$, *g*) $\frac{1}{1000}$, *h*) $\frac{4}{25}$

28. Find each power: *a*) $(-1)^5$, *b*) $(-1)^{82}$, *c*) $(-.3)^3$, *d*) $(-.1)^4$, *e*) $(-.12)^2$, *f*) $(-\frac{2}{5})^3$, *g*) $(-\frac{1}{3})^4$ (7.2)
Ans. *a*) -1, *b*) $+1$, *c*) $-.027$, *d*) $+.0001$, *e*) $+.0144$, *f*) $-\frac{8}{125}$, *g*) $+\frac{1}{81}$

29. Complete each: (7.4)
a) $(-5)^? = -125$ *c*) $(-?)^4 = +.0001$ *e*) $(+\frac{2}{3})^4 = ?$ *g*) $(?)^{171} = -1$
b) $(+10)^? = 10{,}000$ *d*) $(-?)^3 = -.343$ *f*) $(-\frac{1}{2})^5 = ?$ *h*) $(?)^{242} = +1$
Ans. *a*) 3, *b*) 4, *c*) $-.1$, *d*) $-.7$, *e*) $\frac{16}{81}$, *f*) $-\frac{1}{32}$, *g*) -1, *h*) $+1$ or -1

30. Divide: *a*) (+24) by (+3) *c*) (−49) by (−7) *e*) (+4.8) by (+.2) *g*) (−18) by (−4) (8.1)
b) (+88) by (+8) *d*) (−78) by (−6) *f*) (−95) by (−.5) *h*) (+8) by (+12)
Ans. *a*) +8, *b*) +11, *c*) +7, *d*) +13, *e*) +24, *f*) +190, *g*) $+4\frac{1}{2}$, *h*) $+\frac{2}{3}$

31. Divide: *a*) (+30) by (−5) *c*) (+13) by (−2) *e*) (+.2) by (−.04) *g*) (−100) by (+500) (8.2)
b) (−30) by (+5) *d*) (−36) by (+8) *f*) (−30) by (+.1) *h*) (+100) by (−3)
Ans. *a*) −6, *b*) −6, *c*) $-6\frac{1}{2}$, *d*) $-4\frac{1}{2}$, *e*) −5, *f*) −300, *g*) $-\frac{1}{5}$, *h*) $-33\frac{1}{3}$

32. Divide +12 by *a*) +24, *b*) +12, *c*) +4, *d*) −1, *e*) −3, *f*) −48 (8.3)
Divide each by −20 *g*) +60, *h*) +20, *i*) +5, *j*) −225, *k*) −10, *l*) −100
Ans. *a*) $+\frac{1}{2}$, *b*) +1, *c*) +3, *d*) −12, *e*) −4, *f*) $-\frac{1}{4}$, *g*) −3, *h*) −1, *i*) $-\frac{1}{4}$, *j*) $+11\frac{1}{4}$, *k*) $+\frac{1}{2}$, *l*) +5

33. Divide: *a*) $\frac{+25}{+5}$ *b*) $\frac{-25}{+5}$ *c*) $\frac{-2.5}{-.5}$ *d*) $\frac{-.25}{+.5}$ *e*) $\frac{-25}{-.5}$ *f*) $\frac{-.025}{+.5}$ (8.3)
Ans. *a*) +5 *b*) −5 *c*) +5 *d*) $-\frac{1}{2}$ *e*) +50 *f*) −.05

34. Multiply and divide as indicated: (8.4,8.5)
a) $\frac{(+20)(+12)}{(+3)(+5)}$ *c*) $\frac{(+3)(+6)(+10)}{(+12)(-3)}$ *e*) $\frac{0}{(-17)(-24)}$ *g*) $\frac{(+1)(+2)(+3)}{(-4)(-6)}$
b) $\frac{(+20)(-12)}{(-3)(-5)}$ *d*) $\frac{(-3)(-6)(+18)}{(+12)(+3)}$ *f*) $\frac{(-120)(+31)\,0}{(-5)(-8)(-9)}$ *h*) $\frac{(+.1)(+.2)(-30)}{(-.4)(-.1)}$
Ans. *a*) +16, *b*) −16, *c*) −5, *d*) +9, *e*) 0, *f*) 0, *g*) $+\frac{1}{4}$, *h*) −15

35. Evaluate if $y = -3$: (9.1)
a) $3y + 1$ *c*) $2y^2$ *e*) $(-y)^2$ *g*) $\frac{7y}{3}$ *i*) $\frac{3y + 15}{y}$
b) $20 - 4y$ *d*) $3y^3$ *f*) $2 - y^3$ *h*) $y(y^2 - 2)$ *j*) $2y^2 - 5y + 27$
Ans. *a*) −8, *b*) 32, *c*) 18, *d*) −81, *e*) 9, *f*) 29, *g*) −7, *h*) −21, *i*) −2, *j*) 60

36. Evaluate if $x = -1$ and $y = +3$: (9.2)
a) $x + y$ *c*) $x^2 + y^2$ *e*) $4xy - x^2$ *g*) $3xy^2$ *i*) $\frac{y^2}{6x}$
b) $y - 2x$ *d*) $3xy$ *f*) $x^2y + 10$ *h*) $x^3 + 10y$ *j*) $\frac{y + x}{y - x}$
Ans. *a*) 2, *b*) 5, *c*) 10, *d*) −9, *e*) −13, *f*) 13, *g*) −27, *h*) 29, *i*) $-\frac{3}{2}$, *j*) $\frac{1}{2}$

37. Evaluate if $x = -2$, $y = -1$, and $z = +3$: (9.3)
a) $x + y + z$ *c*) $x^2 + y^2 + z^2$ *e*) $2xyz$ *g*) $xy + z^2$ *i*) $\frac{2x - 3y}{4z}$
b) $2x + 2y - 2z$ *d*) $x^3 - y^2 + z$ *f*) $xy + yz$ *h*) $y^2 - 5xz$ *j*) $\frac{x^2 - y^2}{z^2}$
Ans. *a*) 0, *b*) −12, *c*) 14, *d*) −6, *e*) 12, *f*) −1, *g*) 11, *h*) 31, *i*) $-\frac{1}{12}$, *j*) $\frac{1}{3}$

Chapter 4

MONOMIALS and POLYNOMIALS

1. UNDERSTANDING MONOMIALS AND POLYNOMIALS

A **term** is a number or the product of numbers. Each of the numbers being multiplied is a factor of the term.

Thus, the term $-5xy$ consists of three factors, -5, x and y.

In $-5xy$, xy is the literal coefficient and -5 is the numerical coefficient.

An **expression** consists of one or more terms. Expressions may be monomials or polynomials.

1. A **monomial** is an expression of one term.
2. A **polynomial** is an expression of two or more terms.
   - *a*) A **binomial** is a polynomial of two terms.
   - *b*) A **trinomial** is a polynomial of three terms.

Thus, $3x^2$ is a monomial.

The polynomial $3x^2 + 5x$ is a binomial.

The polynomial $3x^2 + 5x - 2$ is a trinomial.

Like terms are terms having the same literal factors, each with the same exponent. Like terms have the same literal coefficient.

Thus, $5x^2y$ and $3x^2y$ are like terms with the same coefficient, x^2y.

Unlike terms are terms that do not have the same literal coefficient.

Thus, $5x^2y$ and $-3xy^2$ are unlike terms.

1.1. Selecting Like Terms

In each polynomial select like terms, if any:

a) $10 - 5x + 2$	*Ans. a*) 10 and $+2$
b) $x + 3y - z + y$	*Ans. b*) $3y$ and $+y$
c) $2ab - 3ac + 4bc - bc$	*Ans. c*) $+4bc$ and $-bc$
d) $2abc + 3acd + 5bcd + 7acd$	*Ans. d*) $3acd$ and $7acd$
e) $y^2 - 2x^2 + 4z^2 - 5y^2$	*Ans. e*) y^2 and $-5y^2$
f) $xy^2 + x^2y^2 - 3x^2y + 7x^2y$	*Ans. f*) $-3x^2y$ and $7x^2y$
g) $2ab^2c^2 - 5a^2b^2c + 8a^2bc^2$	*Ans. g*) no like terms
h) $3(x+y) - 5(x+y) + 2(x-y)$	*Ans. h*) $3(x+y)$ and $-5(x+y)$

2. ADDING MONOMIALS

To add like terms:

1. Add their numerical coefficients.
2. Keep the common literal coefficient.

Thus, $3a^2 - 5a^2 + 4a^2 = (3-5+4)a^2$ or $2a^2$.

To simplify monomials being added:

(*1*) Parentheses may be omitted.

(*2*) "+" meaning "add" may be omitted.

(*3*) If the first monomial is positive, its + sign may be omitted.

Thus, to simplify $(+8x) + (+9x) + (-10x)$ write $8x + 9x - 10x$.

2.1. Adding Like Terms

Procedure:	Add: *a*) $5a^2$ and $3a^2$	*b*) $13(x+y)$ and $-7(x+y)$
1. Add the numerical coefficients:	**1.** $5+3=8$	**1.** $13-7=6$
2. Keep common literal coefficient.	**Ans.** $8a^2$	**Ans.** $6(x+y)$

2.2. Adding Like Terms Horizontally

Simplify and add:

a) $(+5a)+(+2a)+(-4a)$ *b*) $(-8x^2)+(-12x^2)$ *c*) $(-3wy)+(+11wy)$ *d*) $(+16r^2s)+(-13r^2s)$

Solutions: (*First, simplify.*)

a) $(+5a)+(+2a)+(-4a)$	*b*) $(-8x^2)+(-12x^2)$	*c*) $(-3wy)+(+11wy)$	*d*) $(+16r^2s)+(-13r^2s)$
$5a + 2a - 4a$	$-8x^2 - 12x$	$-3wy + 11wy$	$16r^2s - 13r^2s$
Ans. $3a$	**Ans.** $-20x^2$	**Ans.** $8wy$	**Ans.** $3r^2s$

2.3. Adding Like Terms Vertically

Add:	*a*)	*b*)	*c*)	*d*)	*e*)
	$8xy$	$-2abc$	$12c^3d^2$	$15(x-y)$	$-3(a^2+b^2)$
	$-3xy$	$5abc$	$-10c^3d^2$	$-18(x-y)$	$-(a^2+b^2)$
	$12xy$	$-8abc$	$-c^3d^2$	$14(x-y)$	$-8(a^2+b^2)$
Ans.	$17xy$	$-5abc$	c^3d^2	$11(x-y)$	$-12(a^2+b^2)$

2.4. Adding Like and Unlike Terms

Simplify and add: (*Combine only like terms.*)

a) $(+10x)+(+5x)+(-7)$	*b*) $(+3a^2)+(-4a^2)+(-3b^2)$	*c*) $(-8a^2c)+(-6ac^2)+(-3ac^2)$
$10x + 5x - 7$	$3a^2 - 4a^2 - 3b^2$	$-8a^2c - 6ac^2 - 3ac^2$
Ans. $15x-7$	*Ans.* $-a^2-3b^2$	*Ans.* $-8a^2c - 9ac^2$

3. ARRANGING AND ADDING POLYNOMIALS

Arranging the Terms of a Polynomial in Descending or Ascending Order

A polynomial may be arranged as follows:

1. In **descending order**, by having the exponents of the same letter decrease in successive terms.
2. In **ascending order**, by having the exponents of the same letter increase in successive terms.

Thus, $2x^2+3x^3-5x+8$ becomes
$3x^3+2x^2-5x+8$ in descending order
or $8-5x+2x^2+3x^3$ in ascending order.

To Add Polynomials

Procedure:	Add: $5x-2y$ and $y+3x$ **Solution:**
1. Arrange polynomials in order, placing like terms in same column:	**1.** $5x-2y$ $3x+y$
2. Add like terms:	**2.** $8x-y$ **Ans.**

To check addition of polynomials, substitute any convenient value, except 1 or 0, for each of the letters.

Note. If 1 is substituted for a letter, an incorrect exponent will not be detected.
If $x=1$, then $x^2=1$, $x^3=1$, $x^4=1$, etc.

In examples 3.4 and 3.5, checking is shown.

3.1. Arranging Polynomials and Combining Like Terms

Rearrange in descending order and combine:

a) $3a^2 + 2a^2 - 10a - 5a$ *b*) $5x + 8 - 6x + 10x^2$ *c*) $-5y^2 + 8x^2 + 5y^2 + x$

a) No rearrangement needed.	*b*) $10x^2 + 5x - 6x + 8$	*c*) $8x^2 + x - 5y^2 + 5y^2$
Ans. $5a^2 - 15a$	*Ans.* $10x^2 - x + 8$	*Ans.* $8x^2 + x$

3.2. Adding Arranged Polynomials

Add:

a)	*b*)	*c*)	*d*)
$3x - 10$	$5x^2 + 5x$	$5x^3 + 7x^2 - 4$	$10x + 3y$
$2x + 4$	$-x^2 - x$	$-6x^3 \quad - 10$	$7x - 6y + 1$
Ans. a) $5x - 6$	*Ans. b*) $4x^2 + 4x$	*Ans. c*) $-x^3 + 7x^2 - 14$	*Ans. d*) $17x - 3y + 1$

3.3. Adding Polynomials

Add: *a*) $3a + 5b$, $6b - 2a$ and $10b - 25$ *b*) $x^2 + x^3 - 3x$, $4 - 5x^2 + 3x^3$ and $10 - 8x^2 - 5x$

<u>Procedure:</u>	Solutions: *a*)	*b*)
1. Rearrange polynomials with like terms in the same column:	$3a + 5b$ $-2a + 6b$ $10b - 25$	$x^3 + x^2 - 3x$ $3x^3 - 5x^2 \quad + 4$ $-8x^2 - 5x + 10$
2. Add like terms:	**Ans.** *a*) $a + 21b - 25$	**Ans.** *b*) $4x^3 - 12x^2 - 8x + 14$

3.4 and 3.5. Checking the Addition of Polynomials

3.4. Check by letting $x = 2$ and $y = 3$: $(25x - 10y) + (5x - 2y) = 30x - 12y$

Check: Let $x = 2$ and $y = 3$.

$$25x - 10y \rightarrow 50 - 30 = 20$$
$$5x - 2y \rightarrow 10 - 6 = 4$$
$$30x - 12y \rightarrow 60 - 36 = \boxed{24}$$

By adding vertically and horizontally, the sum is 24.

3.5. Check by letting $a = 2$, $b = 4$ and $c = 3$: $(a^2 + b^2 - c^2) + (3a^2 - b^2 + 4c^2) = 4a^2 + 3c^2$

Check: Let $a = 2$, $b = 4$ and $c = 3$.

$$a^2 + b^2 - c^2 \rightarrow 4 + 16 - 9 = 11$$
$$3a^2 - b^2 + 4c^2 \rightarrow 12 - 16 + 36 = 32$$
$$4a^2 \quad + 3c^2 \rightarrow 16 \quad + 27 \quad \boxed{43}$$

By adding vertically and horizontally, the sum is 43.

4. SUBTRACTING MONOMIALS

To **subtract** a term, **add** its opposite. (*Opposite terms differ only in sign.*)

Thus, to subtract $-3x$, add $+3x$, or to subtract $+5x^2$, add $-5x^2$.

4.1. Subtracting Like Terms

Subtract: *a*) $(+2a) - (-5a)$ *b*) $(-3x^2) - (+5x^2)$ *c*) $(-15cd) - (-10cd)$ *d*) $(+3a^2b) - (+8a^2b)$

a) $(+2a) - (-5a)$	*b*) $(-3x^2) - (+5x^2)$	*c*) $(-15cd) - (-10cd)$	*d*) $(+3a^2b) - (+8a^2b)$
$(+2a) + (+5a)$	$(-3x^2) + (-5x^2)$	$(-15cd) + (+10cd)$	$(+3a^2b) + (-8a^2b)$
$2a + 5a$	$-3x^2 - 5x^2$	$-15cd + 10cd$	$3a^2b - 8a^2b$
Ans. $7a$	*Ans.* $-8x^2$	*Ans.* $-5cd$	*Ans.* $-5a^2b$

4.2. Different Subtraction Forms

a) Subtract $(-5a)$ from $(+2a)$

b) From $(-3x^2)$ take $(+5x^2)$

c) Reduce $(+7mn)$ by $(-2mn)$

d) How much does $(-8cd)$ exceed $(+3cd)$?

Solutions:

a) $(+2a)-(-5a)$
$(+2a)+(+5a)$
$2a + 5a$
Ans. $7a$

b) $(-3x^2)-(+5x^2)$
$(-3x^2)+(-5x^2)$
$-3x^2 - 5x^2$
Ans. $-8x^2$

c) $(+7mn)-(-2mn)$
$(+7mn)+(+2mn)$
$7mn + 2mn$
Ans. $9mn$

d) $(-8cd)-(+3cd)$
$(-8cd)+(-3cd)$
$-8cd - 3cd$
Ans. $-11cd$

4.3. Combining Adding and Subtracting of Like Terms

Combine:

a) $(+6x)+(+8x)-(-3x)$ *b*) $(-3a^2)-(-a^2)-(+10a^2)$ *c*) $(+13ab)-(-ab)-(-2ab)+(-3ab)$

Solutions:

a) $(+6x)+(+8x)-(-3x)$
$(+6x)+(+8x)+(+3x)$
$6x + 8x + 3x$
Ans. $17x$

b) $(-3a^2)-(-a^2)-(+10a^2)$
$(-3a^2)+(+a^2)+(-10a^2)$
$-3a^2 + a^2 - 10a^2$
Ans. $-12a^2$

c) $(+13ab)-(-ab)-(-2ab)+(-3ab)$
$(+13ab)+(+ab)+(+2ab)+(-3ab)$
$13ab + ab + 2ab - 3ab$
Ans. $13ab$

5. *SUBTRACTING POLYNOMIALS*

To Subtract Polynomials

Procedure:	Subtract $5x-2y$ from $8x-4y$
	Solution: $(8x-4y)-(5x-2y)$
1. Arrange polynomials in order, placing like terms in same column:	1. $8x - 4y$ **Minuend** (S) $5x - 2y$ **Subtrahend**
2. Subtract like terms:	2. $3x - 2y$ **Difference (Ans.)**

Notes. *1*) Use **(S)** to indicate subtraction.
2) To subtract a polynomial, change each of its signs **mentally**, then add.

To check subtraction, add the difference obtained to the subtrahend. Their sum should be the minuend.

5.1. Subtracting Arranged Polynomials

Subtract:

a) $\begin{array}{rl} & 5b+7 \\ \text{(S)} & 3b-2 \\ \hline \textbf{Ans.} & 2b+9 \end{array}$

b) $\begin{array}{rl} & y^2-3y \\ \text{(S)} & -2y^2-4y \\ \hline \textbf{Ans.} & 3y^2+y \end{array}$

c) $\begin{array}{rl} & 8x^3-2x^2+5x \\ \text{(S)} & 6x^3+9x \\ \hline \textbf{Ans.} & 2x^3-2x^2-4x \end{array}$

d) $\begin{array}{rl} & 3x^2y+4xy^2 \\ \text{(S)} & x^2y+xy^2+y^3 \\ \hline \textbf{Ans.} & 2x^2y+3xy^2-y^3 \end{array}$

5.2. Arranging Polynomials and Subtracting

Arrange and subtract:

a) $2x-y+8$ from $13x+4y+9$

b) $10+3a-5b$ from $7b+2a-8$

c) x^3-10 from $25x^2+3x^3$

d) m^2-18m from $12+3m^2$

Solutions:

a) $\begin{array}{rl} & 13x+4y+9 \\ \text{(S)} & 2x-y+8 \\ \hline \textbf{Ans.} & 11x+5y+1 \end{array}$

b) $\begin{array}{rl} & 2a+7b-8 \\ \text{(S)} & 3a-5b+10 \\ \hline \textbf{Ans.} & -a+12b-18 \end{array}$

c) $\begin{array}{rl} & 3x^3+25x^2 \\ \text{(S)} & x^3-10 \\ \hline \textbf{Ans.} & 2x^3+25x^2+10 \end{array}$

d) $\begin{array}{rl} & 3m^2+12 \\ \text{(S)} & m^2-18m \\ \hline \textbf{Ans.} & 2m^2+18m+12 \end{array}$

5.3. Different Subtraction Forms

a) From $9+x^2$ take $-3x^2-5x+12$

b) Reduce $17ab$ by $-8ab+15$

c) Find $8x^2-15x$ less $20x-7$

d) Subtract $10-4x+x^2$ from 0.

Solutions:

a)
$$\begin{array}{rrrr} & x^2 & & +9 \\ \text{(S)} & -3x^2 & -5x & +12 \\ \hline \textbf{Ans.} & 4x^2 & +5x & -3 \end{array}$$

b)
$$\begin{array}{rrr} & 17ab & \\ \text{(S)} & -8ab & +15 \\ \hline \textbf{Ans.} & 25ab & -15 \end{array}$$

c)
$$\begin{array}{rrrr} & 8x^2 & -15x & \\ \text{(S)} & & 20x & -7 \\ \hline \textbf{Ans.} & 8x^2 & -35x & +7 \end{array}$$

d)
$$\begin{array}{rrrr} & 0 & & \\ \text{(S)} & 10 & -4x & +x^2 \\ \hline \textbf{Ans.} & -10 & +4x & -x^2 \end{array}$$

5.4. Checking the Subtraction of Polynomials

Check each subtraction, using addition:

a)
$$\begin{array}{rrr} & 8c & -3 \\ \text{(S)} & 5c & -8 \\ \hline & 3c & +5\ (?) \end{array}$$

b)
$$\begin{array}{rrr} & 5x^2 & +20 \\ \text{(S)} & x^2 & +15 \\ \hline & 4x^2 & +5\ (?) \end{array}$$

c)
$$\begin{array}{rrrr} & 3y^2 & +5y & \\ \text{(S)} & y^2 & -8y & +5 \\ \hline & 2y^2 & +13y & +5\ (?) \end{array}$$

Checking: (Difference + Subtrahend = Minuend)

a)
$$\begin{array}{rrr} & 3c & +5 \\ \text{(A)} & 5c & -8 \\ \hline & 8c & -3 \end{array}$$
(*Correct*)

b)
$$\begin{array}{rrr} & 4x^2 & +5 \\ \text{(A)} & x^2 & +15 \\ \hline & 5x^2 & +20 \end{array}$$
(*Correct*)

c)
$$\begin{array}{rrrr} & 2y^2 & +13y & +5 \\ \text{(A)} & y^2 & -8y & +5 \\ \hline & 3y^2 & +5y & +10 \end{array}$$
(*Incorrect*)

Correct difference = $2y^2+13y-5$

6. *USING PARENTHESES AND OTHER GROUPING SYMBOLS TO ADD OR SUBTRACT POLYNOMIALS*

Symbols of grouping include:

1. **Parentheses,** ( ), as in $\frac{1}{2}(8-4x)$ or $5-(10x-4)$.
2. **Brackets,** [ ], to include parentheses as in $8-[5+(x-2)]$.
3. **Braces,** { }, to include brackets as in $x-\{3+[x-(y+4)]\}$.
4. **Bar,** ——, as in the fraction $\dfrac{8-4x}{2}$ or $5-\overline{10x-4}$.

Symbols of grouping may be used to show the addition or subtraction of polynomials. Thus:

1. To add $3a+4b$ and $5a-b$, write $(3a+4b)+(5a-b)$.
2. To subtract $2x-5$ from x^2+10x, write $(x^2+10x)-(2x-5)$.
3. To subtract the sum of $8x^2+9$ and $6-x^2$ from $3x^2-12$,
   write $(3x^2-12)-[(8x^2+9)+(6-x^2)]$.

Rules for Removing Parentheses and Grouping Symbols

Rule 1. When removing parentheses preceded by a plus sign, **do not change** the signs of the enclosed terms.

Thus,
$$\begin{array}{l} 3a+(+5a-10) \\ 3a+5a-10 \\ 8a-10 \quad \textbf{Ans.} \end{array} \qquad \begin{array}{l} 3x^2+(-x^2-5x+8) \\ 3x^2-x^2-5x+8 \\ 2x^2-5x+8 \quad \textbf{Ans.} \end{array}$$

Rule 2. When removing parentheses preceded by a minus sign, **change** the sign of each enclosed term.

Thus,
$$\begin{array}{l} 3a-(+5a-10) \\ 3a-5a+10 \\ -2a+10 \quad \textbf{Ans.} \end{array} \qquad \begin{array}{l} 3x^2-(-x^2-5x+8) \\ 3x^2+x^2+5x-8 \\ 4x^2+5x-8 \quad \textbf{Ans.} \end{array}$$

Rule 3. When more than one set of grouping symbols is used, remove one set at a time beginning with the innermost one.

Thus,

$2 + [r-(3-r)]$	$6s - \{5 - [3 + (7s-8)]\}$
$2 + [r-3+r]$	$6s - \{5 - [3 + 7s-8]\}$
$2 + r - 3 + r$	$6s - \{5 - 3 - 7s + 8\}$
$2r - 1$ **Ans.**	$6s - 5 + 3 + 7s - 8$
	$13s - 10$ **Ans.**

6.1. Rule 1. Removing Parentheses Preceded by a Plus Sign

Simplify: (*Do not change the sign of the enclosed terms.*)

a) $13x + (5x-2)$	*b*) $(15a^2-3a) + (-7-2a)$	*c*) $3 + (6-5x) - 2x^2$
$13x + 5x - 2$	$15a^2 - 3a - 7 - 2a$	$3 + 6 - 5x - 2x^2$
Ans. *a*) $18x - 2$	**Ans.** *b*) $15a^2 - 5a - 7$	**Ans.** *c*) $9 - 5x - 2x^2$

6.2. Rule 2. Removing Parentheses Preceded by a Minus Sign

Simplify: (*Change the sign of each enclosed term.*)

a) $13x - (5x-2)$	*b*) $-(15a^2-3a) - (-7-2a)$	*c*) $3 - (5-5x) - 2x^2$
$13x - 5x + 2$	$-15a^2 + 3a + 7 + 2a$	$3 - 5 + 5x - 2x^2$
Ans. *a*) $8x + 2$	**Ans.** *b*) $-15a^2 + 5a + 7$	**Ans.** *c*) $-2 + 5x - 2x^2$

6.3. Rule 3. Brackets Containing Parentheses

Simplify:

Procedure:		
1. Remove ( ):	*a*) $2 - [r - (3-r)]$	*b*) $5 - [(8+4x) - (3x-2)]$
2. Remove [ ]:	$2 - [r - 3 + r]$	$5 - [8 + 4x - 3x + 2]$
	$2 - r + 3 - r$	$5 - 8 - 4x + 3x - 2$
3. Combine:	**Ans.** *a*) $-2r + 5$	**Ans.** *b*) $-x - 5$

6.4. and 6.5. Using Grouping Symbols to Add or Subtract Polynomials

6.4. Subtract $2x-5$ from the sum of $15x+10$ and $3x-5$.

Solution Using Parentheses

$[(15x+10) + (3x-5)] - (2x-5)$

$15x + 10 + 3x - 5 - 2x + 5$

Ans. $16x + 10$

Solution Without Parentheses

$$\begin{array}{rr} & 15x + 10 \\ \text{(A)} & 3x - 5 \\ \hline & 18x + 5 \end{array} \quad \longrightarrow \quad \begin{array}{rr} & 18x + 5 \\ \text{(S)} & 2x - 5 \\ \hline \textbf{Ans.} & 16x + 10 \end{array}$$

6.5. From the sum of $3a-5b$ and $8a+10$ subtract the sum of $2b-8$ and $7a+9$.

Solution Using ( ) and [ ]

$[(3a-5b) + (8a+10)] - [(2b-8) + (7a+9)]$

$[3a-5b+8a+10] - [2b-8+7a+9]$

$[11a - 5b + 10] - [7a + 2b + 1]$

$11a - 5b + 10 - 7a - 2b - 1$

Ans. $4a - 7b + 9$

Solution Without ( ) and [ ]

$$\begin{array}{rr} & 3a - 5b \qquad \\ \text{(A)} & 8a \qquad + 10 \\ \hline & 11a - 5b + 10 \end{array} \qquad \begin{array}{rr} & 2b - 8 \\ \text{(A)} & 7a \qquad + 9 \\ \hline & 7a + 2b + 1 \end{array}$$

$$\begin{array}{rr} & 11a - 5b + 10 \\ \text{(S)} & 7a + 2b + 1 \\ \hline \textbf{Ans.} & 4a - 7b + 9 \end{array}$$

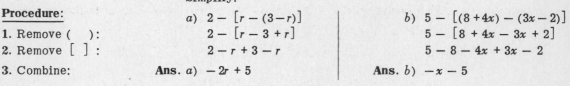

6.6. Removing Symbols of Grouping

Simplify: (*Remove innermost symbols first.*)

Procedure:	
1. Remove $\overline{\quad}$:	$5x - \{8x - [7 - (4x - \overline{8 - 2x})]\} - 5$
	$5x - \{8x - [7 - (4x - 8 + 2x)]\} - 5$
2. Remove ( ):	$5x - \{8x - [7 - 4x + 8 - 2x]\} - 5$
3. Remove [ ] :	$5x - \{8x - 7 + 4x - 8 + 2x\} - 5$
4. Remove { }:	$5x - 8x + 7 - 4x + 8 - 2x - 5$
5. Combine:	**Ans.** $-9x + 10$

7. MULTIPLYING MONOMIALS AND POWERS OF THE SAME BASE

Rule 1. To multiply the powers of the same base, keep the base and add the exponents.

Thus, $x^4 \cdot x^3 = x^7$ $\qquad a^5 \cdot a \cdot b^2 \cdot b^3 = a^6 b^5$

$3^{10} \cdot 3^2 = 3^{12}$ $\qquad 2^4 \cdot 2^5 \cdot 10^2 \cdot 10^4 = 2^9 \cdot 10^6$

$$x^a x^b = x^{a+b}$$

Rule 2. To find the power of a power of a base, keep the base and multiply the exponents.

Thus, $(x^4)^3 = x^{12}$ since $(x^4)^3 = (x^4)(x^4)(x^4)$

$(5^2)^4 = 5^8$ since $(5^2)^4 = (5^2)(5^2)(5^2)(5^2)$

$$(x^a)^b = x^{ab}$$

Rule 3. Changing the order of factors does not change their product.

[This fundamental law is known as the **commutative law of multiplication.**]

Thus, $2 \cdot 3 \cdot 5 = 2 \cdot 5 \cdot 3 = 5 \cdot 2 \cdot 3 = 30$

$3x \cdot 4x^2 = 3 \cdot 4 \cdot xx^2 = 12x^3$

To Multiply Monomials

Multiply $2x$ by $-3x^2$.

Procedure:	Solution:
	$(2x)(-3x^2)$
1. Multiply numerical coefficients:	1. $(2)(-3) = -6$
2. Multiply literal coefficients:	2. $(x)(x^2) = x^3$
3. Multiply results:	3. $-6x^3$ **Ans.**

7.1. Rule 1. Multiplying Powers of the Same Base

Multiply: (*Keep base and add exponents.*)

a) $b \cdot b^2$	*Ans.* b^3	*d*) $x^2 \cdot x^3 \cdot y$	*Ans.* $x^5 y$	*g*) $4^5 \cdot 4^2$	*Ans.* 4^7
b) $x^2 \cdot x^3 \cdot x$	*Ans.* x^6	*e*) $c^4 \cdot c \cdot d^5 \cdot d$	*Ans.* $c^5 d^6$	*h*) $2^3 \cdot 2 \cdot 5$	*Ans.* $2^4 \cdot 5$
c) $x^b x^c$	*Ans.* x^{b+c}	*f*) $x^3 y^2 x^4 y$	*Ans.* $x^7 y^3$	*i*) $3^2 3^3 y^4 y^5$	*Ans.* $3^5 y^9$

7.2. Rule 2. Finding the Power of a Power of a Base

Raise to a power: (*Keep base and multiply exponents.*)

a) $(a^4)^2$	*Ans.* a^8	*d*) $(3^5)^4$	*Ans.* 3^{20}	*g*) $(x^2)^3 (y^3)^2$	*Ans.* $x^6 y^6$
b) $(b^3)^5$	*Ans.* b^{15}	*e*) $(10^4)^{10}$	*Ans.* 10^{40}	*h*) $(4^2)^3 (x^3)^4$	*Ans.* $4^6 x^{12}$
c) $(x^d)^e$	*Ans.* x^{de}	*f*) $(5^m)^n$	*Ans.* 5^{mn}	*i*) $(y^3)^4 y^3 y^4$	*Ans.* y^{19}

7.3. Multiplying Monomials

Multiply:

a) $2(-3b)$	*Ans.* $-6b$	*d*) $(4x^2)(5x^3)$	*Ans.* $20x^5$	*g*) $(3a)(2b)(-10c)$	*Ans.* $-60abc$
b) $(-2b)(5b)$	*Ans.* $-10b^2$	*e*) $(-\frac{1}{2}y^3)(-10y^5)$	*Ans.* $5y^8$	*h*) $(2x^2)(3x^3)(4x^4)$	*Ans.* $24x^9$
c) $(4x)(-7y)$	*Ans.* $-28xy$	*f*) $(10r^4)(-\frac{2}{5}rs)$	*Ans.* $-4r^5s$	*i*) $(-3)(-r^4)(-s^5)$	*Ans.* $-3r^4s^5$

8. MULTIPLYING A POLYNOMIAL BY A MONOMIAL

Rule. To multiply a polynomial by a monomial, multiply each term of the polynomial by the monomial.

[This fundamental law is known as the **distributive law**.]

Thus, $4(a+b-c) = 4a+4b-4c$.

8.1. Multiplying a Polynomial by a Monomial Horizontally

Multiply:

a) $10(a+b)$	*Ans.* $10a+10b$	*d*) $\pi r(r+h)$	*Ans.* $\pi r^2+\pi rh$	*g*) $3(x^2-2x+8)$	*Ans.* $3x^2-6x+24$
b) $x(y-7)$	*Ans.* $xy-7x$	*e*) $-ab(a+b)$	*Ans.* $-a^2b-ab^2$	*h*) $a(a^2+a+5)$	*Ans.* a^3+a^2+5a
c) $-c(d-5e)$	*Ans.* $-cd+5ce$	*f*) $\frac{2}{3}(9k-30m)$	*Ans.* $6k-20m$	*i*) $-x^2(3-2x+x^2)$	*Ans.* $-3x^2+2x^3-x^4$

8.2. Multiplying a Polynomial by a Monomial Vertically

Multiply:

a) y^2+8y-7	*b*) d^3-2d^2-20	*c*) $5a+2b-3c+8$
$3y$	$-d^2$	$-2abc$
Ans. $3y^3+24y^2-21y$	*Ans.* $-d^5+2d^4+20d^2$	*Ans.* $-10a^2bc-4ab^2c+6abc^2-16abc$

8.3. Removing Parentheses

Simplify:

a) $3x+2(x-3)$	*b*) $3x^2-x(5-x)$	*c*) $a(4a-5)-8(a^2-10)$
$3x+2x-6$	$3x^2-5x+x^2$	$4a^2-5a-8a^2+80$
Ans. $5x-6$	*Ans.* $4x^2-5x$	*Ans.* $-4a^2-5a+80$

8.4. Removing Brackets Containing Parentheses

Simplify: (*Remove parentheses, first.*)

a) $3[5x-2(x-4)]$	*b*) $3+5[2-4(a-6)]$	*c*) $a[ab-a(b-c)]$
$3[5x-2x+8]$	$3+5[2-4a+24]$	$a[ab-ab+ac]$
$3[3x+8]$	$3+5[26-4a]$	$a[ac]$
Ans. $9x+24$	*Ans.* $133-20a$	*Ans.* a^2c

8.5. Removing Symbols of Grouping

Simplify: (*Remove innermost symbols, first.*)

	$120y-2\{y+8[-7y-5(y-\overline{3y+4})]\}+320$
1. Remove $\overline{\quad}$:	$120y-2\{y+8[-7y-5(y-3y-4)]\}+320$
2. Remove ( ) :	$120y-2\{y+8[-7y-5y+15y+20]\}+320$
3. Combine :	$120y-2\{y+8[3y+20]\}+320$
4. Remove [ ] :	$120y-2\{y+24y+160\}+320$
5. Remove { } :	$120y-2y-48y-320+320$
6. Combine :	$70y$ *Ans.*

9. MULTIPLYING POLYNOMIALS

To Multiply Polynomials

Procedure:	Multiply $3x + 4$ by $1 + 2x$ Solution:
1. Arrange each polynomial in order:	$3x + 4$ $2x + 1$
2. Multiply each term of one polynomial by each term of the other:	$6x^2 + 8x$ $+ 3x + 4$
3. Add like terms:	**Ans.** $6x^2 + 11x + 4$

To check multiplication of polynomials, interchange the polynomials and multiply again, or substitute any convenient values for the letters, except 1 or 0.

9.1. Multiplying Polynomials

Multiply:

a)	*b*)	*c*)	*d*)
$3x + 4$	$8 + c$	$4r + 7s$	$a^2 - 3a + 5$
$2x - 1$	$3 - 2c$	$4r - 7s$	$5a - 2$
$6x^2 + 8x$	$24 + 3c$	$16r^2 + 28rs$	$5a^3 - 15a^2 + 25a$
$-3x - 4$	$-16c - 2c^2$	$-28rs - 49s^2$	$-2a^2 + 6a - 10$
Ans. $6x^2 + 5x - 4$	*Ans.* $24 - 13c - 2c^2$	*Ans.* $16r^2 \quad -49s^2$	*Ans.* $5a^3 - 17a^2 + 31a - 10$

9.2. Checking Multiplication

Multiply and check: $(5r - 8)(3r - 2)$

Solution:	Check by multiplication:	Check by substitution:
$5r - 8$	$3r - 2$	Let $r = 10$.
$3r - 2$	$5r - 8$	$5r - 8 \rightarrow 42$
$15r^2 - 24r$	$15r^2 - 10r$	$3r - 2 \rightarrow 28$
$-10r + 16$	$-24r + 16$	$15r^2 - 34r + 16 \rightarrow 1500 - 340 + 16 = 1176$
Ans. $15r^2 - 34r + 16$	$15r^2 - 34r + 16$	Correct since $1176 = (42 \times 28)$

9.3. Extended Multiplication

Multiply $5p + 2$, $2p - 1$ and $p + 1$ and check.

Solution:		Check by substitution:
$5p + 2$	$10p^2 - p - 2$	Let $p = 10$.
$2p - 1$	$p + 1$	$5p + 2 \rightarrow 50 + 2 = 52$
$10p^2 + 4p$	$10p^3 - p^2 - 2p$	$2p - 1 \rightarrow 20 - 1 = 19$
$-5p - 2$	$+10p^2 - p - 2$	$p + 1 \rightarrow 10 + 1 = 11$
$10p^2 - p - 2$	$10p^3 + 9p^2 - 3p - 2$ **Ans.**	$10p^3 + 9p^2 - 3p - 2 \rightarrow 10000 + 900 - 30 - 2 = 10868$ Correct since $10868 = (52 \times 19 \times 11)$

10. DIVIDING POWERS AND MONOMIALS

When dividing powers having the same base, arrange them into a fraction and apply the following rules:

Rule 1. If the exponent of the numerator is **larger than** the exponent of the denominator, keep the base and subtract the smaller exponent from the larger.

Thus, $\frac{x^7}{x^4} = x^3$

> If a is larger than b, $\frac{x^a}{x^b} = x^{a-b}$

Rule 2. If the exponents **are equal**, then we have a number divided by itself; the quotient is 1.

Thus, $\frac{x^4}{x^4} = 1$

> $\frac{x^a}{x^a} = 1$

Rule 3. If the exponent of the denominator is larger, make the numerator of the quotient 1, and to obtain its denominator, keep the base and subtract the smaller exponent from the larger.

Thus, $\frac{x^4}{x^7} = \frac{1}{x^3}$

> If a is smaller than b, $\frac{x^a}{x^b} = \frac{1}{x^{b-a}}$

To Divide Monomials

Divide $21ab^2$ by $-7a^2b$.

Procedure:	Solution:	
1. Arrange in fractional form:	1. $\frac{21ab^2}{-7a^2b}$	**Dividend** / **Divisor**
2. Divide numerical coefficients:	2. $\frac{21}{-7} = -3$	
3. Divide literal coefficients:	3. $\frac{ab^2}{a^2b} = \frac{b}{a}$	
4. Multiply the results:	**Ans.** 4. $-\frac{3b}{a}$	**Quotient**

To Check Division of Monomials

Multiply the quotient by the divisor. The result should be the dividend.

Thus, to check $\frac{15ab}{3a} = 5b$, multiply $3a$ by $5b$ to obtain $15ab$.

10.1. Rules 1 to 3. Dividing Powers of the Same Base

Divide:

(Rule 1)		(Rule 2)		(Rule 3)	
a) $\frac{x^8}{x^2}$	*Ans.* x^6	*e*) $\frac{x^2}{x^2}$	*Ans.* 1	*i*) $\frac{x^2}{x^8}$	*Ans.* $\frac{1}{x^6}$
b) $\frac{8^7}{8^3}$	*Ans.* 8^4	*f*) $\frac{-8^3}{8^3}$	*Ans.* -1	*j*) $\frac{8^3}{8^7}$	*Ans.* $\frac{1}{8^4}$
c) $\frac{a^3b^4}{a^2b^2}$	*Ans.* ab^2	*g*) $\frac{a^2b^2}{-a^2b^2}$	*Ans.* -1	*k*) $\frac{a^2b^2}{a^3b^4}$	*Ans.* $\frac{1}{ab^2}$
d) $\frac{x^{2a}}{x^a}$	*Ans.* x^a	*h*) $\frac{-x^a}{-x^a}$	*Ans.* 1	*l*) $\frac{x^a}{x^{2a}}$	*Ans.* $\frac{1}{x^a}$

10.2. Dividing Monomials

Divide:

a) $\frac{24b^2}{3b}$ *Ans.* $8b$

b) $\frac{24b}{8b^2}$ *Ans.* $\frac{3}{b}$

c) $\frac{-24b^2}{-24b^2}$ *Ans.* 1

d) $\frac{28x^2y^4}{28x^2y^2}$ *Ans.* y^2

e) $\frac{28x^2y^2}{-28x^2y^2}$ *Ans.* -1

f) $\frac{-25u^5w}{-25uw^5}$ *Ans.* $\frac{u^4}{w^4}$

g) $\frac{-14abc}{7abcd}$ *Ans.* $-\frac{2}{d}$

h) $\frac{-7a^2bc}{14ab^2c^2}$ *Ans.* $-\frac{a}{2bc}$

i) $\frac{a^3b^4c^5}{a^3b^7c^2}$ *Ans.* $\frac{c^3}{b^3}$

10.3. Checking Division

Check each division, using multiplication:

a) Does $\frac{24b^2}{3b} = 8b$? b) Does $\frac{28x^2y^4}{28x^2y^2} = y^2$? c) Does $\frac{a^3b^4c^6}{-ab^2c^2} = -a^2b^2c^3$?

Check: (Quotient × Divisor = Dividend)

a) Multiply:
$(8b)(3b) = 24b^2$
Correct.

b) Multiply:
$y^2(28x^2y^2) = 28x^2y^4$
Correct.

c) Multiply:
$(-a^2b^2c^3)(-ab^2c^2) = a^3b^4c^5$
Incorrect. The quotient should be $-a^2b^2c^4$.

11. DIVIDING A POLYNOMIAL BY A MONOMIAL

To divide a polynomial by a monomial, divide each term of the polynomial by the monomial.

Thus, $\frac{10x+15}{5} = \frac{10x}{5} + \frac{15}{5} = 2x + 3$. Also, $\frac{ax+bx}{x} = \frac{ax}{x} + \frac{bx}{x} = a + b$.

To check the division, multiply the quotient by the divisor. The result should be the dividend.

Thus, to check $\frac{10x+15}{5} = 2x+3$, multiply $2x+3$ by 5 to obtain $10x+15$.

11.1. Dividing a Polynomial by a Monomial

Divide:

a) $\frac{3a+6b}{3}$ *Ans.* $a+2b$

b) $\frac{r-rt}{r}$ *Ans.* $1-t$

c) $\frac{pq+pr}{-p}$ *Ans.* $-q-r$

d) $\frac{2\pi r + 2\pi R}{2\pi}$ *Ans.* $r+R$

e) $\frac{ab-abc}{ab}$ *Ans.* $1-c$

f) $\frac{9x^2y - 36xy^2}{9xy}$ *Ans.* $x-4y$

g) $\frac{7x-14y+56}{7}$ *Ans.* $x-2y+8$

h) $\frac{x^3+2x^2+5x}{x}$ *Ans.* x^2+2x+5

i) $\frac{3x^5-x^3+5x^2}{-x^2}$ *Ans.* $-3x^3+x-5$

11.2. Checking Division

Check each, using multiplication:

a) Does $\frac{r-rt}{r} = 1-t$? b) Does $\frac{x^2y-2xy^2}{xy} = x-y$? c) Does $\frac{2x-4y+10}{-2} = -x+2y-5$?

Check: (Quotient × Divisor = Dividend)

a) Multiply:
$r(1-t) = r-rt$
Correct.

b) Multiply:
$xy(x-y) = x^2y - xy^2$
Incorrect. The quotient should be $x-2y$.

c) Multiply:
$-2(-x+2y-5) = 2x-4y+10$
Correct.

11.3. Multiplying and Dividing Polynomials by a Monomial

Simplify:	a)	b)	c)
	$5(x-2)+\dfrac{3x-12}{3}$	$\dfrac{x^2-5x}{x}-x(3-x)$	$\dfrac{a^3-a^2bc}{a}-b(b-ac)$
	$5x-10+x-4$	$x-5-3x+x^2$	$a^2-abc-b^2+abc$
Ans.	$6x-14$	x^2-2x-5	a^2-b^2

12. DIVIDING A POLYNOMIAL BY A POLYNOMIAL

To Divide Polynomials

Divide x^2-5x+6 by $x-2$.

Procedure:	**Solution:** (*By Steps*)
1. Set up as a form of long division in which the polynomials are arranged in descending order, leaving space for missing terms:	1. $x-2\overline{)x^2-5x+6}$
2. Divide the first term of the divisor into the first term of the dividend to obtain the first term of the quotient:	2. $\begin{array}{r} x \\ x-2\overline{)x^2-5x+6} \end{array}$
3. Multiply the first term of the quotient by each term of the divisor:	3. $\begin{array}{r} x \\ x-2\overline{)x^2-5x+6} \\ x^2-2x \end{array}$
4. Subtract like terms and bring down one or more terms as needed:	4. $-3x+6$
5. Repeat steps 2 to 4, using the remainder as the new dividend: that is, divide, multiply, subtract and bring down:	5. $\begin{array}{r} -3 \\ x-2\overline{)} \\ \hline -3x+6 \\ \underline{-3x+6} \end{array}$
6. Continue repeating steps 2 to 4 as long as it is possible.	6. No further steps needed.

Full Solution:

$$\begin{array}{r} x-3 \\ x-2\overline{)x^2-5x+6} \\ \underline{x^2-2x} \\ -3x+6 \\ \underline{-3x+6} \end{array}$$

Ans. $x-3$

To Check the Division

1) If no final remainder exists, then multiply the quotient by the divisor. The result should equal the dividend.

   Thus, to check $\dfrac{18}{3}=6$, multiply 3 by 6 to obtain 18.

2) If there is a final remainder, add this to the product of the quotient and divisor. The result should equal the dividend.

   Thus, to check $\dfrac{19}{3}=6\frac{1}{3}$, multiply 3 by 6 and then add 1 to obtain 19.

12.1. Dividing a Polynomial by a Polynomial (*No Remainder*)

Divide: *a*) $x^2-9x+14$ by $x-7$ *b*) $x^3-6x^2+11x-6$ by $x-3$ and check.

Solutions: *a*)

$$\begin{array}{r} x-2 \\ x-7\overline{)x^2-9x+14} \\ \underline{x^2-7x} \\ -2x+14 \\ \underline{-2x+14} \end{array} \qquad \textbf{Check: } \begin{array}{r} x-2 \\ \underline{x-7} \\ x^2-2x \\ \underline{-7x+14} \\ x^2-9x+14 \\ \textbf{(Dividend)} \end{array}$$

Ans. $x-2$

b)

$$\begin{array}{r} x^2-3x+2 \\ x-3\overline{)x^3-6x^2+11x-6} \\ \underline{x^3-3x^2} \\ -3x^2+11x \\ \underline{-3x^2+9x} \\ 2x-6 \\ \underline{2x-6} \end{array} \qquad \textbf{Check: } \begin{array}{r} x^2-3x+2 \\ \underline{x-3} \\ x^3-3x^2+2x \\ \underline{-3x^2+9x-6} \\ x^3-6x^2+11x-6 \\ \textbf{(Dividend)} \end{array}$$

Ans. x^2-3x+2

12.2. Arranging Polynomials and Dividing

Divide: $20a^2 - 3b^2 + 7ab$ by $-b + 4a$ and check:

Solution: Arrange both dividend and divisor in descending powers of a:

$$\begin{array}{r@{\,}l} & 5a + 3b \\ 4a - b\,) & \overline{20a^2 + 7ab - 3b^2} \\ & \underline{20a^2 - 5ab} \\ & \qquad 12ab - 3b^2 \\ & \qquad \underline{12ab - 3b^2} \end{array}$$

Ans. $5a + 3b$

Check:

$$\begin{array}{l} 5a + 3b \\ \underline{4a - \ \ b} \\ 20a^2 + 12ab \\ \underline{\qquad - 5ab - 3b^2} \\ 20a^2 + \ 7ab - 3b^2 \ \textbf{(Dividend)} \end{array}$$

12.3. Terms Missing in Dividend

Divide $x^3 - 64$ by $x - 4$ and check:

Solution: Leave spaces for missing x^2 and x terms:

$$\begin{array}{r@{\,}l} & x^2 + 4x + 16 \\ x - 4\,) & \overline{x^3 \qquad\qquad\quad - 64} \\ & \underline{x^3 - 4x^2} \\ & \quad 4x^2 \\ & \quad \underline{4x^2 - 16x} \\ & \qquad\quad 16x - 64 \\ & \qquad\quad \underline{16x - 64} \end{array}$$

Ans. $x^2 + 4x + 16$

Check:

$$\begin{array}{l} x^2 + 4x + 16 \\ \underline{\qquad\quad x - 4} \\ x^3 + 4x^2 + 16x \\ \underline{\quad - 4x^2 - 16x - 64} \\ x^3 \qquad\qquad\quad - 64 \ \textbf{(Dividend)} \end{array}$$

12.4. Dividing Polynomials (*With Remainder*)

Divide $8x^2 - 10x + 8$ by $2x - 4$ and check:

Solution:

$$\begin{array}{r@{\,}l} & 4x + 3 + \frac{20}{2x-4} \\ 2x - 4\,) & \overline{8x^2 - 10x + 8} \\ & \underline{8x^2 - 16x} \\ & \qquad + 6x + 8 \\ & \qquad \underline{+ 6x - 12} \\ \textit{Remainder:} & \qquad\qquad 20 \end{array}$$

Complete quotient by adding $\frac{20}{2x-4}$

Ans. $4x + 3 + \frac{20}{2x-4}$

Check:

$$\begin{array}{l} 4x + 3 \\ \underline{2x - 4} \\ 8x^2 + 6x \\ \underline{\qquad - 16x - 12} \\ 8x^2 - 10x - 12 \\ \textit{Add Remainder:} \ \underline{+ 20} \\ 8x^2 - 10x + \ 8 \ \textbf{(Dividend)} \end{array}$$

SUPPLEMENTARY PROBLEMS

1. In each polynomial, select like terms, if any: **(1.1)**

a) $16 - 4y + 3y - y^2$

b) $16y^2 - 4y + 20y^2 + 8$

c) $5x^2 - 8xy + 7y^2 - 2$

d) $7a - 3b + 12ab + 10$

e) $a^2b + ab^2 - 2ab + 3ab^2$

f) $4(x+y) - 5(x+y) + 2(x^2+y^2)$

g) $3xy + xz - 3x - 5z$

h) $3x^2y^2 + 2(x^2+y^2) - 4(x^2+y^2)$

Ans. *a*) $-4y$ and $+3y$

b) $16y^2$ and $20y^2$

c) *No Like Terms*

d) *No Like Terms*

e) $+ab^2$ and $+3ab^2$

f) $4(x+y)$ and $-5(x+y)$

g) *No Like Terms*

h) $2(x^2+y^2)$ and $-4(x^2+y^2)$

2. Add: (2.1 to 2.3)

a) $(+5b) + (+16b)$

b) $(-10y^3) + (-7y^3)$

c) $(+7rs) + (-10rs)$

d) $(-20abc) + (-abc)$

e) $\begin{array}{r} +\ 8x^2y^2 \\ +11x^2y^2 \\ \underline{+30x^2y^2} \end{array}$

f) $\begin{array}{r} +3.2h \\ -2.2h \\ \underline{+7.5h} \end{array}$

g) $\begin{array}{r} -3(a+b) \\ +5(a+b) \\ \underline{+2(a+b)} \end{array}$

Ans. a) $21b$, b) $-17y^3$, c) $-3rs$, d) $-21abc$, e) $49x^2y^2$, f) $8.5h$, g) $4(a+b)$

3. Simplify and add: (2.2)

a) $(+13a) + (-2a) + (-a)$

b) $(-2x^2) + (-8x^2) + (-15x^2)$

c) $(+a^3) + (+3a^3) + (-7a^3)$

d) $(+xy^2) + (+xy^2) + (-xy^2)$

e) $(+2.3ab) + (+7.1ab) + (-3.7ab)$

f) $(+\frac{3}{4}r^2s) + (-\frac{1}{2}r^2s) + (1\frac{1}{8}r^2s)$

Ans. a) $13a - 2a - a = 10a$

b) $-2x^2 - 8x^2 - 15x^2 = -25x^2$

c) $a^3 + 3a^3 - 7a^3 = -3a^3$

d) $xy^2 + xy^2 - xy^2 = xy^2$

e) $2.3ab + 7.1ab - 3.7ab = 5.7ab$

f) $\frac{3}{4}r^2s - \frac{1}{2}r^2s + 1\frac{1}{8}r^2s = +1\frac{3}{8}r^2s$

4. Simplify and add: (2.4)

a) $(+12a) + (-3a) + (+10)$

b) $(+12) + (-3a) + (+10a)$

c) $(+12b) + (-3) + (+10b)$

d) $(-10x^2) + (-3x^2) + (-5x)$

e) $(-10x^2) + (-3x) + (-5x^2)$

f) $(+5r^2s) + (-2r^2s) + (+rs^2)$

g) $(+5rs^2) + (-2r^2s) + (+rs^2)$

h) $(+5rs^2) + (-2rs^2) + (+rs^2)$

Ans. a) $12a - 3a + 10 = 9a + 10$

b) $12 - 3a + 10a = 12 + 7a$

c) $12b - 3 + 10b = 22b - 3$

d) $-10x^2 - 3x^2 - 5x = -13x^2 - 5x$

e) $-10x^2 - 3x - 5x^2 = -15x^2 - 3x$

f) $5r^2s - 2r^2s + rs^2 = 3r^2s + rs^2$

g) $5rs^2 - 2r^2s + rs^2 = 6rs^2 - 2r^2s$

h) $5rs^2 - 2rs^2 + rs^2 = 4rs^2$

5. Rearrange in descending order and combine: (3.1)

a) $-6y + 2y^2 + y^3 + 3y^2$

b) $2x^3 + 3x^4 - x^3 - 10 + 5x^2$

c) $6y^2 - x^2 + 10xy + 8x^2$ (*order in terms of x*)

a) $y^3 + 2y^2 + 3y^2 - 6y$
Ans. $y^3 + 5y^2 - 6y$

b) $3x^4 + 2x^3 - x^3 + 5x^2 - 10$
Ans. $3x^4 + x^3 + 5x^2 - 10$

c) $8x^2 - x^2 + 10xy + 6y^2$
Ans. $7x^2 + 10xy + 6y^2$

6. Add: (3.2)

a) $\begin{array}{r} 5y + 12 \\ \underline{-3y - 10} \end{array}$

b) $\begin{array}{r} 6x - 2y \\ \underline{3x + 10y} \end{array}$

c) $\begin{array}{r} 8x^3 + 5x^2 - 10x \\ \underline{x^3 - 6x^2 + 11x} \end{array}$

d) $\begin{array}{r} x^2 - 3x + 25 \\ \underline{-3x^2 - 10x - 30} \end{array}$

Ans. a) $2y + 2$ b) $9x + 8y$ c) $9x^3 - x^2 + x$ d) $-2x^2 - 13x - 5$

7. Add: (3.3)

a) $6x^2 - 7x$ and $-2x^2 - x$

b) $5a + 2$, $3a - 7$ and $-6a + 4$

c) $x^2 - x + 1$ and $7x - 4x^2 + 9$

d) $5y - 3x + 6$ and $-8 - 4x + y$

Ans.

a) $\begin{array}{r} 6x^2 - 7x \\ \underline{-\ 2x^2 - x} \\ 4x^2 - 8x \end{array}$

b) $\begin{array}{r} 5a + 2 \\ 3a - 7 \\ \underline{-6a + 4} \\ 2a - 1 \end{array}$

c) $\begin{array}{r} x^2 - x + 1 \\ \underline{-4x^2 + 7x + 9} \\ -3x^2 + 6x + 10 \end{array}$

d) $\begin{array}{r} -3x + 5y + 6 \\ \underline{-4x + y - 8} \\ -7x + 6y - 2 \end{array}$

8. Check, letting $x = 2$, $y = 3$ and $z = 4$: (3.4, 3.5)

a) $\begin{array}{r} 5x - 3y + z \\ \underline{2x - 2y - 10z} \\ 7x - 5y - 9z \end{array}$

b) $\begin{array}{r} x^2 + 2y^2 - z^2 \\ \underline{x^2 - y^2 + 2z^2} \\ 2x^2 + y^2 + z^2 \end{array}$

Check: a)

$5x - 3y + z \rightarrow 10 - 9 + 4 = 5$

$2x - 2y - 10z \rightarrow 4 - 6 - 40 = -42$

$7x - 5y - 9z \rightarrow 14 - 15 - 36 = \boxed{-37}$

By adding vertically and horizontally, the sum is -37.

b)

$x^2 + 2y^2 - z^2 \rightarrow 4 + 18 - 16 = 6$

$x^2 - y^2 + 2z^2 \rightarrow 4 - 9 + 32 = 27$

$2x^2 + y^2 + z^2 \rightarrow 8 + 9 + 16 = \boxed{33}$

By adding vertically and horizontally, the sum is 33.

9. Subtract: $a)$ $(+5c) - (-3c)$ $c)$ $(-3rs) - (+2rs)$ $e)$ $(+2x^2y) - (+12x^2y)$ **(4.1)**

         $b)$ $(-x^2) - (-4x^2)$ $d)$ $(+cd^2) - (+7cd^2)$ $f)$ $(+5abc) - (-7abc)$

     $Ans.$ $a)$ $8c$, $b)$ $3x^2$, $c)$ $-5rs$, $d)$ $-6cd^2$, $e)$ $-10x^2y$, $f)$ $12abc$

10. Subtract: $a)$ $+3cd$ $b)$ $+5x^2$ $c)$ $-xy^2$ $d)$ $-2pqr$ $e)$ $-4(m-n)$ **(4.1)**

         (S) $\underline{+\ cd}$ (S) $\underline{-7x^2}$ (S) $\underline{-4xy^2}$ (S) $\underline{+7pqr}$ (S) $\underline{-14(m-n)}$

     $Ans.$ $a)$ $2cd$ $b)$ $12x^2$ $c)$ $3xy^2$ $d)$ $-9pqr$ $e)$ $10(m-n)$

11. $a)$ From $(5y^2)$ take $(-2y^2)$ $Ans.$ $5y^2 - (-2y^2) = 7y^2$ **(4.2)**

     $b)$ Reduce $(-2ab)$ by $(-5ab)$ $Ans.$ $-2cb - (-5ab) = 3ab$

     $c)$ How much less than $+x^2$ is $+3x^2$? $Ans.$ $x^2 - (+3x^2) = -2x^2$

     $d)$ How much does $-17y$ exceed $-30y$? $Ans.$ $-17y - (-30y) = 13y$

12. Combine: $a)$ $(+4b) + (-2b) - (-3b)$ $c)$ $(+2cd) - (-3cd) + (-10cd) - (+5cd)$ **(4.3)**

         $b)$ $(-5x^2) - (-x^2) - (+3x^2)$ $d)$ $(-4abc^2) - (-abc^2) - (+3abc^2) - (-12abc^2)$

     $Ans.$ $a)$ $4b - 2b + 3b = 5b$ $c)$ $2cd + 3cd - 10cd - 5cd = -10cd$

            $b)$ $-5x^2 + x^2 - 3x^2 = -7x^2$ $d)$ $-4abc^2 + abc^2 - 3abc^2 + 12abc^2 = 6abc^2$

13. Subtract: $a)$ $3k - 5$ $b)$ $k^2 - 2k$ $c)$ $-2a^2 - 5a + 12$ $d)$ $3r^2 - 5rt$ **(5.1)**

         (S) $\underline{k - 8}$ (S) $\underline{-k^2 - 8k}$ (S) $\underline{-\ a^2\ \ \ \ -\ 7}$ (S) $\underline{-10r^2 - rt + 7t^2}$

     $Ans.$ $a)$ $2k + 3$ $b)$ $2k^2 + 6k$ $c)$ $-a^2 - 5a + 19$ $d)$ $13r^2 - 4rt - 7t^2$

14. Arrange and subtract: **(5.2)**

       $a)$ $5k - 3l - 2$ from $4l - 7 - k$ $c)$ $3x + 5$ from $x^2 - 4x$

       $b)$ $3x + 4y - 5$ from $16 - 2x - y$ $d)$ $m^3 - 18m$ from $8m^2 + 10m$

   $Ans.$ $a)$ $-k + 4l - 7$ $b)$ $-2x - y + 16$ $c)$ $x^2 - 4x$ $d)$ $8m^2 + 10m$

       (S) $\underline{5k - 3l - 2}$ (S) $\underline{3x + 4y - 5}$ (S) $\underline{\ \ \ \ \ \ 3x + 5}$ (S) $\underline{m^3\ \ \ \ \ \ \ \ \ - 18m}$

          $-6k + 7l - 5$ $-5x - 5y + 21$ $x^2 - 7x - 5$ $-m^3 + 8m^2 + 28m$

15. $a)$ Reduce $8x^2$ by $3x^2 + 5$. $Ans.$ $5x^2 - 5$ $c)$ Subtract $a^3 - 2a^2 + 5$ from 0. $Ans.$ $-a^3 + 2a^2 - 5$ **(5.3)**

     $b)$ From $2x - 3y$ take $4y - 7x$. $Ans.$ $9x - 7y$ $d)$ What is $5h^3 - 12h$ less $h^2 + 5h$? $Ans.$ $5h^3 - h^2 - 17h$

16. Check each subtraction, using addition: **(5.4)**

       $a)$ $3x - 2$ $b)$ $8a^2 - 1$ $c)$ $5y^2 - 3y$ $d)$ $2h^2\ \ \ \ \ + 5$

    (S) $\underline{x - 5}$ (S) $\underline{-\ a^2 + 7}$ (S) $\underline{8y^2\ \ \ \ \ + 15}$ (S) $\underline{-3h^2 - h}$

        $2x + 3$ (?) $9a^2 - 8$ (?) $-3y^2 + 3y - 15$ (?) $5h^2 - h + 5$ (?)

Check by addition:

       $a)$ $2x + 3$ $b)$ $9a^2 - 8$ $c)$ $-3y^2 + 3y - 15$ $d)$ $5h^2 - h + 5$

    (A) $\underline{x - 5}$ (A) $\underline{-\ a^2 + 7}$ (A) $\underline{8y^2\ \ \ \ \ + 15}$ (A) $\underline{-3h^2 - h}$

        $3x - 2$ $8a^2 - 1$ $5y^2 + 3y$ $2h^2 - 2h + 5$

       ($Correct$)   ($Correct$)   ($Incorrect$)   ($Incorrect$)

                                        Difference should be   Difference should be

                                         $-3y^2 - 3y - 15$ $5h^2 + h + 5$

17. Simplify: $a)$ $2r + (3 - 7r)$ $b)$ $(7x^2 + 4) + (10 - 2x^2)$ $c)$ $8y + (7y - 20) + (40 - 3y)$ **(6.1)**

     $Ans.$ $a)$ $-5r + 3$ $b)$ $5x^2 + 14$ $c)$ $12y + 20$

18. Simplify: $a)$ $17 - (3a + 4)$ $c)$ $(5x^2 - 2) - (3 - 4x^2)$ $e)$ $24h - (3h - 2) - 6(h - 1)$ **(6.2)**

         $b)$ $17a - (6 - 2a)$ $d)$ $3 - (7y + 10) - (18 - y)$ $f)$ $80 - (30 - 5y) - (3y - y^2)$

     $Ans.$ $a)$ $13 - 3a$ $c)$ $9x^2 - 5$ $e)$ $15h + 8$

            $b)$ $19a - 6$ $d)$ $-6y - 25$ $f)$ $y^2 + 2y + 50$

19. Simplify: **(6.3, 6.6)**

$a)\ 5a - [2a - (3a+4)]$ $b)\ 12x + 4 - [(3-x) - (5x+7)]$ $c)\ 20 - \{[-(d-1) + 3d] - 5d\}$

Ans. $a)\ 6a + 4$ $b)\ 18x + 8$ $c)\ 3d + 19$

20. $a)$ Subtract $3b - 8$ from the sum of $5b - 2$ and $-6b + 5$.   **(6.4, 6.5)**

     $b)$ From the sum of $8x^2 + 5$ and $7x^2 - 2$, subtract the sum of $20x - 8$ and $-x^2 + 5x$.

Ans. $a)\ [(5b - 2) + (-6b + 5)] - (3b - 8) = -4b + 11$

         $b)\ [(8x^2 + 5) + (7x^2 - 2)] - [(20x - 8) + (-x^2 + 5x)] = 16x^2 - 25x + 11$

21. Multiply: $a)\ c^2 \cdot c \cdot c$ $d)\ x^2 \cdot y^3 \cdot x^5$ $g)\ 7^2\,7$ **(7.1)**

               $b)\ y^3 \cdot y^4 \cdot y$ $e)\ m^2\,pm^8\,p$ $h)\ 8^3 \cdot 8 \cdot 10^2$

               $c)\ a^x a^y$ $f)\ a^3\,bab^4$ $i)\ 3^5\,3^4\,x^2\,x^9$

Ans. $a)\ c^4$, $b)\ y^8$, $c)\ a^{x+y}$, $d)\ x^7 y^3$, $e)\ m^{10} p^2$, $f)\ a^4 b^5$, $g)\ 7^3$, $h)\ 8^4 10^2$, $i)\ 3^9 x^{11}$

22. Raise to a power: $a)\ (x^2)^3$ $c)\ (b^2)^6$ $e)\ (3^3)^3$ $g)\ (r^3)^6 (t^2)^2$ **(7.2)**

                           $b)\ (x^3)^2$ $d)\ (a^x)^y$ $f)\ (5^2)^5$ $h)\ (4^7)^2 (10^2)^4$

Ans. $a)\ x^6$, $b)\ x^6$, $c)\ b^{12}$, $d)\ a^{xy}$, $e)\ 3^9$, $f)\ 5^{10}$, $g)\ r^{18} t^4$, $h)\ 4^{14} 10^8$

23. Multiply: $a)\ 3(-5a)$ $d)\ 5y(-y)(-y)$ $g)\ (-7ab)(-3a)(-b)$ **(7.3)**

               $b)\ (-6x)(-x)$ $e)\ (.2x^2)(.3x^3)$ $h)\ (3r^3)(5s^5)$

               $c)\ (-4y)(+3y^2)$ $f)\ (\frac{2}{3}x)(3x)(-5y)$ $i)\ (-5xy^2)^2$

     Ans. $a)\ -15a$ $d)\ 5y^3$ $g)\ -21a^2 b^2$

                $b)\ 6x^2$ $e)\ .06x^5$ $h)\ 15r^3 s^5$

               $c)\ -12y^3$ $f)\ -10x^2 y$ $i)\ 25x^2 y^4$

24. Multiply: $a)\ 2(a-b)$ $d)\ 3a(b-2c)$ $g)\ \frac{1}{2}(a^2 + 6b^2)$ $j)\ -5(x^2 + y^2 - z^2)$ **(8.1)**

               $b)\ -3(x-5)$ $e)\ -5x(x-4)$ $h)\ ab(a-2b)$ $k)\ -a(a^3 + a^2 - 5a)$

               $c)\ a(c-3d)$ $f)\ \frac{1}{2}(6y-8z)$ $i)\ -4r^2(r^3 + 2r^2)$ $l)\ 4x(5 - x - 10x^2)$

     Ans. $a)\ 2a - 2b$ $d)\ 3ab - 6ac$ $g)\ \frac{a^2}{2} + 3b^2$ $j)\ -5x^2 - 5y^2 + 5z^2$

               $b)\ -3x + 15$ $e)\ -5x^2 + 20x$ $h)\ a^2 b - 2ab^2$ $k)\ -a^4 - a^3 + 5a^2$

               $c)\ ac - 3ad$ $f)\ 3y - 4z$ $i)\ -4r^5 - 8r^4$ $l)\ 20x - 4x^2 - 40x^3$

25. Multiply: $a)\ s^2 - s - 1$ $b)\ x^3 - 3x^2 + 11$ $c)\ 7h + 5k - 3m + 5$ **(8.2)**

                    $-3s$ $7x^2$ $2hkm$

     Ans. $a)\ -3s^3 + 3s^2 + 3s$ $b)\ 7x^5 - 21x^4 + 77x^2$ $c)\ 14h^2 km + 10hk^2 m - 6hkm^2 + 10hkm$

26. Simplify: $a)\ 5 + 3(y-2)$ $c)\ 3x + 10(2x-1)$ $e)\ -a(a-2) + 6(a^2 - 5)$ **(8.3)**

              $b)\ x + 2(3-4x)$ $d)\ -4(a-3) - 2(5-a)$ $f)\ \frac{1}{2}(6c - 10d) - \frac{2}{3}(9d - 30c)$

     Ans. $a)\ 3y - 1$ $c)\ 23x - 10$ $e)\ 5a^2 + 2a - 30$

               $b)\ -7x + 6$ $d)\ -2a + 2$ $f)\ 23c - 11d$

27. Simplify: $a)\ 2[3b + 5(b-2)]$ $c)\ 4(5-c) - 3[8 - 7(c-2)]$ **(8.4, 8.5)**

              $b)\ -a[2(a-3) + 6a]$ $d)\ 10\{25 - [5(y+4) - 3(y-1)]\}$

     Ans. $a)\ 16b - 20$, $b)\ -8a^2 + 6a$, $c)\ 17c - 46$, $d)\ -20y + 20$

28. Multiply: $a)\ (5y+2)(3y+1)$ $d)\ (a+2b)(a-b)$ $g)\ (x^2 + x + 1)(x+2)$ **(9.1)**

              $b)\ (x+7)(x-4)$ $e)\ (x-4y)(x-y)$ $h)\ (r^2 - 2r + 3)(r-1)$

              $c)\ (1+y)(2-y)$ $f)\ (ab-5)(ab+2)$ $i)\ (t^2 + 5t - 6)(2t+1)$

     Ans. $a)\ 15y^2 + 11y + 2$ $d)\ a^2 + ab - 2b^2$ $g)\ x^3 + 3x^2 + 3x + 2$

               $b)\ x^2 + 3x - 28$ $e)\ x^2 - 5xy + 4y^2$ $h)\ r^3 - 3r^2 + 5r - 3$

               $c)\ 2 + y - y^2$ $f)\ a^2 b^2 - 3ab - 10$ $i)\ 2t^3 + 11t^2 - 7t - 6$

29. Check each multiplication, letting $x = 10$. **(9.2)**

a) $3x + 1$
(M) $2x + 2$
$6x^2 + 8x + 2$

b) $x^2 + x - 1$
(M) $3x - 2$
$3x^3 + x^2 - 5x + 2$

Check: a) $3x + 1 \rightarrow 30 + 1 = 31$
$2x + 2 \rightarrow 20 + 2 = 22$
$6x^2 + 8x + 2 \rightarrow 600 + 80 + 2 = 682$
Correct since $682 = (31 \times 22)$

b) $x^2 + x - 1 \rightarrow 100 + 10 - 1 = 109$
$3x - 2 \rightarrow 30 - 2 = 28$
$3x^3 + x^2 - 5x + 2 \rightarrow 3000 + 100 - 50 + 2 = 3052$
Correct since $3052 = (109 \times 28)$

30. Multiply: a) $(x+2)(x+3)(x+4)$ b) $(y-1)(y-2)(y+5)$ c) $(2a+3)(3a-4)(a-5)$ **(9.3)**

Ans. a) $(x^2+5x+6)(x+4)$ b) $(y^2-3y+2)(y+5)$ c) $(6a^2+a-12)(a-5)$
$x^3 + 9x^2 + 26x + 24$ $y^3 + 2y^2 - 13y + 10$ $6a^3 - 29a^2 - 17a + 60$

31. Divide: a) $\frac{x^5}{x^3}$ b) $\frac{x^3}{x^3}$ c) $\frac{x^3}{x^5}$ d) $\frac{a^4b^5}{ab^2}$ e) $\frac{ab^2}{a^4b^5}$ f) $\frac{10^5}{10^9}$ g) $\frac{10^9}{10^5}$ h) $\frac{10^5}{10^5}$ **(10.1)**

Ans. a) x^2 b) 1 c) $\frac{1}{x^2}$ d) a^3b^3 e) $\frac{1}{a^3b^3}$ f) $\frac{1}{10^4}$ g) 10^4 h) 1

32. Divide: a) $\frac{32a}{4a}$ b) $\frac{4a}{32a}$ c) $\frac{32a^4}{-8a^3}$ d) $\frac{-a^2b^2c^2}{-ab^3}$ e) $\frac{7xy}{-7xy}$ f) $\frac{-56c^4d}{7cd^4}$ **(10.2)**

Ans. a) 8 b) $\frac{1}{8}$ c) $-4a$ d) $\frac{ac^2}{b}$ e) -1 f) $-\frac{8c^3}{d^3}$

33. Divide: a) r^{10} by r^3 d) u^2 by u^6 g) $(-8x^2y^2) \div (-2xy)$ j) $a^5 \div a^b$ **(10.1, 10.2)**
b) s^5 by s e) $2ab \div ab^2$ h) $10r^4s^4 \div 4r^4s^4$ k) $3^a \div 3^5$
c) $-t^2$ by t^2 f) $-a^2b \div ab^2$ i) $x^a \div x^2$ l) $x^a \div x^b$

Ans. a) r^7, b) s^4, c) -1, d) $\frac{1}{u^4}$, e) $\frac{2}{b}$, f) $-\frac{a}{b}$, g) $4xy$, h) $2\frac{1}{2}$, i) x^{a-2}, j) a^{5-b}, k) 3^{a-5}, l) x^{a-b}

34. Check each division, using multiplication: **(10.3)**

a) Does $\frac{18x^3}{3x} = 6x^2$ (?) b) Does $\frac{6a^2b^2}{-2ab} = -3ab$ (?) c) Does $\frac{-27a^2}{-9a^6} = \frac{3}{a^3}$ (?)

Check, using multiplication:
a) $(3x)(6x^2) = 18x^3$ b) $(-2ab)(-3ab) = 6a^2b^2$ c) $(-9a^6)(\frac{3}{a^3}) = -27a^3$
(*Correct*) (*Correct*) (*Incorrect*). Quotient should be $\frac{3}{a^4}$.

35. Divide: a) $\frac{5a+15c}{5}$ c) $\frac{xy^2+x^2y}{xy}$ e) $\frac{P+Prt}{P}$ g) $\frac{20r+30s-40}{-10}$ **(11.1)**
b) $\frac{xy+x}{x}$ d) $\frac{\pi R^2 - \pi r^2}{\pi}$ f) $\frac{5x^3-5x}{5x}$ h) $\frac{y^5+2y^4-y^3}{y^3}$

Ans. a) $a+3c$, b) $y+1$, c) $y+x$, d) R^2-r^2, e) $1+rt$, f) x^2-1, g) $-2r-3s+4$, h) y^2+2y-1

36. Divide: a) $2x-4$ by 2 d) $(45x^2-15) \div 15$ g) $4\overline{)8a^2-4a+12}$ **(11.1)**
b) $3a-6$ by -3 e) $(15x^2-45x) \div 15x$ h) $-x\overline{)x^3-7x^2+10x}$
c) $-7x^3+7x^2$ by $-7x$ f) $(27xy-18y^2) \div 9y$ i) $3a^2\overline{)30a^4-33a^3+3a^2}$

Ans. a) $x-2$, b) $-a+2$, c) x^2-x, d) $3x^2-1$, e) $x-3$, f) $3x-2y$,
g) $2a^2-a+3$, h) $-x^2+7x-10$, i) $10a^2-11a+1$

37. Simplify: a) $2(x+4) + \frac{6x-8}{2}$ b) $\frac{r^2-10r}{r} + r(3+r)$ c) $\frac{a^4-a^3+a^2}{a} - 2a(a^2-1)$ **(11.3)**

a) $2x + 8 + 3x - 4$ b) $r - 10 + 3r + r^2$ c) $a^3 - a^2 + a - 2a^3 + 2a$
Ans. $5x + 4$ *Ans.* $r^2 + 4r - 10$ *Ans.* $-a^3 - a^2 + 3a$

38. Divide: (12.1*a*, 12.2)

a) $d^2 + 9d + 14$ by $d + 7$
b) $d^2 - 8d + 7$ by $d - 7$
c) $r^2 - 13r + 30$ by $r - 10$
d) $x^2 - 3x - 10$ by $x - 5$
e) $2r^2 + 7r + 6$ by $r + 2$
f) $6y^2 - 29y + 28$ by $2y - 7$
g) $x^2 + xy - 20y^2$ by $x - 4y$
h) $45 - 14b + b^2$ by $9 - b$
i) $15a^2b^2 - 8ab + 1$ by $3ab - 1$
j) $3x^2 - 40 - 2x$ by $3x + 10$
k) $3 + 7y^2 - 22y$ by $3 - y$
l) $x^4 + 4x^2 - 45$ by $x^2 + 9$

Ans. *a*) $d + 2$ *b*) $d - 1$ *c*) $r - 3$ *d*) $x + 2$ *e*) $2r + 3$ *f*) $3y - 4$ *g*) $x + 5y$ *h*) $5 - b$ *i*) $5ab - 1$ *j*) $x - 4$ *k*) $1 - 7y$ *l*) $x^2 - 5$

39. Divide: (12.1b)

a) $x^3 + 5x^2 + 7x + 2$ by $x + 2$
b) $x^3 - 2x^2 - 5x + 6$ by $x - 1$
c) $x^3 - 2x^2 - 5x + 6$ by $x - 3$
d) $6a^3 + 7a^2 + 12a - 5$ by $3a - 1$
e) $6a^3 + 17a^2 + 27a + 20$ by $3a + 4$
f) $21y^3 - 38y^2 + 29y - 40$ by $3y - 5$

Ans. *a*) $x^2 + 3x + 1$ *b*) $x^2 - x - 6$ *c*) $x^2 + x - 2$ *d*) $2a^2 + 3a + 5$ *e*) $2a^2 + 3a + 5$ *f*) $7y^2 - y + 8$

40. Divide: (12.3)

a) $x^2 - 9$ by $x - 3$
b) $x^3 - 4x$ by $x + 2$
c) $7x^2 - 63$ by $x + 3$
d) $x^3 - 8$ by $x - 2$
e) $x^4 - 1$ by $x + 1$
f) $2x^3 - 30x - 8$ by $x - 4$

Ans. *a*) $x + 3$ *b*) $x^2 - 2x$ *c*) $7x - 21$ *d*) $x^2 + 2x + 4$ *e*) $x^3 - x^2 + x - 1$ *f*) $2x^2 + 8x + 2$

41. Divide: (12.4)

a) $x^2 - 5x + 2$ by $x - 3$
b) $2x^2 + 5x - 15$ by $x - 2$
c) $x^2 + 1$ by $x + 1$
d) $x^2 + 9$ by $x + 3$
e) $x^3 - 5x^2 + 6x + 5$ by $x - 3$
f) $2x^3 + 3x^2 - x - 2$ by $2x - 3$

Ans. *a*) $x - 2 + \dfrac{-4}{x - 3}$ *b*) $2x + 9 + \dfrac{3}{x - 2}$ *c*) $x - 1 + \dfrac{2}{x + 1}$ *d*) $x - 3 + \dfrac{18}{x + 3}$ *e*) $x^2 - 2x + \dfrac{5}{x - 3}$ *f*) $x^2 + 3x + 4 + \dfrac{10}{2x - 3}$

42. Check each division, using multiplication: (12.2, 12.4)

a) Does $\dfrac{21x^2 + 11xy - 2y^2}{3x + 2y} = 7x - y$?

b) Does $\dfrac{x^3 - 6}{x - 2} = x^2 + 2x + 4 + \dfrac{2}{x - 2}$?

Check, using multiplication:

a)
$$\begin{array}{lr} & 3x + 2y \\ & \underline{7x - y} \\ \textit{Correct} & 21x^2 + 11xy - 2y^2 \end{array}$$

b)
$$\begin{array}{lr} & x^2 + 2x + 4 \\ & \underline{x - 2} \\ & x^3 \quad - 8 \\ \textit{Add remainder:} & \underline{+ 2} \\ \textit{Correct} & x^3 \quad - 6 \end{array}$$

Chapter 5 EQUATIONS of the FIRST DEGREE in ONE UNKNOWN

1. REVIEWING THE SOLUTION OF FIRST DEGREE EQUATIONS HAVING POSITIVE ROOTS

A first degree equation in one unknown is one which, after it has been simplified, contains only 1 unknown having the exponent 1.

Thus, $2x+5=9$ is a first degree equation in one unknown but $x^2+5=9$ and $\frac{2}{x}+x=3$ are not. When $\frac{2}{x}+x=3$ is simplified, it becomes $2+x^2=3x$.

In Chapter 2, simple equations were solved using the equality rules and inverse operations. In this chapter, these methods of solution will be extended to equations having signed numbers and to more difficult equations.

Inverse Operations

Since addition and subtraction are inverse operations:

1. Use subtraction (**S**) to undo addition (**A**).
2. Use addition (**A**) to undo subtraction (**S**).

Since multiplication and division are inverse operations:

3. Use division (**D**) to undo multiplication (**M**).
4. Use multiplication (**M**) to undo division (**D**).

Order of Inverse Operations

Generally, undo addition and subtraction before undoing multiplication and division.

1.1. Using Subtraction (S) to Undo Addition (A)

Solve: *a)*	$n+2=8$	*b)*	$n+5=2n$	*c)*	$2n+6=n+10$	*d)*	$n^2+8=n^2+n$
$\mathbf{S}_2$	$2=2$	$\mathbf{S}_n$	$n=n$	$\mathbf{S}_{n+6}$	$n+6=n+6$	$\mathbf{S}_{n^2}$	$n^2=n^2$
Ans.	$n=6$	*Ans.*	$5=n$	*Ans.*	$n=4$	*Ans.*	$8=n$

1.2. Using Addition (A) to Undo Subtraction (S)

Solve: *a)*	$x-2=8$	*b)*	$5-x=0$	*c)*	$5-n=-7$	*d)*	$8-n^2=n-n^2$
$\mathbf{A}_2$	$2=2$	$\mathbf{A}_x$	$x=x$	$\mathbf{A}_{7+n}$	$7+n=7+n$	$\mathbf{A}_{n^2}$	$n^2=n^2$
Ans.	$x=10$	*Ans.*	$5=x$	*Ans.*	$12=n$	*Ans.*	$8=n$

1.3. Using Division (D) to Undo Multiplication (M)

Solve: *a)*	$2y=10$	*b)*	$10w=5$	*c)*	$.7a=77$	*d)*	$3\frac{1}{4}b=32\frac{1}{2}$
$\mathbf{D}_2$	$\frac{2y}{2}=\frac{10}{2}$	$\mathbf{D}_{10}$	$\frac{10w}{10}=\frac{5}{10}$	$\mathbf{D}_{.7}$	$\frac{.7a}{.7}=\frac{77}{.7}$	$\mathbf{D}_{3.25}$	$\frac{3.25b}{3.25}=\frac{32.5}{3.25}$
Ans.	$y=5$	*Ans.*	$w=\frac{1}{2}$	*Ans.*	$a=110$	*Ans.*	$b=10$

1.4. Using Multiplication (M) to Undo Division (D)

Solve:

a) $\frac{n}{3} = 6$

M_3 $3\left(\frac{n}{3}\right) = 6 \cdot 3$

Ans. $n = 18$

b) $\frac{6}{n} = 1$

M_n $n\left(\frac{6}{n}\right) = 1 \cdot n$

Ans. $6 = n$

c) $7 = \frac{1}{8}z$

M_8 $8 \cdot 7 = 8\left(\frac{1}{8}z\right)$

Ans. $56 = z$

d) $\frac{b}{.5} = 100$

$M_{.5}$ $.5\left(\frac{b}{.5}\right) = 100(.5)$

Ans. $b = 50$

1.5. Using Two Operations to Solve Equations

Solve:

a)
$$\begin{array}{lrcl} & 5x + 2 & = & 17 \\ S_2 & 2 & = & 2 \\ & 5x & = & 15 \\ D_5 & \frac{5x}{5} & = & \frac{15}{5} \\ Ans. & x & = & 3 \end{array}$$

b)
$$\begin{array}{lrcl} & \frac{x}{6} - 2 & = & 8 \\ A_2 & 2 & = & 2 \\ & \frac{x}{6} & = & 10 \\ M_6 & 6\left(\frac{x}{6}\right) & = & 10(6) \\ Ans. & x & = & 60 \end{array}$$

c)
$$\begin{array}{lrcl} & 12 - x & = & 8 \\ A_x & x & = & +x \\ & 12 & = & 8 + x \\ S_8 & 8 & = & 8 \\ Ans. & 4 & = & x \end{array}$$

d)
$$\begin{array}{lrcl} & \frac{12}{x} & = & 8 \\ M_x & x\left(\frac{12}{x}\right) & = & 8x \\ & 12 & = & 8x \\ D_8 & \frac{12}{8} & = & \frac{8x}{8} \\ Ans. & 1\frac{1}{2} & = & x \end{array}$$

1.6. Equations Containing More Than One Term of the Unknown

Solve:

a)
$$\begin{array}{lrcl} & 8y & = & 3y + 35 \\ S_{3y} & 3y & = & 3y \\ & 5y & = & 35 \\ D_5 & \frac{5y}{5} & = & \frac{35}{5} \\ Ans. & y & = & 7 \end{array}$$

b)
$$\begin{array}{lrcl} & 3y + y + 70 & = & 11y \\ Combine: & 4y + 70 & = & 11y \\ S_{4y} & 4y & = & 4y \\ & 70 & = & 7y \\ D_7 & \frac{70}{7} & = & \frac{7y}{7} \\ Ans. & 10 & = & y \end{array}$$

c)
$$\begin{array}{lrcl} & 5y - 2 & = & 2y + 22 \\ S_{2y} & 2y & = & 2y \\ & 3y - 2 & = & 22 \\ A_2 & 2 & = & 2 \\ & 3y & = & 24 \\ D_3 & \frac{3y}{3} & = & \frac{24}{3} \\ Ans. & y & = & 8 \end{array}$$

1.7. Equations Having Fractional Coefficients

Solve:

a)
$$\begin{array}{lrcl} & \frac{1}{8}a & = & 3 \\ M_8 & 8\left(\frac{1}{8}a\right) & = & 3(8) \\ Ans. & a & = & 24 \end{array}$$

b)
$$\begin{array}{lrcl} & \frac{3}{8}b & = & 12 \\ M_{8/3} & \frac{8}{3}\left(\frac{3}{8}b\right) & = & 12\left(\frac{8}{3}\right) \\ Ans. & b & = & 32 \end{array}$$

c)
$$\begin{array}{lrcl} & \frac{2}{3}c - 3 & = & 7 \\ A_3 & 3 & = & 3 \\ & \frac{2}{3}c & = & 10 \\ M_{3/2} & \frac{3}{2}\left(\frac{2}{3}c\right) & = & \frac{3}{2} \cdot 10 \\ Ans. & c & = & 15 \end{array}$$

d)
$$\begin{array}{lrcl} & \frac{8}{5}d + 12 & = & 36 \\ S_{12} & 12 & = & 12 \\ & \frac{8}{5}d & = & 24 \\ M_{5/8} & \frac{5}{8}\left(\frac{8}{5}d\right) & = & 24\left(\frac{5}{8}\right) \\ Ans. & d & = & 15 \end{array}$$

1.8. Number Problems Leading to First Degree Equations

a) Twice a number is equal to 81 less than five times the same number. Find the number.

b) Three times a number decreased by 8 equals the number increased by 12. Find the number.

Solutions:

a) Let n = the number.

$$\begin{array}{lrcl} A_{81} & 2n & = & 5n - 81 \\ S_{2n} & 2n + 81 & = & 5n \\ D_3 & 81 & = & 3n \\ & 27 & = & n \end{array}$$

Ans. The number is 27.

b) Let n = the number.

$$\begin{array}{lrcl} A_8 & 3n - 8 & = & n + 12 \\ S_n & 3n & = & n + 20 \\ D_2 & 2n & = & 20 \\ & n & = & 10 \end{array}$$

Ans. The number is 10.

2. SOLVING FIRST DEGREE EQUATIONS HAVING NEGATIVE ROOTS

To change $-x$ to $+x$, multiply by -1 or divide by -1.

$$\begin{array}{lrcl} \text{Thus, if} & -x & = & +5 \\ \mathbf{M}_{-1} & (-1)(-x) & = & (-1)(+5) \\ & +x & = & -5 \end{array}$$

2.1. Using Addition or Subtraction to Solve Equations

Solve: *a*)
$$\begin{array}{lrcl} & n+8 & = & 2 \\ \mathbf{S}_8 & 8 & = & 8 \\ \hline \textit{Ans.} & n & = & -6 \end{array}$$

b)
$$\begin{array}{lrcl} & n-8 & = & -13 \\ \mathbf{A}_8 & 8 & = & 8 \\ \hline \textit{Ans.} & n & = & -5 \end{array}$$

c)
$$\begin{array}{lrcl} & n-15 & = & 2n-9 \\ \mathbf{S}_{n-9} & n-9 & = & n-9 \\ \hline \textit{Ans.} & -6 & = & n \end{array}$$

d)
$$\begin{array}{lrcl} & 14 & = & -n+10 \\ \mathbf{A}_{n-14} & n-14 & = & n-14 \\ \hline \textit{Ans.} & n & = & -4 \end{array}$$

2.2. Using Multiplication or Division to Solve Equations

Solve: *a*)
$$\begin{array}{lrcl} & 2x & = & -8 \\ \mathbf{D}_2 & \frac{2x}{2} & = & \frac{-8}{2} \\ \textit{Ans.} & x & = & -4 \end{array}$$

b)
$$\begin{array}{lrcl} & \frac{x}{2} & = & -8 \\ \mathbf{M}_2 & 2(\frac{x}{2}) & = & (-8)2 \\ \textit{Ans.} & x & = & -16 \end{array}$$

c)
$$\begin{array}{lrcl} & -12x & = & 3 \\ \mathbf{D}_{-12} & \frac{-12x}{-12} & = & \frac{3}{-12} \\ \textit{Ans.} & x & = & -\frac{1}{4} \end{array}$$

d)
$$\begin{array}{lrcl} & -81 & = & 2.7x \\ \mathbf{D}_{2.7} & \frac{-81}{2.7} & = & \frac{2.7x}{2.7} \\ \textit{Ans.} & -30 & = & x \end{array}$$

2.3. Using Two Operations to Solve Equations

Solve: *a*)
$$\begin{array}{lrcl} & 3y+20 & = & 11 \\ \mathbf{S}_{20} & 20 & = & 20 \\ \hline & 3y & = & -9 \\ \mathbf{D}_3 & \frac{3y}{3} & = & \frac{-9}{3} \\ \textit{Ans.} & y & = & -3 \end{array}$$

b)
$$\begin{array}{lrcl} & 8-3y & = & 29 \\ \mathbf{S}_8 & 8 & = & 8 \\ \hline & -3y & = & 21 \\ \mathbf{D}_{-3} & \frac{-3y}{-3} & = & \frac{21}{-3} \\ \textit{Ans.} & y & = & -7 \end{array}$$

c)
$$\begin{array}{lrcl} & \frac{18}{y} & = & -6 \\ \mathbf{M}_y & \frac{18}{y}\cdot y & = & -6y \\ & 18 & = & -6y \\ \mathbf{D}_{-6} & \frac{18}{-6} & = & \frac{-6y}{-6} \\ \textit{Ans.} & -3 & = & y \end{array}$$

d)
$$\begin{array}{lrcl} & 20+5y & = & 2y \\ \mathbf{S}_{5y} & 5y & = & 5y \\ \hline & 20 & = & -3y \\ \mathbf{D}_{-3} & \frac{20}{-3} & = & \frac{-3y}{-3} \\ \textit{Ans.} & -6\frac{2}{3} & = & y \end{array}$$

2.4. Equations Having Fractional Coefficients

Solve: *a*)
$$\begin{array}{lrcl} & 5+\frac{y}{3} & = & 3 \\ \mathbf{S}_5 & 5 & = & 5 \\ \hline & \frac{y}{3} & = & -2 \\ \mathbf{M}_3 & 3(\frac{y}{3}) & = & (-2)3 \\ \textit{Ans.} & y & = & -6 \end{array}$$

b)
$$\begin{array}{lrcl} & \frac{3y}{4}-14 & = & -23 \\ \mathbf{A}_{14} & 14 & = & 14 \\ \hline & \frac{3y}{4} & = & -9 \\ \mathbf{M}_{4/3} & \frac{4}{3}(\frac{3y}{4}) & = & (-9)\frac{4}{3} \\ \textit{Ans.} & y & = & -12 \end{array}$$

c)
$$\begin{array}{lrcl} & 5\frac{3}{4}y & = & -10+5\frac{1}{2}y \\ \mathbf{S}_{5\frac{1}{2}y} & 5\frac{1}{2}y & = & 5\frac{1}{2}y \\ \hline & \frac{1}{4}y & = & -10 \\ \mathbf{M}_4 & 4(\frac{1}{4}y) & = & (-10)4 \\ \textit{Ans.} & y & = & -40 \end{array}$$

2.5. More Difficult Equations

Solve: *a*)
$$\begin{array}{lrcl} & 5y-2y+11 & = & 8y+26 \\ \textit{Combine:} & 3y+11 & = & 8y+26 \\ \mathbf{S}_{3y+26} & 3y+26 & = & 3y+26 \\ \hline & -15 & = & 5y \\ \mathbf{D}_5 & \frac{-15}{5} & = & \frac{5y}{5} \\ \textit{Ans.} & -3 & = & y \end{array}$$

b)
$$\begin{array}{lrcl} & 4\frac{3}{4}a+11 & = & 3a-24 \\ \mathbf{S}_{3a+11} & 3a+11 & = & 3a+11 \\ \hline & 1\frac{3}{4}a & = & -35 \\ \mathbf{M}_{4/7} & \frac{4}{7}(\frac{7}{4}a) & = & (-35)\frac{4}{7} \\ \textit{Ans.} & a & = & -20 \end{array}$$

2.6. Problems Having Negative Roots

a) The sum of three numbers, represented by n, $n-4$ and $3n+7$, is -2. Find the numbers.

b) A merchant's profits in dollars on three articles are represented by p, $p-3$ and $2p-4$. If the total profit is \$1, find the profit on each.

Solutions:

a)
$$n+(n-4)+(3n+7) = -2$$
$$n+n-4+3n+7 = -2$$
S_3 $$5n+3 = -2$$
D_5 $$5n = -5$$
$$n = -1 \quad \begin{cases} n-4 = -5 \\ 3n+7 = 4 \end{cases}$$

Ans. Numbers are, -1, -5 and 4.

b)
$$p+(p-3)+(2p-4) = 1$$
$$p+p-3+2p-4 = 1$$
A_7 $$4p-7 = 1$$
D_4 $$4p = 8$$
$$p = 2 \quad \begin{cases} p-3 = -1 \\ 2p-4 = 0 \end{cases}$$

Ans. The merchant made \$2 on the first article, lost \$1 on the second and made no profit on the third.

3. SOLVING EQUATIONS BY TRANSPOSING

Rule of Transposition

To transpose a term from one side of an equation to another, change its sign.
Thus, by transposing, $3x = 24-x$ is changed into $3x+x = 24$.
In solving equations, "**Tr**" will indicate "transpose" or by "transposing".

Opposites are terms differing only in signs. Thus, $4xy$ and $-4xy$ are opposites.

The rule of transposition is based on the fact that a term may be eliminated from one side of an equation if its opposite is added to the other.

Thus,
$$\begin{aligned} 13x &= 24+5x \\ -5x &= \quad\ -5x \\ \hline 13x-5x &= 24 \end{aligned}$$
Note how $5x$ has been eliminated from the right side and its opposite $-5x$ added to the left.

Using transposition, the second step becomes unnecessary.

To Solve Equations by Transposing Terms

Procedure:	Solve: $8a-55 = 3a-40$ **Solution:**
1. Transpose (**Tr**) so that like terms will be on the same side of equation: (*Change signs of transposed terms.*)	**1.** $8a-55 = 3a-40$ **Tr** $8a-3a = 55-40$
2. Combine like terms:	**2. Combine:** $5a = 15$
3. Divide by the coefficient of the unknown:	**3. D_5** $\frac{5a}{5} = \frac{15}{5}$
	Ans. $a = 3$

3.1. Transposing Terms in an Equation

Solve, using transposition:

a) $31-6x = 25$
Tr $31-25 = 6x$
D_6 $6 = 6x$
Ans. $1 = x$

b) $2y-28 = 42-5y$
Tr $2y+5y = 42+28$
D_7 $7y = 70$
Ans. $y = 10$

c) $11z+3-4z = 5z$
Tr $11z-4z-5z = -3$
D_2 $2z = -3$
Ans. $z = -1\frac{1}{2}$

3.2. More Difficult Equations

Solve and check:

a)

$$
\begin{aligned}
& & 5a - 7 + 4a &= 2a - 28 + 4a \\
&\textbf{Tr} & 5a + 4a - 4a - 2a &= 7 - 28 \\
&\textbf{D}_3 & 3a &= -21 \\
& & a &= -7
\end{aligned}
$$

Check:

$$
\begin{aligned}
5a - 7 + 4a &= 2a - 28 + 4a \\
5(-7) - 7 + 4(-7) &\overset{?}{=} 2(-7) - 28 + 4(-7) \\
-35 - 7 - 28 &\overset{?}{=} -14 - 28 - 28 \\
-70 &= -70
\end{aligned}
$$

Ans. $a = -7$

b)

$$
\begin{aligned}
& & n^2 + 3n - 8 &= n^2 - 5n - 32 \\
&\textbf{Tr} & n^2 - n^2 + 3n + 5n &= -32 + 8 \\
&\textbf{D}_8 & 8n &= -24 \\
& & n &= -3
\end{aligned}
$$

Check:

$$
\begin{aligned}
n^2 + 3n - 8 &= n^2 - 5n - 32 \\
(-3)^2 + 3(-3) - 8 &\overset{?}{=} (-3)^2 - 5(-3) - 32 \\
9 - 9 - 8 &\overset{?}{=} 9 + 15 - 32 \\
-8 &= -8
\end{aligned}
$$

Ans. $n = -3$

4. SOLVING EQUATIONS CONTAINING PARENTHESES

To Solve Equations by Removing Parentheses

Procedure:	Solve: $8 + 2(x-5) = 14$
	Solution:
1. Remove parentheses:	**1.** $8 + 2(x-5) = 14$
	Remove ( ): $8 + 2x - 10 = 14$
2. Solve the resulting equation:	**2.** $2x - 2 = 14$
	Ans. $x = 8$

4.1. Removing Parentheses to Solve Equations

Remove parentheses, then solve:

a) **Remove ( ):**

$$
\begin{aligned}
& & 3(n+2) &= 30 + n \\
&\textbf{Tr} & 3n + 6 &= 30 + n \\
& & 3n - n &= 30 - 6 \\
& & 2n &= 24 \\
&\textit{Ans.} & n &= 12
\end{aligned}
$$

b) **Remove ( ):**

$$
\begin{aligned}
& & -18 &= 10 - 2(3-x) \\
& & -18 &= 10 - 6 + 2x \\
& & -18 &= 4 + 2x \\
& & -22 &= 2x \\
&\textit{Ans.} & -11 &= x
\end{aligned}
$$

4.2. More Difficult Equations with Parentheses

Solve and check:

a) **Remove ( ):**

$$
\begin{aligned}
& & 6y(y-2) &= 3y(2y-1) + 27 \\
&\textbf{Tr} & 6y^2 - 12y &= 6y^2 - 3y + 27 \\
&\textbf{Combine:} & 6y^2 - 6y^2 - 12y + 3y &= 27 \\
& & -9y &= 27 \\
& & y &= -3
\end{aligned}
$$

Check:

$$
\begin{aligned}
6y(y-2) &= 3y(2y-1) + 27 \\
6(-3)(-5) &\overset{?}{=} 3(-3)(-7) + 27 \\
90 &\overset{?}{=} 63 + 27 \\
90 &= 90
\end{aligned}
$$

Ans. $y = -3$

b) **Remove ( ):**

$$
\begin{aligned}
& & (a+1)(a-5) + 7 &= 44 - a(7-a) \\
&\textbf{Tr} & a^2 - 4a - 5 + 7 &= 44 - 7a + a^2 \\
&\textbf{Combine:} & a^2 - a^2 - 4a + 7a &= 44 + 5 - 7 \\
& & 3a &= 42 \\
& & a &= 14
\end{aligned}
$$

Check:

$$
\begin{aligned}
(a+1)(a-5) + 7 &= 44 - a(7-a) \\
(15)(9) + 7 &\overset{?}{=} 44 - 14(-7) \\
135 + 7 &\overset{?}{=} 44 + 98 \\
142 &= 142
\end{aligned}
$$

Ans. $a = 14$

4.3. and 4.4. Problems Leading to Equations with Parentheses

4.3. *a*) Find a number if twice the sum of the number and 7 equals three times the difference of the number and 10.

b) Find a number if 20 minus twice the number equals three times the sum of twice the number and 20.

Solutions:

a)

	Let n = the number
Remove ( ):	$2(n+7) = 3(n-10)$
Tr	$2n + 14 = 3n - 30$
	$14 + 30 = 3n - 2n$
	$44 = n$

Ans. Number is 44.

b)

	Let n = the number
Remove ( ):	$20 - 2n = 3(2n+20)$
Tr	$20 - 2n = 6n + 60$
	$20 - 60 = 6n + 2n$
	$-40 = 8n$ or $n = -5$

Ans. Number is -5.

4.4. John, George and Tom earned \$120 together. George earned \$20 less than John and Tom earned twice as much as George. Find the earnings of each.

Solution:

Let n = John's earnings in \$
$n - 20$ = George's earnings in \$
$2(n-20)$ = Tom's earnings in \$

Remove ( ):

$$n + (n-20) + 2(n-20) = 120$$
$$n + n - 20 + 2n - 40 = 120$$
$$4n - 60 = 120$$
$$4n = 180 \qquad \begin{cases} n - 20 = 25 \\ 2(n-20) = 50 \end{cases}$$
$$n = 45$$

Ans. John, George and Tom earned \$45, \$25 and \$50 respectively.

5. *SOLVING EQUATIONS CONTAINING ONE FRACTION OR FRACTIONS HAVING THE SAME DENOMINATOR*

To Solve Equations Having Same Denominator by Clearing of Fractions

Procedure:	Solve: *a*) $\frac{x}{3} + 5 = 2x$	*b*) $\frac{x}{5} + 6 = \frac{7x}{5}$
	Solutions:	
1. Clear of fractions by multiplying both sides of the equation by the denominator:	**1.** $\frac{x}{3} + 5 = 2x$	**1.** $\frac{x}{5} + 6 = \frac{7x}{5}$
	Multiply by denominator 3:	**Multiply by denominator 5:**
	M_3 $3(\frac{x}{3} + 5) = 3(2x)$	M_5 $5(\frac{x}{5} + 6) = 5(\frac{7x}{5})$
2. Solve the resulting equation:	$x + 15 = 6x$	$x + 30 = 7x$
	Ans. $3 = x$	**Ans.** $5 = x$

5.1. Fractional Equations Having the Same Denominator

Solve:

a)	*b*)	*c*)
$\frac{3x}{7} - 2 = \frac{x}{7}$	$\frac{4}{y} = 5 - \frac{1}{y}$	$\frac{5x}{4} - \frac{3}{4} = -12$
M_7 $7(\frac{3x}{7} - 2) = 7(\frac{x}{7})$	M_y $y(\frac{4}{y}) = y(5 - \frac{1}{y})$	M_4 $4(\frac{5x}{4} - \frac{3}{4}) = 4(-12)$
$3x - 14 = x$	$4 = 5y - 1$	$5x - 3 = -48$
$2x = 14$	$5 = 5y$	$5x = -45$
Ans. $x = 7$	*Ans.* $1 = y$	*Ans.* $x = -9$

5.2. Fractional Equations Having Binomial Numerator or Denominator

Solve:

a) $\frac{2x+7}{4} = x - \frac{3}{4}$

$M_4 \quad 4(\frac{2x+7}{4}) = 4(x - \frac{3}{4})$

$2x + 7 = 4x - 3$

$10 = 2x$

Ans. $5 = x$

b) $10 = \frac{5x}{2x-3}$

$M_{(2x-3)} \quad 10(2x-3) = (\frac{5x}{2x-3})(2x-3)$

$20x - 30 = 5x$

$15x = 30$

Ans. $x = 2$

c) $\frac{x-3}{x} - 5 = \frac{x+7}{x}$

$M_x \quad x(\frac{x-3}{x} - 5) = (\frac{x+7}{x})x$

$x - 3 - 5x = x + 7$

$-5x = 10$

Ans. $x = -2$

6. SOLVING EQUATIONS CONTAINING FRACTIONS HAVING DIFFERENT DENOMINATORS: LOWEST COMMON DENOMINATOR (L.C.D.)

The **lowest common denominator (L.C.D.)** of two or more fractions is the smallest number divisible by their denominators without remainder.

Thus, in $\frac{1}{2} + \frac{x}{3} = \frac{7}{4}$, 12 is the **L.C.D.** since 12 is the smallest number divisible by 2, 3 and 4 without remainder. Larger common denominators of 2, 3 and 4 are 24, 36, 48, etc.

To Solve Equations Having Different Denominators by Clearing of Fractions

Solve: $\frac{x}{2} + \frac{x}{3} = 20$

Procedure:

1. Clear of fractions by multiplying both sides of the equation by the **L.C.D.**:
2. Solve the resulting equation:

Solution:

1. **L.C.D.** $= 6$

$M_6 \quad 6(\frac{x}{2} + \frac{x}{3}) = 6(20)$

2. $3x + 2x = 120, \quad 5x = 120$

Ans. $x = 24$

6.1. Fractional Equations Having Numerical Denominators

Solve:

a) $\frac{x}{2} - \frac{x}{3} = 5$

L.C.D. $= 6$

$M_6 \quad 6(\frac{x}{2} - \frac{x}{3}) = 6 \cdot 5$

$3x - 2x = 30$

Ans. $x = 30$

b) $\frac{a}{2} - \frac{a}{3} - \frac{a}{5} = 2$

L.C.D. $= 30$

$M_{30} \quad 30(\frac{a}{2} - \frac{a}{3} - \frac{a}{5}) = 30(2)$

$15a - 10a - 6a = 60$

Ans. $a = -60$

c) $\frac{3y}{4} - \frac{y}{3} = 10$

L.C.D. $= 12$

$M_{12} \quad 12(\frac{3y}{4} - \frac{y}{3}) = 12(10)$

$9y - 4y = 120$

$5y = 120$

Ans. $y = 24$

6.2. Fractional Equations Having Literal Denominators

Solve:

a) $\frac{10}{x} = \frac{25}{3x} - \frac{1}{3}$

L.C.D. $= 3x$

$M_{3x} \quad 3x(\frac{10}{x}) = 3x(\frac{25}{3x} - \frac{1}{3})$

$30 = 25 - x$

Ans. $x = -5$

b) $\frac{8}{a} - 3 = \frac{7}{2a}$

L.C.D. $= 2a$

$M_{2a} \quad 2a(\frac{8}{a} - 3) = 2a(\frac{7}{2a})$

$16 - 6a = 7$

Ans. $a = 1\frac{1}{2}$

c) $\frac{5}{6} = \frac{7}{3x} + 1$

L.C.D. $= 6x$

$M_{6x} \quad 6x(\frac{5}{6}) = 6x(\frac{7}{3x} + 1)$

$5x = 14 + 6x$

Ans. $x = -14$

6.3. Fractional Equations Having Binomial Numerators

Solve:

a) $\frac{6x+13}{2} + \frac{x+3}{3} = \frac{5}{6}$

L.C.D. = 6

M_6 $6(\frac{6x+13}{2}) + 6(\frac{x+3}{3}) = 6(\frac{5}{6})$

$18x + 39 + 2x + 6 = 5$

Ans. $x = -2$

b) $\frac{5y+4}{9} = 2 + \frac{2y+4}{6}$

L.C.D. = 18

M_{18} $18(\frac{5y+4}{9}) = 18(2) + 18(\frac{2y+4}{6})$

$10y + 8 = 36 + 6y + 12$

Ans. $y = 10$

6.4. Fractional Equations Having Binomial Denominators

Solve:

a) $\frac{3}{8} = \frac{6}{5-y}$

L.C.D. = $8(5-y)$

$M_{L.C.D.}$ $(\frac{3}{8})8(5-y) = 8(5-y)(\frac{6}{5-y})$

$3(5-y) = 48$

Ans. $y = -11$

b) $\frac{6}{5} + \frac{3}{w-3} = \frac{9}{5(w-3)}$

L.C.D. = $5(w-3)$

$M_{L.C.D.}$ $(\frac{6}{5})(5)(w-3) + 5(w-3)(\frac{3}{w-3}) = \frac{9}{5(w-3)}(5)(w-3)$

$6(w-3) + 15 = 9$

Ans. $w = 2$

7. SOLVING EQUATIONS CONTAINING DECIMALS

A **decimal** may be written as a fraction whose denominator is 10, 100 or a power of 10. Thus, the denominator of .003 or $\frac{3}{1000}$ is 1000.

To Solve an Equation Having Decimals

Solve: $.15x + 7 = .5x$

Procedure:	**Solution:**
1. Clear of decimals by multiplying both sides of the equation by the denominator of the decimal having the greatest number of decimal places:	1. $.15x$ has more decimal places than $.5x$. The denominator of $.15x$ is 100. M_{100} $100(.15x+7) = 100(.5x)$
2. Solve the resulting equation:	2. $15x + 700 = 50x$ **Ans.** $x = 20$

In some cases, it may be better not to clear an equation of decimals. Thus, if $3a = .54$, simply divide by 3 to obtain $a = .18$. Or, if $2a - .28 = .44$, then $2a = .72$ and $a = .36$. These are cases where the coefficient of the unknown is not a decimal.

7.1. Equations with One Decimal

First clear each equation of decimals, then solve:

a) $.3a = 6$

M_{10} $10(.3a) = 60$

$3a = 60$

Ans. $a = 20$

b) $8 = .05b$

M_{100} $100(8) = 100(.05b)$

$800 = 5b$

Ans. $160 = b$

c) $2.8c = 54 + c$

M_{10} $10(2.8c) = 10(54+c)$

$28c = 540 + 10c$

Ans. $c = 30$

7.2. Solving Equations Without Clearing of Decimals

Solve, without clearing of decimals:

a)	b)	c)	d)
$3a = .6$	$5r - 5 = .05$	$\frac{x}{4} = .28$	$\frac{x}{5} + 2 = 3.5$
D_3 $\frac{3a}{3} = \frac{.6}{3}$	D_5 $\frac{5r}{5} = \frac{5.05}{5}$	M_4 $4(\frac{x}{4}) = 4(.28)$	M_5 $5(\frac{x}{5}) = (1.5)5$
Ans. $a = .2$	*Ans.* $r = 1.01$	*Ans.* $x = 1.12$	*Ans.* $x = 7.5$

7.3. Equations With Two or More Decimals

Solve (*Multiply by the denominator of the decimal with the largest number of decimal places.*):

a)	b)	c)	d)
$.05x = 2.5$	$-.9 = .003y$	$.2a = a - .8$	$.5a - 3.5 = .75$
M_{100} $5x = 250$	M_{1000} $-900 = 3y$	M_{10} $2a = 10a - 8$	M_{100} $50a - 350 = 75$
Ans. $x = 50$	*Ans.* $-300 = y$	*Ans.* $a = 1$	*Ans.* $a = 8.5$

7.4. Equations Containing Percents

Solve:

a)	b)	c)
25% of $x = 10$	$x + 40\% x = 56$	$x - 16\% x = 420$
$(25\% = \frac{1}{4})$	$(40\% = .4)$	$(16\% = .16)$
$\frac{x}{4} = 10$	$x + .4x = 56$	$x - .16x = 420$
Ans. $x = 40$	*Ans.* $x = 40$	*Ans.* $x = 500$

7.5. Equations Containing Decimals and Parentheses

Solve:

a)	b)	c)
$.3(50-x) = 6$	$.8 = .02(x-35)$	$5(x+.8) = -16$
M_{10} $3(50-x) = 60$	M_{100} $80 = 2(x-35)$	M_{10} $50(x+.8) = -160$
$150 - 3x = 60$	$80 = 2x - 70$	$50x + 40 = -160$
Ans. $x = 30$	*Ans.* $75 = x$	*Ans.* $x = -4$

7.6. More Difficult Decimal Equations

Solve:

a)	b)
$.04x + .03(5000-x) = 190$	$.3(x-200) + .03(1000-x) = 105$
M_{100} $4x + 3(5000-x) = 19{,}000$	M_{100} $30(x-200) + 3(1000-x) = 10{,}500$
$4x + 15{,}000 - 3x = 19{,}000$	$30x - 6000 + 3000 - 3x = 10{,}500$
Ans. $x = 4000$	*Ans.* $x = 500$

8. *SOLVING LITERAL EQUATIONS*

Literal equations contain two or more letters.

Thus, $x + y = 20$, $5x = 15a$, $D = RT$ and $2x + 3y - 5z = 12$ are literal equations.

To solve a literal equation for any letter, follow the same procedure used in solving any equation for an unknown.

Thus, to solve for x in $5x = 25a$, divide both sides by 5 to obtain $x = 5a$.

All Formulas are Literal Equations !

Thus, $D = RT$ and $P = 2L + 2W$ are literal equations.

8.1. Solving Literal Equations Using One Operation

Solve for x:

a)
$$\begin{aligned} x - y &= 8 \\ \mathbf{A}_y \quad y &= y \\ \text{Ans. } x &= y + 8 \end{aligned}$$

b)
$$\begin{aligned} x + 10 &= h \\ \mathbf{S}_{10} \quad 10 &= 10 \\ \text{Ans. } x &= h - 10 \end{aligned}$$

c)
$$\begin{aligned} ax &= b \\ \mathbf{D}_a \quad \frac{ax}{a} &= \frac{b}{a} \\ \text{Ans. } x &= \frac{b}{a} \end{aligned}$$

d)
$$\begin{aligned} \frac{x}{a} &= b \\ \mathbf{M}_a \quad a\left(\frac{x}{a}\right) &= a(b) \\ \text{Ans. } x &= ab \end{aligned}$$

8.2. Solving for One of the Letters in a Formula

Solve for the letter indicated:

a) $RT = D$

Solve for R:

$\mathbf{D}_T \quad \frac{RT}{T} = \frac{D}{T}$

Ans. $R = \frac{D}{T}$

b) $S = C + P$

Solve for C:

$\mathbf{Tr} \quad S - P = C$

Ans. $S - P = C$

c) $C = \frac{5}{9}(F - 32)$

Solve for F: ($\mathbf{M}_{9/5}$, first)

$\mathbf{Tr} \quad \frac{9}{5}C = F - 32$

Ans. $\frac{9}{5}C + 32 = F$

d) $A = \frac{1}{2}bh$

Solve for b: ($\mathbf{M}_2$, first)

$\mathbf{D}_h \quad 2A = bh$

Ans. $\frac{2A}{h} = b$

8.3. Solving and Checking Literal Equations

Solve for y and check:

a)
$$\begin{aligned} 2y - 4a &= 8a \\ 2y &= 8a + 4a \\ 2y &= 12a \\ y &= 6a \end{aligned}$$

Check for $y = 6a$:
$$\begin{aligned} 2y - 4a &= 8a \\ 2(6a) - 4a &\stackrel{?}{=} 8a \\ 12a - 4a &\stackrel{?}{=} 8a \\ 8a &= 8a \end{aligned}$$

b)
$$\begin{aligned} 2y - 24a &= 8y \\ 2y - 8y &= 24a \\ -6y &= 24a \\ y &= -4a \end{aligned}$$

Check for $y = -4a$:
$$\begin{aligned} 2y - 24a &= 8y \\ 2(-4a) - 24a &\stackrel{?}{=} 8(-4a) \\ -8a - 24a &\stackrel{?}{=} -32a \\ -32a &= -32a \end{aligned}$$

c)
$$\begin{aligned} 4b + 3y &= 12b + y \\ 3y - y &= 12b - 4b \\ 2y &= 8b \\ y &= 4b \end{aligned}$$

Check for $y = 4b$:
$$\begin{aligned} 4b + 3y &= 12b + y \\ 4b + 3(4b) &\stackrel{?}{=} 12b + 4b \\ 4b + 12b &\stackrel{?}{=} 16b \\ 16b &= 16b \end{aligned}$$

8.4. Solving a Literal Equation for Each Letter

Solve for the letter indicated:

a) $2x = 3y - 4z$

Solve for x:

$\mathbf{D}_2 \quad 2x = 3y - 4z$

Ans. $x = \frac{3y - 4z}{2}$

b) $2x = 3y - 4z$

Solve for y, transposing first:

$\mathbf{D}_3 \quad 2x + 4z = 3y$

Ans. $\frac{2x + 4z}{3} = y$

c) $2x = 3y - 4z$

Solve for z, transposing first:

$\mathbf{D}_4 \quad 4z = 3y - 2x$

Ans. $z = \frac{3y - 2x}{4}$

8.5. Solving More Difficult Literal Equations

Solve for x or y:

a)
$$\begin{aligned} 3(x - 2b) &= 9a - 15b \\ 3x - 6b &= 9a - 15b \\ \mathbf{D}_3 \quad 3x &= 9a - 9b \\ \text{Ans. } x &= 3a - 3b \end{aligned}$$

b) $\frac{y}{5} - h = f$

Transpose first:

$\mathbf{M}_5 \quad \frac{y}{5} = f + h$

Ans. $y = 5f + 5h$

c) $\frac{x}{a} - \frac{b}{5} = \frac{c}{10}$

L.C.D. $= 10a$

$$\begin{aligned} \mathbf{M}_{10a} \quad 10a\left(\frac{x}{a} - \frac{b}{5}\right) &= \left(\frac{c}{10}\right)10a \\ \mathbf{Tr} \quad 10x - 2ab &= ac \\ \mathbf{D}_{10} \quad 10x &= 2ab + ac \\ \text{Ans. } x &= \frac{2ab + ac}{10} \end{aligned}$$

SUPPLEMENTARY PROBLEMS

1. Solve: **(1.1)**

a) $n+3=11$ *Ans.* $n=8$
b) $n+7=20$ *Ans.* $n=13$
c) $12+n=30$ *Ans.* $n=18$
d) $42=n+13$ *Ans.* $n=29$
e) $n+.5=.9$ *Ans.* $n=.4$
f) $n+3.4=5.1$ *Ans.* $n=1.7$
g) $n+1\frac{1}{2}=4\frac{1}{2}$ *Ans.* $n=3$
h) $n+2\frac{1}{8}=5\frac{1}{2}$ *Ans.* $n=3\frac{3}{8}$
i) $2n+5=n+12$ *Ans.* $n=7$
j) $n+18=2n+10$ *Ans.* $n=8$
k) $n^2+n=n^2+15$ *Ans.* $n=15$
l) $2n^2+34=2n^2+n$ *Ans.* $n=34$

2. Solve: **(1.2)**

a) $x-9=15$ *Ans.* $x=24$
b) $x-20=50$ *Ans.* $x=70$
c) $17=x-13$ *Ans.* $x=30$
d) $100=x-41$ *Ans.* $x=141$
e) $x-.3=1.7$ *Ans.* $x=2$
f) $x-5=8.3$ *Ans.* $x=13.3$
g) $x-1\frac{1}{3}=3\frac{2}{3}$ *Ans.* $x=5$
h) $2\frac{1}{4}=x-7\frac{1}{4}$ *Ans.* $x=9\frac{1}{2}$
i) $8-x=0$ *Ans.* $x=8$
j) $10-x=3$ *Ans.* $x=7$
k) $10-x=-4$ *Ans.* $x=14$
l) $12-x^2=x-x^2$ *Ans.* $x=12$

3. Solve: **(1.3)**

a) $3y=15$ *Ans.* $y=5$
b) $4y=30$ *Ans.* $y=7\frac{1}{2}$
c) $36=12y$ *Ans.* $y=3$
d) $6y=3$ *Ans.* $y=\frac{1}{2}$
e) $5y=6.5$ *Ans.* $y=1.3$
f) $.3y=9$ *Ans.* $y=30$
g) $1.1y=8.8$ *Ans.* $y=8$
h) $4=.4y$ *Ans.* $y=10$
i) $1\frac{1}{2}y=1\frac{1}{2}$ *Ans.* $y=1$
j) $2\frac{1}{2}y=10$ *Ans.* $y=4$
k) $15=1\frac{1}{4}y$ *Ans.* $y=12$
l) $14=2\frac{1}{3}y$ *Ans.* $y=6$

4. Solve **(1.4)**

a) $\frac{a}{4}=3$ *Ans.* $a=12$
b) $\frac{a}{5}=1$ *Ans.* $a=5$
c) $12=\frac{a}{2}$ *Ans.* $a=24$
d) $25=\frac{a}{10}$ *Ans.* $a=250$
e) $\frac{1}{3}a=20$ *Ans.* $a=60$
f) $\frac{1}{7}a=12$ *Ans.* $a=84$
g) $30=\frac{1}{8}a$ *Ans.* $a=240$
h) $5.7=\frac{1}{10}a$ *Ans.* $a=57$
i) $\frac{5}{a}=1$ *Ans.* $a=5$
j) $1=\frac{78}{a}$ *Ans.* $a=78$
k) $\frac{a}{.3}=50$ *Ans.* $a=15$
l) $200=\frac{a}{.7}$ *Ans.* $a=140$

5. Solve: **(1.5)**

a) $3x+1=13$ *Ans.* $x=4$
b) $5x+3=33$ *Ans.* $x=6$
c) $8+4x=44$ *Ans.* $x=9$
d) $11+7x=88$ *Ans.* $x=11$
e) $76=10x+6$ *Ans.* $x=7$
f) $45=15+12x$ *Ans.* $x=2\frac{1}{2}$
g) $3.4=1.4+2x$ *Ans.* $x=1$
h) $7.9=3.1+4x$ *Ans.* $x=1.2$
i) $8x+3\frac{1}{4}=19\frac{1}{4}$ *Ans.* $x=2$
j) $20x+5\frac{7}{8}=15\frac{7}{8}$ *Ans.* $x=\frac{1}{2}$
k) $3x+2\frac{1}{4}=6$ *Ans.* $x=1\frac{1}{4}$
l) $4x+3=5\frac{2}{3}$ *Ans.* $x=\frac{2}{3}$

6. Solve (*Use* **A** *and* **D** *or* **A** *and* **M**): **(1.5)**

a) $5b-2=33$ *Ans.* $b=7$
b) $6b-8=25$ *Ans.* $b=5\frac{1}{2}$
c) $89=9b-1$ *Ans.* $b=10$
d) $42=10b-3$ *Ans.* $b=4\frac{1}{2}$
e) $3b-.4=.8$ *Ans.* $b=.4$
f) $11b-.34=.21$ *Ans.* $b=.05$
g) $7b-2\frac{1}{2}=4\frac{1}{2}$ *Ans.* $b=1$
h) $2b-\frac{2}{5}=\frac{4}{5}$ *Ans.* $b=\frac{3}{5}$
i) $\frac{b}{3}-5=9$ *Ans.* $b=42$
j) $\frac{b}{6}-5=10$ *Ans.* $b=90$
k) $\frac{1}{4}b-9=12$ *Ans.* $b=84$
l) $20=\frac{1}{8}b-11$ *Ans.* $b=248$

7. Solve (*Use* **M** *and* **D** *or* **A** *and* **S**): **(1.5)**

a) $\frac{21}{r}=3$ *Ans.* $r=7$
b) $\frac{14}{r}=4$ *Ans.* $r=3\frac{1}{2}$
c) $8=\frac{24}{r}$ *Ans.* $r=3$
d) $12=\frac{3}{r}$ *Ans.* $r=\frac{1}{4}$
e) $\frac{5.4}{r}=3$ *Ans.* $r=1.8$
f) $.4=\frac{10}{r}$ *Ans.* $r=25$
g) $\frac{5}{r}=2\frac{1}{2}$ *Ans.* $r=2$
h) $2\frac{1}{4}=\frac{9}{r}$ *Ans.* $r=4$
i) $30-r=17$ *Ans.* $r=13$
j) $8.4-r=5.7$ *Ans.* $r=2.7$
k) $5.24=8.29-r$ *Ans.* $r=3.05$
l) $17\frac{3}{4}=20\frac{1}{4}-r$ *Ans.* $r=2\frac{1}{2}$

8. Solve: (1.6)

a) $6y = 2y + 16$ *Ans.* $y = 4$
b) $10y = 30 + 5y$ *Ans.* $y = 6$
c) $18 + 2y = 11y$ *Ans.* $y = 2$
d) $3y + 40 = 8y$ *Ans.* $y = 8$
e) $2y + 7y = 72$ *Ans.* $y = 8$
f) $3y + 5y + 6 = 26$ *Ans.* $y = 2\frac{1}{2}$
g) $21 = 10y - 4y$ *Ans.* $y = 3\frac{1}{2}$
h) $24y - 21y = 14$ *Ans.* $y = 4\frac{2}{3}$
i) $13y - 4 = 10y + 2$ *Ans.* $y = 2$
j) $20y + 16 = 30y - 44$ *Ans.* $y = 6$
k) $2y - 1.7 = y + 1.4$ *Ans.* $y = 3.1$
l) $7y - .8 = 4y - .2$ *Ans.* $y = .2$

9. Solve: (1.7)

a) $\frac{1}{5}t = 14$ *Ans.* $t = 70$
b) $20 = \frac{1}{10}t$ *Ans.* $t = 200$
c) $\frac{2}{5}t = 16$ *Ans.* $t = 40$
d) $30 = \frac{6}{7}t$ *Ans.* $t = 35$
e) $\frac{1}{3}t + 9 = 14$ *Ans.* $t = 15$
f) $\frac{1}{4}t - 6 = 8$ *Ans.* $t = 56$
g) $\frac{2}{5}t + 7 = 13$ *Ans.* $t = 15$
h) $\frac{3}{7}t - 6 = 6$ *Ans.* $t = 28$
i) $24 + \frac{1}{8}t = 31$ *Ans.* $t = 56$
j) $17 + \frac{3}{7}t = 29$ *Ans.* $t = 28$
k) $40 = \frac{3}{2}t - 11$ *Ans.* $t = 34$
l) $\frac{8}{7}t + 9 = \frac{3}{7}t + 39$ *Ans.* $t = 42$

10. Four times a number increased by 45 equals seven times the number. Find the number. (1.8)
Ans. 15

11. Ten times a number decreased by 7 equals eight times the number increased by 21. Find the number. *Ans.* 14 (1.8)

12. Two-thirds of a number increased by 10 equals 24. Find the number. *Ans.* 21 (1.8)

13. Solve: (2.1)

a) $n + 11 = 3$ *Ans.* $n = -8$
b) $n + 25 = -5$ *Ans.* $n = -30$
c) $30 + n = 24$ *Ans.* $n = -6$
d) $40 = n + 70$ *Ans.* $n = -30$
e) $n - 5 = -12$ *Ans.* $n = -7$
f) $n - .9 = -2$ *Ans.* $n = -1.1$
g) $n - 3\frac{1}{3} = -8\frac{1}{3}$ *Ans.* $n = -5$
h) $n - \frac{4}{5} = -8$ *Ans.* $n = -7\frac{1}{5}$
i) $n - 20 = 2n - 3$ *Ans.* $n = -17$
j) $2n + 6 = n - 10$ *Ans.* $n = -16$
k) $-n + 2 = 5$ *Ans.* $n = -3$
l) $.34 = .29 - n$ *Ans.* $n = -.05$

14. Solve: (2.2)

a) $3x = -66$ *Ans.* $x = -22$
b) $13x = -130$ *Ans.* $x = -10$
c) $-50 = 15x$ *Ans.* $x = -3\frac{1}{3}$
d) $-7 = 21x$ *Ans.* $x = -\frac{1}{3}$
e) $\frac{x}{5} = -4$ *Ans.* $x = -20$
f) $\frac{x}{2} = -3\frac{1}{2}$ *Ans.* $x = -7$
g) $-4.8 = \frac{x}{10}$ *Ans.* $x = -48$
h) $-.13 = \frac{x}{6}$ *Ans.* $x = -.78$
i) $-7x = 35$ *Ans.* $x = -5$
j) $-\frac{x}{4} = 120$ *Ans.* $x = -480$
k) $3.1x = -31$ *Ans.* $x = -10$
l) $-3\frac{1}{4}x = 130$ *Ans.* $x = -40$

15. Solve: (2.3)

a) $2y + 16 = 2$ *Ans.* $y = -7$
b) $10y + 42 = 37$ *Ans.* $y = -\frac{1}{2}$
c) $3y - 5 = -17$ *Ans.* $y = -4$
d) $20 + 11y = 9$ *Ans.* $y = -1$
e) $6y + 35 = y$ *Ans.* $y = -7$
f) $8y - 20 = 10y$ *Ans.* $y = -10$
g) $13y = 6y - 84$ *Ans.* $y = -12$
h) $3y = 9y + 78$ *Ans.* $y = -13$
i) $\frac{40}{y} = -5$ *Ans.* $y = -8$
j) $-3 = \frac{18}{y}$ *Ans.* $y = -6$
k) $-\frac{20}{y} = 4$ *Ans.* $y = -5$
l) $6 = -\frac{3}{y}$ *Ans.* $y = -\frac{1}{2}$

16. Solve: (2.4)

a) $\frac{y}{4} + 6 = 5$ *Ans.* $y = -4$
b) $8 + \frac{y}{5} = -1$ *Ans.* $y = -45$
c) $30 = 25 - \frac{y}{3}$ *Ans.* $y = -15$
d) $\frac{2y}{3} + 7 = -7$ *Ans.* $y = -21$
e) $12 + \frac{2y}{5} = -8$ *Ans.* $y = -50$
f) $24 = \frac{7y}{5} + 31$ *Ans.* $y = -5$
g) $y - \frac{1}{2}y = -20$ *Ans.* $y = -40$
h) $y + \frac{2}{3}y = -45$ *Ans.* $y = -27$
i) $8\frac{1}{2}y + 6 = 7\frac{3}{4}y$ *Ans.* $y = -8$

17. Solve: (2.5)

a) $4y - 9y + 22 = 3y + 30$ *Ans.* $y = -1$
b) $12y - 10 = 8 + 3y - 36$ *Ans.* $y = -2$
c) $2\frac{1}{2}a + 10 = 4\frac{1}{4}a + 52$ *Ans.* $a = -24$
d) $5.4b - 14 = 8b + 38$ *Ans.* $b = -20$

18. The sum of two numbers represented by n and $2n+8$ is -7. Find the numbers. **(2.6)**
Ans. -5 and -2

19. The sum of three numbers represented by x, $3x$ and $3-2x$ is 1. Find the numbers. **(2.6)**
Ans. -1, -3 and 5

20. Solve, using transposition: **(3.1)**

a) $7-2r=3$	*Ans.* $r=2$	d) $4s-8=16-2s$	*Ans.* $s=4$	g) $6t+t=10+11t$	*Ans.* $t=-2\frac{1}{2}$
b) $27=30-6r$	*Ans.* $r=\frac{1}{2}$	e) $12+s=6s+7$	*Ans.* $s=1$	h) $4t+40-65=-t$	*Ans.* $t=5$
c) $10r+37=-23$	*Ans.* $r=-6$	f) $40-9s=3s+64$	*Ans.* $s=-2$	i) $20+8t=40-22$	*Ans.* $t=-\frac{1}{4}$

21. Solve: **(3.2)**
a) $8a+1+3a=7+9a-12$ *Ans.* $a=-3$ b) $n^2-6n+1=n^2-8n-9$ *Ans.* $n=-5$

22. Solve: **(4.1)**

a) $4(x+1)=20$	*Ans.* $x=4$	e) $6(y-1)=7y-12$	*Ans.* $y=6$	i) $3(z+1)=4(6-z)$	*Ans.* $z=3$
b) $3(x-2)=-6$	*Ans.* $x=0$	f) $30-2(y-1)=38$	*Ans.* $y=-3$	j) $10(2-z)=4(z-9)$	*Ans.* $z=4$
c) $5(7-x)=25$	*Ans.* $x=2$	g) $20+8(2-y)=44$	*Ans.* $y=-1$	k) $6(3z-1)=-7(8+z)$	*Ans.* $z=-2$
d) $42=7(2x-1)$	*Ans.* $x=3\frac{1}{2}$	h) $12y-3=5(2y+1)$	*Ans.* $y=4$	l) $2(z+1)-3(4z-2)=6z$	*Ans.* $z=\frac{1}{2}$

23. Solve: **(4.2)**
a) $3r(2r+4)=2r(3r+8)-12$ *Ans.* $r=3$ b) $(s+3)(s+5)+s(10-s)=11s+1$ *Ans.* $s=-2$

24. Find a number if twice the sum of the number and 4 equals 11 more than the number. **(4.3)**
Ans. 3

25. Find a number such that three times the sum of the number and 2 equals four times the number decreased by 3. *Ans.* 9 **(4.3)**

26. Find a number if 25 minus three times the number equals eight times the difference obtained when 1 is subtracted from the number. *Ans.* 3 **(4.3)**

27. Three boys earned \$60 together. Henry earned \$2 less than Ed and Jack earned twice as much as Henry. Find their earnings. **(4.4)**
Ans. Henry, Ed and Jack earned \$14.50, \$16.50 and \$29 respectively.

28. Solve: **(5.1)**

a) $\frac{3x}{4}=9$	*Ans.* $x=12$	d) $\frac{12}{x}=-3$	*Ans.* $x=-4$	g) $\frac{y}{3}+10=y$	*Ans.* $y=15$
b) $\frac{2x}{5}+8=6$	*Ans.* $x=-5$	e) $7=\frac{84}{x}$	*Ans.* $x=12$	h) $20-\frac{3y}{5}=y-12$	*Ans.* $y=20$
c) $\frac{x}{3}-5=5$	*Ans.* $x=30$	f) $\frac{10}{x}-2=18$	*Ans.* $x=\frac{1}{2}$	i) $\frac{15}{3y}+3=18$	*Ans.* $y=\frac{1}{3}$

29. Solve: **(5.1)**

a) $\frac{2x}{5}+6=\frac{x}{5}$	*Ans.* $x=-30$	c) $\frac{4h}{3}-\frac{5h}{3}=-2$	*Ans.* $h=6$	e) $\frac{3}{r}=2-\frac{7}{r}$	*Ans.* $r=5$
b) $10-\frac{x}{7}=\frac{4x}{7}$	*Ans.* $x=14$	d) $\frac{6h}{5}+6=\frac{2h}{5}$	*Ans.* $h=-7\frac{1}{2}$	f) $\frac{1}{r}+3=\frac{9}{r}-\frac{2}{r}$	*Ans.* $r=2$

30. Solve: **(5.2)**

a) $\frac{3x-1}{7}=2x+3$ b) $\frac{y}{2y-9}=2$ c) $\frac{z-6}{5}+z=\frac{4z+16}{5}$

Ans. a) $x=-2$ b) $y=6$ c) $z=11$

31. Solve: **(6.1)**

a) $\frac{x}{2} - \frac{x}{3} = 7$ *Ans.* $x = 42$ d) $\frac{y}{4} + \frac{y}{3} + \frac{y}{2} = 26$ *Ans.* $y = 24$ g) $\frac{3x}{4} - \frac{2x}{3} = \frac{3}{4}$ *Ans.* $x = 9$

b) $\frac{x}{5} + \frac{x}{6} = 11$ *Ans.* $x = 30$ e) $\frac{y}{5} + \frac{y}{3} - \frac{y}{2} = 3$ *Ans.* $y = 90$ h) $\frac{x}{2} = \frac{3x}{7} - 5$ *Ans.* $x = -70$

c) $\frac{x}{2} = 12 - \frac{x}{4}$ *Ans.* $x = 16$ f) $10 + \frac{y}{6} = \frac{y}{3} - 4$ *Ans.* $y = 84$ i) $\frac{5x}{2} - \frac{2x}{3} = -\frac{11}{6}$ *Ans.* $x = -1$

32. Solve: **(6.2)**

a) $\frac{5}{x} - \frac{2}{x} = 3$ *Ans.* $x = 1$ c) $\frac{1}{x} + \frac{1}{2} = \frac{5}{x}$ *Ans.* $x = 8$ e) $\frac{2}{3x} + \frac{1}{x} = 5$ *Ans.* $x = \frac{1}{3}$

b) $\frac{7}{a} = 2 + \frac{1}{a}$ *Ans.* $a = 3$ d) $\frac{3}{b} + \frac{1}{4} = \frac{2}{b}$ *Ans.* $b = -4$ f) $\frac{3}{4c} = \frac{1}{c} - \frac{1}{4}$ *Ans.* $c = 1$

33. Solve: **(6.1, 6.2)**

a) $\frac{r}{6} = \frac{1}{2}$ *Ans.* $r = 3$ d) $\frac{b+3}{b} = \frac{2}{5}$ *Ans.* $b = -5$ g) $\frac{2c+4}{12} = \frac{c+4}{7}$ *Ans.* $c = 10$

b) $\frac{r}{12} = \frac{3}{4}$ *Ans.* $r = 9$ e) $\frac{b+6}{b} = \frac{7}{5}$ *Ans.* $b = 15$ h) $\frac{6c+3}{11} = \frac{3c}{5}$ *Ans.* $c = 5$

c) $\frac{8}{r} = \frac{4}{3}$ *Ans.* $r = 6$ f) $\frac{2b-12}{b} = \frac{10}{7}$ *Ans.* $b = 21$ i) $\frac{c}{10} = \frac{c-12}{6}$ *Ans.* $c = 30$

34. Solve: **(6.3)**

a) $\frac{x-2}{3} - \frac{x+1}{4} = 4$ *Ans.* $x = 59$ c) $\frac{w-2}{4} - \frac{w+4}{3} = -\frac{5}{6}$ *Ans.* $w = -12$

b) $\frac{y-3}{5} - 1 = \frac{y-5}{4}$ *Ans.* $y = -7$ d) $1 - \frac{2m-5}{3} = \frac{m+3}{2}$ *Ans.* $m = 1$

35. Solve: **(6.4)**

a) $\frac{d+8}{d-2} = \frac{9}{4}$ *Ans.* $d = 10$ c) $\frac{8}{y-2} - \frac{13}{2} = \frac{3}{2y-4}$ *Ans.* $y = 3$

b) $\frac{4}{x-4} = \frac{7}{x+2}$ *Ans.* $x = 12$ d) $\frac{10}{r-3} + \frac{4}{3-r} = 6$ *Ans.* $r = 4$

36. Solve: **(7.1, 7.3)**

a) $.5d = 3.5$ *Ans.* $d = 7$ d) $3.1c = .42 + c$ *Ans.* $c = .2$ g) $x - 4.2x = .8x - 12$ *Ans.* $x = 3$

b) $.05e = 4$ *Ans.* $e = 80$ e) $6d - 10 = 3.5d$ *Ans.* $d = 4$ h) $x + .4x + 8 = -20$ *Ans.* $x = -20$

c) $60 = .3f$ *Ans.* $f = 200$ f) $8.6m + 3 = 7.1m$ *Ans.* $m = -2$ i) $x - .125x - 1.2 = 19.8$ *Ans.* $x = 24$

37. Solve: **(7.2)**

a) $6a = 3.3$ *Ans.* $a = .55$ d) $\frac{2}{3}b = 7.8$ *Ans.* $b = 11.7$ g) $4c - 6.6 = c - .18$ *Ans.* $c = 2.14$

b) $7a = 7.217$ *Ans.* $a = 1.031$ e) $8b - 4 = .32$ *Ans.* $b = .54$ h) $3c - 2.6 = c + 5$ *Ans.* $c = 3.8$

c) $\frac{a}{2} = 12.45$ *Ans.* $a = 24.9$ f) $\frac{b}{5} + .05 = 1.03$ *Ans.* $b = 4.9$ i) $4c + .8 = c + .44$ *Ans.* $c = -.12$

38. Solve: **(7.4)**

a) $33\frac{1}{3}\%$ of $x = 24$ *Ans.* $x = 72$ d) $x + 20\%x = 30$ *Ans.* $x = 25$ g) $x - 75\%x = 70$ *Ans.* $x = 280$

b) $16\frac{2}{3}\%$ of $x = 3.2$ *Ans.* $x = 19.2$ e) $2x + 10\%x = 7$ *Ans.* $x = 3\frac{1}{3}$ h) $2x - 50\%x = -9$ *Ans.* $x = -6$

c) 70% of $x = 140$ *Ans.* $x = 200$ f) $x = 40.5 - 35\%x$ *Ans.* $x = 30$ i) $x + 10 = 4 + 87\frac{1}{2}\%x$ *Ans.* $x = -48$

39. Solve: **(7.5, 7.6)**

a) $.2(x+5) = 10$ *Ans.* $x = 45$

b) $4(x - .3) = 12$ *Ans.* $x = 3.3$

c) $.03(x + 200) = 45$ *Ans.* $x = 1300$

d) $50 - .05(x - 100) = 20$ *Ans.* $x = 700$

e) $.03y + .02(5000 - y) = 140$ *Ans.* $y = 4000$

f) $.05y - .03(600 - y) = 14$ *Ans.* $y = 400$

g) $.1(1000 - x) + .07(2000 - x) = 104$ *Ans.* $x = 800$

h) $250 - .3(x + 100) = .5(600 - x) - 40$ *Ans.* $x = 200$

40. Solve for x: (8.1)

a) $x - b = 3b$ d) $ax = -10a$ g) $\frac{x}{a} = 10b$

b) $x - 5a = 20$ e) $bx = b^3$ h) $\frac{x}{4b} = \frac{b}{2}$

c) $x + 10c = c + 8$ f) $2cx = -8cd$ i) $a + b = \frac{x}{3}$

Ans. a) $x = 4b$ d) $x = -10$ g) $x = 10ab$

b) $x = 5a + 20$ e) $x = b^2$ h) $x = 2b^2$

c) $x = -9c + 8$ f) $x = -4d$ i) $x = 3a + 3b$

41. Solve for the letter indicated: (8.2)

a) $LW = A$ for L d) $A = \frac{1}{2}bh$ for h g) $A = \frac{1}{2}h(b+b')$ for h

b) $RP = I$ for P e) $V = \frac{1}{3}Bh$ for B h) $s = \frac{1}{2}at^2$ for a

c) $P = S - C$ for S f) $S = 2\pi rh$ for r i) $F = \frac{9}{5}C + 32$ for C

Ans. a) $L = \frac{A}{W}$ d) $h = \frac{2A}{b}$ g) $h = \frac{2A}{b + b'}$

b) $P = \frac{I}{R}$ e) $B = \frac{3V}{h}$ h) $a = \frac{2s}{t^2}$

c) $S = P + C$ f) $r = \frac{S}{2\pi h}$ i) $C = \frac{5}{9}(F - 32)$

42. Solve for x or y: (8.3)

a) $2x = 6a + 22a$ d) $ax - 3a = 5a$ g) $ay - 6b = 3ay$

b) $3y - a = -10a$ e) $\frac{x}{3} + b = 7b$ h) $\frac{y}{m} + 3m = 4m$

c) $\frac{x}{4} + b = -b$ f) $\frac{2x}{3} - 4c = 10c$ i) $\frac{y}{r} - 5 = r + 2$

Ans. a) $x = 14a$ d) $x = 8$ g) $y = -\frac{3b}{a}$

b) $y = -3a$ e) $x = 18b$ h) $y = m^2$

c) $x = -8b$ f) $x = 21c$ i) $y = r^2 + 7r$

43. Solve for the letter indicated: (8.4)

a) $x - 10 = y$ for x d) $2x + 3y = 10$ for y g) $x + y = z$ for x

b) $2y = 6x + 8$ for y e) $2x + 3y = 12$ for x h) $\frac{x}{3} - 2y = 3z$ for x

c) $x + 2y = 20$ for x f) $\frac{x}{2} + 8 = y$ for x i) $ax - by = cz$ for x

Ans. a) $x = y + 10$ d) $y = \frac{10 - 2x}{3}$ g) $x = z - y$

b) $y = 3x + 4$ e) $x = \frac{12 - 3y}{2}$ h) $x = 6y + 9z$

c) $x = 20 - 2y$ f) $x = 2y - 16$ i) $x = \frac{by + cz}{a}$

44. Solve for x or y: (8.5)

a) $5(x+a) = 10(x-2a)$ *Ans.* $x = 5a$ d) $\frac{x}{6} - \frac{a}{3} = \frac{b}{2}$ *Ans.* $x = 2a + 3b$

b) $\frac{x}{3} + b = c - 4b$ *Ans.* $x = 3c - 15b$ e) $3(5-y) = 7(y-b)$ *Ans.* $y = \frac{7b + 15}{10}$

c) $\frac{x}{12} = \frac{a}{6} + b$ *Ans.* $x = 2a + 12b$ f) $\frac{y}{3} = \frac{y}{4} + c$ *Ans.* $y = 12c$

Chapter 6

FORMULAS

1. UNDERSTANDING POLYGONS, CIRCLES AND SOLIDS

A. Understanding Polygons in General

A **polygon** is a closed figure in a plane (flat surface) bounded by straight lines.

Names of Polygon According to the Number of Sides

No. of Sides	Polygon
3	Triangle
4	Quadrilateral
5	Pentagon
6	Hexagon

No. of Sides	Polygon
8	Octagon
10	Decagon
12	Dodecagon
n	n-gon

An **equilateral polygon** is a polygon having equal sides.
Thus, a square is an equilateral polygon.

An **equiangular polygon** is a polygon having equal angles.
Thus, a rectangle is an equiangular polygon.

A **regular polygon** is an equilateral and equiangular polygon.
Thus, a regular pentagon is a 5-sided equilateral and equiangular polygon.

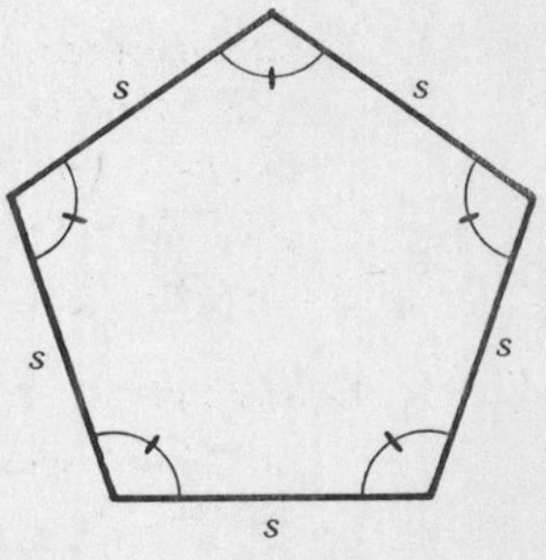

Regular Pentagon

Symbols for Equal Sides and Equal Angles

Using the same letter for sides indicates they are equal. Equal angles may be shown by using arcs crossed by the same number of strokes, as in the adjacent diagram.

B. Understanding Circles

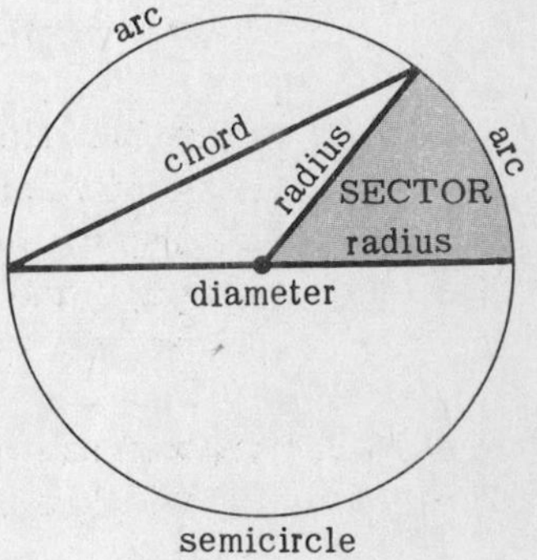

The **circumference** is the distance around a circle.

A **radius** is a line joining the center to a point on the circumference. All radii of a circle are equal.

A **chord** is a line joining any two points on the circumference.

A **diameter** is a chord through the center. A diameter is twice a radius. It is the largest chord.

An **arc** is a part of the circumference of a circle.

A **sector** is a part of the area of a circle bounded by an arc and two radii.

C. Understanding Triangles

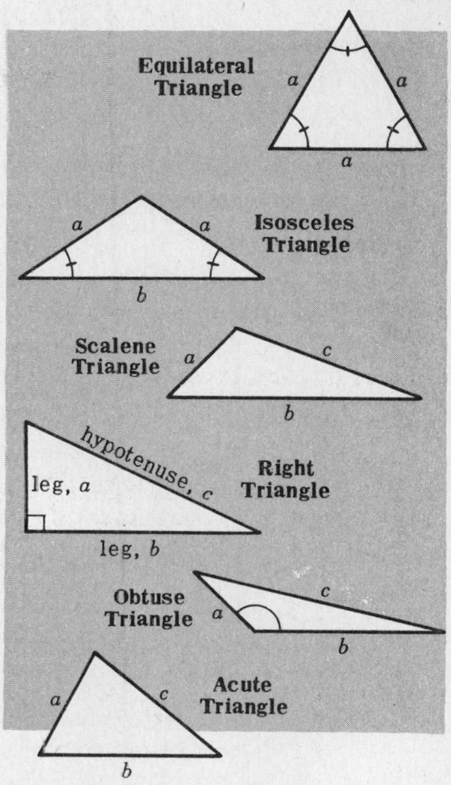

1. An **equilateral triangle** has three equal sides. It also has three equal angles, each 60°.

2. An **isosceles triangle** has at least two equal sides. It also has at least two equal angles. The equal angles shown lie along the base (b) and are called the base angles.

3. A **scalene triangle** has no equal sides.

4. A **right triangle** has one right angle.
   Its **hypotenuse** is opposite the right angle.
   Its **legs** (or **arms**) are the other two sides.
   The symbol for the right angle is a square corner.

5. An **obtuse triangle** has one obtuse angle (more than 90° and less than 180°.)

6. An **acute triangle** has all acute angles (less than 90°).

D. Understanding Quadrilaterals

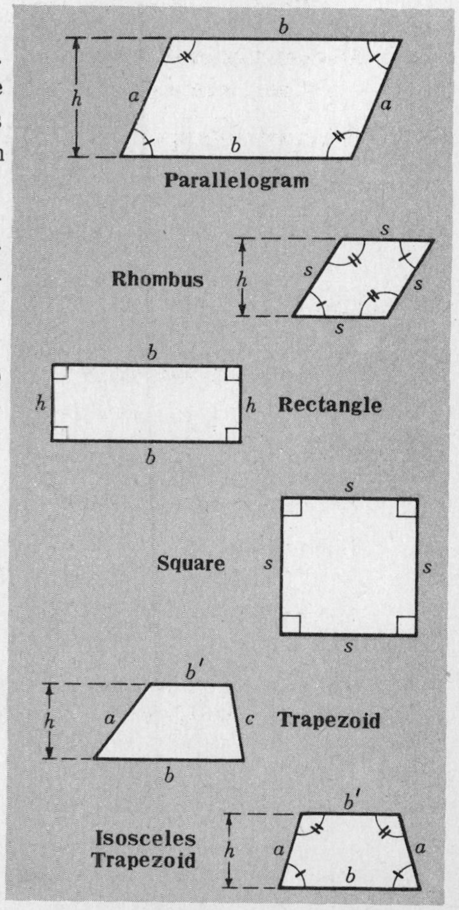

1. A **parallelogram** has two pairs of parallel sides. Its opposite sides and its opposite angles are equal. The distance between the two bases is h. This distance is at right angles to both bases.

2. A **rhombus** has four equal sides. Its opposite angles are equal. It is an equilateral parallelogram.

3. A **rectangle** has four right angles. Its bases (b) and its heights (h) are equal. It is an equiangular parallelogram.

4. A **square** has four equal sides and four right angles. It is an equilateral and equiangular parallelogram. A **square unit** is a square whose side is 1 unit. Thus, a square foot is a square whose side is 1 foot.

5. A **trapezoid** has one and only one pair of parallel sides. The unequal bases are represented by b and b'.

6. An **isosceles trapezoid** has two equal legs (nonparallel sides). Note the equal angles.

E. Understanding Solids

1. A **solid** is an enclosed portion of space bounded by plane and curved surfaces.

   Thus, the **pyramid**  , the **cube**  , the **cone**  , the **cylinder**  and the **sphere**  are solids.

2. A **polyhedron** is a solid bounded by plane (flat) surfaces only. Thus, the pyramid and cube are polyhedrons. The cone, cylinder and sphere are not polyhedrons since each has a curved surface. The faces of a polyhedron are its bounding polygons. The edges of a polyhedron are the sides of its faces.

3. A **prism** is a polyhedron two of whose faces are parallel polygons and whose remaining faces are parallelograms. The bases of a prism are its parallel polygons. These may have any number of sides. The lateral (side) faces are its parallelograms. The distance between the two bases is h. This line is at right angles to each base.

   A **right prism** is a prism whose lateral faces are rectangles. The distance, h, is the height of any of the lateral faces.

Prism

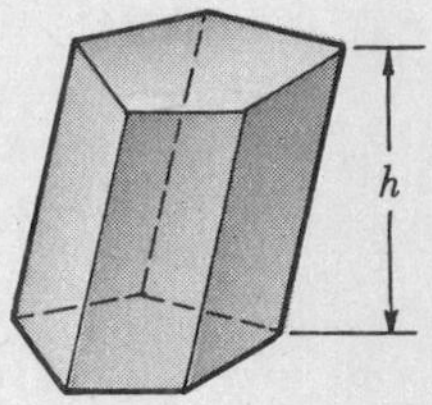

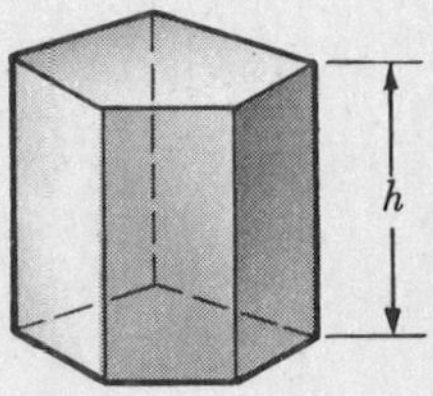

Right Prism

4. A **rectangular solid** (box) is a prism bounded by six rectangles. The rectangular solid can be formed from the pattern of six rectangles folded along the dotted lines. The length (l), the width (w) and the height (h) are its dimensions.

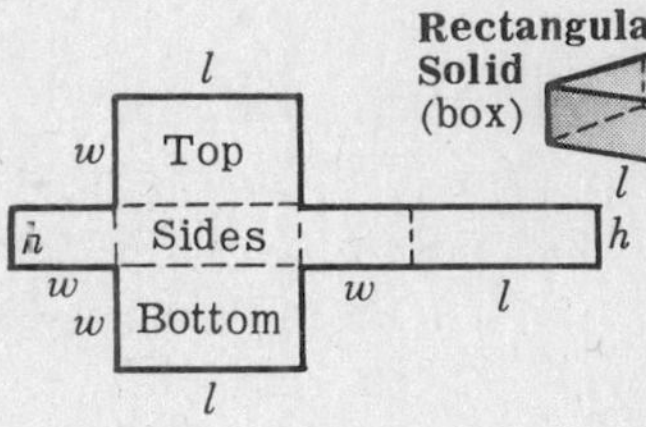

Rectangular Solid (box)

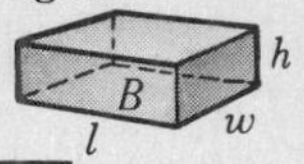

5. A **cube** is a rectangular solid bounded by six squares. The cube can be formed from the pattern of six squares folded along the dotted lines. Each equal dimension is represented by e in the diagram.

   A **cubic unit** is a cube whose edge is 1 unit. Thus, a cubic inch is a cube whose edge is 1 inch.

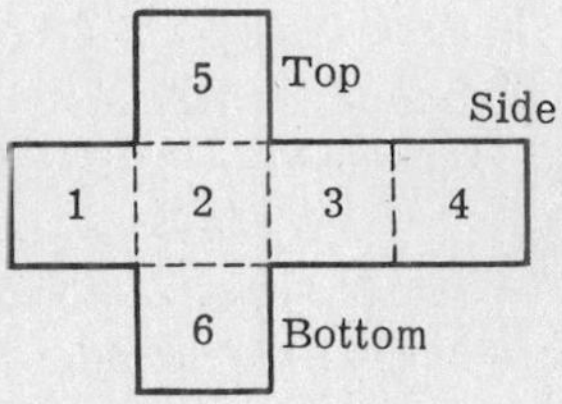

Cube

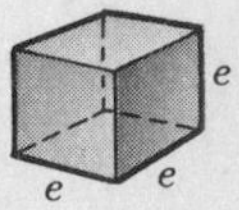

6. A **pyramid** is a polyhedron whose base is a polygon and whose other faces meet at a point, its vertex. The base (B) may have any number of sides. However, the other faces must be triangles. The distance from the vertex to the base is equal to the altitude or height (h), a line from the vertex at right angles to the base.

   A **regular pyramid** is a pyramid whose base is a regular polygon and whose altitude joins the vertex and the center of the base.

Pyramid

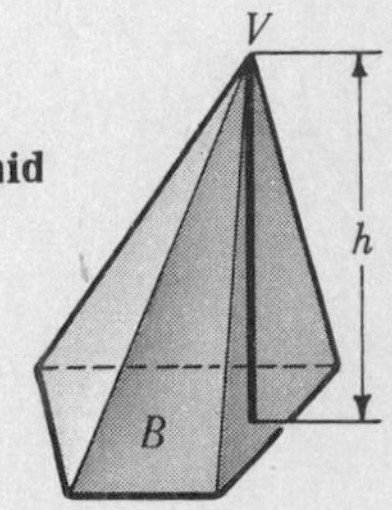

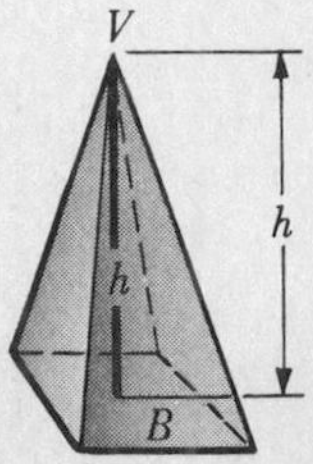

Regular Pyramid (square base)

7. A **circular cone** is a solid whose base is a circle and whose lateral surface comes to a point. (*A circular cone will be referred to as a cone.*)

A **cone of revolution** is formed by revolving a right triangle about one of its legs. This leg becomes the altitude or height (h) of the cone and the other becomes the radius (r) of the base.

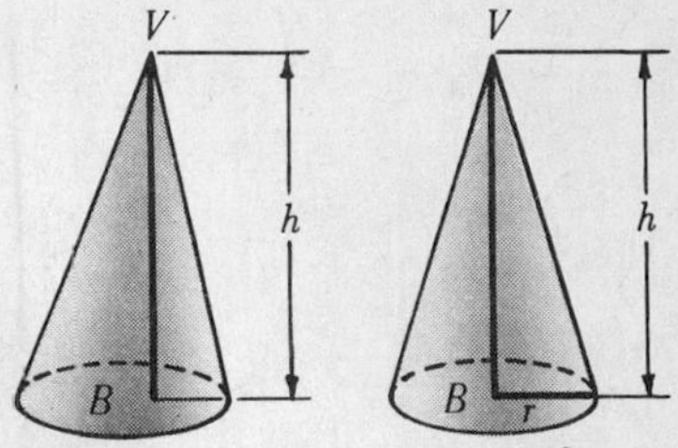

Cone   Cone of Revolution

8. A **sphere** is a solid such that every point on its surface is at an equal distance from the same point, its center.

A sphere is formed by revolving a semicircle about its diameter.

Sphere
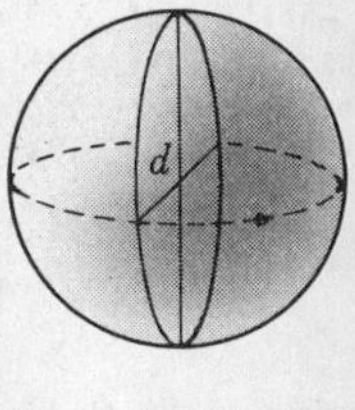

9. A **circular cylinder** is a solid whose bases are parallel circles. Any cross-section parallel to the bases is also a circle. (*A circular cylinder will be referred to as a cylinder.*)

A **cylinder of revolution** is formed by revolving a rectangle about one of its two dimensions. This dimension becomes the height (h) of the cylinder and the other becomes the radius (r) of the base.

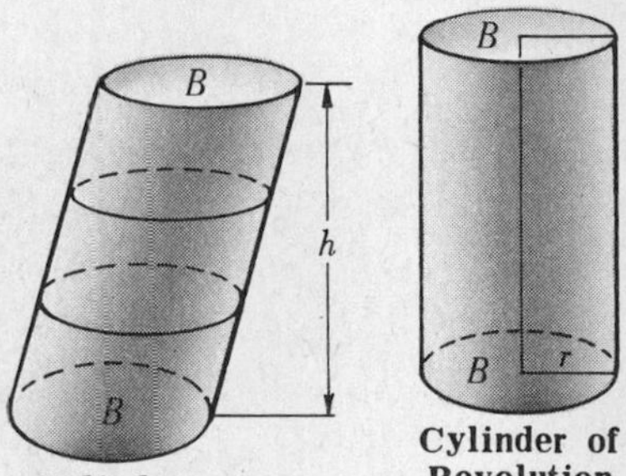

Cylinder   Cylinder of Revolution

2. FORMULAS FOR PERIMETERS AND CIRCUMFERENCES: LINEAR MEASURE

A formula is an equality expressing in mathematical symbols a numerical rule or relationship among quantities.

Thus, $S = C + P$ is a formula for the rule: **selling price equals cost plus profit.**

In perimeter formulas, the perimeter is in the same unit as the dimensions. Thus, if the side of a square is 3 yards, the perimeter is 12 yards.

The **perimeter of a polygon** is the distance around it. Thus, the perimeter (p) of the triangle shown is the sum of its three sides; that is,

$$p = a + b + c$$

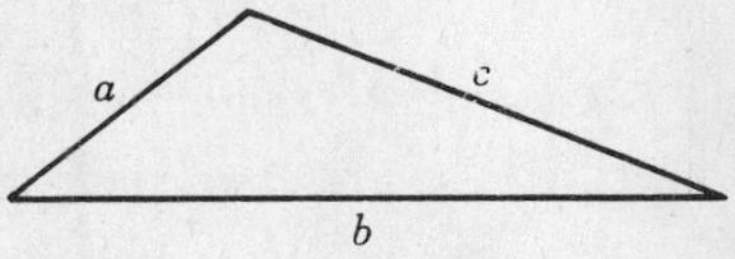

The **circumference of a circle** is the distance around it. For any circle, the circumference (c) is π times the diameter (d); that is, $c = \pi d$.

The value of π, using 5 digits, is 3.1416. If less accuracy is needed, 3.142, 3.14 or $\frac{22}{7}$ may be used.

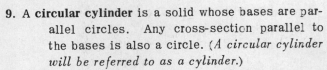

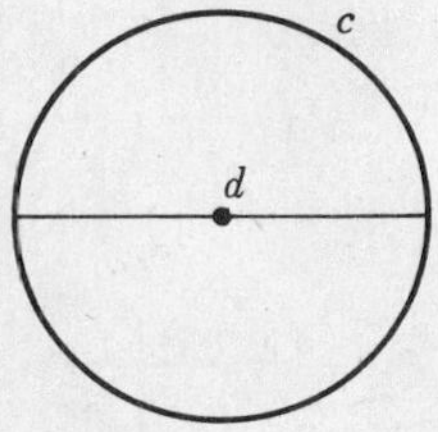

2.1. Perimeter Formulas for Triangles and Quadrilaterals

State the formula for the perimeter (p) of each, using the letters shown on page 87:

a) equilateral triangle	*Ans.* $p = 3a$	*d*) square	*Ans.* $p = 4s$
b) isosceles triangle	*Ans.* $p = 2a + b$	*e*) rectangle	*Ans.* $p = 2b + 2h$
c) scalene triangle	*Ans.* $p = a + b + c$	*f*) obtuse triangle	*Ans.* $p = a + b + c$

2.2. Perimeter Formulas for Polygons

State the name of the equilateral polygon to which each perimeter formula applies:

a) $p = 3s$ *Ans.* equilateral triangle

b) $p = 4s$ *Ans.* square or rhombus

c) $p = 10s$ *Ans.* equilateral decagon

d) $p = 12s$ *Ans.* equilateral dodecagon

2.3. Finding a Side of an Equilateral Polygon

If each of the following polygons has a perimeter of 36 in., find a side:

a) square

b) equilateral triangle

c) equilateral hexagon

d) equilateral decagon

Solutions:

a) $p = 4s$
$36 = 4s$
$9 = s$
Ans. 9 in.

b) $p = 3s$
$36 = 3s$
$12 = s$
Ans. 12 in.

c) $p = 6s$
$36 = 6s$
$6 = s$
Ans. 6 in.

d) $p = 10s$
$36 = 10s$
$3.6 = s$
Ans. 3.6 in.

2.4. Finding Perimeters of Regular Polygons

Find the perimeter of

a) an equilateral triangle with a side of 5 in.

b) a square with a side of $4\frac{1}{2}$ ft.

c) a regular hexagon with a side of 2 yd. 1 ft.

Solutions:

a)

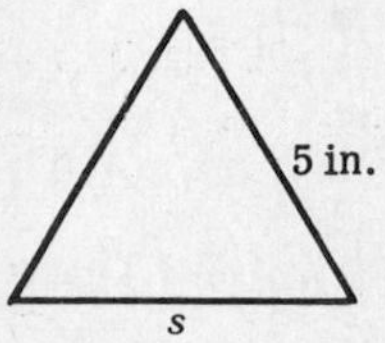

$p = 3s$

Let s = no. of in. in side = 5

$p = 3(5) = 15$

Ans. 15 in.

b)

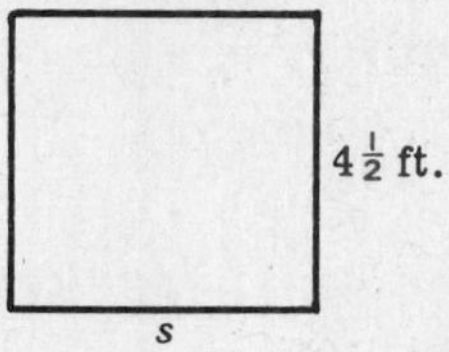

$p = 4s$

Let s = no. of ft. in side = $4\frac{1}{2}$

$p = 4(4\frac{1}{2}) = 18$

Ans. 18 ft.

c)

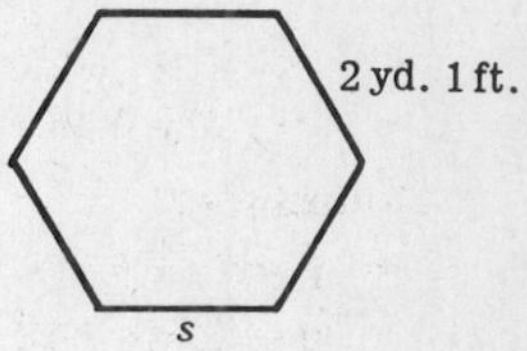

$p = 6s$

Let s = no. of yd. in 1 side = $2\frac{1}{3}$

$p = 6(2\frac{1}{3}) = 14$

Ans. 14 yd. or 42 ft.

2.5. Finding Perimeters of Quadrilaterals

Find the perimeter of

a) a rectangle with sides of 4 ft. and $1\frac{1}{2}$ ft.

b) a parallelogram with sides of 4 yd. and 2 ft.

Solutions:

a)

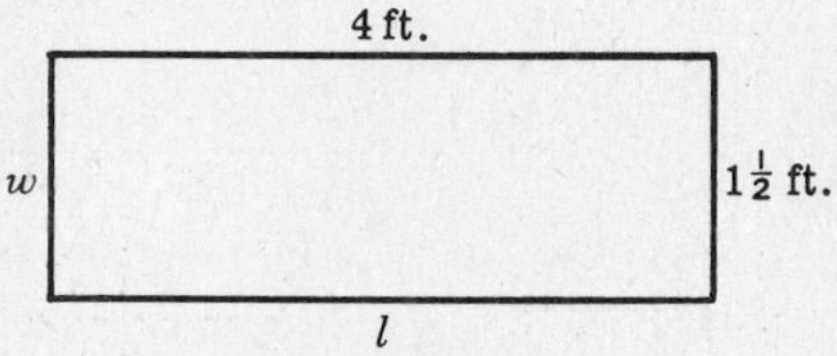

$p = 2l + 2w$

Let l = no. of ft. in length = 4

and w = no. of ft. in width = $1\frac{1}{2}$

$p = 2(4) + 2(1\frac{1}{2}) = 11$

Ans. 11 ft.

b)

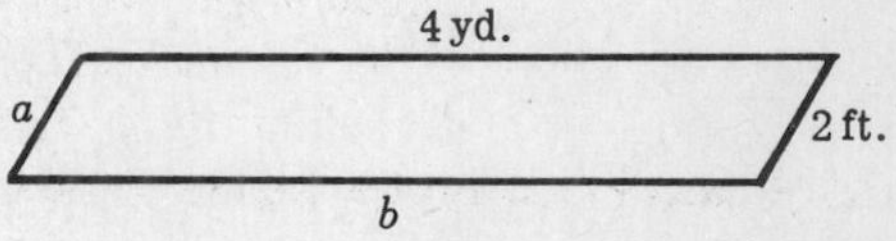

$p = 2a + 2b$

Let a = no. of yd. in 1 side = $\frac{2}{3}$

and b = no. of yd. in other side = 4

$p = 2(\frac{2}{3}) + 2(4) = 9\frac{1}{3}$

Ans. $9\frac{1}{3}$ yd. or 28 ft.

2.6. Finding the Perimeter of a Rectangle

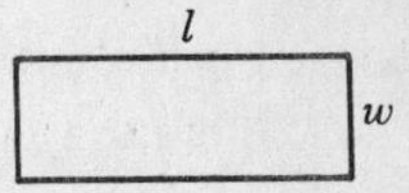

If l, w and p are in inches, find the perimeter of the rectangle shown if

a) $l = 10$, $w = 3$ b) $l = 3\frac{1}{2}$, $w = 4\frac{1}{4}$ c) $l = 3.1$, $w = 2.6$

Solutions:

a) $p = 2l + 2w$
$p = 2(10) + 2(3)$
$p = 26$
Ans. 26 in.

b) $p = 2l + 2w$
$p = 2(3\frac{1}{2}) + 2(4\frac{1}{4})$
$p = 15\frac{1}{2}$
Ans. $15\frac{1}{2}$ in.

c) $p = 2l + 2w$
$p = 2(3.1) + 2(2.6)$
$p = 11.4$
Ans. 11.4 in.

2.7. Finding the Length or Width of a Rectangle

For a rectangle, if l, w and p are in ft., find

a) the length if $p = 20$, $w = 3$
b) the width if $p = 27$, $l = 5\frac{1}{2}$
c) the width if $p = 40$ and $l = w + 5$
d) the length if $p = 30$ and $w = 3l$

Solutions:

a) $p = 2l + 2w$
$20 = 2l + 6$
$7 = l$
Ans. 7 ft.

b) $p = 2l + 2w$
$27 = 2(5\frac{1}{2}) + 2w$
$8 = w$
Ans. 8 ft.

c) $p = 2l + 2w$
$40 = 2(w+5) + 2w$
$7\frac{1}{2} = w$
Ans. $7\frac{1}{2}$ ft.

d) $p = 2l + 2w$
$30 = 2l + 2(3l)$
$3\frac{3}{4} = l$
Ans. $3\frac{3}{4}$ ft.

2.8. Perimeter of an Isosceles Triangle

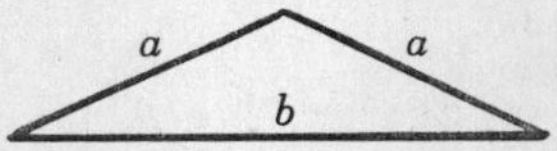

For the isosceles triangle shown, if a, b and p are in inches find

a) the perimeter if $a = 10$ and $b = 12$
b) the base if $p = 25$ and $a = 8$
c) the base if $p = 33$ and $a = b - 3$
d) an equal side if $p = 35$ and $b = 1.5a$

Solutions:

a) $p = 2a + b$
$p = 2(10) + 12$
$p = 32$
Ans. 32 in.

b) $p = 2a + b$
$25 = 2(8) + b$
$9 = b$
Ans. 9 in.

c) $p = 2a + b$
$33 = 2(b - 3) + b$
$13 = b$
Ans. 13 in.

d) $p = 2a + b$
$35 = 2a + 1.5a$
$10 = a$
Ans. 10 in.

2.9. Circumference and Arc Formulas

For any circle, state a formula which relates

a) the diameter (d) and radius (r) *Ans.* $d = 2r$
b) the circumference (c) and the diameter (d) *Ans.* $c = \pi d$
c) the circumference (c) and the radius (r) *Ans.* $c = 2\pi r$
d) the circumference (c) and an arc of 90° (a) *Ans.* $c = 4a$
e) an arc of 60° (a') and the radius (r) *Ans.* $a' = \frac{\pi r}{3}$
(Read a' as "a-prime".)

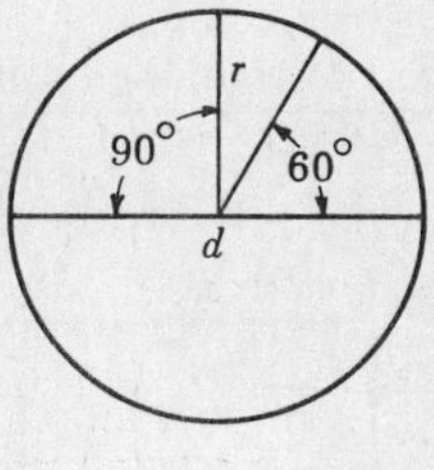

2.10. Circumference of a Circle

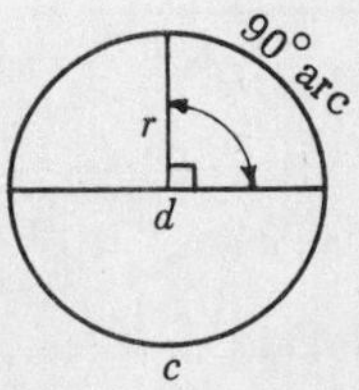

For a circle, if r, d and c are in ft., find, using $\pi = 3.14$,

a) the circumference if $r = 5$
b) a 90° arc if $d = 8$
c) the radius if $c = 942$
d) the diameter if $c = 1570$

Solutions:

a) $c = 2\pi r$
$= 2(3.14)5$
$= 31.4$
Ans. 31.4 ft.

b) 90° arc $= \frac{c}{4} = \frac{\pi d}{4}$
$= \frac{(3.14)8}{4}$
$= 6.28$
Ans. 6.28 ft.

c) $c = 2\pi r$
$942 = 2(3.14)r$
$150 = r$
Ans. 150 ft.

d) $c = \pi d$
$1570 = 3.14d$
$500 = d$
Ans. 500 ft.

2.11. Perimeters of Combined Figures

State the formula for the perimeter, p, for each figure:

(sq.= square, *each curved figure is a semi-circle*)

a)

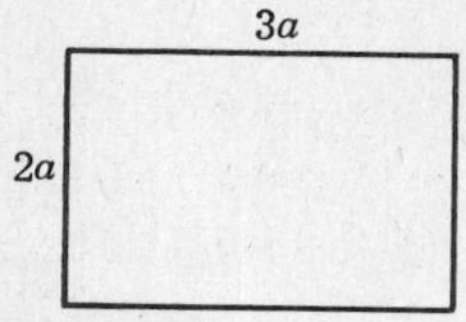

b)

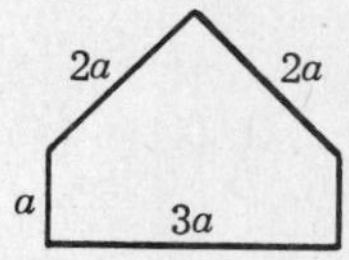

c)

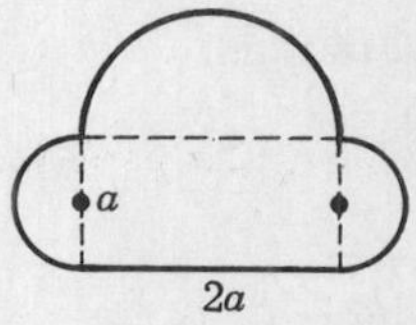

d)

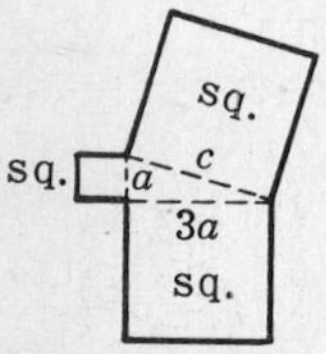

e)

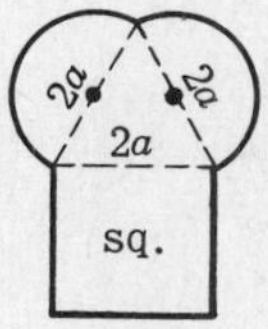

f)

(equilateral triangles around a regular hexagon)

Ans. *a*) $p = 10a$ *b*) $p = 9a$ *c*) $p = 2a + 2\pi a$

d) $p = 12a + 3c$ *e*) $p = 6a + 2\pi a$ *f*) $p = 12a$

3. FORMULAS FOR AREAS: SQUARE MEASURE

A **square unit** is a square whose side is 1 unit. Thus, a square inch is a square whose side is 1 inch.

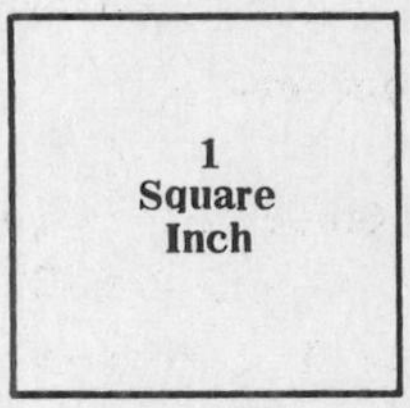

1 inch

The **area of a polygon or circle** is the number of square units contained in its surface. Thus, the area of a rectangle 5 units long and 4 units wide contains 20 square units.

In area formulas, the area is in **square units**, the unit being the same as that used for the dimensions. Thus, if the side of a square is 3 yards, its area is 9 square yards.

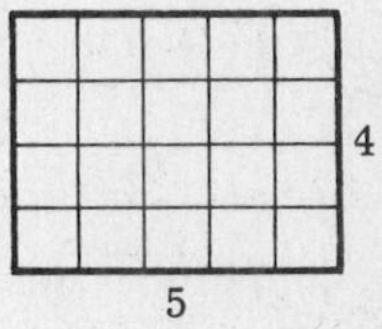

Area Formulas, using A for the area of the figure:

1. **Rectangle:** $A = bh$

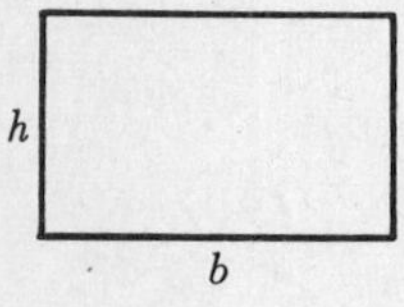

2. **Parallelogram:** $A = bh$

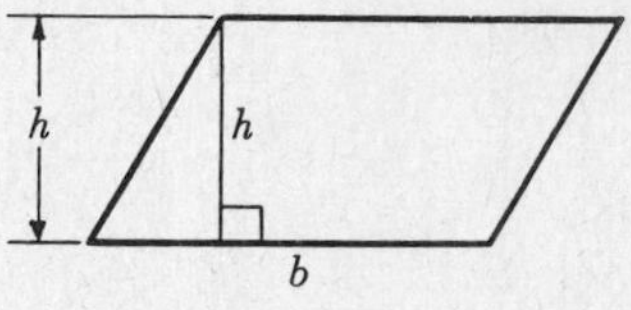

3. **Triangle:** $A = \frac{bh}{2}$

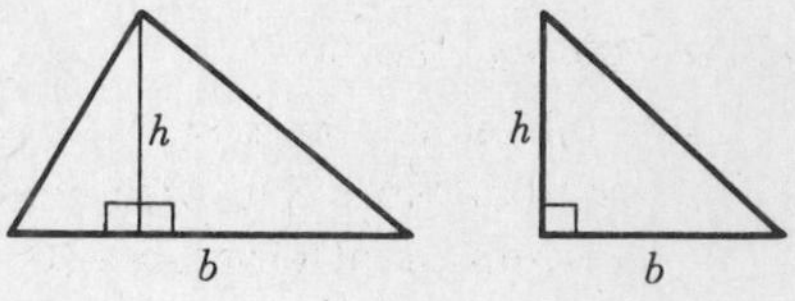

4. **Square:** $A = s^2$

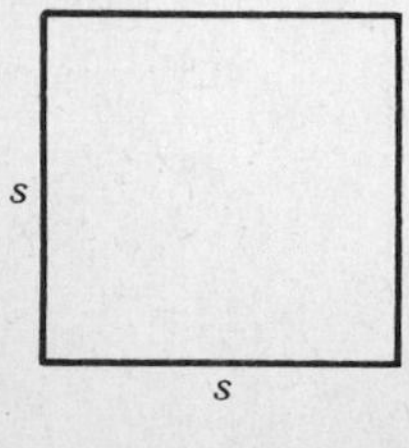

5. **Trapezoid:** $A = \frac{h}{2}(b+b')$

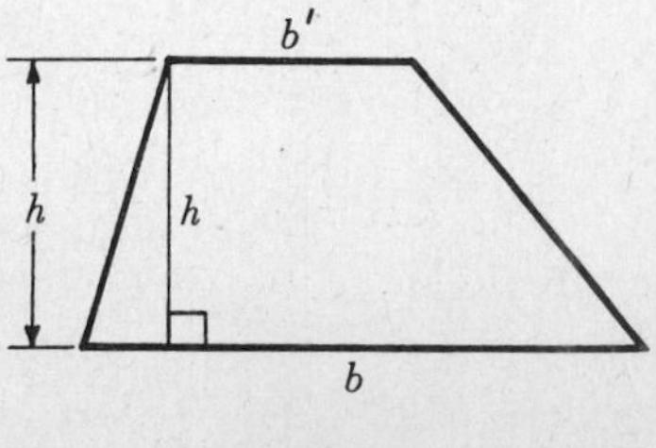

6. **Circle:** $A = \pi r^2$. Also, $A = \frac{\pi d^2}{4}$

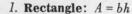

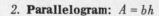

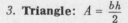

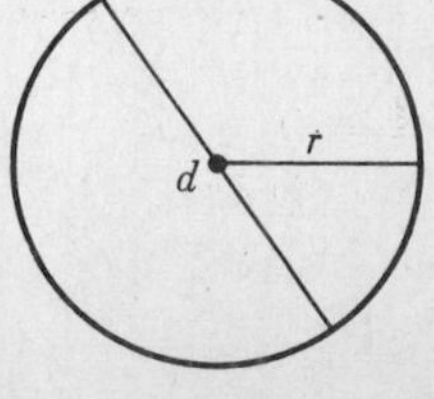

Area Formulas for Solids, using T for the total area of the solid:

1. Total area of the six squares of a **cube**:

$$T = 6e^2$$

$A = e^2$ for each face.

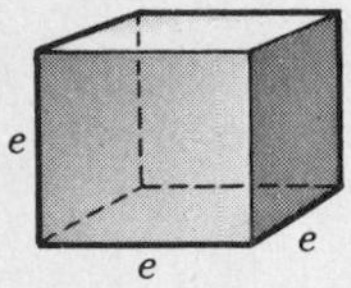

2. Total area of the six rectangles of a **rectangular solid**: $T = 2lw + 2lh + 2wh$

$A = lw$ for top or bottom faces
$A = lh$ for front or back faces
$A = wh$ for left or right faces

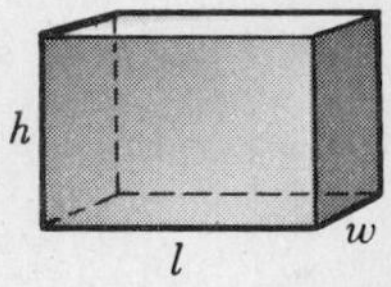

3. Total area of a **sphere**:

$$A = 4\pi r^2$$

4. Total area of a **cylinder of revolution**:

$$T = 2\pi rh + 2\pi r^2$$
$$T = 2\pi r(r+h)$$

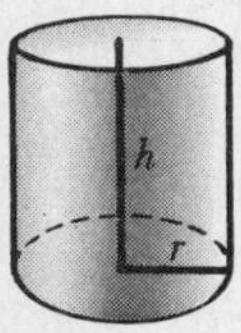

3.1. Relations Among Square Units

Find the area (A) of (*Hint*: A square unit is a square whose side is 1 unit.)
a) a square foot in square inches,
b) a square yard in square feet,
c) a square meter in square centimeters.

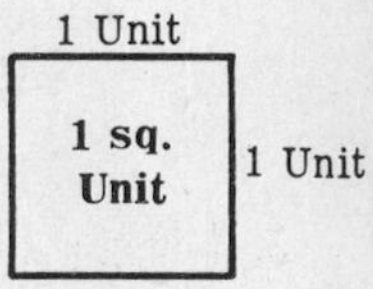

Solutions:

a) $A = s^2$
Since 1 ft. = 12 in.,
$A = 12^2 = 144$
Ans. 1 sq. ft. = 144 sq. in.

b) $A = s^2$
Since 1 yd. = 3 ft.,
$A = 3^2 = 9$
Ans. 1 sq. yd. = 9 sq. ft.

c) $A = s^2$
Since 1 meter = 100 cm.,
$A = 100^2 = 10{,}000$
Ans. 1 sq. meter = 10,000 sq. cm.

3.2. Finding the Area of Squares

Find the area of a square (A) in sq. ft. whose side is
a) 6 in. *b*) 6 ft. *c*) 6 yd. *d*) 6 rd.

Solutions: (*To find area in sq. ft., express side in ft.*)

a) $A = s^2$
Since 6 in. = $\frac{1}{2}$ ft.,
$A = (\frac{1}{2})^2 = \frac{1}{4}$
Ans. $\frac{1}{4}$ sq. ft.

b) $A = s^2$
$A = 6^2$
$A = 36$
Ans. 36 sq. ft.

c) $A = s^2$
Since 6 yd. = 18 ft.,
$A = 18^2 = 324$
Ans. 324 sq. ft.

d) $A = s^2$
Since 6 rd. = $6(16\frac{1}{2})$ or 99 ft.,
$A = 99^2 = 9801$
Ans. 9801 sq. ft.

3.3. Finding Areas

Find the area of
a) a rectangle with sides of 4 ft. and $2\frac{1}{2}$ ft.,
b) a parallelogram with a base of 5.8 in. and a height of 2.3 in.,
c) a triangle with a base of 4 ft. and an altitude to the base of 3 ft. 6 in.

Solutions:

a)

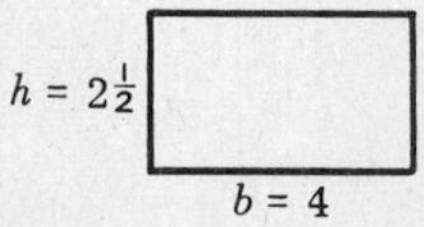

$A = bh$
$A = 4(2\frac{1}{2}) = 10$
Ans. 10 sq. ft.

b)

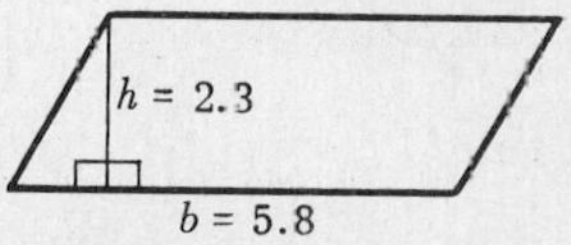

$A = bh$
$A = (5.8)(2.3) = 13.34$
Ans. 13.34 sq. in.

c)

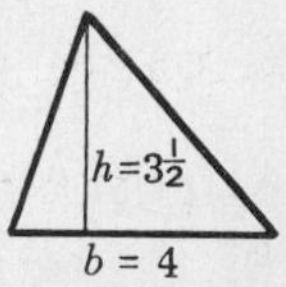

$A = \frac{1}{2}bh$
$A = \frac{1}{2}(4)(3\frac{1}{2}) = 7$
Ans. 7 sq. ft.

3.4. Area of Circle Formulas

For a circle, state a formula which relates

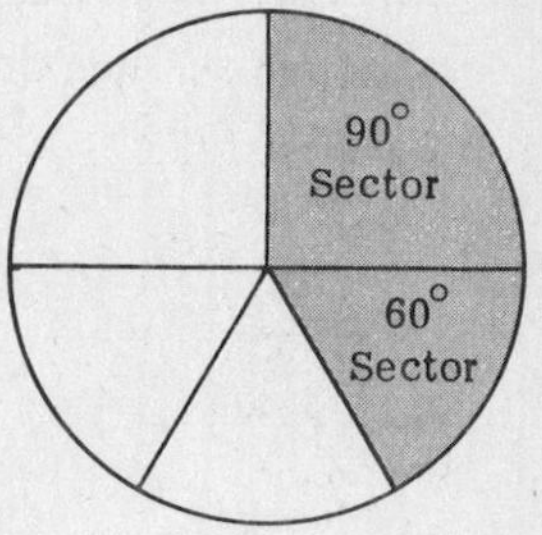

a) the area (A) and the radius (r) *Ans.* $A = \pi r^2$

b) the area (A) and the diameter (d) *Ans.* $A = \dfrac{\pi d^2}{4}$

c) the area (A) and a sector of 90° (S) *Ans.* $A = 4S$

d) a sector of 60° (S') and the area (A) *Ans.* $S' = \dfrac{A}{6}$

e) a sector of 40° (S'') and the radius (r) *Ans.* $S'' = \dfrac{\pi r^2}{9}$
(Read S'' as "S-double prime".)

3.5. Area of a Circle

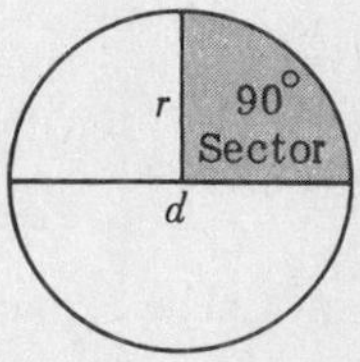

For a circle, if r and d are in inches, find, *to the nearest integer*, the area of

a) the circle if $r = 20$ *c*) a 90° sector if $r = 4$

b) the circle if $d = 10$ *d*) a 30° sector if $d = 12.2$

Solutions: Let $\pi = 3.14$.

a)	*b*)	*c*)	*d*)
$A = \pi r^2$	$A = \pi r^2$	$A = \pi r^2$	$A = \pi r^2$
$A = 3.14(20^2)$	$A = 3.14(5^2)$	$A = (3.14)4^2$	$A = (3.14)(6.1^2)$
$= 3.14(400)$	$= 3.14(25)$	90° sector $= \frac{1}{4}(3.14)(16)$	30° sector $= \frac{1}{12}(3.14)(37.21)$
$= 1256$	$= 78.5$	$= 12.56$	$= 9.7366$
Ans. 1256 sq. in.	*Ans.* 79 sq. in.	*Ans.* 13 sq. in.	*Ans.* 10 sq. in.

3.6. Formulas for Combined Areas

State the formula for the area (A) of each shaded figure:

a)

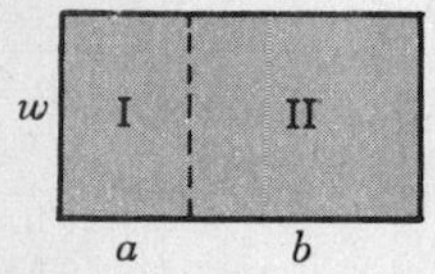

b)

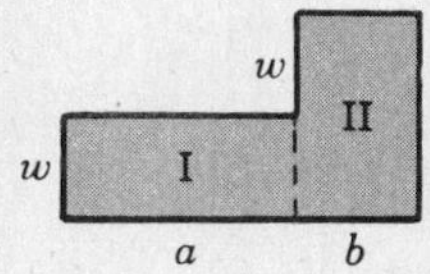

c)

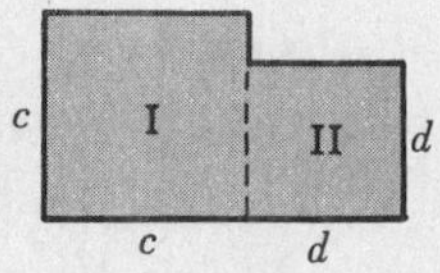

d)

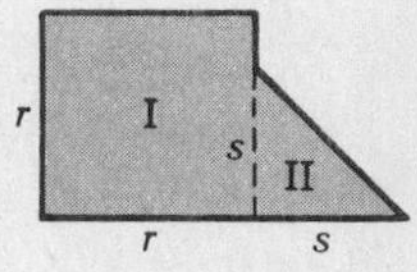

Solutions: (*Abbreviations used are* rect. *for* rectangle, sq. *for* square *and* Δ *for* triangle.)

a)	*b*)	*c*)	*d*)
$A =$ rect.I + rect.II	$A =$ rect.I + rect.II	$A =$ sq.I + sq.II	$A =$ sq.I + ΔII
Ans. $A = aw + bw$	*Ans.* $A = aw + 2bw$	*Ans.* $A = c^2 + d^2$	*Ans.* $A = r^2 + \dfrac{s^2}{2}$

3.7. Formulas for Reduced Areas

State the formula for the area (A) of each shaded figure:

a)

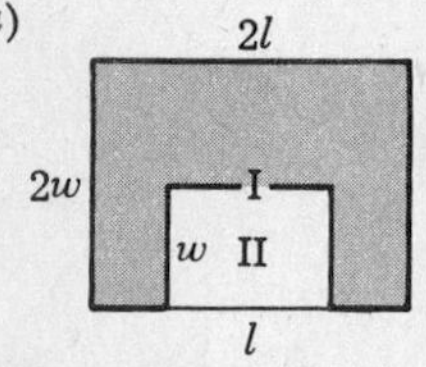

b)

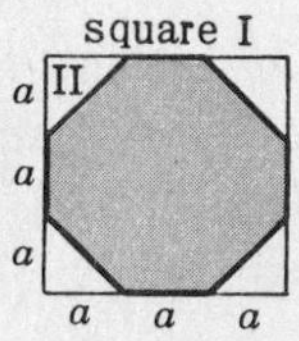

c)

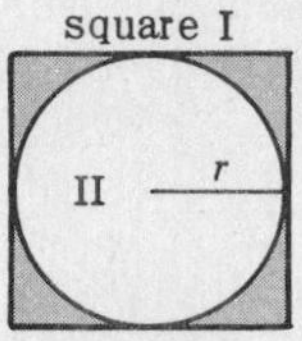

d)

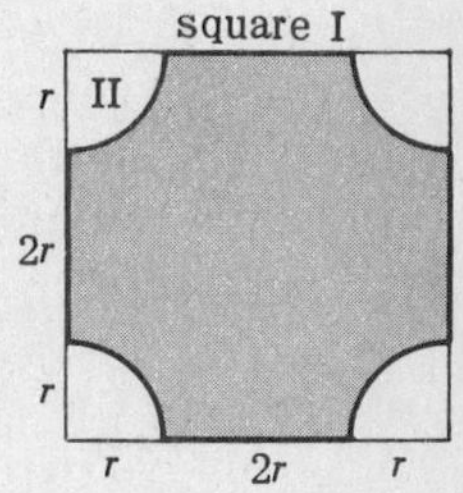

Solutions: (⊙ is used for circle.)

a)	*b*)	*c*)	*d*)
$A =$ rect.I − rect.II	$A =$ sq.I − 4ΔII	$A =$ sq.I − ⊙II	$A =$ sq.I − 4 sector II
$A = 4lw - lw$	$A = (3a)^2 - 4\left(\dfrac{a^2}{2}\right)$	$A = (2r)^2 - \pi r^2$	$A = (4r)^2 - 4\left(\dfrac{\pi r^2}{4}\right)$
Ans. $A = 3lw$	*Ans.* $A = 7a^2$	*Ans.* $A = 4r^2 - \pi r^2$	*Ans.* $A = 16r^2 - \pi r^2$

3.8. Finding Total Areas of Solids

Find, to the nearest integer, the total area of

a) a cube with an edge of 5 in.

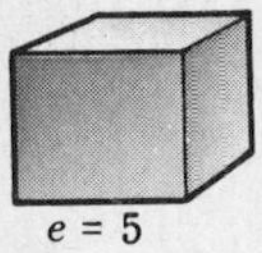

b) a rectangular solid with dimensions of 10 ft., 7 ft. and $4\frac{1}{2}$ ft.

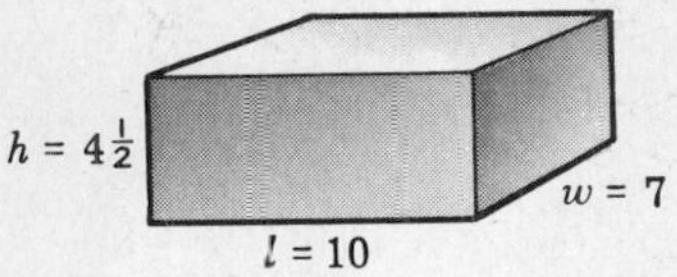

c) a sphere with a radius of 1.1 yd.

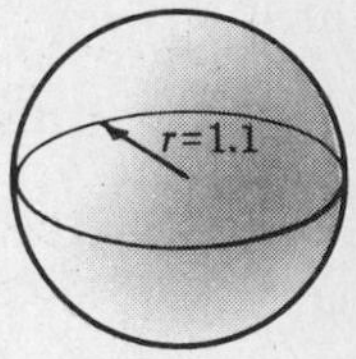

Solutions:

a) $T = 6e^2$
$T = 6(5^2)$
$= 150$
Ans. 150 sq. in.

b) $T = 2lw + 2lh + 2wh$
$T = 2(10)(7) + 2(10)(4\frac{1}{2}) + 2(7)(4\frac{1}{2})$
$= 293$
Ans. 293 sq. ft.

c) $T = 4\pi r^2$
$T = 4(3.14)(1.1^2)$
$= 15.1976$
Ans. 15 sq. yd.

4. FORMULAS FOR VOLUMES: CUBIC MEASURE

A **cubic unit** is a cube whose edge is 1 unit. Thus, a cubic inch is a cube whose side is 1 inch.

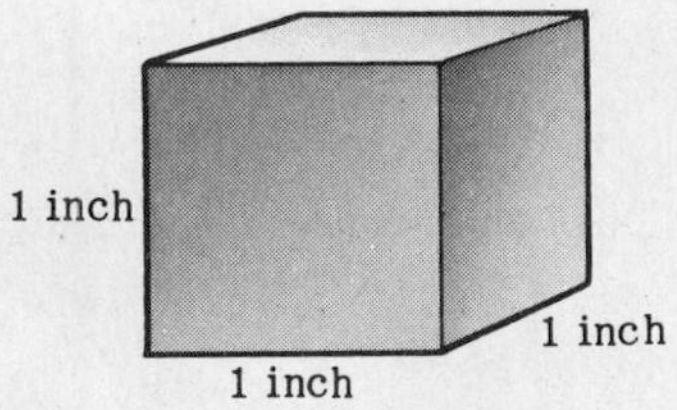

1 Cubic Inch

The **volume of a solid** is the number of cubic units that it contains. Thus, a box 5 units long, 3 units wide and 4 units high has a volume of 60 cubic units; that is, it has a capacity or space large enough to contain 60 cubes, 1 unit on a side.

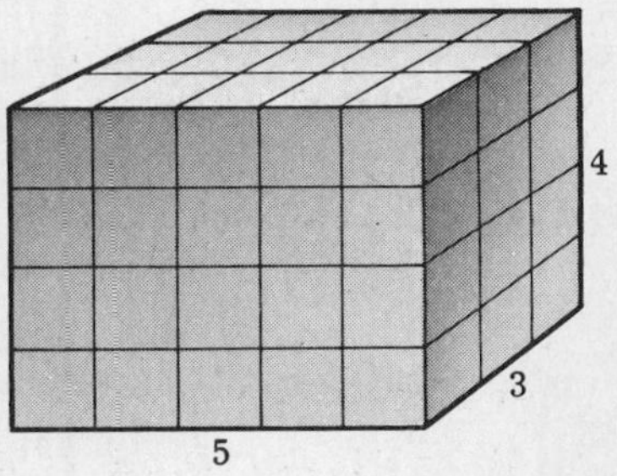

In volume formulas, the volume is in **cubic units**, the unit being the same as that used for the dimensions. Thus, if the edge of a cube is 3 yards, its volume is 27 cubic yards.

Volume Formulas, using V for the volume of the solid, B for the area of a base and h for the distance between the bases or between the vertex and a base:

1. **Rectangular Solid:** $V = lwh$

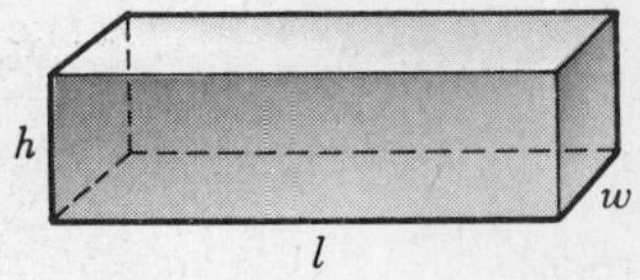

2. **Prism:** $V = Bh$

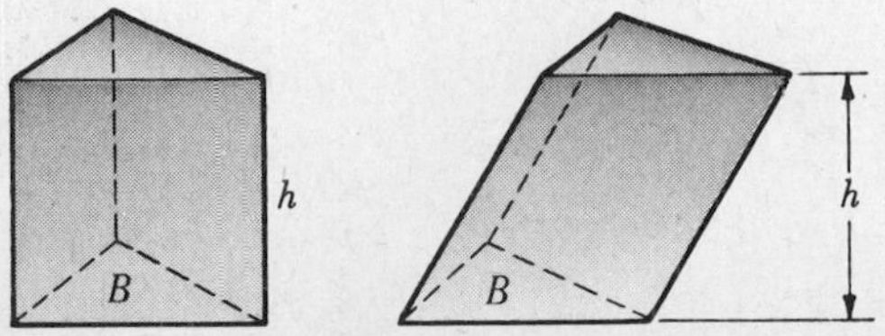

3. **Cylinder:** $V = Bh$ or $V = \pi r^2 h$

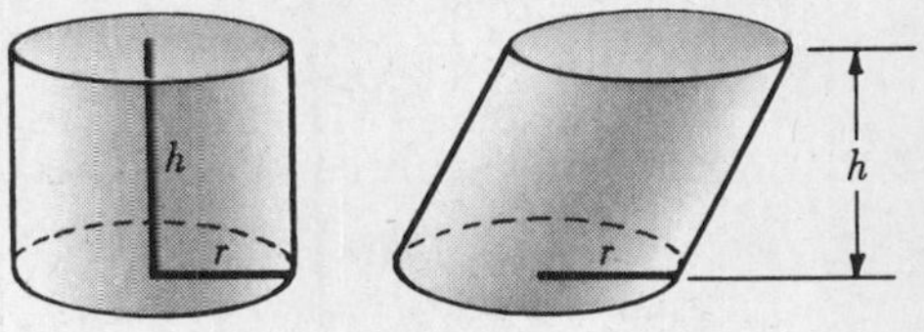

4. **Cube:** $V = e^3$

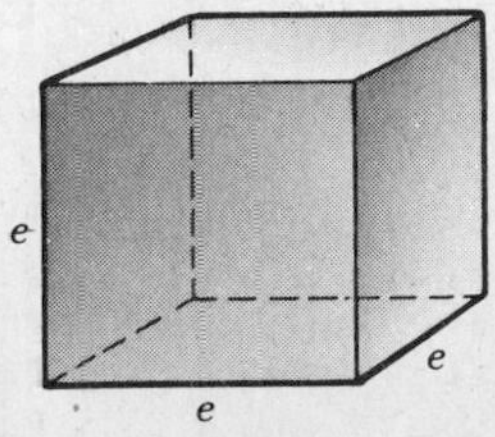

5. **Pyramid:** $V = \frac{1}{3}Bh$

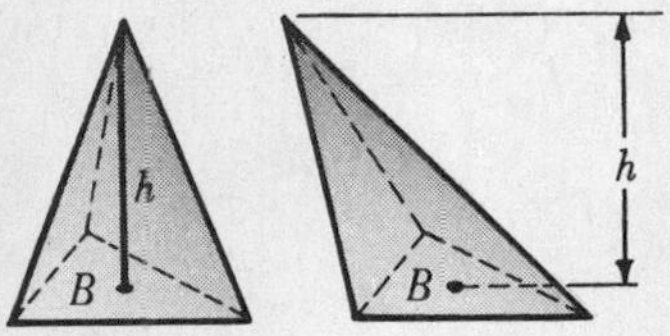

6. **Cone:** $V = \frac{1}{3}Bh$ or $V = \frac{1}{3}\pi r^2 h$

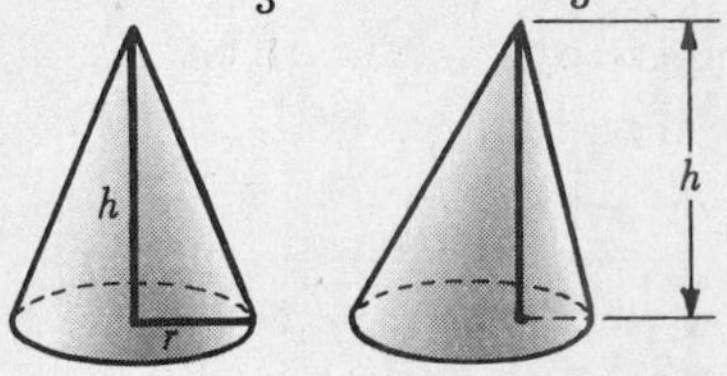

7. **Sphere:** $V = \frac{4}{3}\pi r^3$

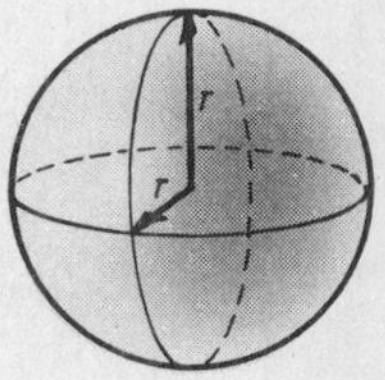

4.1. Relations Among Cubic Units

Find the volume V of (*Hint*: A cubic unit is a cube whose edge is 1 unit.)

a) a cubic foot in cubic inches
b) a cubic yard in cubic feet
c) a liter (cubic decimeter) in cubic centimeters.

1 Cubic Unit

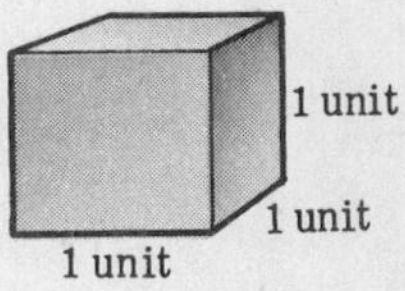

Solutions:

a) $V = e^3$
Since 1 ft. = 12 in.,
$V = 12^3 = 1728$
Ans. 1 cu. ft. = 1728 cu. in.

b) $V = e^3$
Since 1 yd. = 3 ft.,
$V = 3^3 = 27$
Ans. 1 cu. yd. = 27 cu. ft.

c) $V = e^3$
Since 1 cu. dm. = 10 cu. cm.,
$V = 10^3 = 1000$
Ans. 1 liter = 1000 cu. cm.

4.2. Finding Volumes of Cubes

Find the volume of a cube (V) in cu. ft. whose edge is *a*) 4 in., *b*) 4 ft., *c*) 4 yd., *d*) 4 rd.

Solutions: (To find volume in cu. ft., express side in ft.)

a) $V = e^3$
Since 4 in. = $\frac{1}{3}$ ft.,
$V = (\frac{1}{3})^3 = \frac{1}{27}$
Ans. $\frac{1}{27}$ cu. ft.

b) $V = e^3$
$V = 4^3 = 64$
Ans. 64 cu. ft.

c) $V = e^3$
Since 4 yd. = 12 ft.,
$V = 12^3 = 1728$
Ans. 1728 cu. ft.

d) $V = e^3$
Since 4 rd. = $4(16\frac{1}{2})$ or 66 ft.
$V = 66^3 = 287{,}496$
Ans. 287,496 cu. ft.

4.3. Finding Volumes of Rectangular Solid, Prism and Pyramid

Find the volume of

a) a rectangular solid having a length of 6 in., a width of 4 in. and a height of 1 ft.
b) a prism having a height of 15 yd. and a triangular base of 120 sq. ft.
c) a pyramid having a height of 8 yd. and a square base whose side is $4\frac{1}{2}$ yd.

Solutions:

a) $V = lwh$
$V = 6(4)(12) = 288$
Ans. 288 cu. in.

b) $V = Bh$
$V = 120(45) = 5400$
Ans. 5400 cu. ft. or 200 cu. yd.

c) $V = \frac{1}{3}Bh$
$V = \frac{1}{3}(\frac{9}{2})^2(8) = 54$
Ans. 54 cu. yd.

4.4. Finding Volumes of Sphere, Cylinder and Cone

Find the volume, to the nearest integer, of

a) a sphere with a radius of 10 in.
b) a cylinder with a height of 4 yd. and a base whose radius is 2 ft.
c) a cone with a height of 2 ft. and a base whose radius is 2 yd.

Solutions: (Let $\pi = 3.14$)

a) $V = \frac{4}{3}\pi r^3$
$= \frac{4}{3}(3.14)10^3$
$= 4186\frac{2}{3}$
Ans. 4187 cu. in.

b) $V = \pi r^2 h$
$= (3.14)(2^2)12$
$= 150.72$
Ans. 151 cu. ft.

c) $V = \frac{1}{3}\pi r^2 h$
$= \frac{1}{3}(3.14)(6^2)(2)$
$= 75.36$
Ans. 75 cu. ft.

4.5. Deriving Formulas from V = Bh

From $V = Bh$, the volume formula for a prism or cylinder, derive the volume formulas for each of the following:

a)

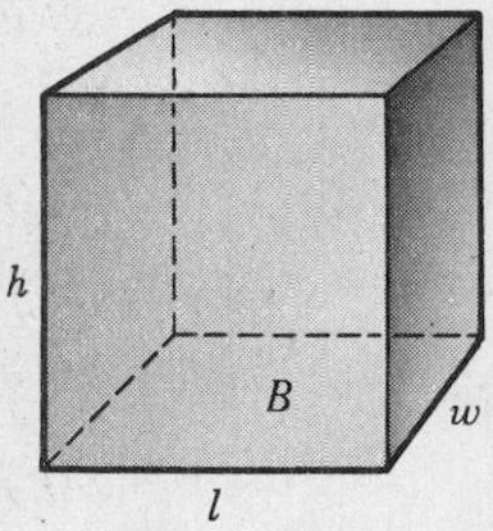

Rectangular Solid

b)

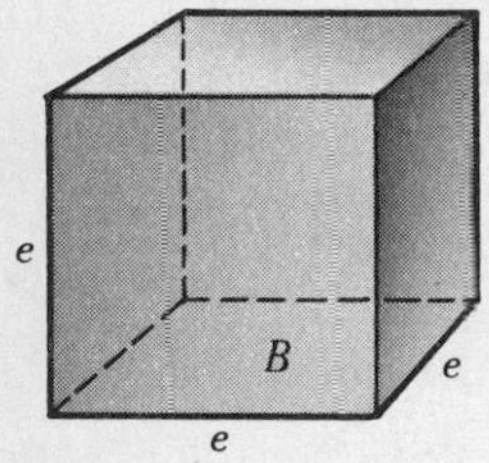

Cube

c)

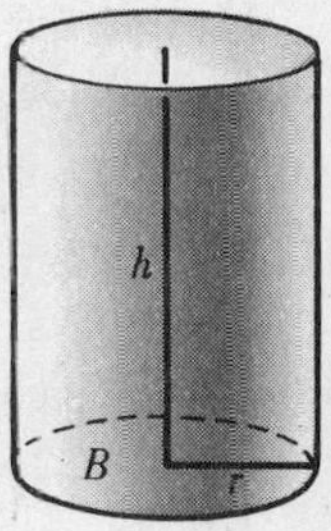

Cylinder of Revolution

d)

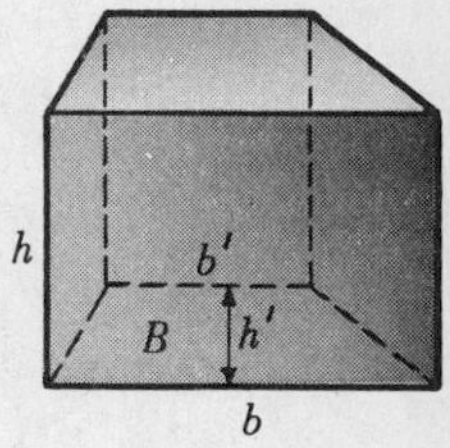

Right Prism with a Trapezoid for a Base

Solutions:

a) $V = Bh$
Since $B = lw$,
$V = (lw)h$
Ans. $V = lwh$

b) $V = Bh$
Since $B = e^2$,
and $h = e$,
$V = (e^2)e$
Ans. $V = e^3$

c) $V = Bh$
Since $B = \pi r^2$,
$V = (\pi r^2)h$
Ans. $V = \pi r^2 h$

d) $V = Bh$
Since $B = \frac{h'}{2}(b+b')$,
$V = \frac{h'}{2}(b+b')h$
Ans. $V = \frac{hh'}{2}(b+b')$

4.6. Formulas for Combined Volumes

State the formula for the volume of each solid:

a)

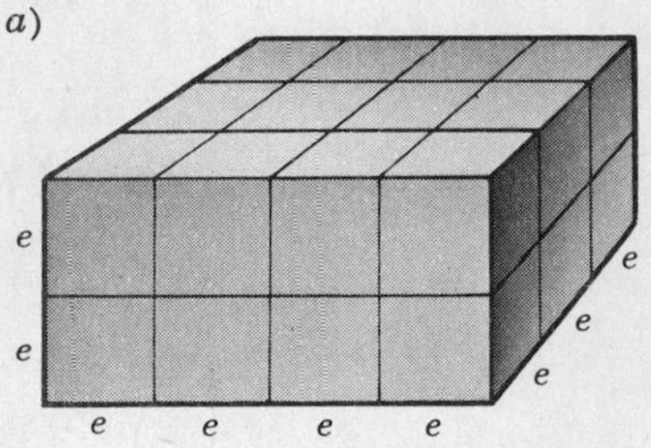

b)

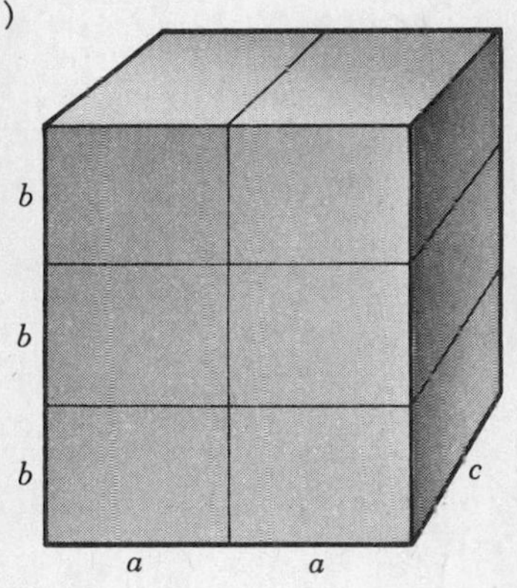

c)

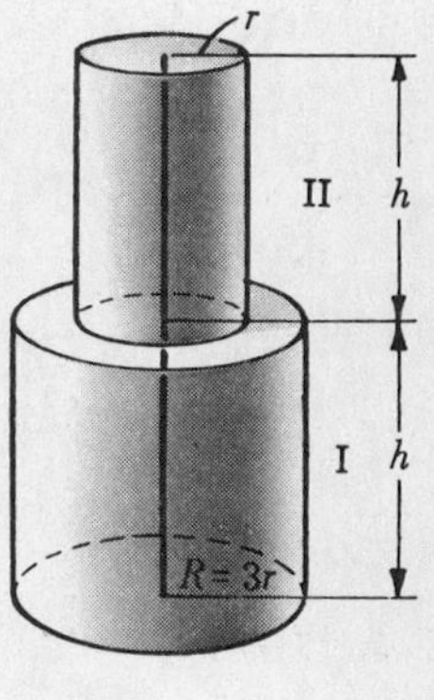

Solutions:

a) $V = lwh$
Now, $l = 4e$, $w = 3e$, $h = 2e$
Hence, $V = (4e)(3e)(2e)$
Ans. $V = 24e^3$

b) $V = lwh$
Now, $l = 2a$, $w = c$, $h = 3b$
Hence, $V = (2a)(c)(3b)$
Ans. $V = 6abc$

c) $V = \text{cyl.I} + \text{cyl.II}$
$V = \pi R^2 h + \pi r^2 h$
Now, $R = 3r$
Hence, $V = \pi(3r)^2 h + \pi r^2 h$
Ans. $V = 10\pi r^2 h$

5. DERIVING FORMULAS

To Derive a Formula for Related Quantities

Derive a formula relating the **distance** (D) traveled in a **time** (T) at a **rate of speed** (R).

Procedure:

1. **Obtain sets of values** for these quantities, using convenient numbers:
2. **Make a table of values** for those sets of values:
3. **State the rule** that follows:
4. **State the formula** that expresses the rule:

Solution:

1. **Sets of Values:**
   At 50 mph for 2 hr, 100 mi. will be traveled.
   At 25 mph for 10 hr, 250 mi. will be traveled.
   At 40 mph for 3 hr, 120 mi. will be traveled.
2. **Table of Values:** (*Place units above quantities.*)

(mph) Rate (R)	(hr.) Time (T)	(mi.) Distance (D)
50	2	$50 \cdot 2 = 100$
25	10	$25 \cdot 10 = 250$
40	3	$40 \cdot 3 = 120$

3. **Rule:** The product of the rate and time equals the distance.
4. **Formula:**

$$RT = D$$

Note: If D is in mi. and T in hr., then R must be in mi. per hr. (mph);
or, if D is in ft. and T in sec., then R must be in ft. per sec. (fps).
Rate *must be in* **distance** *units per* **time** *unit.*

Obtaining Formulas from a More General Formula

A formula, such as $RT = D$, relates **three** quantities: time, rate and distance. Each of these quantities may vary in value; that is, they may have many values. However, in a problem, situation or discussion, one of these quantities may have a fixed or unchanging value. When such is the case, this constant value may be used to obtain a formula relating the other **two** quantities.

Thus, $D = RT$ leads to $D = 30T$ if the rate of speed is fixed at 30 mph or 30 feet per min. Or, $D = RT$ leads to $D = 3R$ when the time of travel is fixed at 3 hr. or 3 min.

5.1. Deriving a Coin Formula

Derive a formula for the number of nickels (n) equivalent to q quarters:
(equivalent *means equal in value*)

Solution:

1. Sets of values:

2 quarters equal 10 nickels
4 quarters equal 20 nickels
10 quarters equal 50 nickels
q quarters equal $5q$ nickels

2. Table of Values:

No. of Quarters (q)	No. of Nickels (n)
2	$5 \cdot 2$ or 10
4	$5 \cdot 4$ or 20
10	$5 \cdot 10$ or 50
q	$5q$

3. Rule: The number of nickels equivalent to a number of quarters is five times that number.

4. Formula: $n = 5q$ *Ans.*

5.2. Deriving a Coin Formula

Derive a formula for the value in cents (c) of d dimes and n nickels.

Solution:

1. Sets of values:

In 3 dimes and 4 nickels, there are 50¢.
In 4 dimes and 2 nickels, there are 50¢.
In 5 dimes and 3 nickels, there are 65¢.
In d dimes and n nickels, there are $(10d+5n)$¢.

2. Table of Values:

No. of **Dimes (d)**	No. of **Nickels (n)**	(cents) **Value of Dimes & Nickels (c)**
3	4	$10\cdot3+5\cdot4$ or 50
4	2	$10\cdot4+5\cdot2$ or 50
5	3	$10\cdot5+5\cdot3$ or 65
d	n	$10d+5n$

3. Rule: The value in cents of dimes and nickels is ten times the number of dimes plus five times the number of nickels.

4. Formula: $c = 10d + 5n$ *Ans.*

5.3. Deriving Coin Formulas

Derive a formula for each relationship:

a) For the number of pennies (p) equivalent to q quarters. *Ans.* $p = 25q$

b) For the number of nickels (n) equivalent to d dimes. *Ans.* $n = 2d$

c) For the number of quarters (q) equivalent to D dollars. *Ans.* $q = 4D$

d) For the number of pennies (p) equivalent to n nickels and q quarters. *Ans.* $p = 5n + 25q$

e) For the number of nickels (n) equivalent to q quarters and d dimes. *Ans.* $n = 5q + 2d$

5.4. Deriving Time Formulas

Derive a formula for each relationship:

a) For the number of seconds (s) in m minutes. *Ans.* $s = 60m$

b) For the number of hours (h) in d days. *Ans.* $h = 24d$

c) For the number of weeks (w) in d days. *Ans.* $w = \frac{d}{7}$

d) For the number of days (d) in w weeks and 5 days. *Ans.* $d = 7w + 5$

e) For the number of minutes (m) in h hours and 30 sec. *Ans.* $m = 60h + \frac{1}{2}$

5.5. Deriving Length Formulas

Derive a formula for each relationship:

a) For the number of in. (i) in f feet. *Ans.* $i = 12f$

b) For the number of ft. (f) in y yards. *Ans.* $f = 3y$

c) For the number of yd. (y) in i inches. *Ans.* $y = \frac{i}{36}$

d) For the number of ft. (f) in m miles and 50 yd. *Ans.* $f = 5280m + 150$

5.6. Obtaining Formulas from D = RT

From $D = RT$, derive a formula for each relationship:

a) For the distance in mi. and time in hr. when the rate is 35 mph. *Ans.* $D = 35T$

b) For the distance in ft. and the time in sec. when sound travels at 1100 ft. per sec. *Ans.* $D = 1100T$

c) For the distance in mi. and the time in sec. when light travels at 186,000 mi. per sec. *Ans.* $D = 186{,}000T$

d) For the distance in mi. and the rate in mph when the time of travel is 1 hr. and 30 min. *Ans.* $D = 1\frac{1}{2}R$

e) For the rate in mph and the time in hr. when the distance traveled is 125 mi. *Ans.* $125 = RT$

6. TRANSFORMING FORMULAS

The **subject of a formula** is the letter that has been isolated and expressed in terms of the other letters.

Thus, in $p = 4s$, p is the subject of the formula.

Transforming a formula is the process of changing the subject.

Thus, $p = 4s$ becomes $\frac{p}{4} = s$ when both sides are divided by 4. In the transforming of the formula, the subject has changed from p to s.

In solving a formula for a letter, the formula is transformed in order that the letter be made the subject.

Thus, to solve $D = 5T$ for T, transform it into $T = \frac{D}{5}$.

Use Inverse Operations to Transform Formulas, as follows:

1. Use **division** to undo **multiplication.**
   Thus, $c = 25q$ becomes $\frac{c}{25} = q$ by division.

2. Use **multiplication** to undo **division.**
   Thus, $w = \frac{d}{7}$ becomes $7w = d$ by multiplication.

3. Use **subtraction** to undo **addition.**
   Thus, $S = P+C$ becomes $S-P = C$ by subtraction.

4. Use **addition** to undo **subtraction.**
   Thus, $S = C-L$ becomes $S+L = C$ by addition.

Formulas may be transformed by transposing terms. In transposing a term, change its sign. Thus, $a+b = 180$ becomes $a = 180-b$ when $+b$ is transposed. Actually, $+b$ has been subtracted from both sides to undo addition.

6.1. Transformations Requiring Division

Solve for the letter indicated:

a) $D = RT$ for R	b) $D = RT$ for T	c) $V = LWH$ for L	d) $c = 10d$ for d	e) $C = 2\pi r$ for r
Solutions:				
a) $D = RT$	b) $D = RT$	c) $V = LWH$	d) $c = 10d$	e) $C = 2\pi r$
$\mathbf{D}_T$ $\frac{D}{T} = \frac{RT}{T}$	$\mathbf{D}_R$ $\frac{D}{R} = \frac{RT}{R}$	$\mathbf{D}_{WH}$ $\frac{V}{WH} = \frac{LWH}{WH}$	$\mathbf{D}_{10}$ $\frac{c}{10} = \frac{10d}{10}$	$\mathbf{D}_{2\pi}$ $\frac{C}{2\pi} = \frac{2\pi r}{2\pi}$
Ans. $\frac{D}{T} = R$	*Ans.* $\frac{D}{R} = T$	*Ans.* $\frac{V}{WH} = L$	*Ans.* $\frac{c}{10} = d$	*Ans.* $\frac{C}{2\pi} = r$

6.2. Transformations Requiring Multiplication

Solve for the letter indicated:

a) $\frac{i}{12} = f$ for i	b) $f = \frac{n}{d}$ for n	c) $\frac{V}{LW} = H$ for V	d) $\frac{b}{2} = \frac{A}{h}$ for A
Solutions:			
a) $\frac{i}{12} = f$	b) $f = \frac{n}{d}$	c) $\frac{V}{LW} = H$	d) $\frac{b}{2} = \frac{A}{h}$
$\mathbf{M}_{12}$ $12(\frac{i}{12}) = 12f$	$\mathbf{M}_d$ $fd = d(\frac{n}{d})$	$\mathbf{M}_{LW}$ $LW(\frac{V}{LW}) = LWH$	$\mathbf{M}_h$ $(\frac{b}{2})h = (\frac{A}{h})h$
Ans. $i = 12f$	*Ans.* $fd = n$	*Ans.* $V = LWH$	*Ans.* $\frac{bh}{2} = A$

6.3. Transformations Requiring Addition or Subtraction

(Transposing is the result of adding or subtracting.)

Solve for the letter indicated:

a) $a+b=90$ for a	b) $a=b-180$ for b	c) $a+c=b+100d$ for b	d) $a-b-25=c$ for a
Solutions:			
a) $a+b=90$	b) $a=b-180$	c) $a+c=b+100d$	d) $a-b-25=c$
Transpose b:	**Transpose** -180:	**Transpose** $100d$:	**Transpose** $-b-25$:
Ans. $a=90-b$	*Ans.* $a+180=b$	*Ans.* $a+c-100d=b$	*Ans.* $a=b+c+25$

6.4. Transformations Requiring Two Operations

Solve each formula for the letter indicated:

a) $P=2a+b$ for a	b) $c=10d+25q$ for q	c) $F=\frac{9}{5}C+32$ for C	d) $V=\frac{1}{3}Bh$ for B
Solutions:			
a) $P=2a+b$	b) $c=10d+25q$	c) $F=\frac{9}{5}C+32$	d) $V=\frac{1}{3}Bh$
Transpose b:	**Transpose** $10d$:	**Transpose** 32:	$\mathbf{M}_3$ $3V=3(\frac{1}{3}Bh)$
$\mathbf{D}_2$ $P-b=2a$	$\mathbf{D}_{25}$ $c-10d=25q$	$\mathbf{M}_{5/9}$ $F-32=\frac{9}{5}C$	$\mathbf{D}_h$ $3V=Bh$
Ans. $\frac{P-b}{2}=a$	*Ans.* $\frac{c-10d}{25}=q$	*Ans.* $\frac{5}{9}(F-32)=C$	*Ans.* $\frac{3V}{h}=B$

6.5. More Difficult Transformations

Solve each formula for the letter indicated:

a) $A=\frac{h}{2}(b+b')$ for h	b) $S=\frac{n}{2}(a+l)$ for a	c) $l=a+(n-1)d$ for n
Solutions:		
a) $A=\frac{h}{2}(b+b')$	b) $\mathbf{M}_2$ $S=\frac{n}{2}(a+l)$	c) $l=a+(n-1)d$
$\mathbf{M}_2$ $2A=2(\frac{h}{2})(b+b')$	$\mathbf{D}_n$ $2S=n(a+l)$	**Transpose** a:
$\mathbf{D}_{(b+b')}$ $\frac{2A}{b+b'}=\frac{h(b+b')}{b+b'}$	$\frac{2S}{n}=a+l$	$\mathbf{D}_d$ $l-a=(n-1)d$
Ans. $\frac{2A}{b+b'}=h$	**Transpose** $+l$:	$\frac{l-a}{d}=n-1$
	Ans. $\frac{2S}{n}-l=a$	**Transpose** -1:
		Ans. $\frac{l-a}{d}+1=n$

7. FINDING THE VALUE OF AN UNKNOWN IN A FORMULA

To Find an Isolated Unknown, Substitute and Solve

The value of an unknown in a formula may be found if values are given for the other letters. By substitution replace the letters by their given values and then solve for the unknown.

Thus, in $A=bh$ if $b=10$ and $h=5$, then $A=(10)(5)$ or 50.

To Find an Unisolated Unknown, Substitute First or Transform First

When the unknown is not isolated, its value may be found by either of two methods. One method is to substitute first and then solve. The second method is to transform the formula first to isolate the unknown. After the transformation, substitution is then used.

Thus, in $p=3s$ if $p=27$, the value of s may be found

(1) by substituting first: $27=3s$, $s=9$

or (2) by transforming first:

Transform $p=3s$ into $s=\frac{p}{3}$. Then $s=\frac{27}{3}$ or 9.

7.1. Finding the Value of an Isolated Unknown

Find each unknown:

a) Find V if $V = lwh$ and $l = 10$, $w = 2$, $h = 3.2$

b) Find S if $S = \frac{n}{2}(a+l)$ and $n = 8$, $a = 5$, $l = 12$.

c) Find A if $A = p + prt$ and $p = 800$, $r = .04$, $t = 3$.

d) Find S if $S = \frac{1}{2}gt^2$ and $g = 32$, $t = 5$.

Solutions:

a) $V = lwh$

$V = 10(2)(3.2)$

Ans. $V = 64$

b) $S = \frac{n}{2}(a+l)$

$S = \frac{8}{2}(5+12)$

Ans. $S = 68$

c) $A = p + prt$

$A = 800 + 800(.04)3$

Ans. $A = 896$

d) $S = \frac{1}{2}gt^2$

$S = \frac{1}{2} \cdot 32 \cdot 5^2$

Ans. $S = 400$

7.2. Finding the Value of an Unknown that is Not Isolated

Find each unknown, using substitution first or transformation first:

a) Find h if $A = bh$, $b = 13$ and $A = 156$.

Solutions:

(1) By substitution, first:

$A = bh$

$156 = 13h$

Ans. $12 = h$

(2) By transformation, first:

$A = bh$

Transform: $\frac{A}{b} = h$

Substitute: $\frac{156}{13} = h$

Ans. $12 = h$

b) Find a if $p = 2a+b$, $b = 20$ and $p = 74$.

(1) By substitution, first:

$p = 2a + b$

$74 = 2a + 20$

Ans. $27 = a$

(2) By transformation, first:

$p = 2a + b$

Transform: $\frac{p-b}{2} = a$

Substitute: $\frac{74-20}{2} = a$

Ans. $27 = a$

c) Find h if $V = \frac{1}{3}Bh$, $B = 240$ and $V = 960$.

(1) By substitution, first:

$V = \frac{1}{3}Bh$

$960 = \frac{1}{3}(240)h$

Ans. $12 = h$

(2) By transformation, first:

$V = \frac{1}{3}Bh$

Transform: $\frac{3V}{B} = h$

Substitute: $\frac{3(960)}{240} = h$

Ans. $12 = h$

7.3. More Difficult Evaluations Using Transformations

Find each unknown: a) Find h if $V = \pi r^2 h$, $\pi = 3.14$, $V = 9420$ and $r = 10$.

b) Find t if $A = p + prt$, $A = 864$, $p = 800$ and $r = 2$.

Solutions:

a) $V = \pi r^2 h$

Transform: $\frac{V}{\pi r^2} = h$

Substitute: $\frac{9420}{(3.14)(100)} = h$

$\frac{9420}{314} = h$

Ans. $30 = h$

b) $A = p + prt$

Transform: $\frac{A-p}{pr} = t$

Substitute: $\frac{864-800}{800(2)} = t$

$\frac{64}{1600} = t$

Ans. $.04 = t$

7.4. Finding an Unknown in a Problem

A train takes 3 hours and 15 minutes to go a distance of 247 miles. Find its average speed.

Solution: Here, $D = RT$, $T = 3\frac{1}{4}$ hr. and $D = 247$. To find R:

(1) By substitution, first:

$247 = \frac{13}{4}R$

$M_{4/13}$ $\frac{4}{\cancel{13}} \cdot \overset{19}{\cancel{247}} = R$

$76 = R$

(2) By transposition, first:

$D = RT$

Transform: $\frac{D}{T} = R$

$247 \div \frac{13}{4} = R$

$247 \cdot \frac{4}{13} = R$

Ans. Average rate is 76 mph

SUPPLEMENTARY PROBLEMS

1. State the formula for the perimeter (p) of each, using the letters shown on page 87: **(2.1)**

a) right triangle *Ans.* $p = a + b + c$

b) acute triangle *Ans.* $p = a + b + c$

c) parallelogram *Ans.* $p = 2a + 2b$

d) trapezoid *Ans.* $p = a + b + b' + c$

e) rhombus *Ans.* $p = 4s$

f) isosceles trapezoid *Ans.* $p = 2a + b + b'$

2. State the name of the equilateral polygon to which each perimeter formula applies: **(2.2)**

a) $p = 5s$ *Ans.* equilateral pentagon

b) $p = 8s$ *Ans.* equilateral octagon

c) $p = 6s$ *Ans.* equilateral hexagon

d) $p = ns$ *Ans.* equilateral n-gon

3. Stating the formula used, find the perimeter of each equilateral polygon having a side of 6 in. Express each answer in inches, feet and yards. **(2.1, 2.4)**

a) equilateral triangle

b) regular hexagon

e) equilateral centagon (100 sides)

b) square or rhombus

d) equilateral decagon

f) regular n-gon

Ans. *a*) $p = 3s$; 18 in., $1\frac{1}{2}$ ft., $\frac{1}{2}$ yd. *c*) $p = 6s$; 36 in., 3 ft., 1 yd. *e*) $p = 100s$; 600 in., 50 ft., $16\frac{2}{3}$ yd.

b) $p = 4s$; 24 in., 2 ft., $\frac{2}{3}$ yd. *d*) $p = 10s$; 60 in., 5 ft., $1\frac{2}{3}$ yd. *f*) $p = ns$; $6n$ in., $\frac{n}{2}$ ft., $\frac{n}{6}$ yd.

4. A piece of wire 12 ft. long is to be bent into the form of an equilateral polygon. How many feet will be available for each side of **(2.3)**

a) an equilateral triangle?

b) a square?

c) a regular pentagon?

d) an equilateral octagon?

e) an equilateral dodecagon?

f) an equilateral centagon?

Ans. *a*) each of 3 sides = 4 ft. *c*) each of 5 sides = 2.4 ft. *e*) each of 12 sides = 1 ft.

b) each of 4 sides = 3 ft. *d*) each of 8 sides = $1\frac{1}{2}$ ft. *f*) each of 100 sides = .12 ft.

5. Find the perimeter of **(2.5)**

a) a rectangle with sides of 3 yd. and $5\frac{1}{2}$ yd. *Ans.* *a*) 17 yd.

b) a rhombus with side of 4 yd. 1 ft. *Ans.* *b*) $17\frac{1}{3}$ yd.

c) a parallelogram with sides of 3.4 ft. and 4.7 ft. *Ans.* *c*) 16.2 ft.

d) an isosceles trapezoid with bases of 13 yd. 1 ft. and 6 yd. 1 ft., and each remaining side 5 yd. 2 ft. *Ans.* *d*) 31 yd.

6. For a rectangle shown, if l, w and p are in inches, find **(2.6, 2.7)**

a) the perimeter if $l = 12$ and $w = 7$

b) the perimeter if $l = 5\frac{1}{2}$ and $w = 2\frac{3}{8}$

c) the perimeter if $l = 7.6$ and $w = 4.3$

d) the length if $p = 30$ and $w = 9\frac{1}{2}$

e) the length if $p = 52$ and $w = l - 4$

f) the width if $p = 60$ and $l = 2w + 3$

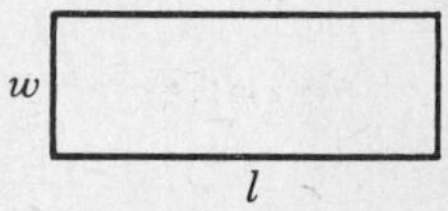

Ans. *a*) 38 in., *b*) $15\frac{3}{4}$ in., *c*) 23.8 in., *d*) $5\frac{1}{2}$ in.,
e) Since $52 = 2l + 2(l-4)$, length is 15 in., *f*) Since $60 = 2(2w+3) + 2w$, width is 9 in.

7. For the isosceles triangle shown, if a, b and p are in yd., find **(2.8)**

a) the perimeter if $a = 12$ and $b = 16$

b) the perimeter if $a = 3\frac{1}{4}$ and $b = 4\frac{3}{8}$

c) the perimeter if $a = 1.35$ and $b = 2.04$

d) the base if $p = 20$ and $a = 7$

e) each equal side if $p = 32$ and $b = a + 5$

f) the base if $p = 81$ and $a = b - 3$

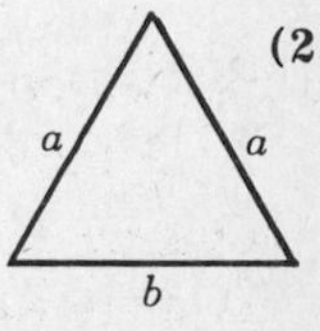

Ans. *a*) 40 yd., *b*) $10\frac{7}{8}$ yd., *c*) 4.74 yd., *d*) Since $20 = 14 + b$, base is 6 yd.,
e) Since $32 = 2a + (a+5)$, equal side is 9 yd., *f*) Since $81 = 2(b-3) + b$, base is 29 yd.

8. For any circle, state a formula which relates **(2.9)**

a) the circumference (c) and an arc of 45° (a) *Ans.* *a*) $c = 8a$

b) the semi-circumference (s) and the radius (r) *Ans.* *b*) $s = \pi r$

c) an arc of 120° (a') and the diameter (d) *Ans.* *c*) $a' = \frac{\pi d}{3}$

d) an arc of 20° (a'') and semi-circumference (s) *Ans.* *d*) $a'' = \frac{s}{9}$

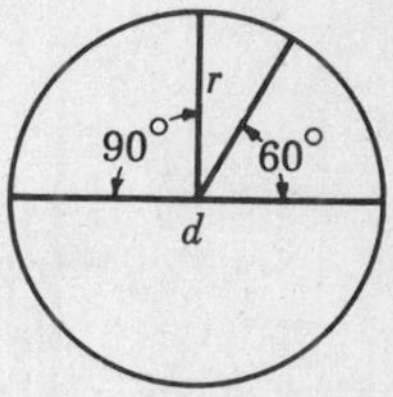

9. For a circle, if r, d and c (c = circumference) are in mi., find, using $\pi = 3.14$, **(2.10)**

a) circumference if $r = 6$
b) semi-circumference if $d = 20$
c) 60° arc if $r = 12$
d) radius if $c = 157$
e) diameter if $c = 314$
f) radius, if a 90° arc = 9.42 mi.

Ans. *a*) 37.68 mi., *b*) 31.4 mi., *c*) 12.56 mi., *d*) 25 mi., *e*) 100 mi.,
f) Since 90° arc = $\frac{1}{4}$ of circumference, $9.42 = \frac{1}{4} \times 2(3.14)r$. Radius is 6 mi.

10. State the formula for the perimeter of each figure: **(2.11)**

a)
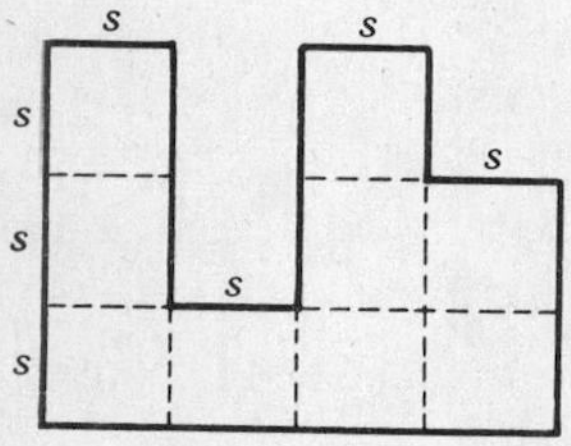

b)
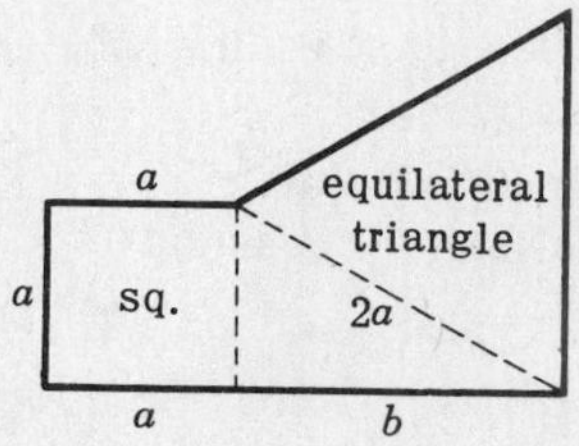

c)
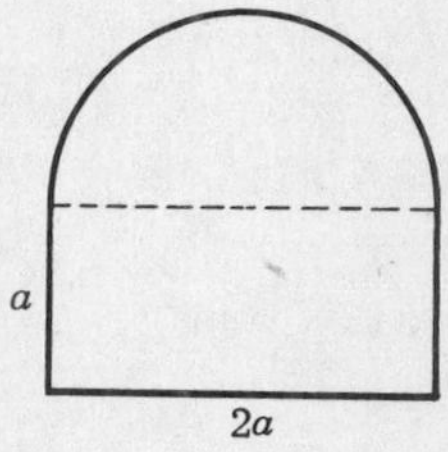

Ans. *a*) $p = 18s$ *b*) $p = 7a + b$ *c*) $4a + \pi a$

11. Find the area of **(3.1)**

a) a square yard in square inches
b) a square rod in square yards (1 rd. = $5\frac{1}{2}$ yd.)
c) a square rod in square feet (1 rd. = $16\frac{1}{2}$ ft.)

Ans. *a*) 36^2 or 1296 sq. in., *b*) $(5\frac{1}{2})^2$ or $30\frac{1}{4}$ sq. yd., *c*) $(16\frac{1}{2})^2$ or $272\frac{1}{4}$ sq. ft.

12. Find the area of a square in sq. in. whose side is **(3.2)**

a) 5 in. *b*) $6\frac{1}{2}$ in. *c*) 7.2 in. *d*) 1 ft. 3 in. *e*) 1.3 ft.

Ans. *a*) 25 sq. in., *b*) $42\frac{1}{4}$ sq. in., *c*) 51.84 sq. in., *d*) 225 sq. in., *e*) 243.36 sq. in.

13. Find the area in sq. ft. of **(3.3)**

a) a rectangle with sides of $8\frac{1}{2}$ ft. and 12 ft.
b) a parallelogram with a base of $5\frac{1}{2}$ ft. and an altitude of 4 ft.
c) a triangle with a base of 10.4 ft. and an altitude of 8 ft.
d) a trapezoid with bases of 6 ft. and 4 ft. and an altitude of $3\frac{1}{2}$ ft. ($A = \frac{h}{2}(b + b')$)

Ans. *a*) 102 sq. ft., *b*) 22 sq. ft., *c*) 41.6 sq. ft., *d*) $17\frac{1}{2}$ sq. ft.

14. For a circle, state a formula which relates **(3.4)**

a) the area of the circle (A) and a sector of 60° (S)
b) a sector of 120° (S') and the area of the circle (A)
c) a sector of 90° (S'') and the radius (r)
d) a sector of 90° (S'') and the diameter (d).

Ans. *a*) $A = 6S$, *b*) $S' = \frac{A}{3}$, *c*) $S'' = \frac{\pi r^2}{4}$, *d*) $S'' = \frac{1}{4}(\frac{\pi d^2}{4}) = \frac{\pi d^2}{16}$

15. Find, in terms of π and to the nearest integer, using $\pi = 3.14$, the area of **(3.5)**

a) a circle whose radius is 30 in. *Ans.* 900π or 2826 sq. in.
b) a circle whose diameter is 18 in. *Ans.* 81π or 254 sq. in.
c) a semi-circle whose radius is 14 ft. *Ans.* 98π or 308 sq. ft.
d) a 45° sector whose radius is 12 yd. *Ans.* 18π or 57 sq. yd.

16. State the formula for the area (A) of each shaded figure: **(3.6, 3.7)**

a)

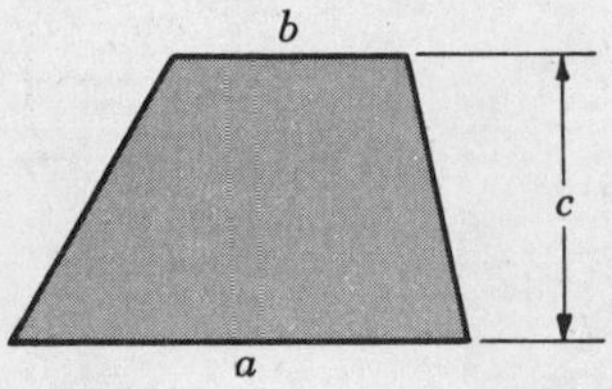

b)

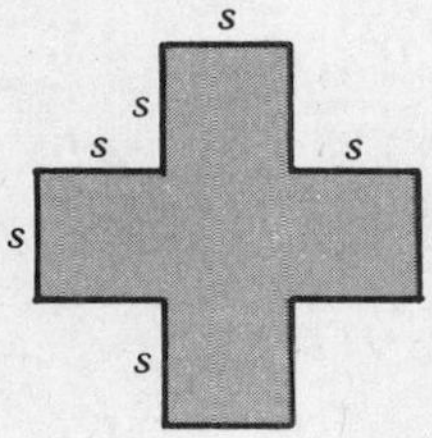

c)

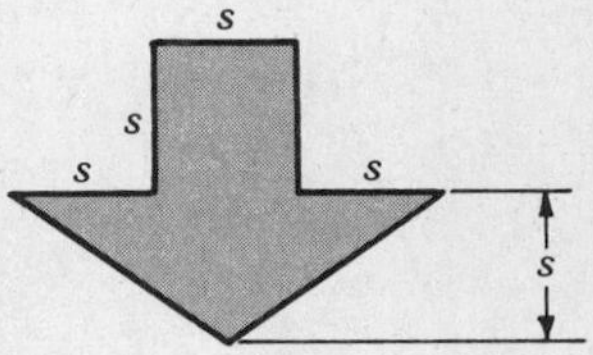

d)

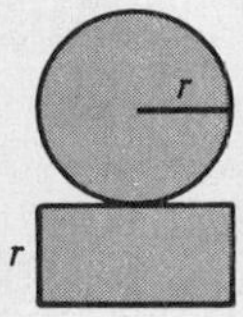

e)

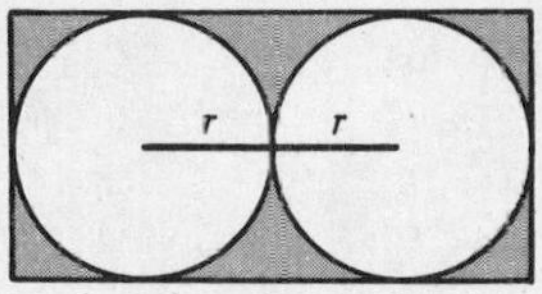

f)

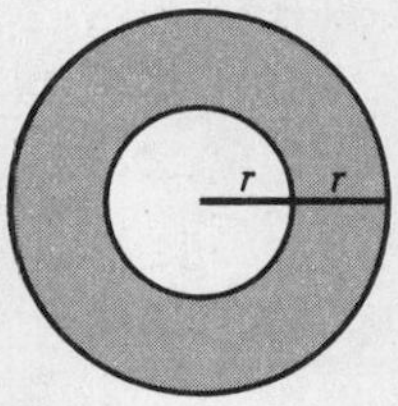

Ans. *a*) $A = \frac{c}{2}(a+b)$ *b*) $A = 5s^2$ *c*) $A = s^2 + 1\frac{1}{2}s^2$ or $2\frac{1}{2}s^2$

d) $A = \pi r^2 + 2r^2$ *e*) $A = 8r^2 - 2\pi r^2$ *f*) $A = 4\pi r^2 - \pi r^2$ or $3\pi r^2$

17. Find, to the nearest integer, using $\pi = 3.14$. the total area of **(3.8)**

a) a cube with an edge of 7 yd.

b) a rectangular solid with dimensions of 8 ft., $6\frac{1}{2}$ ft. and 14 ft.

c) a sphere with a radius of 30 in.

d) a cylinder of revolution with a radius of 10 rd. and a height of $4\frac{1}{2}$ rd. [*Hint.* Use $T = 2\pi r(r+h)$]

Ans. *a*) $6(7^2)$ or 294 sq. yd., *b*) $2(8)(6\frac{1}{2}) + 2(8)(14) + 2(6\frac{1}{2})(14)$ or 510 sq. ft.,

c) $4(3.14)30^2$ or 11,304 sq. in., *d*) $2(3.14)(10)(10 + 4\frac{1}{2})$ or 911 sq. rd.

18. Find the volume of **(4.1)**

a) a cubic yard in cubic inches

b) a cubic rod in cubic yards

c) a cubic meter in cubic centimeters (1 meter = 100 cm.).

Ans. *a*) 36^3 or 46,656 cu. in., *b*) $(5\frac{1}{2})^3$ or $166\frac{3}{8}$ cu. yd., *c*) 100^3 or 1,000,000 cu. cm.

19. Find, to the nearest cubic inch, the volume of a cube whose edge is **(4.2)**

a) 3 in. *b*) $4\frac{1}{2}$ in. *c*) 7.5 in. *d*) .3 ft. *e*) 1 ft. 2 in.

Ans. *a*) 27 cu. in., *b*) 91 cu. in., *c*) 422 cu. in., *d*) 47 cu. in., *e*) 2744 cu. in.

20. Find, to the nearest integer, the volume of **(4.3)**

a) a rectangular solid whose length is 3 in., width $8\frac{1}{2}$ in. and height 8 in.

b) a prism having a height of 2 ft. and a square base whose side is 3 yd.

c) a pyramid having a height of 2 yd. and a base whose area is 6.4 sq. ft.

Ans. *a*) $3(8\frac{1}{2})(8)$ or 204 cu. in., *b*) $2(9)(9)$ or 162 cu. ft., *c*) $\frac{1}{3}(6)(6.4)$ or 13 cu. ft.

21. Find, to the nearest integer, the volume of **(4.4)**

a) a sphere with a radius of 6 in.

b) a cylinder having a height of 10 ft. and a base whose radius is 2 yd.

c) a cone having a height of 3 yd. and a base whose radius is 1.4 ft.

a) $\frac{4}{3}(3.14)6^3$ or 904.32 *b*) $(3.14)(6^2)10$ or 1130.4 *c*) $\frac{1}{3}(3.14)(1.4^2)(9)$ or 18.4632

Ans. 904 cu. in. *Ans.* 1130 cu. ft. *Ans.* 18 cu. ft.

22. From $V = \frac{1}{3}Bh$, the volume formula for a pyramid or cone, derive volume formulas for each of the following: **(4.5)**

a)

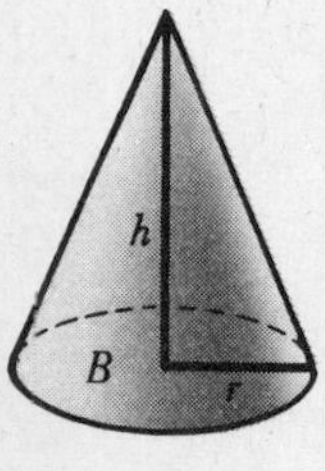

Cone

b)

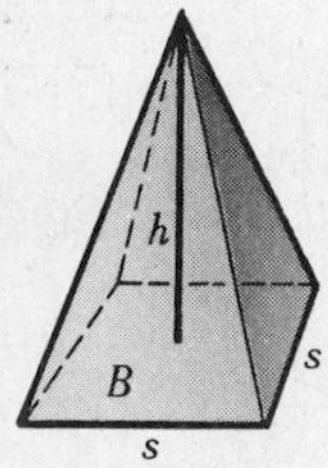

Pyramid with a Square Base

c)

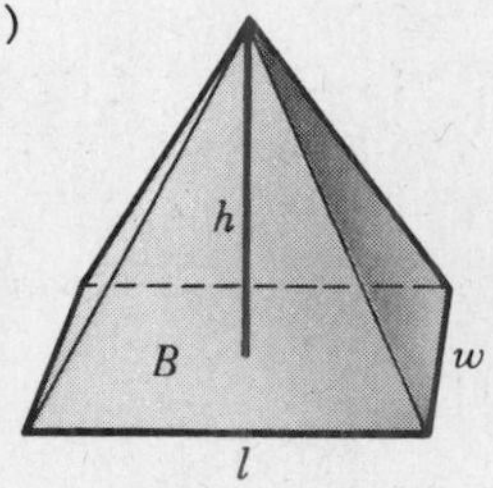

Pyramid with a Rectangular Base

d)

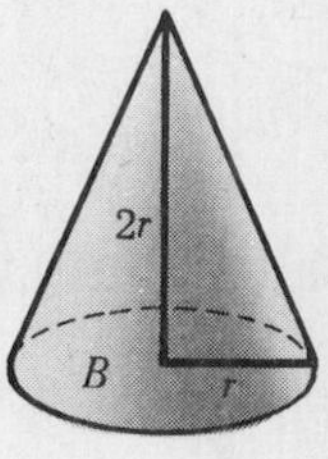

Cone where $h = 2r$

a) $V = \frac{1}{3}Bh$
Since $B = \pi r^2$,
$V = \frac{1}{3}(\pi r^2)h$
Ans. $V = \frac{1}{3}\pi r^2 h$

b) $V = \frac{1}{3}Bh$
Since $B = s^2$,
$V = \frac{1}{3}(s^2)h$
Ans. $V = \frac{1}{3}s^2 h$

c) $V = \frac{1}{3}Bh$
Since $B = lw$,
$V = \frac{1}{3}(lw)h$
Ans. $V = \frac{1}{3}lwh$

d) $V = \frac{1}{3}Bh$
Since $B = \pi r^2$
and $h = 2r$,
$V = \frac{1}{3}(\pi r^2)(2r)$
Ans. $V = \frac{2}{3}\pi r^3$

23. State a formula for the volume of each solid: **(4.6)**

a)

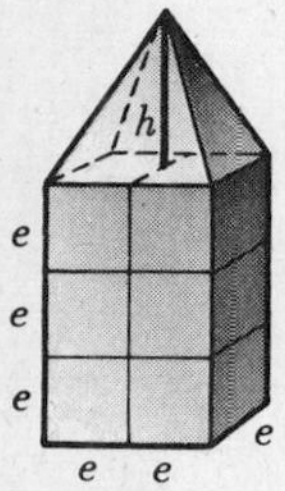

b)

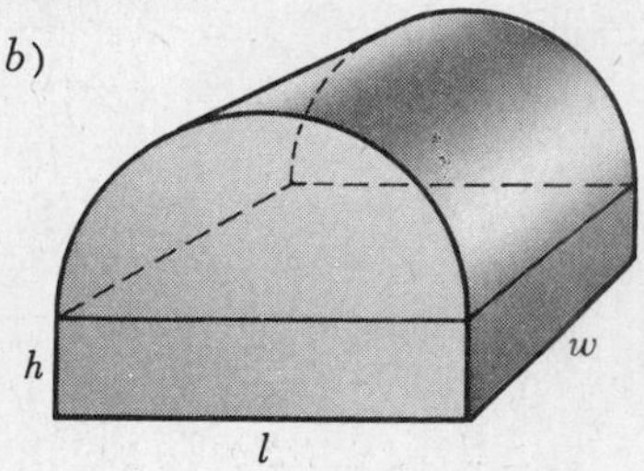

c)

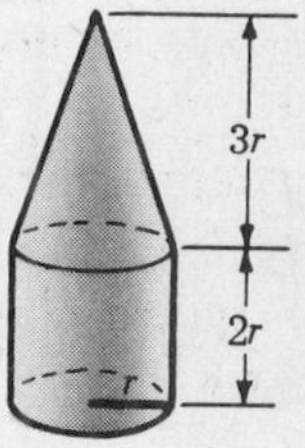

a) $(2e)(3e)e + \frac{1}{3}(2e^2)h$
Ans. $6e^3 + \frac{2e^2 h}{3}$

b) $lwh + \frac{1}{2}\cdot\pi(\frac{l}{2})^2 w$
Ans. $lwh + \frac{\pi l^2 w}{8}$

c) $\pi r^2(2r) + \frac{1}{3}\pi r^2(3r)$
Ans. $3\pi r^3$

24. Derive a formula for each relationship: **(5.1)**

a) the no. of pennies (p) equivalent to d dimes
b) the no. of dimes (d) equivalent to p pennies
c) the no. of nickels (n) equivalent to D dollars
d) the no. of half-dollars (h) equivalent to q quarters
e) the no. of quarters (q) equivalent to d dimes.

Ans. *a*) $p = 10d$, *b*) $d = \frac{p}{10}$, *c*) $n = 20D$, *d*) $h = \frac{1}{2}q$, *e*) $q = \frac{2}{5}d$ or $\frac{2d}{5}$

25. Derive a formula for each relationship: **(5.2, 5.3)**

a) the value in cents (c) of d dimes and q quarters
b) the value in cents (c) of n nickels and D dollars
c) the no. of nickels (n) equivalent to d dimes and h half-dollars
d) the no. of dimes (d) equivalent to n nickels and p pennies
e) the no. of quarters (q) equivalent to D dollars and n nickels.

Ans. *a*) $c = 10d + 25q$, *b*) $c = 5n + 100D$, *c*) $n = 2d + 10h$, *d*) $d = \frac{n}{2} + \frac{p}{10}$, *e*) $q = 4D + \frac{n}{5}$

26. Derive a formula for each relationship: (5.4)

$a)$ The no. of sec. (s) in h hr.
$b)$ The no. of hr. (h) in m min.
$c)$ The no. of hr. (h) in w wk.
$d)$ The no. of da. (d) in M mo. of 30 days.
$e)$ The no. of da. (d) in M mo. of 30 days, w wk. and 5 da.
$f)$ The no. of min. (m) in h hr. and 30 sec.
$g)$ The no. of da. (d) in y yr. of 365 da. and 3 weeks.

Ans. a) $s = 3600h$, b) $h = \frac{m}{60}$, c) $h = 168w$, d) $d = 30M$, e) $d = 30M + 7w + 5$, f) $m = 60h + \frac{1}{2}$, g) $d = 365y + 21$

27. Derive a formula for each relationship: (5.5)

$a)$ The no. of in. (i) in y yd.
$b)$ The no. of yd. (y) in f ft.
$c)$ The no. of yd. (y) in r rd.
$d)$ The no. of mi. (m) in f ft.
$e)$ The no. of meters (m) in c centimeters. (1 meter = 100 cm.)
$f)$ The no. of centimeters (c) in d decimeters. (1 decimeter = 10 cm.)

Ans. a) $i = 36y$, b) $y = \frac{f}{3}$, c) $y = 5\frac{1}{2}r$, d) $m = \frac{f}{5280}$, e) $m = \frac{c}{100}$, f) $c = 10d$

28. From $D = RT$, obtain a formula for each relationship: (5.6)

$a)$ Distance in mi. and rate in mph for a time of 5 hr.
$b)$ Distance in mi. and rate in mph for a time of 30 min.
$c)$ Time in hr. and rate in mph for a distance of 25 mi.
$d)$ Time in sec. and rate in ft. per sec. for a distance of 100 yd.
$e)$ Distance in ft. and time in min. for a rate of 20 ft. per min.
$f)$ Distance in ft. and time in min. for a rate of 20 ft. per sec.

Ans. a) $D = 5R$, b) $D = \frac{1}{2}R$, c) $RT = 25$, d) $RT = 300$, e) $D = 20T$, f) $D = 1200T$ (20 ft. per sec = 1200 ft. per min.)

29. Solve for the letter indicated: (6.1)

$a)$ $d = 2r$ for r
$b)$ $p = 5s$ for s
$c)$ $D = 30T$ for T
$d)$ $25W = A$ for W
$e)$ $c = \pi d$ for d
$f)$ $c = \pi d$ for π
$g)$ $NP = C$ for N
$h)$ $I = PR$ for R
$i)$ $V = LWH$ for H
$j)$ $V = 2\pi r^2 h$ for h
$k)$ $9C = 5(F-32)$ for C
$l)$ $2A = h(b+b')$ for h

Ans. a) $\frac{d}{2} = r$ b) $\frac{p}{5} = s$ c) $\frac{D}{30} = T$ d) $W = \frac{A}{25}$ e) $\frac{c}{\pi} = d$ f) $\pi = \frac{c}{d}$ g) $N = \frac{C}{P}$ h) $\frac{I}{P} = R$ i) $\frac{V}{LW} = H$ j) $\frac{V}{2\pi r^2} = h$ k) $C = \frac{5(F-32)}{9}$ l) $\frac{2A}{b+b'} = h$

30. Solve for the letter indicated: (6.2)

$a)$ $\frac{p}{10} = s$ for p
$b)$ $R = \frac{D}{15}$ for D
$c)$ $W = \frac{A}{8}$ for A
$d)$ $w = \frac{d}{7}$ for d
$e)$ $\pi = \frac{c}{2r}$ for c
$f)$ $\frac{M}{D} = F$ for M
$g)$ $P = \frac{A}{2F}$ for A
$h)$ $\frac{T}{Q} = 5R$ for T
$i)$ $\frac{V}{3LW} = H$ for V
$j)$ $\frac{T}{14RS} = \frac{1}{2}$ for T
$k)$ $\frac{L}{KA} = V^2$ for L
$l)$ $\frac{V}{\pi r^2} = \frac{h}{3}$ for V

Ans. a) $p = 10s$ b) $15R = D$ c) $8W = A$ d) $7w = d$ e) $2\pi r = c$ f) $M = FD$ g) $2PF = A$ h) $T = 5RQ$ i) $V = 3LWH$ j) $T = 7RS$ k) $L = KAV^2$ l) $V = \frac{\pi r^2 h}{3}$

31. Solve for the letter indicated: (6.3)

$a)$ $a + b = 60$ for a
$b)$ $3c + g = 85$ for g
$c)$ $h - 10r = l$ for h
$d)$ $3m = 4n + p$ for p
$e)$ $10r = s - 5t$ for s
$f)$ $4g + h - 12 = j$ for h
$g)$ $5a + b = c - d$ for b
$h)$ $5a - 4c = 3e + f$ for f
$i)$ $\frac{b}{2} - 10 + c = 100p$ for c

Ans. a) $a = 60 - b$ b) $g = 85 - 3c$ c) $h = l + 10r$ d) $3m - 4n = p$ e) $10r + 5t = s$ f) $h = j + 12 - 4g$ g) $b = c - d - 5a$ h) $5a - 4c - 3e = f$ i) $c = 100p + 10 - \frac{b}{2}$

32. Solve for the letter indicated: (6.4)

a) $4P - 3R = 40$ for P
b) $36 - 5i = 12f$ for i
c) $\frac{P}{2} + R = S$ for P
d) $A = \frac{1}{2}bh$ for b
e) $V = \frac{1}{3}\pi r^2 h$ for h
f) $A = \frac{1}{2}h(b+b')$ for h
g) $\frac{R}{2} - 4S = T$ for R
h) $8h - \frac{k}{5} = 12$ for k
i) $20p - \frac{2}{3}q = 8t$ for q

Ans. *a)* $P = \frac{3R+40}{4}$
b) $\frac{36-12f}{5} = i$
c) $P = 2S - 2R$
d) $\frac{2A}{h} = b$
e) $\frac{3V}{\pi r^2} = h$
f) $\frac{2A}{b+b'} = h$
g) $R = 8S + 2T$
h) $5(8h-12) = k$ or $40h - 60 = k$
i) $\frac{3}{2}(20p-8t) = q$ or $30p - 12t = q$

33. Solve for the letter indicated: (6.5)

a) $l = a+(n-1)d$ for d
b) $S = \frac{n}{2}(a+l)$ for l
c) $F = \frac{9}{5}C + 32$ for C

Ans. *a)* $\frac{l-a}{n-1} = d$
b) $l = \frac{2s}{n} - a$ or $\frac{2s-an}{n}$
c) $\frac{5}{9}(F-32) = C$

34. Find each unknown: (7.1)

a) Find I if $I = prt$ and $p = 3000$, $r = .05$, $t = 2$. *Ans.* *a)* 300
b) Find t if $t = \frac{I}{pr}$ and $I = 40$, $p = 2000$, $r = .01$. *Ans.* *b)* 2
c) Find F if $F = \frac{9}{5}C+32$ and $C = 55$. *Ans.* *c)* 131
d) Find F if $F = \frac{9}{5}C+32$ and $C = -40$. *Ans.* *d)* -40
e) Find C if $C = \frac{5}{9}(F-32)$ and $F = 212$. *Ans.* *e)* 100
f) Find S if $S = \frac{1}{2}gt^2$ and $g = 32$, $t = 8$. *Ans.* *f)* 1024
g) Find g if $g = \frac{2S}{t^2}$ and $S = 800$, $t = 10$. *Ans.* *g)* 16
h) Find S if $S = \frac{a-lr}{1-r}$ and $a = 5$, $l = 40$, $r = -1$. *Ans.* *h)* $22\frac{1}{2}$
i) Find A if $A = p+prt$ and $p = 500$, $r = .04$, $t = 2\frac{1}{2}$. *Ans.* *i)* 550

35. Find each unknown: (7.2, 7.3)

a) Find R if $D = RT$ and $D = 30$, $T = 4$. *Ans.* *a)* $7\frac{1}{2}$
b) Find b if $A = bh$ and $A = 22$, $h = 2.2$. *Ans.* *b)* 10
c) Find a if $P = 2a+b$ and $P = 12$, $b = 3$. *Ans.* *c)* $4\frac{1}{2}$
d) Find w if $P = 2l + 2w$ and $P = 68$, $l = 21$. *Ans.* *d)* 13
e) Find c if $p = a+2b+c$ and $p = 33$, $a = 11$, $b = 3\frac{1}{2}$. *Ans.* *e)* 15
f) Find h if $2A = h(b+b')$ and $A = 70$, $b = 3.3$, $b' = 6.7$. *Ans.* *f)* 14
g) Find B if $V = \frac{1}{3}Bh$ and $V = 480$, $h = 12$. *Ans.* *g)* 120
h) Find a if $l = a+(n-1)d$ and $l = 140$, $n = 8$, $d = 3$. *Ans.* *h)* 119
i) Find E if $I = \frac{E}{R+r}$ and $I = 40$, $R = 3$, $r = 1\frac{1}{2}$. *Ans.* *i)* 180
j) Find E if $C = \frac{nE}{R+nr}$ if $C = 240$, $n = 8$, $R = 14$, $r = 2$. *Ans.* *j)* 900

36. *a)* A train takes 5 hours and 20 minutes to go a distance of 304 miles. Find its average speed.
Ans. 57 mph ($R = \frac{D}{T}$) (7.4)

b) A rectangle has a perimeter of 2 yd. and a length of 22 in. Find its width. *Ans.* 14 in.

Chapter 7

GRAPHS of LINEAR EQUATIONS

1. UNDERSTANDING GRAPHS

Reviewing Number Scales

A **number scale** is a line on which distances from a point are numbered in equal units, positively in one direction and negatively in the other.

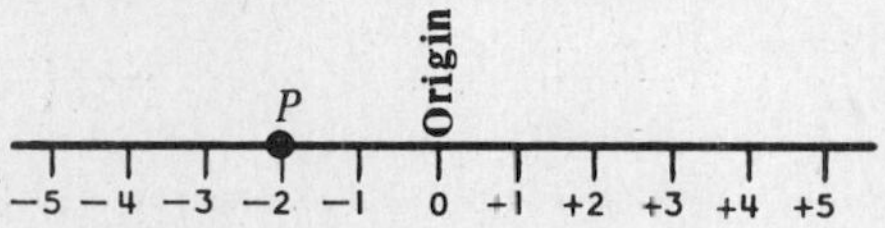

Horizontal Number Scale

Vertical Number Scale

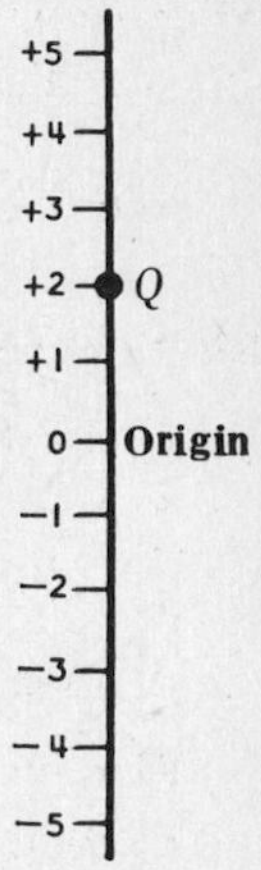

The **origin** is the zero point from which distances are numbered.

Note on the horizontal number scale, how **positive numbers** are to the **right** of the origin, while on the vertical number scale they are **above** the origin.

The position of any point on a number scale is found by determining its distance from the origin. Thus, P on the horizontal scale is -2 units from the origin, and Q on the vertical scale is $+2$ units from the origin.

Forming a Graph by Combining Number Scales

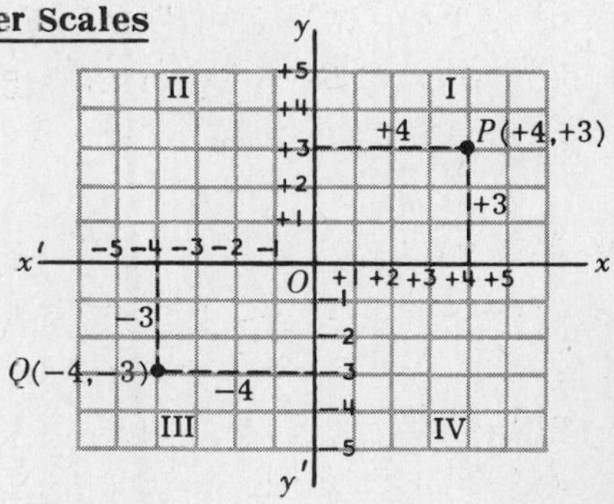

The graph shown is formed by combining two number scales at right angles to each other so that their zero points coincide.

The horizontal number scale is usually called the x-**axis**. To show this, x and x' are placed at each end. (x' is read "x-prime".)

The vertical number scale is called the y-**axis**. Its end letters are y and y'.

The origin, O, is the point where the two scales cross each other.

To locate a point, determine its **distance from each axis**. Note on the graph that for P, the distances are $+4$ and $+3$.

The **coordinates of a point** are its distances from the axes.

(1) The x-**coordinate of a point**, its **abscissa**, is its distance from the y-axis. This distance is measured along the x-axis. For P, this is $+4$; for Q, -4.

(2) The y-**coordinate of a point**, its **ordinate**, is its distance from the x-axis. This distance is measured along the y-axis. For P, this is $+3$; for Q, -3.

(3) In stating the coordinates of a point, the x-coordinate precedes the y-coordinate, just as x in the alphabet precedes y. Place the coordinates in parentheses. Thus the coordinates of P are written $(+4, +3)$ or $(4, 3)$, those for Q, $(-4, -3)$.

The **quadrants of a graph** are the four parts cut off by the axes. Note on the graph how these are numbered I, II, III and IV in a counterclockwise direction.

Comparing a Map and a Graph

A map is a special kind of graph. Each position on a map, such as the map of Graphtown shown, may be located using a street number and an avenue number. Each point on a graph is located using an x number and a y number. The points B, M, L, S and T on the graph correspond to the map position of the bank, museum, library, school and town hall, respectively.

Note how quadrant I corresponds to the northeast map section, quadrant II to the northwest, quadrant III to the southwest and quadrant IV to the southeast.

Use the map and graph in exercises **1.1.** and **1.2.**

MAP of GRAPHTOWN

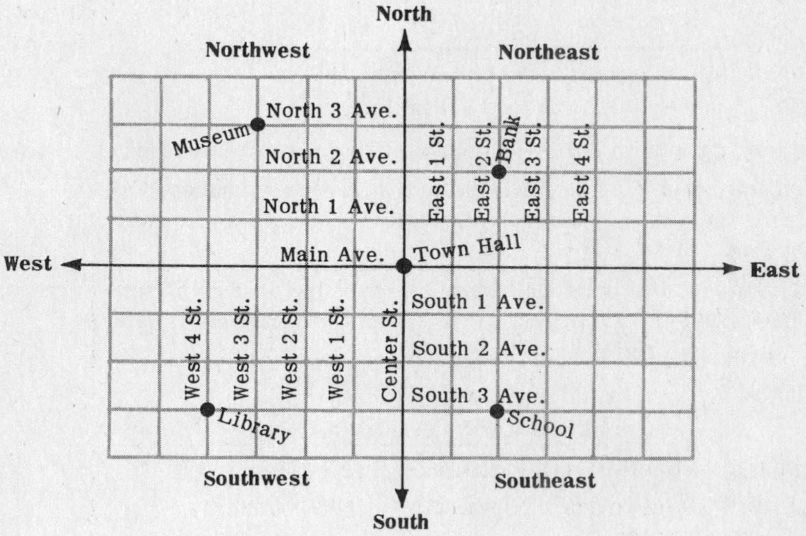

GRAPH

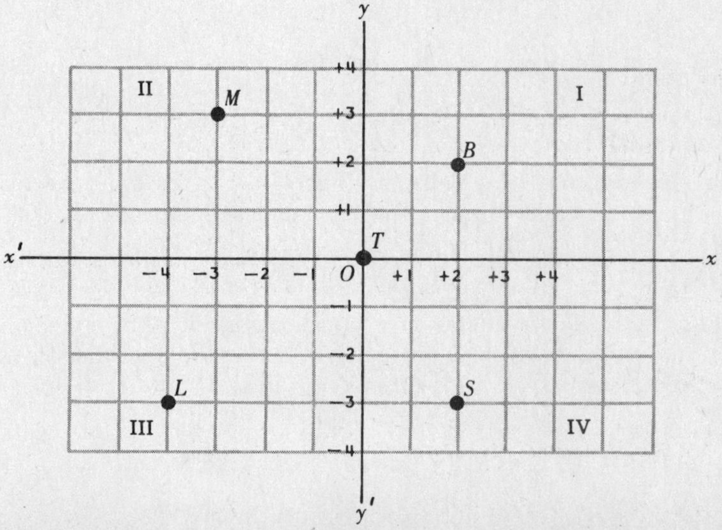

1.1. Locating Points on a Graph

On the graph shown above, locate each point, placing the x-coordinate before the y-coordinate:

	Point	Coordinates (Ans.)
a)	B	$(+2, +2)$ or $(2,2)$
b)	M	$(-3, +3)$ or $(-3,3)$
c)	T	$(0,0)$, the origin

	Point	Coordinates (Ans.)
d)	L	$(-4, -3)$
e)	S	$(+2, -3)$ or $(2, -3)$

1.2. Locating Positions on a Map

On the map of Graphtown, locate the position of each indicated building, placing the street before the avenue.

	Building	Position (Ans.)
a)	Bank	(E 2 St., N 2 Ave.)
b)	Museum	(W 3 St., N 3 Ave.)
c)	Townhall	(Center St., Main Ave.)

	Building	Position (Ans.)
d)	Library	(W 4 St., S 3 Ave.)
e)	School	(E 2 St., S 3 Ave.)

1.3. Coordinates of Points in the Four Quadrants

State the signs of the coordinates of

a) any point A in I
b) any point B in II
c) any point C in III
d) any point D in IV.

Ans. a) $A: (+,+)$ $\quad$ c) $C: (-,-)$
$\quad$ b) $B: (-,+)$ $\quad$ d) $D: (+,-)$

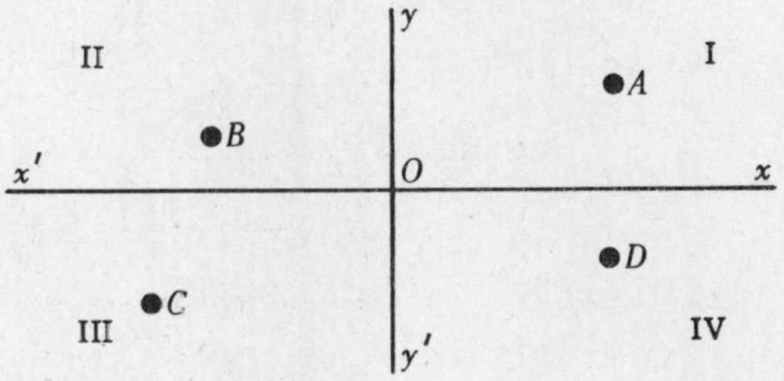

1.4. Coordinates of Points Between the Quadrants

State the zero value of one coordinate and the sign of the other for

a) any point P between I and II
b) any point Q between II and III
c) any point R between III and IV
d) any point S between IV and I.

Ans. a) $P: (0,+)$ $\quad$ c) $R: (0,-)$
$\quad$ b) $Q: (-,0)$ $\quad$ d) $S: (+,0)$

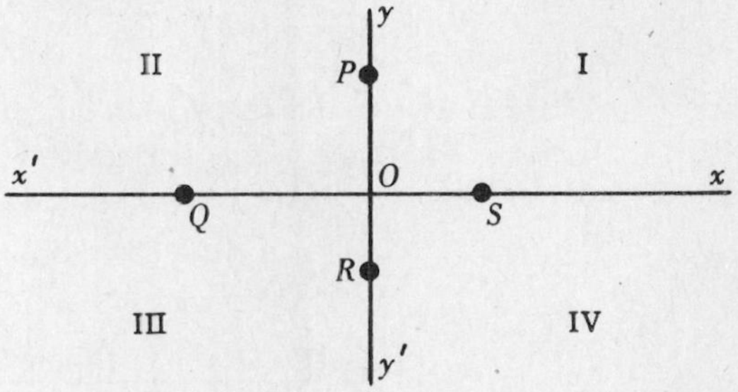

1.5. Graphing a Quadrilateral

If $A(3,1)$, $B(-5,1)$, $C(-5,-3)$ and $D(3,-3)$ are the vertices of the rectangle shown, find its perimeter and area.

Solution:

The base and the height of the rectangle $ABCD$ are 8 and 4. Hence the perimeter is 24 units and the area is 32 sq. units. *Ans.* 24, 32

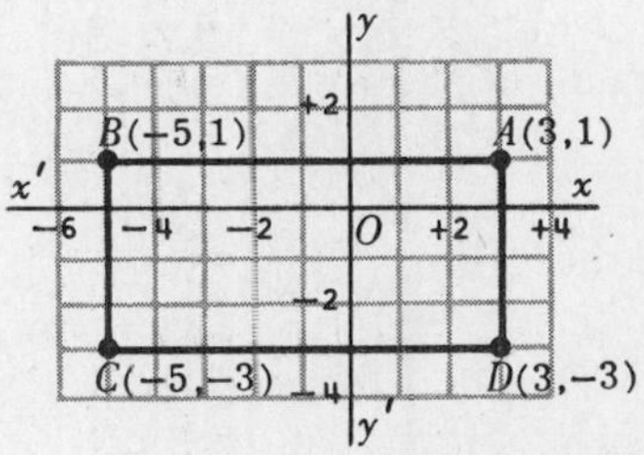

1.6. Graphing a Triangle

If $A(4\frac{1}{2},-2)$, $B(-2\frac{1}{2},-2)$ and $C(1,5)$ are the vertices of the triangle shown, find its area.

Solution:

The base, $BA = 7$. The height, $CD = 7$. Since $A = \frac{1}{2}bh$, $A = \frac{1}{2}(7)(7) = 24\frac{1}{2}$. *Ans.* $24\frac{1}{2}$

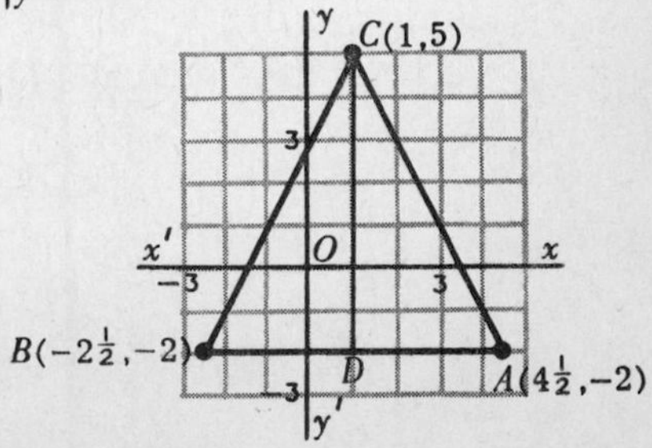

2. GRAPHING LINEAR EQUATIONS

A **linear equation** is an equation whose graph is a straight line.
Note "line" in "linear".

To Graph a Linear Equation

Procedure:

1. **Make a table of coordinates for three pairs of values as follows:** Let x have convenient values such as 2, 0 and -2. Substitute each of these for x and find the corresponding value of y.

2. **Plot the points and draw the straight line joining them:**

   Note. If correct, two points determine a line. The third point serves as a check point to ensure correctness.

Graph: $y = x+4$

Solution:

1. **Table of Coordinate Values:**

Since $y = x+4$,	Point	Coordinates (x,y)
(1) if $x = 2$, $y = 2+4 = 6 \rightarrow$	A	(2,6)
(2) if $x = 0$, $y = 0+4 = 4 \rightarrow$	B	(0,4)
(3) if $x = -2$, $y = -2+4 = 2 \rightarrow$	C	(−2,2)

2. **Join the Plotted Points:**

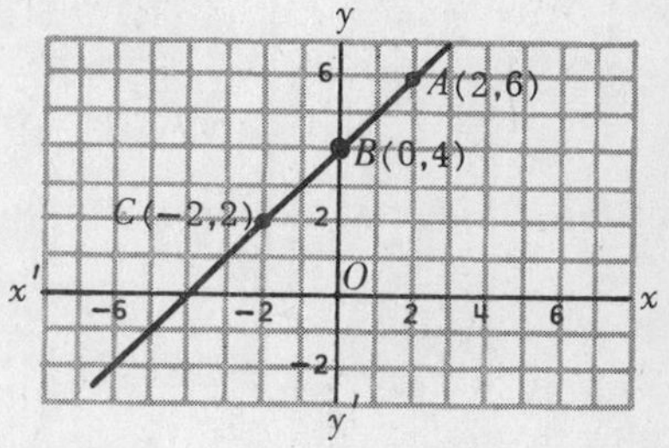

An intercept of a graph is the distance from the origin to the point where the graph crosses either axis.

(1) The **x-intercept** of a graph is the value of x for the point where the graph crosses the x-axis. At this point, $y = 0$.
Thus, for $2x+3y = 6$, the x-intercept $= 3$.

(2) The **y-intercept** of a graph is the value of y for the point where the graph crosses the y-axis. At this point, $x = 0$.
Thus, for $2x+3y = 6$, the y-intercept $= 2$.

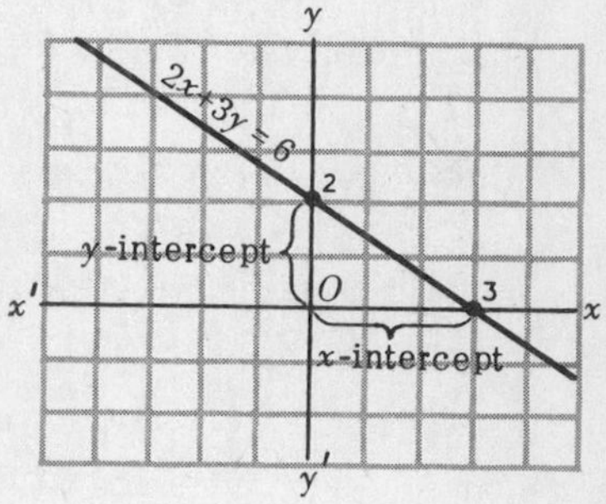

To Graph a Linear Equation Using Intercepts

Procedure:

1. **Make a table of pairs of values as follows:**

   a) Let $x = 0$ to obtain the y-intercept.
   b) Let $y = 0$ to obtain the x-intercept.
   c) Obtain a third or check point, using any convenient value for either unknown.

2. **Plot the points and join them with a straight line:**

Graph: $2x+3y = 6$

Solution:

1. **Table of Coordinate Values:**

	Point	Coordinates (x,y)
a) If $x = 0$, $y = 2$.	A	(0,2)
b) If $y = 0$, $x = 3$.	B	(3,0)
c) If $x = -3$, $y = 4$.	C	(−3,4)

2. **Join the Plotted Points:**

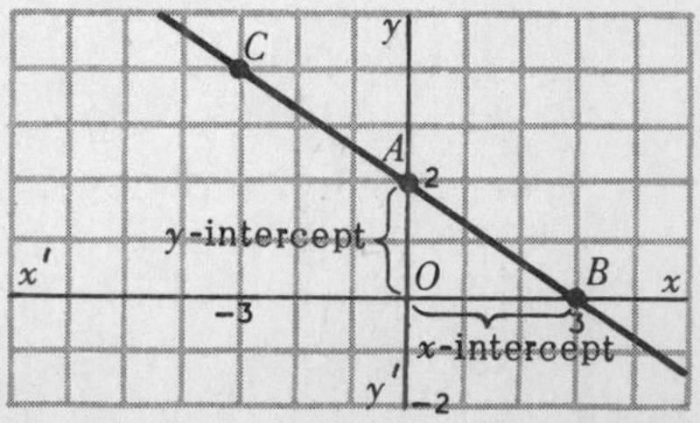

Equations of the First Degree

(1) An equation of the first degree in one unknown is one which, after it has been simplified, contains only one unknown having the exponent 1.

Thus, $2x = 7$ is an equation of the first degree in one unknown.

(2) An equation of the first degree in two unknowns is one which, after it has been simplified, contains only 2 unknowns, each of them in a separate term and having the exponent 1.

Thus, $2x = y + 7$ is an equation of the first degree in two unknowns, but $2xy = 7$ is not since x and y are not in separate terms.

Rule 1. The graph of a first degree equation in one or two unknowns is a straight line. Hence, such equations are **linear equations.**

Thus, the graphs of the equations $y = x + 4$ and $2x + 3y = 6$ are straight lines.

Rule 2. The graph of a first degree equation in only one unknown is either the x-axis, the y-axis or a line parallel to one of the axes. Thus,

(1) The graph of $y = 0$ is the x-axis and the graph of $x = 0$ is the y-axis.

(2) The graphs of $y = 3$ and $y = -3$ are lines parallel to the x-axis.

(3) The graphs of $x = 4$ and $x = -4$ are lines parallel to the y-axis.

Note. The coordinates of any point of intersection of two such lines are obtainable from the equations of these lines. Thus, (4,3) is the intersection of $x = 4$ and $y = 3$.

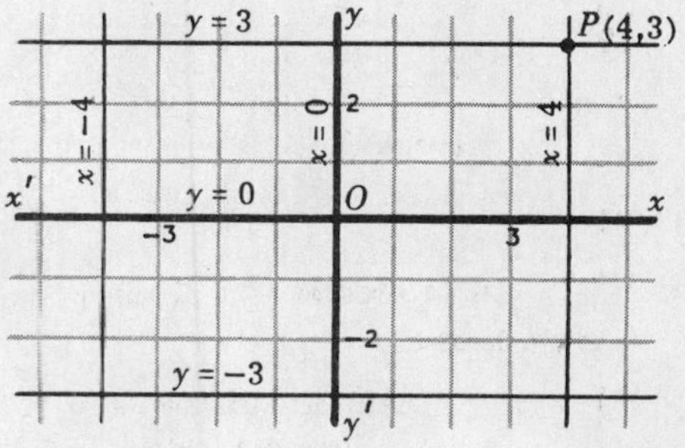

Rule 3. If a point is **on the graph** of an equation, its coordinates **satisfy** the equation. Thus, $x = 1$ and $y = 4$, the coordinates of P on the graph of $y = x + 3$, satisfy the equation $y = x + 3$.

Rule 4. If a point is **not on the graph** of an equation, its coordinates **do not satisfy** the equation.

Thus, $x = 3$ and $y = 4$, the coordinates of B which is not on the graph of $y = x + 3$, do not satisfy the equation $y = x + 3$.

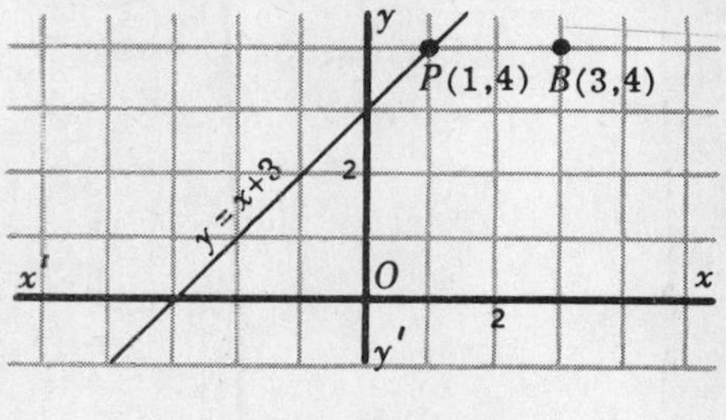

2.1. Making Tables of Coordinate Values

Complete the table of values for each equation:

a) $y = 2x - 3$

	(x,y)
(1)	$(-2,?)$
(2)	$(0,?)$
(3)	$(2,?)$

b) $x = 3y + 1$

	(x,y)
(1)	$(?,-2)$
(2)	$(?,0)$
(3)	$(?,2)$

(Vertically Arranged Tables)

c) $x + 2y = 10$

	(1)	(2)	(3)
x	?	?	?
y	-2	0	2

d) $y = 3$

	(1)	(2)	(3)
x	-2	0	2
y	?	?	?

(Horizontally Arranged Tables)

Solutions:

If one of two coordinate values is given, the corresponding value is found by substituting the given value in the equation and solving.

a)

	(x,y)
(1)	$(-2,-7)$
(2)	$(0,-3)$
(3)	$(2,1)$

b)

	(x,y)
(1)	$(-5,-2)$
(2)	$(1,0)$
(3)	$(7,2)$

c)

	(1)	(2)	(3)
x	14	10	6
y	-2	0	2

d)

	(1)	(2)	(3)
x	-2	0	2
y	3	3	3

($y = 3$ for all values of x.)

2.2. Rule 1. Graphing Linear Equations

Using the same set of axes, graph:

a) $y = \frac{x}{2}$ and $y = 5 - x$ | b) $y = \frac{x-2}{2}$ and $y = 3$

Procedure:

1. Make a table of values:

Solutions:

(a)

Point	(x,y)
A	$(-2,-1)$
B	$(0,0)$
C	$(2,1)$

Point	(x,y)
D	$(-2,7)$
E	$(0,5)$
F	$(2,3)$

(b)

	G	H	I
x	-2	0	2
y	-2	-1	0

	J	K	L
x	-2	0	2
y	3	3	3

2. Join the plotted points:

(a)

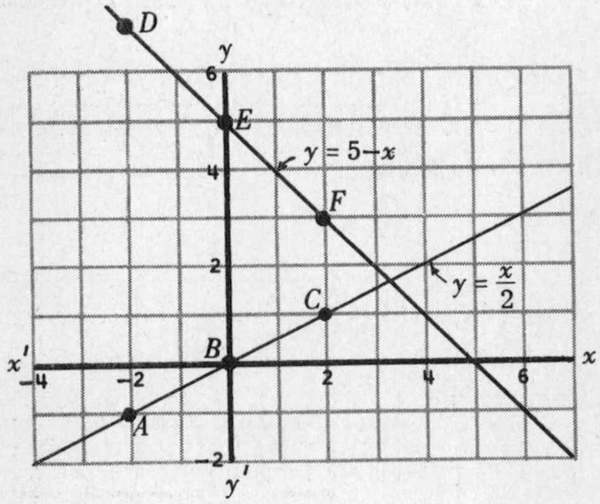

(b)

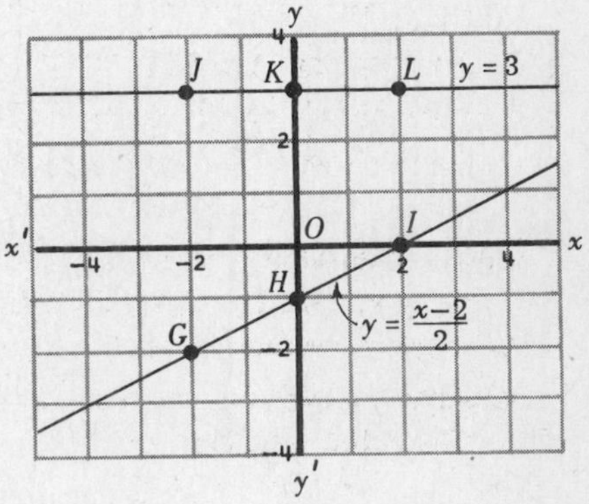

2.3. Graphing Linear Equations, Using Intercepts

Graph, using intercepts:

a) $2x + 5y = 10$ | b) $3x - 4y = 6$

Procedure:

1. Make a table of values:
   a) Let $x = 0$ to find y-intercept:
   b) Let $y = 0$ to find x-intercept:
   c) Obtain a check point:

Solutions:

(a)

Point	(x,y)
A	$(0,2)$
B	$(5,0)$
C	$(2\frac{1}{2},1)$

(b)

Point	(x,y)
D	$(0,-1\frac{1}{2})$
E	$(2,0)$
F	$(4,1\frac{1}{2})$

2. Join the plotted points:

(a)

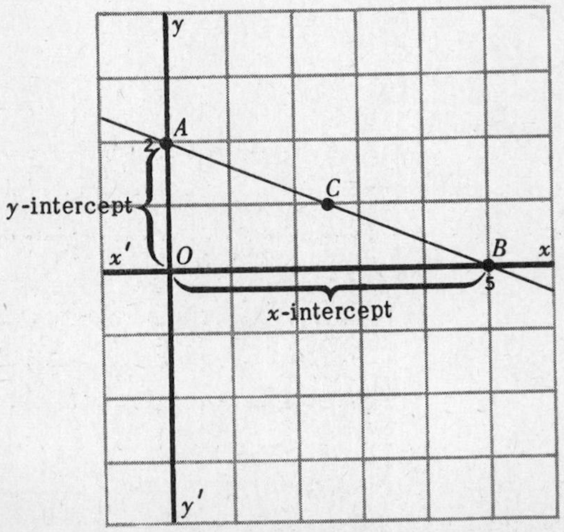

(b)

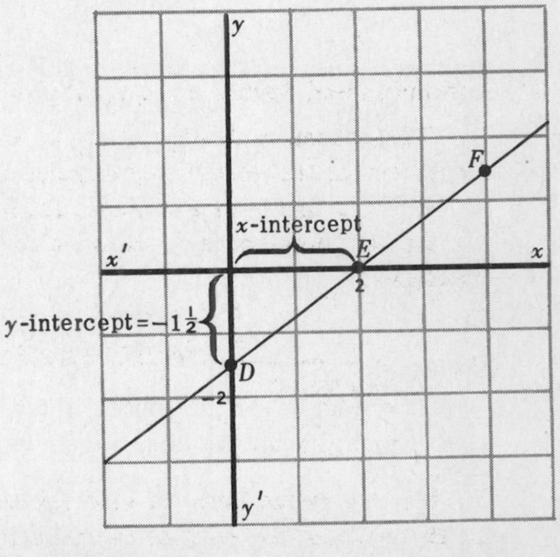

2.4. Rule 2. Graphing Equations of First Degree in Only One Unknown

Using a separate set of axes for each, graph

a) $x=-4$, $x=0$, $x=2\frac{1}{2}$ *b*) $y=3\frac{1}{2}$, $y=0$, $y=-4$

Solutions: (*Each graph is an axis or a line parallel to an axis.*)

a) Graphs of $x=-4$ and $x=2\frac{1}{2}$ are parallel to y-axis. Graph of $x=0$ is the y-axis.

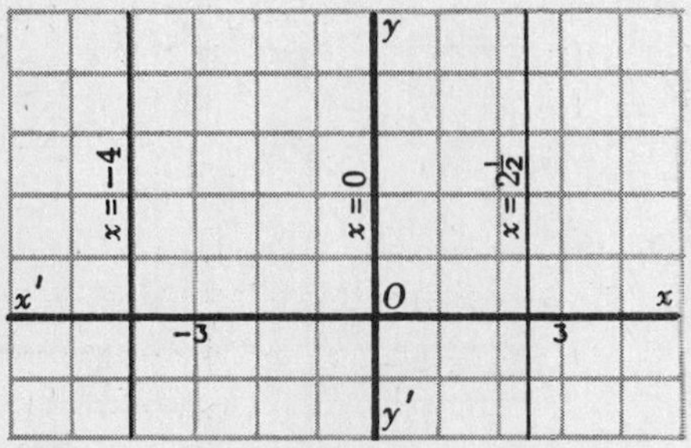

b) Graphs of $y=3\frac{1}{2}$ and $y=-4$ are parallel to x-axis. Graph of $y=0$ is the x-axis.

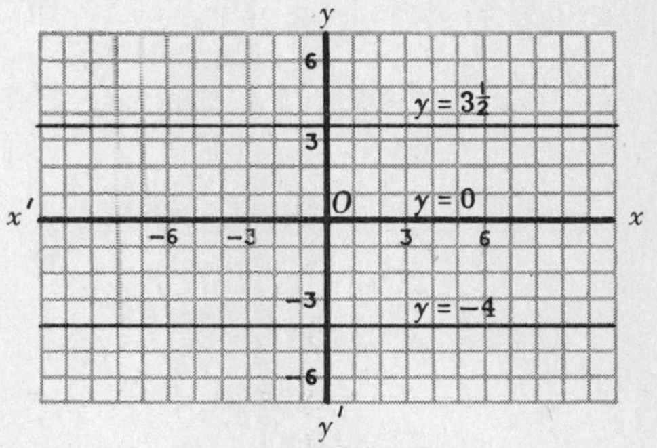

2.5. Rule 3. Coordinate Values of Any Point on a Line

$A(4,3)$ and $B(-5,6)$ are on the graph of $x+3y=13$. Show that their coordinates satisfy the equation.

Solution: (*Substitute to test each pair of coordinate values.*)

Test for $A(4,3)$:	Test for $B(-5,6)$:
$x + 3y = 13$	$x + 3y = 13$
$4 + 3(3) \stackrel{?}{=} 13$	$-5 + 3(6) \stackrel{?}{=} 13$
$13 = 13$	$13 = 13$

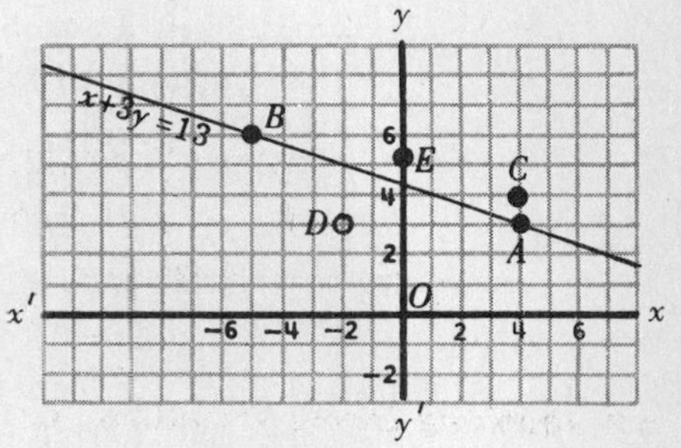

2.6. Rule 4. Coordinate Values of Any Point Not on a Line

$C(4,4)$, $D(-2,3)$ and $E(0,5\frac{1}{3})$ are not on the graph of $x+3y=13$ used in **2.5**. Show that their coordinates do not satisfy the equation.

Solutions: (*Substitute to test each pair of coordinate values.*)

Test for $C(4,4)$:	Test for $D(-2,3)$:	Test for $E(0,5\frac{1}{3})$:
$x + 3y = 13$	$x + 3y = 13$	$x + 3y = 13$
$4 + 3(4) \stackrel{?}{=} 13$	$-2 + 3(3) \stackrel{?}{=} 13$	$0 + 3(5\frac{1}{3}) \stackrel{?}{=} 13$
$4 + 12 \stackrel{?}{=} 13$	$-2 + 9 \stackrel{?}{=} 13$	$16 \neq 13$
$16 \neq 13$	$7 \neq 13$	

2.7. Intercepts and Points of Intersection

The graphs of $3x+2y=6$, $x+2y=-2$ and $2y-3x=6$ are shown in the adjoining diagram. Find

a) the x and y-intercepts of each line,

b) the coordinates of their points of intersection.

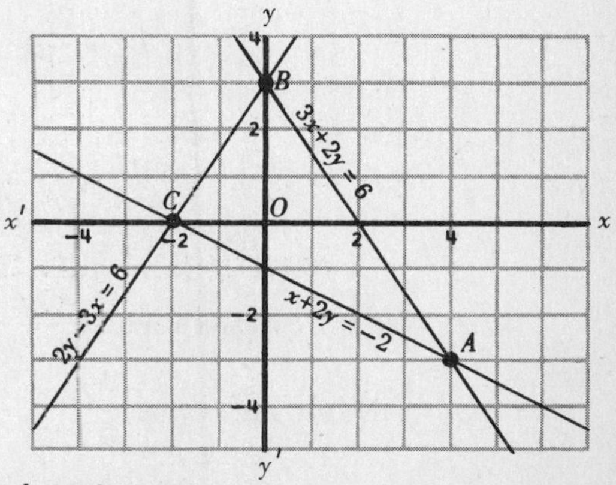

Solutions:

a)

	(Let $y=0$) x-intercept	(Let $x=0$) y-intercept
$3x+2y=6$	2	3
$x+2y=-2$	−2	−1
$2y-3x=6$	−2	3

b) $A(4,-3)$ is the point of intersection of $3x+2y = 6$ and $x+2y = -2$.

$B(0,3)$ is the point of intersection of $3x+2y = 6$ and $2y-3x = 6$.

$C(-2,0)$ is the point of intersection of $x+2y = -2$ and $2y-3x = 6$.

3. SOLVING A PAIR OF LINEAR EQUATIONS GRAPHICALLY

The common solution of two linear equations is the **one and only one** pair of values that satisfies both equations.

Thus, $x=3$, $y=7$ is the common solution of the equations $x+y=10$ and $y=x+4$. Note that the graphs of these equations meet at the point, $x=3$ and $y=7$. Since two intersecting lines meet in one and only one point, this pair of values is the common solution.

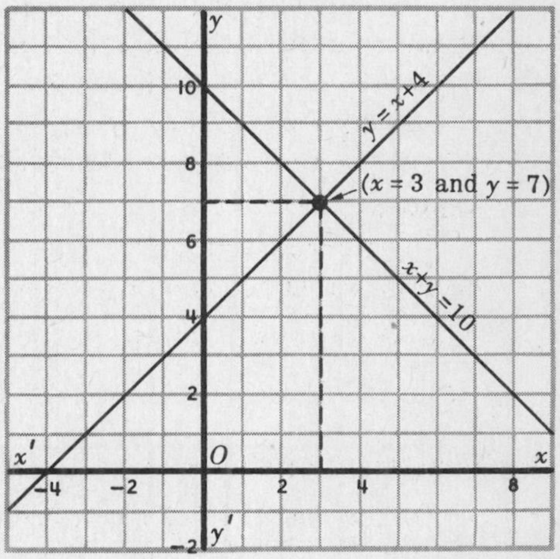

Consistent, Inconsistent and Dependent Equations

(1) Equations are **consistent** if one and only one pair of values satisfies both equations. Thus, $x+y=10$ and $y=x+4$, shown in the adjoining graph, are consistent equations.

(2) Equations are **inconsistent** if no pair of values satisfies both equations. Thus, $x+y=3$ and $x+y=5$ are inconsistent equations. Note that the graphs of these equations are parallel lines and cannot meet. There is no common solution. Two numbers cannot have a sum of 3 and also 5.

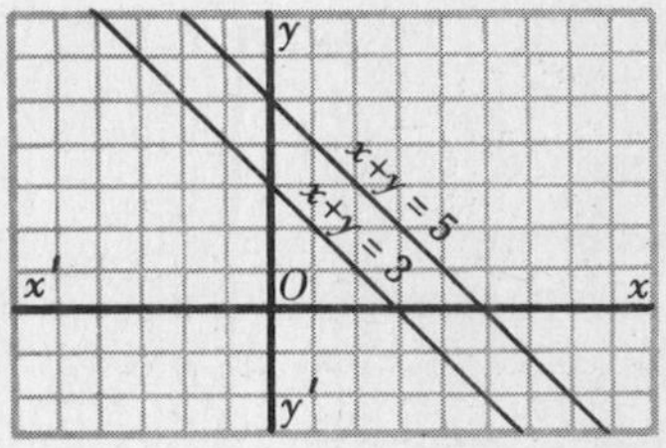

(3) Equations are **dependent** if any pair of values that satisfies one also satisfies the other. Thus, $y=x+2$ and $3y=3x+6$ are dependent equations. Note that the same line is the graph of either equation. Hence, any pair of values of a point on this line satisfies either equation.

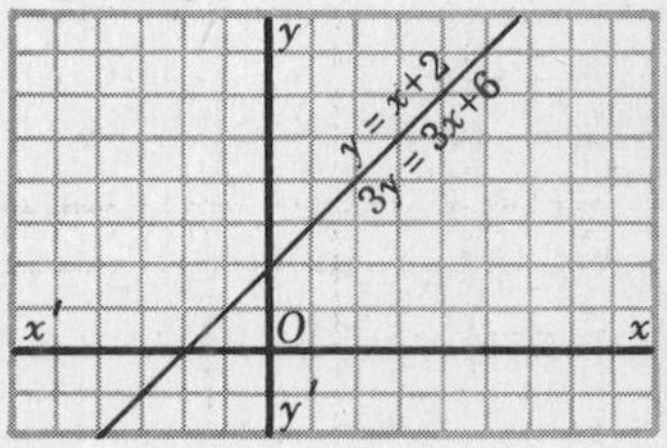

If equations are dependent, one equation can be obtained from the other by performing the same operation with the same number on both sides. Thus, $3y=3x+6$ can be obtained from $y=x+2$ by multiplying both sides of $y=x+2$ by 3.

To Solve a Pair of Linear Equations Graphically

Solve $x+2y=7$ and $y=2x+1$ graphically and check their common solution.

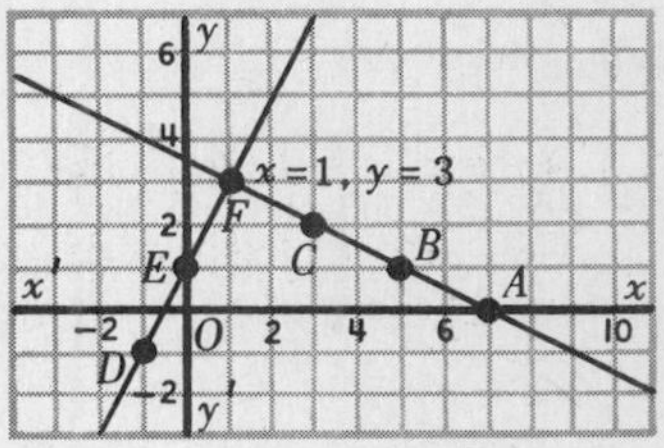

Procedure: | **Solution:**

1. Graph each equation, using the same set of axes:

1.

$x+2y=7$	$y=2x+1$
(x,y)	(x,y)
$A(7,0)$	$D(-1,-1)$
$B(5,1)$	$E(0,1)$
$C(3,2)$	$F(1,3)$

2. Find the common solution:

2. The common solution is the pair of coordinate values of the point of intersection: $x=1$, $y=3$.

3. Check the values found:

3. Check, using $x=1$ and $y=3$:

$x+2y=7$	$y=2x+1$
$1+6 \stackrel{?}{=} 7$	$3 \stackrel{?}{=} 2+1$
$7=7$	$3=3$

3.1. Finding Common Solutions Graphically

From the graphs of $3y+x=5$, $x+y=3$ and $y=x+3$ shown, find the common solution of

a) $3y + x = 5$ and $x + y = 3$,
b) $3y + x = 5$ and $y = x + 3$,
c) $x + y = 3$ and $y = x + 3$.

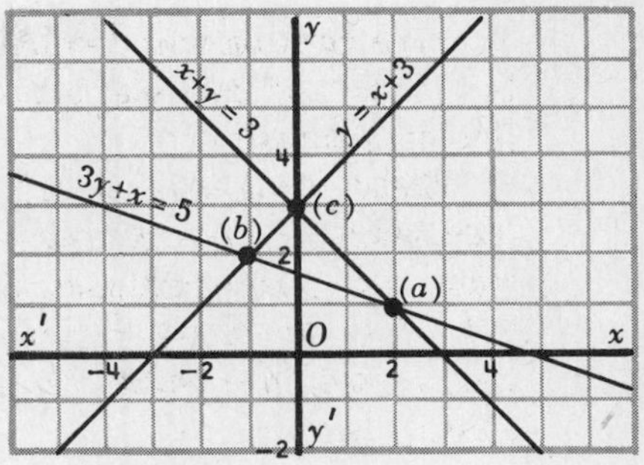

Solutions:

The common solution is the pair of values of the coordinates at the points of intersection.

Ans. *a*) $x=2,\ y=1$ *b*) $x=-1,\ y=2$ *c*) $x=0,\ y=3$

3.2. Consistent Equations and Their Graphic Solution

Solve each pair of equations graphically:

a) $y = 3$
$x = 4$

Solutions:

Graph of $y=3$ is a line parallel to x-axis | Graph of $x=4$ is a line parallel to y-axis

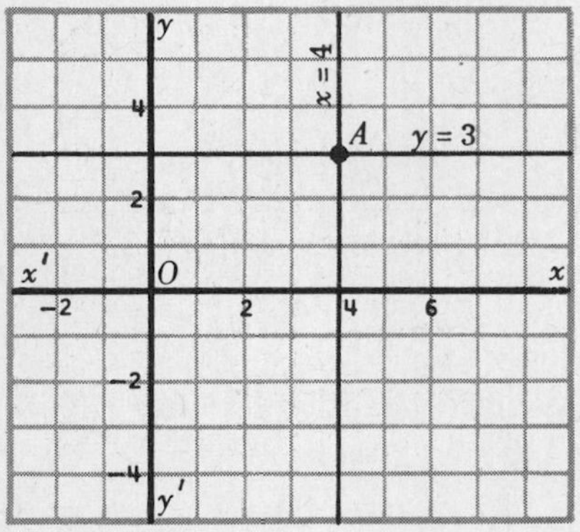

Using intersection point A,
$x=4,\ y=3$ *Ans.*

b) $y = 1$
$y = x - 2$

Graph of $y=1$ is a line parallel to x-axis

$y = x - 2$
(x,y)
$(-2,-4)$
$(0,-2)$
$(2,0)$

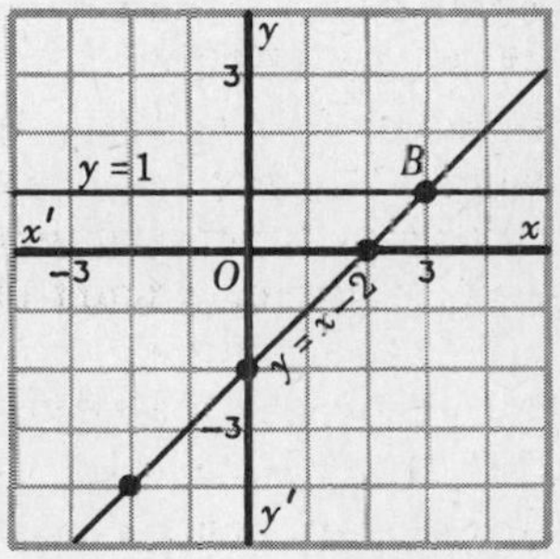

Using intersection point B,
$x=3,\ y=1$ *Ans.*

c) $x+y = 8$
$y = x+6$

$x+y = 8$	$y = x+6$
(x,y)	(x,y)
$(0,8)$	$(-2,4)$
$(2,6)$	$(0,6)$
$(4,4)$	$(2,8)$

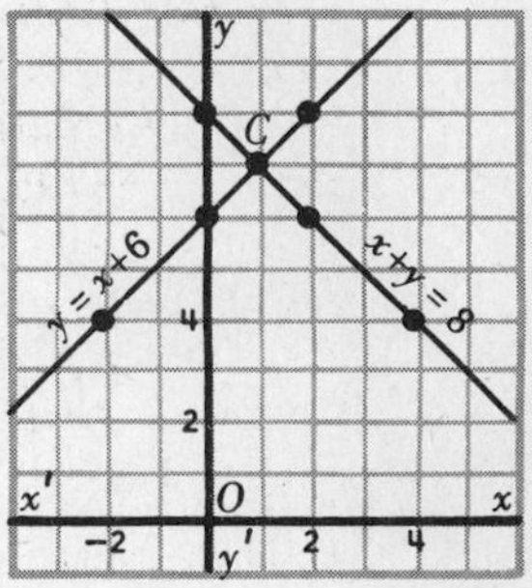

Using intersection point C,
$x=1,\ y=7$ *Ans.*

3.3. Inconsistent Equations

Show graphically that there is no common solution for the following pair of equations:

$$x + y = 4, \quad 2x = 6 - 2y$$

Solution:

$x + y = 4$	$2x = 6 - 2y$
(x,y)	(x,y)
$(0,4)$	$(0,3)$
$(4,0)$	$(3,0)$
$(2,2)$	$(2,1)$

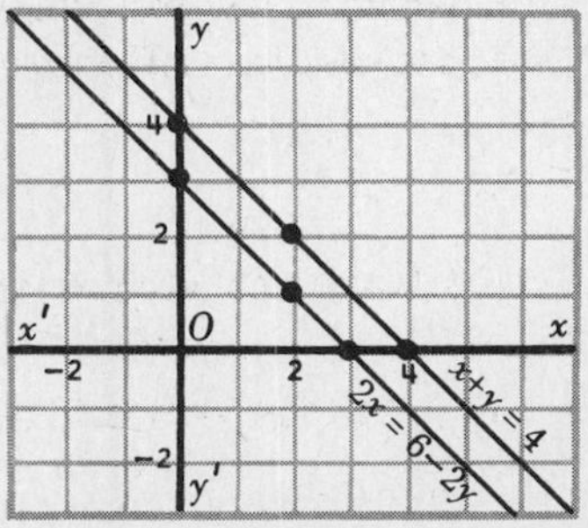

The graphs are parallel lines. Hence, the equations are inconsistent and there is no common solution.

3.4. Dependent Equations

For the equations $x-2y=2$ and $2x=4y+4$, show graphically that any pair of values satisfying one equation satisfies the other equation also.

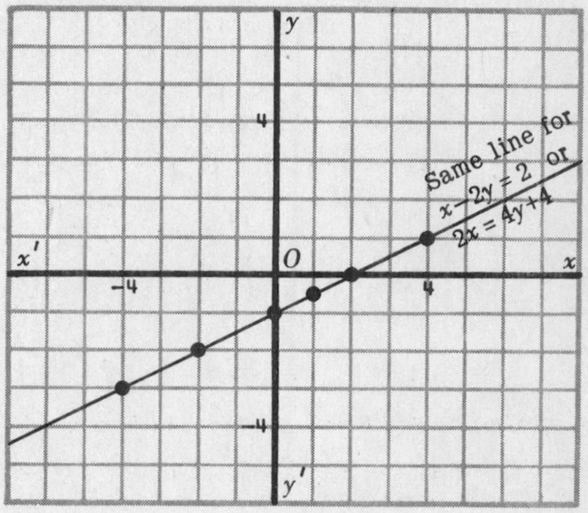

Solution:

$x-2y=2$	$2x=4y+4$
(x,y)	(x,y)
$(0,-1)$	$(-2,-2)$
$(2,0)$	$(-4,-3)$
$(4,1)$	$(1,-\frac{1}{2})$

The same line is the graph of the two equations. Hence, any pair of values satisfying one equation satisfies the other also.

4. DERIVING A LINEAR EQUATION FROM A TABLE OF VALUES

Deriving a Simple Linear Equation By Inspection

A simple linear equation such as $y=x+3$ involves only one operation. Any y-value equals 3 **added to** its corresponding x-value. Note this in the following table of values for $y=x+3$:

y	-7	0	3	4	6	13
x	-10	-3	0	1	3	10

A simple equation can be derived from a table of values by inspection. Once the single operation is found, the equation follows. For example, suppose we are given the table

y	4	8	16	40
x	1	2	4	10

An inspection shows that any y-value equals 4 times its corresponding x-value. Hence, its equation is $y=4x$.

Note also that in $y=x+3$ and $y=4x$, y is expressed in terms of x. In such cases, it is useful to place y-values above x-values in a horizontally arranged table.

Deriving a Linear Equation by the Ratio Method

How can we derive a linear equation when two operations are involved as in the case of $y=4x+2$? Here any y-value equals 2 added to four times its corresponding x-value.

Examine the following table of values for $y=4x+2$ and notice that as x increases 1, y increases 4; as x increases 2, y increases 8 and finally as x decreases 1, y decreases 4.

Change in y $\quad +4 \quad +8 \quad -4$

y	6	10	18	14
x	1	2	4	3

Change in x $\quad +1 \quad +2 \quad -1$

Compare each change in y (above the table) with the corresponding change in x (below the table). Notice that the y-change or y-difference equals **4** times the x-change or x-difference. From this, we may conclude that the equation is of the form $y=4x+b$. The value of b may now be found. To find b, substitute any tabular pair of values for x and y in $y=4x+b$.

Thus, in $y = 4x + b$, substitute $x = 1$ and $y = 6$.

$6 = 4(1) + b$

$2 = b$ Since $b = 2$, the equation is $y = 4x + 2$.

Rule. If a linear equation has the form $y = mx + b$, the value of m can be obtained from a table using the ratio of the y-difference to the corresponding x-difference; that is,

$$m = \frac{y\text{-difference}}{x\text{-difference}} = \frac{\text{difference of two } y\text{-values}}{\text{corresponding difference of two } x\text{-values}}$$

(Think of $\underline{m}$ as the $\underline{\text{m}}$ultiplier of x in $y = \underline{m}x + b$.)

4.1. Deriving Simple Linear Equations by Inspection

Derive the linear equation for each table:

a)

y	3	4	10
x	1	2	8

b)

y	2	0	−4
x	5	3	−1

c)

y	7	21	70
x	1	3	10

d)

y	2	5	9
x	6	15	27

Solutions:

a) Since each y-value is 2 more than its corresponding x-value, $y = x + 2$. *Ans.*

b) Since each y-value is 3 less than its corresponding x-value, $y = x - 3$. *Ans.*

c) Since each y-value is 7 times its corresponding x-value, $y = 7x$. *Ans.*

d) Since each y-value is one-third of its corresponding x-value, $y = \frac{x}{3}$. *Ans.*

4.2. Deriving a Linear Equation of Form $y = mx + b$ by Ratio Method

Derive the linear equation for each table:

a)

y	1	4	10
x	0	1	3

b)

y	−12	−2	18
x	−2	0	4

Procedure:

1. Find m:

$$m = \frac{y\text{-difference}}{x\text{-difference}}$$

2. Find b:
Substitute any tabular pair of values in
$y = mx + b$

3. Form equation
$y = mx + b$

Solutions:

a)

1. y-differences: +3, +6

y	1	4	10
x	0	1	3

x-differences: +1, +2

$$m = \frac{+3}{+1} = \frac{+6}{+2} = 3$$

2. Since $m = 3$, $y = 3x + b$.
Substitute $x = 0$, $y = 1$
in $y = 3x + b$
$1 = 3(0) + b$
$1 = b$

3. $y = 3x + 1$ *Ans.*

b)

1. y-differences: +10, +20

y	−12	−2	18
x	−2	0	4

x-differences: +2, +4

$$m = \frac{+10}{+2} = \frac{+20}{+4} = 5$$

2. Since $m = 5$, $y = 5x + b$.
Substitute $y = 18$, $x = 4$
in $y = 5x + b$
$18 = 5(4) + b$
$-2 = b$

3. $y = 5x - 2$ *Ans.*

SUPPLEMENTARY PROBLEMS

1. State the coordinates of each lettered point on the graph. **(1.1)**

Ans.
$A(3,0)$
$B(4,3)$
$C(3,4)$
$D(0,2)$
$E(-2,4)$
$F(-4,2)$
$G(-1,0)$
$H(-3\frac{1}{2},-2)$
$I(-2,-3)$
$J(0,-4)$
$K(1\frac{1}{2},-2\frac{1}{2})$
$L(4,-2\frac{1}{2})$

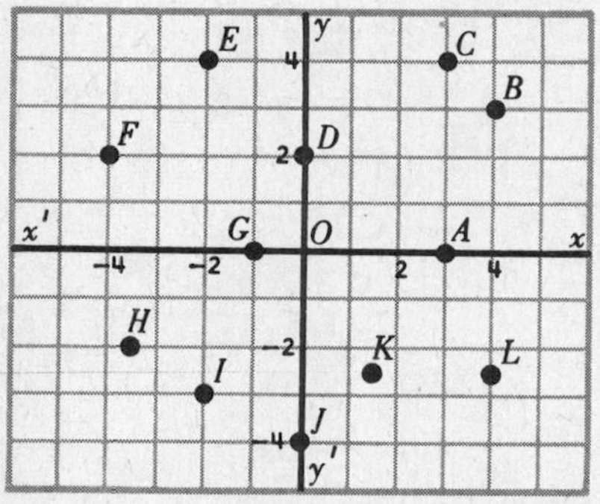

2. Plot each point and locate it with reference to quadrants I to IV: **(1.3, 1.4)**

$A(-2,-3)$ $C(0,-1)$ $E(3,-4)$ $G(0,3)$
$B(-3,2)$ $D(-3,0)$ $F(1\frac{1}{2},2\frac{1}{2})$ $H(3\frac{1}{2},0)$

Ans.

F is in I
B is in II
A is in III
E is in IV
G is between I and II
D is between II and III
C is between III and IV
H is between IV and I

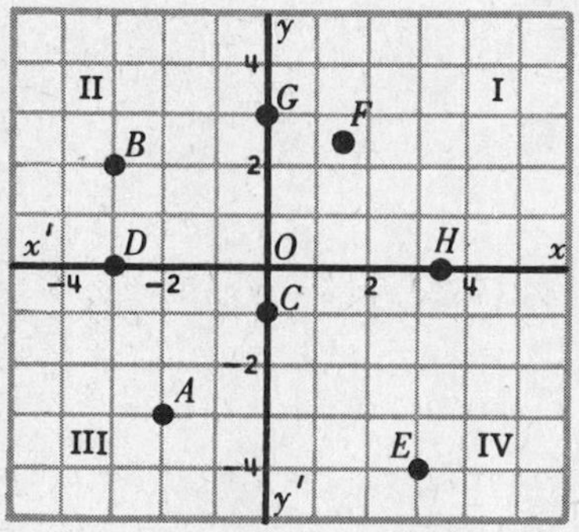

3. Plot each point: $A(2,3)$, $B(-3,3)$, $C(-3,-2)$, $D(2,-2)$. Find the perimeter and area of the square $ABCD$. **(1.5)**

Ans. Perimeter of square formed is 20 units, its area is 25 square units.

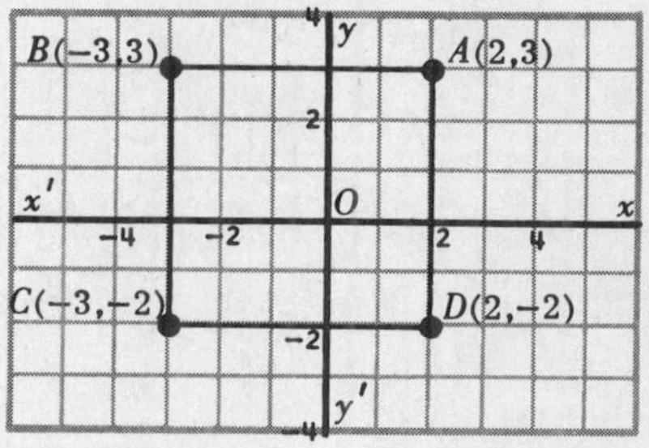

4. Plot each point: $A(4,3)$, $B(-1,3)$, $C(-3,-3)$, $D(2,-3)$. Find the area of parallelogram $ABCD$ and triangle BCD. **(1.5, 1.6)**

Ans. Area of parallelogram = $bh = 5(6)$ or 30 sq. units
Area of $\triangle BCD$ = $\frac{1}{2}bh = \frac{1}{2}(30) = 15$ sq. units

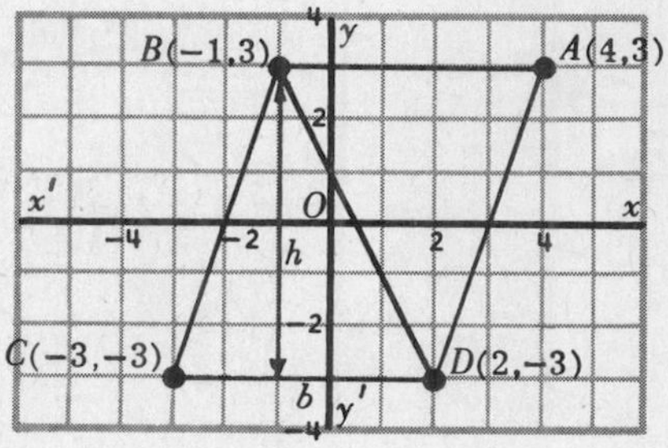

5. Graph each equation after completing each table of values: **(2.1, 2.2)**

a) $y = -4$

	(x,y)
A	$(-2,?)$
B	$(0,?)$
C	$(2,?)$

Ans. a)

	(x,y)
A	$(-2,-4)$
B	$(0,-4)$
C	$(2,-4)$

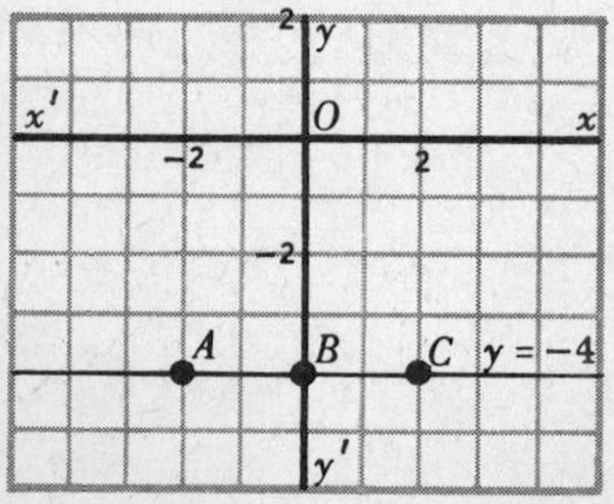

b) $y = 2x$

	(x,y)
D	$(-2,?)$
E	$(0,?)$
F	$(2,?)$

Ans. b)

	(x,y)
D	$(-2,-4)$
E	$(0,0)$
F	$(2,4)$

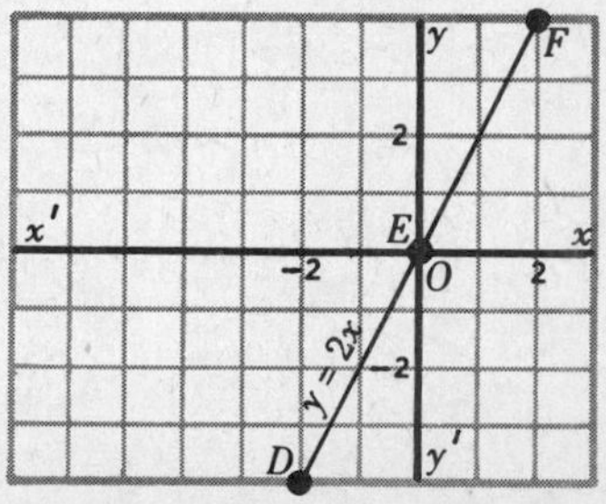

c) $x = 2y + 3$

	(x,y)
G	$(?,-1)$
H	$(?,0)$
I	$(?,1)$

Ans. c)

	(x,y)
G	$(1,-1)$
H	$(3,0)$
I	$(5,1)$

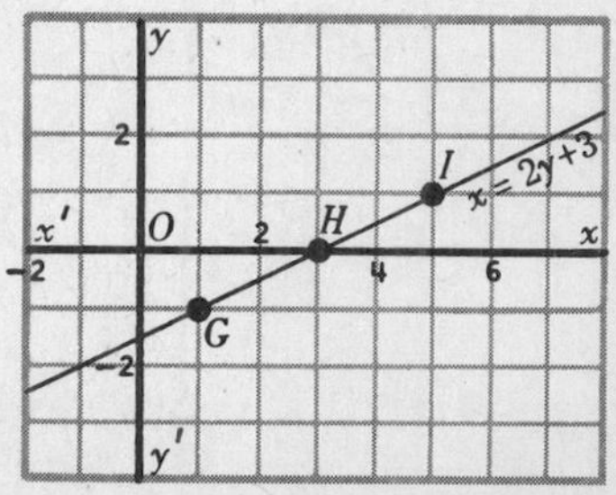

6. Graph each equation, using intercepts: (2.3)

a) $3x + 2y = 6$

Ans. a)

	(x,y)
A	$(0,3)$
B	$(2,0)$
C	$(4,-3)$

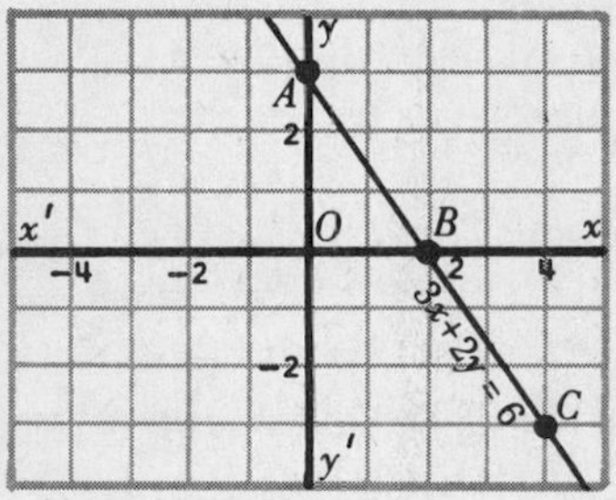

b) $4y - 3x = 6$

Ans. b)

	(x,y)
D	$(0,1\frac{1}{2})$
E	$(-2,0)$
F	$(2,3)$

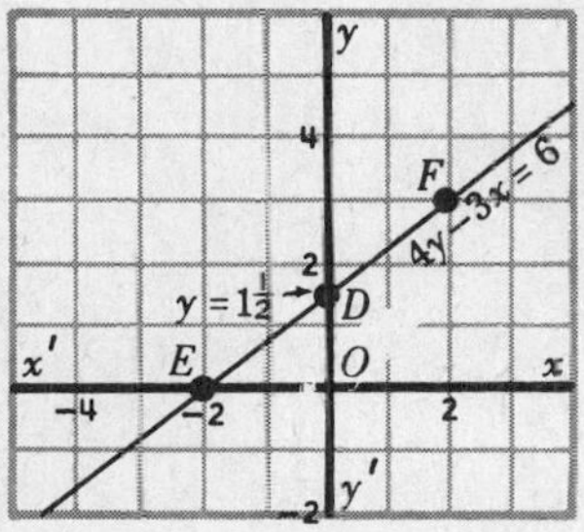

c) $\frac{x}{4} + \frac{y}{3} = 1$

Ans. c)

	(x,y)
G	$(0,3)$
H	$(4,0)$
I	$(-4,6)$

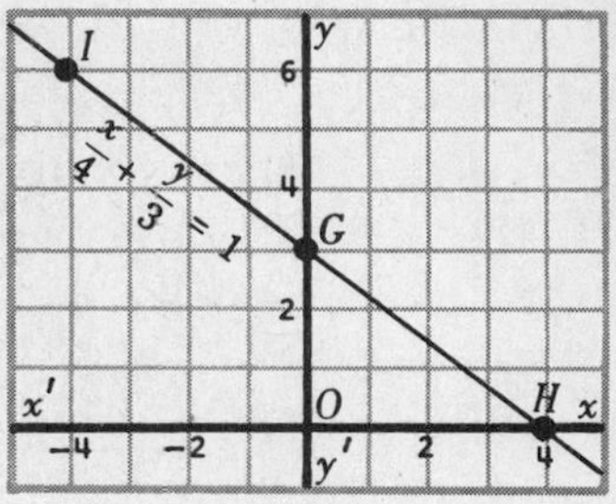

7. Using one set of axes, graph each line and state the coordinates of their nine points of intersection: $x = 1\frac{1}{2}$, $x = 0$, $x = -3$, $y = 5$, $y = 0$, $y = -2\frac{1}{2}$. (2.4)

Ans. The points of intersection are

$A(1\frac{1}{2},0)$	$D(-3,5)$	$G(0,-2\frac{1}{2})$
$B(1\frac{1}{2},5)$	$E(-3,0)$	$H(1\frac{1}{2},-2\frac{1}{2})$
$C(0,5)$	$F(-3,-2\frac{1}{2})$	$O(0,0)$

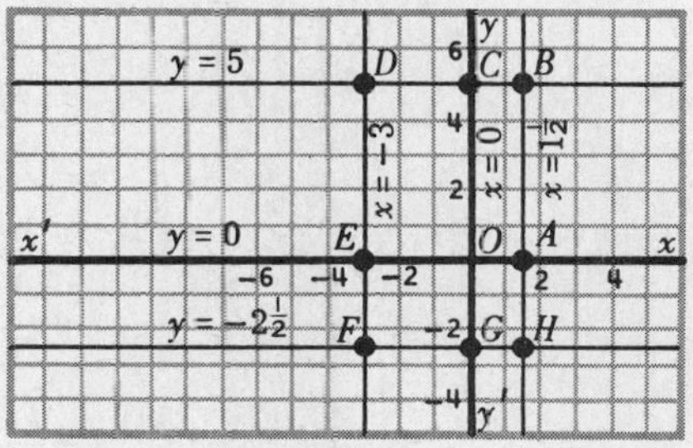

8. Locate $A(1,0)$ and $B(2,-4)$ on the graph of $y + 4x = 4$ and show that their coordinates satisfy the equation. Also, show that $C(2,2)$ and $D(-1,3)$ are not on the graph of $y + 4x = 4$ and show that their coordinates do not satisfy the equation. (2.5, 2.6)

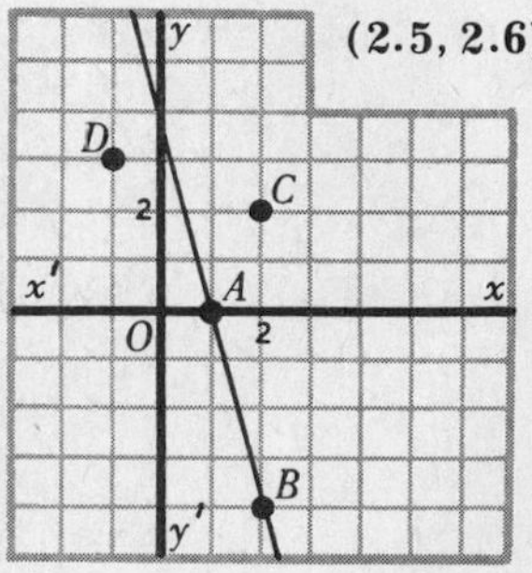

Ans. Test $A(1,0)$	Test $B(2,-4)$	Test $C(2,2)$	Test $D(-1,3)$
$y + 4x = 4$	$y + 4x = 4$	$y + 4x = 4$	$y + 4x = 4$
$0 + 4(1) \stackrel{?}{=} 4$	$-4 + 4(2) \stackrel{?}{=} 4$	$2 + 4(2) \stackrel{?}{=} 4$	$3 + 4(-1) \stackrel{?}{=} 4$
$4 = 4$	$4 = 4$	$10 \neq 4$	$-1 \neq 4$
A is on graph	B is on graph	C is not on graph	D is not on graph

9. The graphs of $2y - x = 6$, $2y - x = 2$ and $2y = 3x - 2$ are shown in the adjacent figure. Find (2.7)

a) the x and y-intercepts of each line,

b) the coordinates of any point of intersection common to two graphs.

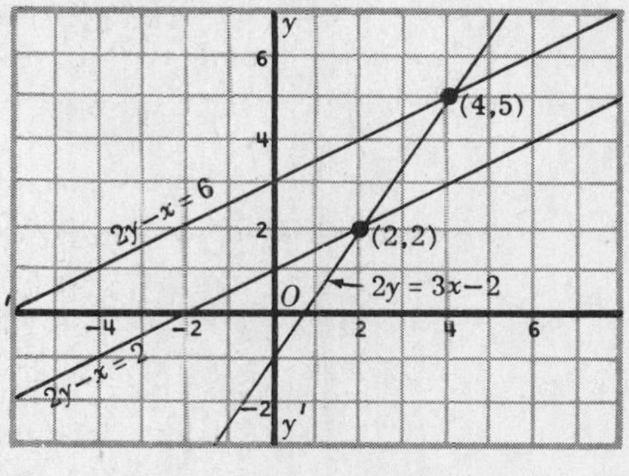

Ans. a)

equation	(Let $y=0$.) x-intercept	(Let $x=0$.) y-intercept
$2y - x = 6$	-6	3
$2y - x = 2$	-2	1
$2y = 3x - 2$	$\frac{2}{3}$	-1

b) $(2,2)$ is the point of intersection of $2y - x = 2$ and $2y = 3x - 2$.
$(4,5)$ is the point of intersection of $2y - x = 6$ and $2y = 3x - 2$.
Since $2y - x = 2$ and $2y - x = 6$ are parallel, they have no point of intersection.

10. From the graph of $3x+4y=12$, $4y+7x=12$ and $3x+4y=-4$, find any solution to any two equations. **(3.1)**

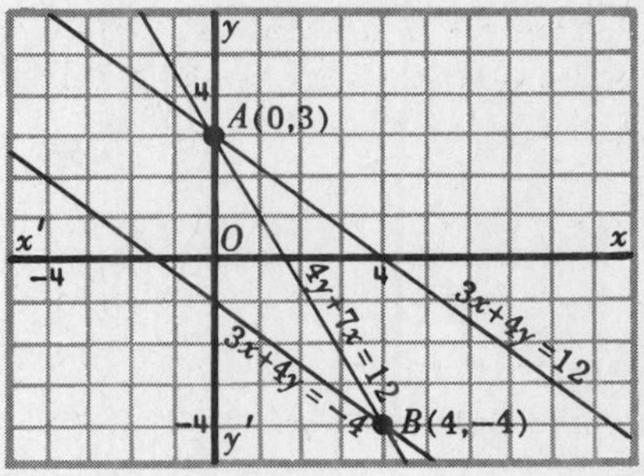

Ans. The common solution of $3x+4y=12$ and $4y+7x=12$ is $x=0$, $y=3$. The common solution of $3x+4y=-4$ and $4y+7x=12$ is $x=4$, $y=-4$. Since $3x+4y=12$ and $3x+4y=-4$ are parallel, there is no common solution.

11. Solve each pair of equations graphically: **(3.2)**

a) $y=-2$
$x=3\frac{1}{2}$

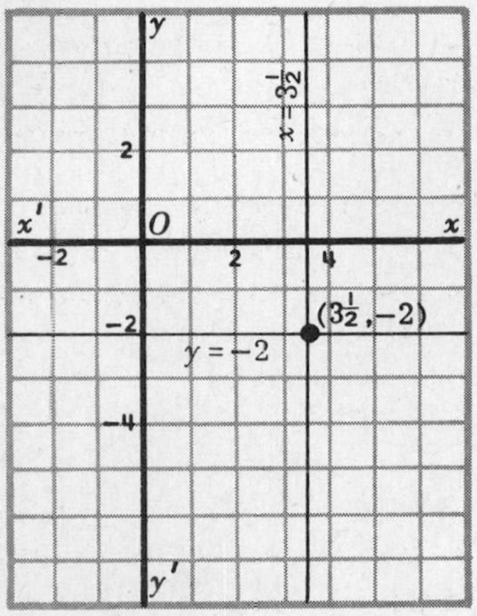

Ans. $x=3\frac{1}{2}$, $y=-2$

b) $y=4$
$x+y=2$

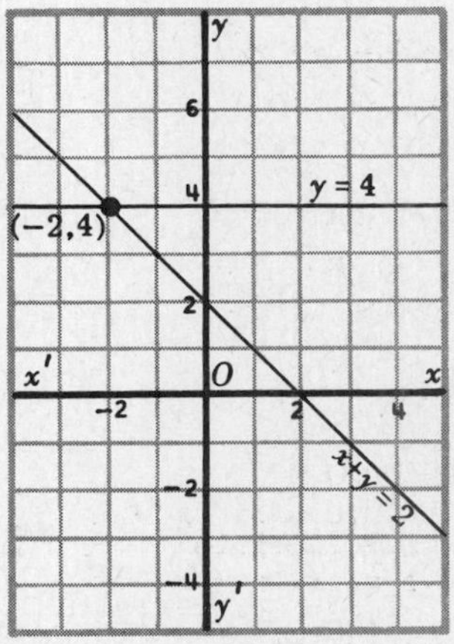

Ans. $x=-2$, $y=4$

c) $3x+4y=-6$
$y=3-3x$

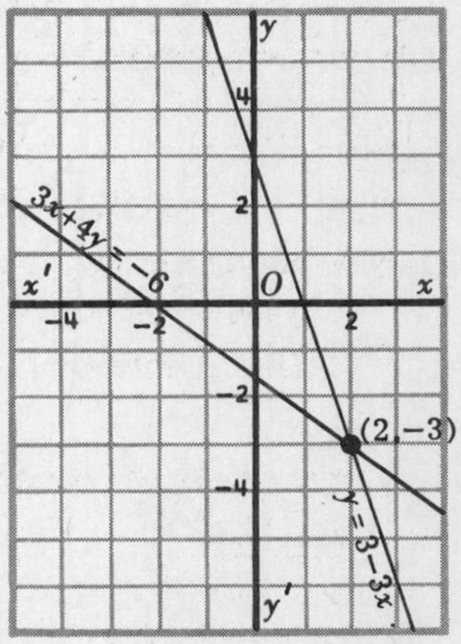

Ans. $x=2$, $y=-3$

12. Graph each equation and determine which equations have no common solution. **(3.3)**

a) $2x+3y=6$
$5x+3y=6$
$2x+3y=-3$

Ans.

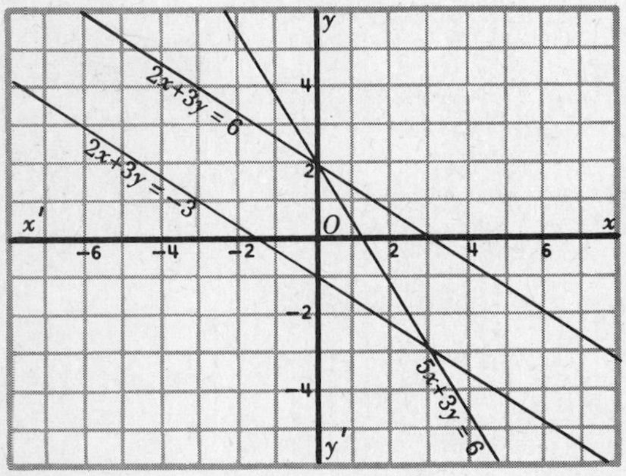

$2x+3y=6$ and $2x+3y=-3$ have no common solution.

b) $y+2x=0$
$2y-3x=6$
$y=3-2x$

Ans.

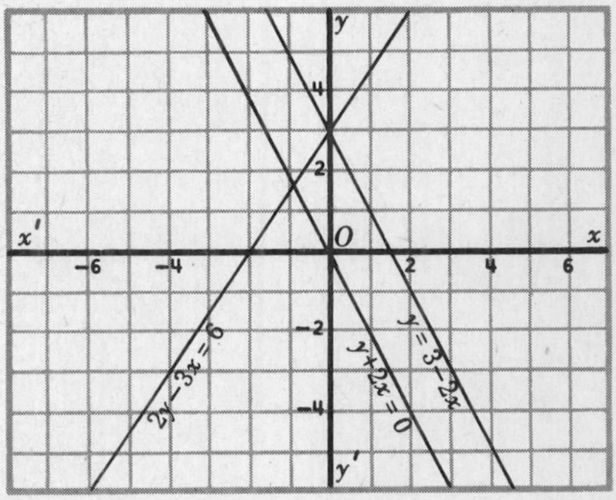

$y=3-2x$ and $y+2x=0$ have no common solution.

13. Graph each equation and determine which equations are such that any pair of values satisfying one equation satisfies the other also: **(3.4)**

a) $2y - x = 2$
$y + 2x = 6$
$3x = 6y - 6$

Ans.

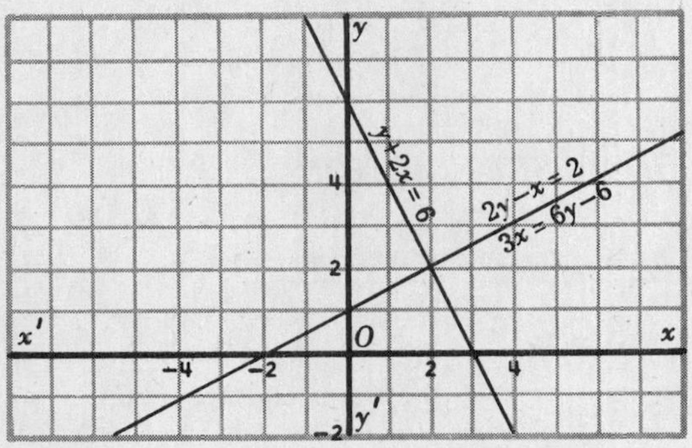

The graph of $2y - x = 2$ and $3x = 6y - 6$ is the same line. A pair of values satisfying one equation satisfies the other.

b) $2y = x - 4$
$x + 4y = 4$
$8y = 8 - 2x$

Ans.

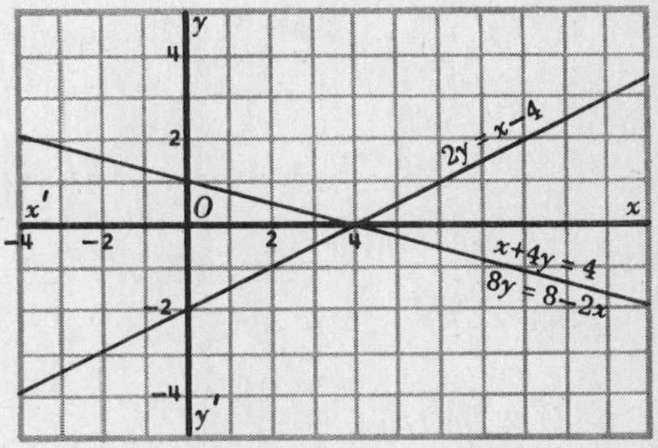

The graph of $x + 4y = 4$ and $8y = 8 - 2x$ is the same line. A pair of values satisfying one equation satisfies the other.

14. Derive the linear equation for each table: **(4.1)**

a)

y	5	8	9
x	2	5	6

b)

y	−4	−6	−10
x	−2	−3	−5

c)

y	−8	−7	−3
x	0	1	5

d)

y	−1	1	4
x	−3	3	12

e)

y	2	2	2
x	7	−7	20

f)

y	4	2	0
x	3	5	7

g)

p	6	9	15
q	8	12	20

h)

r	$4\frac{1}{2}$	$\frac{1}{2}$	−2
s	$1\frac{1}{2}$	$-2\frac{1}{2}$	−5

Ans. *a*) $y = x + 3$ *b*) $y = 2x$ *c*) $y = x - 8$ *d*) $y = \frac{x}{3}$
e) $y = 2$ *f*) $x + y = 7$ *g*) $p = \frac{3}{4}q$ *h*) $r = s + 3$

15. Derive the linear equation for each table: **(4.2)**

a)

y	−1	1	5
x	0	1	3

b)

y	−1	3	15
x	−1	0	3

c)

y	4	3	−2
x	−1	0	5

d)

p	8	0	−4
q	0	4	6

e)

y	2	8	14
x	−2	0	2

f)

y	−19	−4	6
x	−3	0	2

g)

y	6	0	−2
x	0	3	4

h)

s	1	3	7
t	−4	0	8

i)

y	7	16	28
x	1	4	8

j)

y	27	13	−1
x	5	3	1

k)

y	7	−2	−14
x	−1	2	6

l)

a	5	7	11
b	6	9	15

Ans. *a*) $y = 2x - 1$ *b*) $y = 4x + 3$ *c*) $y = -x + 3$ *d*) $p = -2q + 8$
e) $y = 3x + 8$ *f*) $y = 5x - 4$ *g*) $y = -2x + 6$ *h*) $s = \frac{1}{2}t + 3$
i) $y = 3x + 4$ *j*) $y = 7x - 8$ *k*) $y = -3x + 4$ *l*) $a = \frac{2}{3}b + 1$

Chapter 8

EQUATIONS of FIRST DEGREE in TWO UNKNOWNS

1. SOLVING A PAIR OF EQUATIONS BY ADDITION OR SUBTRACTION

To Solve a Pair of Equations by Adding or Subtracting

Solve: (*1*) $3x-7=y$
(*2*) $4x-5y=2$

Procedure:	**Solution:**
1. **Arrange** so that like terms are in the same column:	1. **Arrange:** (*1*) $3x-7=y \longrightarrow 3x-y=7$ (*2*) $4x-5y=2$
2. **Multiply** so that the coefficients of one of the unknowns will have the same absolute value:	2. **Multiply:** M_5 $3x-y=7 \longrightarrow 15x-5y=35$
3. **To eliminate the unknown** whose coefficients have the same absolute value: *a*) **Add** if their signs are **unlike**, *b*) **Subtract** if their signs are **like**.	3. **Eliminate** *y*: $15x-5y=35$ **Subtract:** $4x-5y=2$ $11x=33$
4. **Find the remaining unknown** by solving the resulting equation:	4. **Find** *x*: D_{11} $11x=33$ $x=3$
5. **Find the other unknown** by substituting the value found in any equation having both unknowns:	5. **Find** *y*: $4x-5y=2$ Substitute 3 for *x*: $4(3)-5y=2$ $y=2$
6. **Check** the common solution in each of the original equations:	6. **Check** for $x=3$, $y=2$: (*1*) $3x-7=y$; $3(3)-7\overset{?}{=}2$; $2=2$ (*2*) $4x-5y=2$; $4(3)-5(2)\overset{?}{=}2$; $2=2$

1.1. Using Addition or Subtraction to Eliminate One Unknown

Add or subtract to eliminate one unknown; then find the value of remaining unknown:

a)	*b*)	*c*)	*d*)
$5x+3y=19$ $x+3y=11$	$10x+4y=58$ $13x-4y=57$	$10y=38-6x$ $12y=48-6x$	$3x+10=5y$ $7x+20=-5y$

Solutions:

a)	*b*)	*c*)	*d*)
By subt., $4x=8$ *Ans.* $x=2$	By adding, $23x=115$ *Ans.* $x=5$	By subt., $-2y=-10$ *Ans.* $y=5$	By adding, $10x+30=0$ *Ans.* $x=-3$

1.2. Solutions Not Requiring Multiplication

Solve by addition or subtraction and check:

a) (*1*) $5x + 3y = 17$
(*2*) $x + 3y = 1$

b) (*1*) $10x + 4y = 20$
(*2*) $13x - 4y = -66$

Solutions:

a) By subtracting, $4x = 16$
$x = 4$

Subst. 4 for x in $x + 3y = 1$
$4 + 3y = 1$
$y = -1$

Ans. $x=4$, $y=-1$

Check for $x=4$, $y=-1$:

(*1*) $5x+3y=17$	(*2*) $x+3y=1$
$20-3 \stackrel{?}{=} 17$	$4-3 \stackrel{?}{=} 1$
$17=17$	$1=1$

b) By adding, $23x = -46$
$x = -2$

Subst. -2 for x in $10x + 4y = 20$
$-20 + 4y = 20$
$y = 10$

Ans. $x=-2$, $y=10$

Check for $x=-2$, $y=10$

(*1*) $10x+4y = 20$	(*2*) $13x-4y = -66$
$-20+40 \stackrel{?}{=} 20$	$-26-40 \stackrel{?}{=} -66$
$20 = 20$	$-66 = -66$

1.3. Solutions Requiring Multiplication

Solve by addition or subtraction:

a) (*1*) $13x - 4y = 57$
(*2*) $5x + 2y = 29$

b) (*1*) $3r - 5s = 19$
(*2*) $2r - 4s = 16$

Solutions: [In (*a*) one multiplication is needed, in (*b*) two are needed.]

a) (*1*) $13x - 4y = 57$
(*2*) M_2 $5x+2y = 29 \rightarrow 10x + 4y = 58$

By adding, $23x = 115$
$x = 5$

Subst. 5 for x in $5x + 2y = 29$
$25 + 2y = 29$
$y = 2$

Ans. $x=5$, $y=2$

b) (*1*) M_2 $3r - 5s = 19 \rightarrow 6r - 10s = 38$
(*2*) M_3 $2r - 4s = 16 \rightarrow 6r - 12s = 48$

By subtracting, $2s = -10$
$s = -5$

Subst. -5 for s in $3r - 5s = 19$
$3r + 25 = 19$
$r = -2$

Ans. $s=-5$, $r=-2$

1.4. Solutions Requiring Rearrangement of Terms

Rearrange, then solve by addition or subtraction:

a) (*1*) $2x+y=16$
(*2*) $y = 25-5x$

b) (*1*) $3(x-2) = 8y+1$
(*2*) $8(y+1) = 4x-4$

<u>**Procedure:**</u>	**Solutions:** (*a*)	(*b*)
1. Rearrange:	(*1*) $2x+y = 16$ (*2*) $5x+y = 25$	(*1*) $3x-8y = 7$ (*2*) $-4x+8y = -12$
2. Add or subtract:	By subtr., $-3x = -9$ $x = 3$	By adding, $-x = -5$ $x = 5$
3. Substitute:	Subst. 3 for x in $2x+y = 16$ $6+y = 16$ $y = 10$	Subst. 5 for x in $3(x-2) = 8y+1$ $9 = 8y+1$ $1 = y$
	Ans. $x=3$, $y=10$	*Ans.* $x=5$, $y=1$

1.5. Fractional Pairs of Equations

Solve by addition or subtraction:

a) (1) $\frac{1}{2}x + \frac{1}{4}y = 5$
(2) $\frac{1}{2}x - \frac{3}{4}y = -3$

Solutions:

a)
(1) $\frac{1}{2}x + \frac{1}{4}y = 5$
(2) $\frac{1}{2}x - \frac{3}{4}y = -3$

By subtracting, $y = 8$
Subst. 8 for y in $\frac{1}{2}x + \frac{1}{4}y = 5$
$\frac{1}{2}x + 2 = 5$
$x = 6$

Ans. $x = 6,\ y = 8$

b) (1) $\frac{a}{b} + 5 = -\frac{4}{b}$
(2) $-2a - 3b = -6$

(1) M_b $\frac{a}{b} + 5 = -\frac{4}{b}$
M_2 $a + 5b = -4$
$2a + 10b = -8$
(2) $-2a - 3b = -6$

By adding, $7b = -14,\quad b = -2$
Subst. -2 for b in $-2a - 3b = -6$
$-2a + 6 = -6$

Ans. $a = 6,\ b = -2$

1.6. Decimal Pairs of Equations

Solve by addition or subtraction:

a) (1) $.3x = .6y$
(2) $x + 6y = 8000$

Solutions:

(1) M_{10} $.3x = .6y$
Rearrange $3x = 6y$
$3x - 6y = 0$
(2) $x + 6y = 8000$

By adding, $4x = 8000,\quad x = 2000$
Subst. 2000 for x in $x + 6y = 8000$
$2000 + 6y = 8000$

Ans. $x = 2000,\ y = 1000$

b) (1) $.2x = .05y + 75$
(2) $3x - 5y = 700$

(1) M_{100} $.2x = .05y + 75$
Rearrange $20x = 5y + 7500$
$20x - 5y = 7500$
(2) $3x - 5y = 700$

By subtracting, $17x = 6800,\quad x = 400$
Subst. 400 for x in $3x - 5y = 700$
$1200 - 5y = 700$

Ans. $x = 400,\ y = 100$

2. SOLVING A PAIR OF EQUATIONS BY SUBSTITUTION

To Solve a Pair of Equations by Substitution

Solve: (1) $x - 2y = 7$
(2) $3x + y = 35$

Procedure:

1. **Express one unknown in terms of the other** by transforming one of the equations: **
2. In the other equation, **substitute** for the first unknown the expression containing the other unknown:
3. **Find the remaining unknown** by solving the resulting equation:
4. **Find the other unknown** by substituting the value found in any equation having both unknowns:
5. **Check** the common solution in each of the original equations:

Solution:

1. **Express** x **in terms** of y:
(1) $x - 2y = 7 \longrightarrow x = 2y + 7$
2. **Substitute** for x:
Subst. $(2y+7)$ for x in $3x + y = 35$
$3(2y+7) + y = 35$
3. **Find** y: $6y + 21 + y = 35$
$7y = 14$
$y = 2$
4. **Find** x: Subst. 2 for y in $x = 2y + 7$
$x = 4 + 7$
$x = 11$
5. **Check** for $x = 11,\ y = 2$:

(1)	(2)
$x - 2y \stackrel{?}{=} 7$	$3x + y \stackrel{?}{=} 35$
$11 - 2(2) \stackrel{?}{=} 7$	$3(11) + 2 \stackrel{?}{=} 35$
$7 = 7$	$35 = 35$

**Do this after studying the equations to determine which transformation provides the easier expression for one of the unknowns.

2.1. Using Substitution to Eliminate One Unknown

Substitute to eliminate one unknown; then find the value of the remaining unknown:

a) $y = 4x$, $3x + y = 21$

a) Subst. $4x$ for y in $3x + y = 21$
$$3x + 4x = 21$$
$$7x = 21, \quad x = 3 \quad Ans.$$

b) $x = y + 8$, $x + 3y = 48$

b) Subst. $(y+8)$ for x in $x + 3y = 48$
$$(y+8) + 3y = 48$$
$$4y = 40, \quad y = 10 \quad Ans.$$

c) $x = \frac{y}{2}$, $4x = y + 6$

c) Subst. $\frac{y}{2}$ for x in $4x = y + 6$
$$4(\tfrac{y}{2}) = y + 6$$
$$2y = y + 6, \quad y = 6 \quad Ans.$$

d) $y = 3x + 2$, $2y - 5x = 7$

d) Subst. $(3x+2)$ for y in $2y - 5x = 7$
$$2(3x+2) - 5x = 7,$$
$$6x + 4 - 5x = 7, \quad x = 3 \quad Ans.$$

2.2. Substitution Solutions Requiring Transformation

Solve by substitution:

a) (1) $2x = 3y + 14$
(2) $x - y = 10$

Solutions:

Rearrange (2) → $x = y + 10$

Subst. $y+10$ for x in $2x = 3y + 14$
$$2(y+10) = 3y + 14$$
$$2y + 20 = 3y + 14$$
$$y = 6$$
Since $x = y + 10$, $x = 6 + 10 = 16$.

Ans. $x = 16$, $y = 6$

b) (1) $y - 2x = 9$
(2) $5x = 3y - 26$

Rearrange (1) → $y = 2x + 9$

Subst. $2x+9$ for y in $5x = 3y - 26$
$$5x = 3(2x+9) - 26$$
$$5x = 6x + 27 - 26$$
$$x = -1$$
Since $y = 2x + 9$, $y = -2 + 9 = 7$

Ans. $x = -1$, $y = 7$

2.3. Fractional and Decimal Pairs of Equations

Solve by substitution:

a) (1) $a = 3b$
(2) $\frac{b+5}{3} = 15 - a$

Solutions:

M_3 $\frac{b+5}{3} = 15 - a$

Subst. $3b$ for a in
$$b + 5 = 45 - 3a$$
$$b + 5 = 45 - 9b$$
$$b = 4$$
Since $a = 3b$, $a = 3(4) = 12$.

Ans. $a = 12$, $b = 4$

b) (1) $x = 7 - y$
(2) $\frac{x}{3} = \frac{4-y}{2}$

M_6 $\frac{x}{3} = \frac{4-y}{2}$

Subst. $7-y$ for x in
$$2x = 12 - 3y$$
$$14 - 2y = 12 - 3y$$
$$y = -2$$
Since $x = 7 - y$, $x = 7 + 2 = 9$.

Ans. $x = 9$, $y = -2$

c) (1) $y = 2000 - x$
(2) $.3x = \frac{y}{2} - 40$

M_{10} $.3x = \frac{y}{2} - 40$

Subst. $2000 - x$ for y in
$$3x = 5y - 400$$
$$3x = 5(2000 - x) - 400$$
$$x = 1200$$
Since $y = 2000 - x$, $y = 800$.

Ans. $x = 1200$, $y = 800$

SUPPLEMENTARY PROBLEMS

1. Add or subtract to eliminate one unknown; then find the value of the remaining unknown: **(1.1)**

a) $3x - y = 21$, $2x + y = 4$

b) $7a + b = 22$, $5a + b = 14$

c) $m + 4n = 6$, $m - 2n = 18$

d) $3k + 5p = 9$, $3k - p = -9$

e) $8r = 11 - 5s$, $5r = 15 + 5s$

f) $-10u = 2v + 30$, $-u = -2v + 14$

a) (Add, $5x=25$) *Ans.* $x=5$
b) (Subt., $2a=8$) *Ans.* $a=4$
c) (Subt., $6n=-12$) *Ans.* $n=-2$
d) (Subt., $6p=18$) *Ans.* $p=3$
e) (Add, $13r=26$) *Ans.* $r=2$
f) (Add, $-11u=44$) *Ans.* $u=-4$

2. Solve by addition or subtraction: **(1.2)**

a) $7a + t = 42$, $3a - t = 8$

b) $4r - 2s = -14$, $4r - 5s = -32$

c) $2a + 5s = 44$, $6a - 5s = -8$

d) $7c - 3d = 23$, $-2c - 3d = 5$

e) $8r - 1 = 3t$, $8r + 1 = 9t$

f) $-7g = 31 - 2h$, $-17g = 17 + 2h$

a) (Add) *Ans.* $a=5,\ t=7$
b) (Subt.) *Ans.* $r=-\frac{1}{2},\ s=6$
c) (Add) *Ans.* $a=4\frac{1}{2},\ s=7$
d) (Subt.) *Ans.* $c=2,\ d=-3$
e) (Subt.) *Ans.* $r=\frac{1}{4},\ t=\frac{1}{3}$
f) (Add) *Ans.* $g=-2,\ h=8\frac{1}{2}$

3. Solve by addition or subtraction: **(1.3*a*)**

(*One multiplication is needed before adding or subtracting.*)

a) (*1*) $2x - y = 5$, (*2*) $3x + 3y = 21$

b) (*1*) $A + 4B = 18$, (*2*) $5A + 3B = 5$

c) (*1*) $3c + 5d = 11$, (*2*) $2c - d = 16$

d) (*1*) $7r = 5s + 15$, (*2*) $2r = s + 9$

e) (*1*) $3h - 4j = 13$, (*2*) $-6h + 3j = -21$

f) (*1*) $6t = 3v + 63$, (*2*) $5t = 9v + 85$

a) (M_3 in eq.*1* and add) *Ans.* $x=4,\ y=3$
b) (M_5 in eq.*1* and subt.) *Ans.* $A=-2,\ B=5$
c) (M_5 in eq.*2* and add) *Ans.* $c=7,\ d=-2$
d) (M_5 in eq.*2* and subt.) *Ans.* $r=10,\ s=11$
e) (M_2 in eq.*1* and add) *Ans.* $h=3,\ j=-1$
f) (M_3 in eq.*1* and subt.) *Ans.* $t=8,\ v=-5$

4. Solve by addition or subtraction: **(1.3*b*)**

(*Two multiplications are needed before adding or subtracting.*)

a) (*1*) $3x + 5y = 11$, (*2*) $4x - 3y = 5$

b) (*1*) $5t - 2s = 17$, (*2*) $8t + 5s = 19$

c) (*1*) $8a + 3b = 13$, (*2*) $3a + 2b = 11$

d) (*1*) $6c + 12b = 7$, (*2*) $8c - 15b = -1$

e) (*1*) $-3m + 4p = -6$, (*2*) $5m - 6p = 8$

f) (*1*) $7r + 11s = 35$, (*2*) $6r - 12s = 30$

a) (M_4 in eq.*1*, M_3 in eq.*2* and subtract) *Ans.* $x=2,\ y=1$
b) (M_5 in eq.*1*, M_2 in eq.*2* and add) *Ans.* $t=3,\ s=-1$
c) (M_2 in eq.*1*, M_3 in eq.*2* and subtract) *Ans.* $a=-1,\ b=7$
d) (M_4 in eq.*1*, M_3 in eq.*2* and subtract) *Ans.* $c=\frac{1}{2},\ b=\frac{1}{3}$
e) (M_3 in eq.*1*, M_2 in eq.*2* and add) *Ans.* $m=-2,\ p=-3$
f) (M_6 in eq.*1*, M_7 in eq.*2* and subtract) *Ans.* $r=5,\ s=0$

5. Rearrange; then solve by addition or subtraction: **(1.4)**

a) $3x - y = 17$, $y = 8 - 2x$

b) $2y - 7x = 2$, $3x = 14 - y$

c) $9 = 8p - 5q$, $10 + 7q = 6p$

d) $10C - 9D = 18$, $6D + 2C = 1$

e) $12r - 3s = 3$, $3(r-1) = s - 4$

f) $4(W+3) = 3Z + 7$, $2(Z-5) = W + 5$

Ans. *a*) $x=5,\ y=-2$ *b*) $x=2,\ y=8$ *c*) $p=\frac{1}{2},\ q=-1$ *d*) $C=\frac{3}{2},\ D=-\frac{1}{3}$ *e*) $r=2,\ s=7$ *f*) $W=7,\ Z=11$

6. Solve by addition or subtraction: **(1.5)**

a) $\frac{1}{2}c + \frac{1}{2}d = 4$, $\frac{1}{2}c - \frac{1}{2}d = -2$

b) $\frac{4}{5}r - \frac{1}{4}s = 11$, $\frac{3}{5}r - \frac{1}{4}s = 8$

c) $3K - \frac{1}{3}N = 11$, $2K + \frac{1}{3}N = 4$

d) $x - y = 17$, $\frac{4}{3}x + \frac{3}{2}y = 0$

e) $3r - 2p = 32$, $\frac{r}{5} + 3p = -1$

f) $\frac{3b}{a} - 4 = \frac{3}{a}$, $\frac{5a}{b} + 2 = \frac{25}{b}$

Ans. a) $c=2,\ d=6$ b) $r=15,\ s=4$ c) $K=3,\ N=-6$ d) $x=9,\ y=-8$ e) $r=10,\ p=-1$ f) $a=3,\ b=5$

7. Solve by addition or subtraction: **(1.6)**

a) $3x - y = 500$, $.7x + .2y = 550$

b) $a - 2b = 500$, $.03a + .02b = 51$

c) $y = 4x - 100$, $.06y = .05x + 32$

d) $.03C + .04D = 44$, $.04C + .02D = 42$

e) $.8R - .7T = 140$, $.03R + .05T = 51$

f) $.05(W+2000) = .03(Y+3000)$, $W = \frac{Y}{2} + 500$

Ans. a) $x=500,\ y=1000$ b) $a=1400,\ b=450$ c) $y=700,\ x=200$ d) $C=800,\ D=500$ e) $R=700,\ T=600$ f) $W=4000,\ Y=7000$

8. Substitute to eliminate one unknown, then find the value of the remaining unknown: **(2.1)**

a) $y = 2x$, $7x - y = 35$

b) $a = b + 2$, $3a + 4b = 20$

c) $R = \frac{S}{3}$, $3R + 2S = 36$

d) $r = 4t - 1$, $6t + r = 79$

e) $3p = 27 - q$, $2q = 3p$

f) $5y - 9x = -24$, $5y = 11x$

g) $a = 9 - 3b$, $7b + 5a = 33$

h) $3d - 2g = 27$, $d = 4 - g$

i) $s = \frac{t}{3} - 1$, $6s + t = 21$

a) (Subst. $2x$ for y) *Ans.* $x=7$

b) (Subst. $b+2$ for a) *Ans.* $b=2$

c) (Subst. $\frac{S}{3}$ for R) *Ans.* $S=12$

d) (Subst. $4t-1$ for r) *Ans.* $t=8$

e) (Subst. $2q$ for $3p$) *Ans.* $q=9$

f) (Subst. $11x$ for $5y$) *Ans.* $x=-12$

g) (Subst. $9-3b$ for a) *Ans.* $b=1\frac{1}{2}$

h) (Subst. $4-g$ for d) *Ans.* $g=-3$

i) (Subst. $\frac{t}{3} - 1$ for s) *Ans.* $t=9$

9. Solve by substitution: **(2.2)**

a) $x - y = 12$, $3x = 1 - 4y$

b) $5A - 8B = 8$, $B + A = 12$

c) $r - 3s = 11$, $.5s + 30 = 4r$

d) $p = 2(r-5)$, $4p + 40 = r - 7$

e) $6a = 7c + 7$, $7c - a = 28$

f) $h - 5 = \frac{d}{3}$, $3h - 2d = -6$

a) (Subst. $y+12$ for x) *Ans.* $x=7,\ y=-5$

b) (Subst. $12-B$ for A) *Ans.* $A=8,\ B=4$

c) (Subst. $3s+11$ for r) *Ans.* $r=5,\ s=-2$

d) (Subst. $2r-10$ for p) *Ans.* $p=-12,\ r=-1$

e) (Subst. $7c-28$ for a) *Ans.* $a=7,\ c=5$

f) (Subst. $\frac{d}{3}+5$ for h) *Ans.* $h=12,\ d=21$

10. Solve by substitution: **(2.3)**

a) $x - 9b = 0$, $\frac{x}{3} = 2b + \frac{1}{3}$

b) $r + 5 = 2s$, $\frac{4s+1}{5} = 3r - 3$

c) $\frac{c}{2} + \frac{d}{3} = 9$, $c = 4d + 4$

d) $h + 10m = 900$, $.4h = -2m + 300$

a) (Subst. $9b$ for x) *Ans.* $b=\frac{1}{3},\ x=3$

b) (Subst. $2s-5$ for r) *Ans.* $s=3\frac{1}{2},\ r=2$

c) (Subst. $4d+4$ for c) *Ans.* $c=16,\ d=3$

d) (Subst. $900-10m$ for h) *Ans.* $h=600,\ m=30$

Chapter 9

PROBLEM-SOLVING

The four steps of problem solving are as follows:

1. **Representation** of unknowns by letters.
2. **Translation** of relationships about unknowns into equations.
3. **Solution** of equations to find the values of the unknowns.
4. **Verification** or **check** of the values found to see if they **satisfy the original problem.**

1. NUMBER PROBLEMS HAVING ONE UNKNOWN

In number problems having one unknown, a single relationship about the unknown is needed. After the unknown is represented by a letter such as n or x, this relationship is used to obtain an equation.

The value of the unknown is found by solving the equation. However, the value found must be checked in the original problem. **Do not check in any equation; the equation may be incorrect!**

To Solve a Number Problem Having One Unknown

Solve: Twice a certain number increased by 10 equals 32. Find the number.

Procedure:	**Solution:**
1. **Represent** the unknown by a letter:	1. **Representation:** Let n = the number.
2. **Translate** the relationship about the unknown into an equation:	2. **Translation:** Twice a certain number increased by 10 equals 32. $\underbrace{\text{Twice a certain number}}_{2n}\ \underbrace{\text{increased by 10}}_{+\ 10}\ \text{equals}\ 32$, i.e. $2n + 10 = 32$
3. **Solve** the equation:	3. **Solution:** $2n=22$ or $n=11$. *Ans.* The number is 11.
4. **Check** the value found in the **original problem:** **(Do not check in any equation!)**	4. **Check:** Does 11 satisfy the statement, "Twice the number increased by 10 equals 32."? If it does, then $2(11) + 10 \stackrel{?}{=} 32$, $32 = 32$

1.1. Translation of Statements into Equations

Using n to represent the unknown, express each statement as an equation; then, find the unknown:

		Equations	**Value of n**
a) If a number is decreased by 5, the result is 28.	*Ans.*	$n - 5 = 28$	$n = 33$
b) Three times a number, increased by 8 equals 41.	*Ans.*	$3n + 8 = 41$	$n = 11$
c) Two-fifths a number equals 18.	*Ans.*	$\frac{2n}{5} = 18$	$n = 45$
d) A number added to one-fourth of itself equals 45.	*Ans.*	$n + \frac{n}{4} = 45$	$n = 36$
e) When eight times a number is diminished by 20, the remainder is 28.	*Ans.*	$8n - 20 = 28$	$n = 6$
f) 10 exceeds one-fourth of a number by 3.	*Ans.*	$10 - \frac{n}{4} = 3$	$n = 28$

1.2. Verification or Check in Original Statement

Check each statement for the number indicated:

a) Does 20 check in "one-fifth of the number increased by 3 is 7"?
Check: $\frac{20}{5} + 3 \stackrel{?}{=} 7, \quad 4 + 3 \stackrel{?}{=} 7, \quad 7 = 7$ *Ans.* Yes

b) Does 3 check in "seven times a number less 12 is 9"?
Check: $7(3) - 12 \stackrel{?}{=} 9, \quad 21 - 12 \stackrel{?}{=} 9, \quad 9 = 9$ *Ans.* Yes

c) Does 24 check in "three-fourths of the number decreased by 10 equals 8"?
Check: $\frac{3}{4}(24) - 10 \stackrel{?}{=} 8, \quad 18 - 10 \stackrel{?}{=} 8, \quad 8 = 8$ *Ans.* Yes

d) Does 21 check in "five times the sum of the number and 8 is three times the number less 2"?
Check: $5(21 + 8) \stackrel{?}{=} 3(21) - 2, \quad 5(29) \stackrel{?}{=} 63 - 2, \quad 145 \neq 61$ *Ans.* No

1.3. Complete Solutions of Number Problems

a) Find a number such that three times the number decreased by 5 equals 19.

b) What number added to 40 is the same as five times the number?

c) If eleven times a number is increased by 10, the result is fourteen times the number less 5. Find the number.

Solutions:

a) Let n = the number.

$$3n - 5 = 19$$
$$3n = 24, \quad \underline{n = 8}$$

Ans. The number is 8.

Check:

Does 8 check in "three times the number decreased by 5 equals 19"?

$$24 - 5 \stackrel{?}{=} 19, \quad 19 = 19$$

b) Let n = the number.

$$n + 40 = 5n$$
$$40 = 4n, \quad \underline{n = 10}$$

Ans. The number is 10.

Does 10 check in "the number added to 40 is five times the number"?

$$40 + 10 \stackrel{?}{=} 50, \quad 50 = 50$$

c) Let n = the number.

$$11n + 10 \stackrel{?}{=} 14n - 5$$
$$15 = 3n, \quad \underline{n = 5}$$

Ans. The number is 5.

Does 5 check in "eleven times the number increased by 10 is fourteen times the number less 5"?

$$55 + 10 \stackrel{?}{=} 70 - 5, \quad 65 = 65$$

2. *NUMBER PROBLEMS HAVING TWO UNKNOWNS*

In number problems having two unknowns, two relationships concerning the unknowns are needed. Such problems may be solved by either of two methods:

Method 1. Using One Letter and Obtaining One Equation

One of the relationships is used to represent the two unknowns in terms of one letter. The other relationship is then used to obtain a single equation.

Method 2. Using Two Letters and Obtaining Two Equations

Each of the unknowns is represented by a different letter. Each of the two relationships is then used to obtain a separate equation.

The value found for the unknown by either method must be checked in the **original problem.**

Do Not Check in Any Equation or Equations Since They may be Incorrect!

To Solve a Number Problem Having Two Unknowns

Solve: One positive number is twice another. The larger is 10 more than the smaller. Find the numbers.

Procedure, using one letter:

1. **Represent** one of the unknowns by a letter. Represent the other unknown in terms of the letter, using one of the relationships.

2. **Translate** the other relationship into an equation:

3. **Solve** the equation:

4. **Check** the values found in the **original problem:**

Solution: Method 1.

1. **Representation:** Let s = the smaller number.
       Then $2s$ = the larger number, since the larger is twice the smaller.

2. **Translation:** The larger is 10 more than the smaller.
     $2s$ $=$ $s + 10$

3. **Solution:** $2s - s = 10$, $s = 10$ and $2s = \underline{20}$

Ans. The numbers are 20 and 10.

4. **Check:** Are the numbers 20 and 10?

One number is twice another.	The larger is 10 more
Hence, $20 \overset{?}{=} 2(10)$	than the smaller.
$20 = 20$	Hence, $20 \overset{?}{=} 10 + 10$
	$20 = 20$

Procedure, using two letters:

1. **Represent** each of the two unknowns by a different letter:

2. **Translate** each relationship into a separate equation:

3. **Solve** both equations:

4. **Check** in the **original problem:**

Solution: Method 2.

1. **Representation:** Let l = the larger number, and s = the smaller number.

2. **Translation:** One number is twice another.
       l $=$ $2s$

       The larger is 10 more than the smaller.
       l $=$ $s + 10$

3. **Solution:** Substitute $2s$ for l in $l = s + 10$,
       $2s = s + 10$, $s = 10$.
       Since $l = 2s$, $l = 20$.

Ans. The numbers are 20 and 10.

4. **Check:** (Same as check for Method 1.)

2.1. Representing Two Unknowns Using One Letter

If n represents a number, represent another number that is

a) five more than n	*Ans.* $n+5$	f) the product of n and 15	*Ans.* $15n$
b) ten less than n	*Ans.* $n-10$	g) the quotient of n and 3	*Ans.* $\frac{n}{3}$
c) five times as large as n	*Ans.* $5n$	h) 3 more than twice n	*Ans.* $2n + 3$
d) one-fifth of n	*Ans.* $\frac{n}{5}$	i) 80 reduced by six times n	*Ans.* $80 - 6n$
e) the sum of twice n and 8	*Ans.* $2n+8$	j) ten less than the product of n and 5.	*Ans.* $5n-10$

2.2. Using One Equation for Two Unknowns

Express each statement as an equation; then, find the numbers:

a) Two numbers are represented by n and $(n+5)$.
Their sum is 3 less than three times the smaller.   *Ans.* $n + (n+5) = 3n - 3$, $n = 8$
                           Numbers are 8 and 13.

b) Two numbers are represented by n and $(20-n)$.
The first is three times the second.   *Ans.* $n = 3(20-n)$, $n = 15$
                           Numbers are 15 and 5.

c) Two numbers are represented by n and $(3n-2)$.
The second number is twice the first number less 6.   *Ans.* $(3n-2) = 2n - 6$, $n = -4$
                           Numbers are -4 and -14.

d) Two numbers are represented by n and $6n$.
Ten times the first exceeds the second by 18.   *Ans.* $10n - 6n = 18$, $n = 4\frac{1}{2}$
                           Numbers are $4\frac{1}{2}$ and 27.

2.3. Using Two Equations for Two Unknowns

Using s for the smaller number and l for the larger, obtain two equations for each problem; then find l and s:

a) The sum of two numbers is 15. Their difference is 8.
Ans. a) $l + s = 15$, $l - s = 8$ | $l = 11\frac{1}{2}$, $s = 3\frac{1}{2}$

b) The sum of the larger and 8 is three times the smaller. The larger reduced by 10 equals twice the smaller.
b) $l + 8 = 3s$, $l - 10 = 2s$ | $l = 46$, $s = 18$

c) Separate 40 into two parts such that the larger exceeds twice the smaller by 52.
c) $l + s = 40$, $l - 2s = 52$ | $l = 44$, $s = -4$

2.4. Complete Solution of Number Problem Having Two Unknowns

The larger of two numbers is three times the smaller. Their sum is 8 more than twice the smaller. Find the numbers.

Solution:

Method 1

1. Representation, using one letter:

Let s = smaller number
$3s$ = larger number, since larger is three times as large.

2. Translation, using one equation:

Their sum is 8 more than twice the smaller.

$$\underbrace{3s + s}_{\text{Their sum}} = \underbrace{2s + 8}_{\text{8 more than twice the smaller}}$$

3. Solution:

$4s = 2s + 8, \quad 2s = 8$
$s = 4 \text{ and } 3s = 12$

Ans. Numbers are 12 and 4.

4. Check: (*Do this in the original problem.*)

Method 2

1. Representation, using two letters:

Let s = smaller number
l = larger number

2. Translation, using two equations:

Larger is three times the smaller.

(1) $l = 3s$

Their sum is 8 more than twice the smaller.

(2) $l + s = 2s + 8$

3. Solution: Substitute $3s$ for l in (2):

$3s + s = 2s + 8, \quad 2s = 8$
$s = 4 \text{ and } l = 3s = 12$

Ans. Numbers are 12 and 4.

4. Check: (*Do this in the original problem.*)

3. CONSECUTIVE INTEGER PROBLEMS

An integer is a whole number. An integer may be a positive whole number such as 25, a negative whole number such as −15, or zero.

Each consecutive integer problem involves a set of consecutive integers, a set of consecutive even integers or a set of consecutive odd integers. Each such set involves integers arranged in **increasing order** from left to right.

TABLE OF INTEGERS

	Consecutive Integers	Consecutive Even Integers	Consecutive Odd Integers
Illustrations	4, 5, 6, 7 −4, −3, −2, −1	4, 6, 8, 10 −4, −2, 0, 2	5, 7, 9, 11 −5, −3, −1, 1
Kinds of Integers	**Odd or Even**	**Even Only**	**Odd Only**
Differ by	1	2	2
Representation of First Consecutive No. of Second Consecutive No. of Third Consecutive No.	 n $n + 1$ $n + 2$	 n $n + 2$ $n + 4$	 n $n + 2$ $n + 4$

Note. In the table, n represents the first number of a set. However, n may be used to represent any other number in the set. Thus, a set of three consecutive integers may be represented by $n - 1$, n and $n + 1$.

3.1. Representation Using n for First Integer

Using n for the first integer, represent

	Representation	
a) three consecutive integers and their sum,	*Ans.* *a*) $n,\ n+1,\ n+2$	Sum = $3n+3$
b) three consecutive even integers and their sum,	*Ans.* *b*) $n,\ n+2,\ n+4$	Sum = $3n+6$
c) three consecutive odd integers and their sum,	*Ans.* *c*) $n,\ n+2,\ n+4$	Sum = $3n+6$
d) four consecutive integers and their sum.	*Ans.* *d*) $n,\ n+1,\ n+2,\ n+3$	Sum = $4n+6$

3.2. Representation Using n for Middle Integer

Using n for the middle integer, represent

	Representation	
a) three consecutive integers and their sum,	*Ans.* *a*) $n-1,\ n,\ n+1$	Sum = $3n$
b) three consecutive even integers and their sum,	*Ans.* *b*) $n-2,\ n,\ n+2$	Sum = $3n$
c) five consecutive odd integers and their sum.	*Ans.* *c*) $n-4,\ n-2,\ n,\ n+2,\ n+4$	Sum = $5n$

3.3. Translation in Consecutive Integer Problems

Using n, $n+1$ and $n+2$ for three consecutive integers, express each statement as an equation; then, find the integers:

	Equations	Integers
a) Their sum is 21.	*Ans.* *a*) $3n+3 = 21$	6, 7, 8
b) The sum of the first two is 7 more than the third.	*Ans.* *b*) $2n+1 = (n+2)+7$	8, 9, 10
c) The sum of the second and third is 2 less than three times the first.	*Ans.* *c*) $2n+3 = 3n-2$	5, 6, 7
d) The third added to twice the first is 12 more than twice the second.	*Ans.* *d*) $2n+(n+2) = 2(n+1)+12$	12, 13, 14

3.4. Translation in Consecutive Even Integer Problems

Using n, $n+2$ and $n+4$ for three consecutive even integers, express each statement as an equation; then, find the integers:

	Equations	Integers
a) Their sum is 42.	*Ans.* *a*) $3n+6 = 42$	12, 14, 16
b) The second is half the first.	*Ans.* *b*) $n+2 = \frac{n}{2}$	$-4,\ -2,\ 0$
c) The first equals the sum of the second and third.	*Ans.* *c*) $n = 2n+6$	$-6,\ -4,\ -2$

3.5. Complete Solutions of an Integer Problem

Find five consecutive odd integers whose sum is 45.

Solutions:

Method 1

Represent the five consecutive odd integers, using $n,\ n+2,\ n+4,\ n+6$ and $n+8$.

Then, their sum $= 5n+20 = 45$.

$5n = 25$, $n = 5$ (the first)

Ans. 5, 7, 9, 11, 13

Method 2

Represent the five consecutive odd integers, using $n-4,\ n-2,\ n,\ n+2$ and $n+4$.

Then, their sum $= 5n = 45$.

$n = 9$ (the third)

Ans. 5, 7, 9, 11, 13

4. AGE PROBLEMS

Rule 1. To find a person's future age a number of years hence, **add** that number of years to his present age.

Thus, in 10 yr., a person 17 yr. old will be $17+10$ or 27 yr. old.

Rule 2. To find a person's past age a number of years ago, **subtract** that number of years from his present age.

Thus, 10 yr. ago, a person 17 yr. old was $17-10$ or 7 yr. old.

4.1. Representing Ages on Basis of Present Age

Represent the age of a person in years

a) 10 yr. hence if his present age is x yr. *Ans. a*) $x+10$

b) 10 yr. ago if his present age is x yr. *Ans. b*) $x-10$

c) in y yr. if his present age is 40 yr. *Ans. c*) $40+y$

d) y yr. ago if his present age is 40 yr. *Ans. d*) $40-y$

e) y yr. ago if his present age is p yr. *Ans. e*) $p-y$

4.2. Representing Ages

Find or represent the age of a person (in years)

(In answer, the expression in parentheses is present age.)

a) 5 yr. hence if he was 20 yr. old 10 yr. ago. *Ans. a*) $(20+10)+5 = 35$

b) y yr. hence if he was 30 yr. old 5 yr. ago. *Ans. b*) $(30+5)+y = y+35$

c) 5 yr. ago if he will be 20 yr. old in y yr. *Ans. c*) $(20-y)-5 = 15-y$

4.3. Using One Equation for Two Unknowns in Age Problems

Using $4S$ and S to represent the present ages of a father and his son, express each statement as an equation; then, find their present ages:

		Equations	Ages Now
a) In 14 yr., the father will be twice as old as his son will be then.	*Ans. a*)	$4S+14 = 2(S+14)$ $S = 7$	28, 7
b) In 5 yr., the father will be 21 yr. older than twice the age of his son now.	*Ans. b*)	$4S+5 = 2S+21$ $S = 8$	32, 8
c) 3 yr. ago, the father was five times as old as his son was then.	*Ans. c*)	$4S-3 = 5(S-3)$ $S = 12$	48, 12

4.4. Using Two Equations for Two Unknowns in Age Problems

Obtain two equations for each problem, using F and S to represent the present ages of a father and son, respectively; then find their present ages:

		Equations	Ages Now
a) The sum of their present ages is 40. The father is 20 yr. older than the son.	*Ans. a*)	(*1*) $F+S = 40$ (*2*) $F = S+20$	30, 10
b) The sum of their present ages is 50. In 5 yr., the father will be three times as old as the son will be then.	*Ans. b*)	(*1*) $F+S = 50$ (*2*) $F+5 = 3(S+5)$	40, 10
c) In 8 yr., the father will be twice as old as his son will be then. 3 yr. ago, the father was three times as his son was then.	*Ans. c*)	(*1*) $F+8 = 2(S+8)$ (*2*) $F-3 = 3(S-3)$	36, 14

4.5. Complete Solution of an Age Problem: Two Methods

A father is now 20 yr. older than his son. In 8 yr., the father's age will be 5 yr. more than twice the son's age then. Find their present ages.

Solutions: <u>**Method 1, using two letters.**</u>

Let F = father's present age
S = son's present age

Then (1) $F = S + 20$
(2) $F + 8 = 2(S+8) + 5$

Substitute $S+20$ for F in (2):
$(S+20) + 8 = 2(S+8) + 5, \quad S = 7$

<u>**Method 2, using one letter.**</u>

Let S = son's present age
and $S+20$ = father's present age

Then $(S+20) + 8 = 2(S+8) + 5$
$S + 28 = 2S + 16 + 5$
$7 = S$

Ans. 27 yr. and 7 yr.

5. *RATIO PROBLEMS*

Ratios are used to compare quantities by division.
The ratio of two quantities expressed in the same unit is the first divided by the second. Thus, the ratio of 10 ft. to 5 ft. is 10 ft. ÷ 5 ft. which equals 2.

Ways of Expressing a Ratio

A ratio can be expressed in the following ways:

1. Using a colon, 3:4.
2. Using "to", 3 to 4.
3. As a common fraction, $\frac{3}{4}$.
4. As a decimal fraction, .75.
5. As a percent, 75%.

General Principles of Ratios

1. To find the ratios between quantities, the quantities must have the same unit.
   Thus, to find the ratio of 1 foot to 4 inches, first change the foot to 12 inches. Then, take the ratio of 12 inches to 4 inches. The result is a ratio of 3 to 1 = 3.
2. A ratio is an abstract number; that is, a number without a unit of measure.
   Thus, the ratio of \$7 to \$10 = 7 : 10 = .7. The common unit of \$ must be removed.
3. A ratio should be simplified by reducing to lowest terms and eliminating fractions contained in the ratio.
   Thus, the ratio of 20 to 30 = 2 to 3 = $\frac{2}{3}$.
   Also, the ratio of $2\frac{1}{2}$ to $\frac{1}{2}$ = 5 to 1 = 5.
4. The ratios of three or more quantities may be expressed as a **continued ratio**. This is simply an enlarged ratio statement.
   Thus, the ratio of \$2 to \$3 to \$5 is the continued ratio, 2 : 3 : 5. This enlarged ratio is a combination of three separate ratios. These are 2 : 3, 3 : 5 and 2 : 5, as shown:

   2 : 5
   2 : 3 : 5
   2 : 3 3 : 5

5.1. Ratio of Two Quantities with Same Unit

Express each ratio in lowest terms:

a) \$15 to \$3	*Ans. a*) $\frac{15}{3} = 5$	*e*) \$2.50 to \$1.50	*Ans. e*) $\frac{2.50}{1.50} = \frac{5}{3}$
b) 15 lb. to 3 lb.	*Ans. b*) $\frac{15}{3} = 5$	*f*) \$1.25 to \$5	*Ans. f*) $\frac{1.25}{5} = \frac{1}{4}$
c) 3 oz. to 15 oz.	*Ans. c*) $\frac{3}{15} = \frac{1}{5}$	*g*) $2\frac{1}{2}$ da. to 2 da.	*Ans. g*) $2\frac{1}{2} \div 2 = \frac{5}{4}$
d) 24 sec. to 18 sec.	*Ans. d*) $\frac{24}{18} = \frac{4}{3}$	*h*) $2\frac{1}{4}$ yr. to $\frac{1}{4}$ yr.	*Ans. h*) $2\frac{1}{4} \div \frac{1}{4} = 9$

5.2. Ratio of Two Quantities with Different Units

Express each ratio in lowest terms:

	Change to Same Unit	Ratio	
a) 2 yr. to 3 mo.	*a*) 24 mo. to 3 mo.	$\frac{24}{3} = 8$	*Ans.*
b) 80¢ to \$3.20	*b*) 80¢ to 320¢	$\frac{80}{320} = \frac{1}{4}$	*Ans.*
c) $1\frac{2}{3}$ yd. to 2 ft.	*c*) 5 ft. to 2 ft.	$\frac{5}{2} = 2\frac{1}{2}$	*Ans.*
d) 50 mph to 1 mi. per min.	*d*) 50 mph to 60 mph	$\frac{50}{60} = \frac{5}{6}$	*Ans.*

5.3. Continued Ratio of Three Quantities

Express each continued ratio in lowest terms:

	Change to Same Unit	Continued Ratio
a) 1 gal. to 2 qt. to 2 pt.	*a*) 8 pt. to 4 pt. to 2 pt.	$8:4:2 = 4:2:1$ *Ans.*
b) 1 ton to 1 lb. to 8 oz.	*b*) 2000 lb. to 1 lb. to $\frac{1}{2}$ lb.	$2000:1:\frac{1}{2} = 4000:2:1$
c) \$1 to 1 quarter to 2 dimes	*c*) 100¢ to 25¢ to 20¢	$100:25:20 = 20:5:4$
d) 30 sec. to 2 min. to 1 hr.	*d*) 30 sec. to 120 sec. to 3600 sec.	$30:120:3600 = 1:4:120$

5.4. Numerical Ratios

Express each ratio in lowest terms:

a) 50 to 60 *Ans.* $\frac{50}{60} = \frac{5}{6}$

b) 70 to 55 *Ans.* $\frac{70}{55} = \frac{14}{11}$

c) 175 to 75 *Ans.* $\frac{175}{75} = \frac{7}{3}$

d) 6.3 to .9 *Ans.* $\frac{6.3}{.9} = 7$

e) $1\frac{3}{4}$ to 7 *Ans.* $\frac{7}{4} \div 7 = \frac{1}{4}$

f) 12 to $\frac{3}{8}$ *Ans.* $12 \div \frac{3}{8} = 32$

g) $7\frac{1}{6}$ to $1\frac{1}{3}$ *Ans.* $\frac{43}{6} \div \frac{4}{3} = \frac{43}{8}$

h) 80% to 30% *Ans.* $\frac{80\%}{30\%} = \frac{8}{3}$

5.5. Algebraic Ratios

Express each ratio in lowest terms:

a) $2x$ to $5x$ *Ans.* $\frac{2x}{5x} = \frac{2}{5}$

b) $3a$ to $6b$ *Ans.* $\frac{3a}{6b} = \frac{a}{2b}$

c) πD to πd *Ans.* $\frac{\pi D}{\pi d} = \frac{D}{d}$

d) $4ab$ to $3ab$ *Ans.* $\frac{4ab}{3ab} = \frac{4}{3}$

e) $5s^2$ to s^3 *Ans.* $\frac{5s^2}{s^3} = \frac{5}{s}$

f) x to $5x$ to $7x$ *Ans.* $x:5x:7x = 1:5:7$

5.6. Representation of Numbers in a Fixed Ratio

Using x as their common factor, represent the numbers and their sum if

	Numbers	Sum
a) two numbers have a ratio of 4 to 3,	*Ans.* $4x$ and $3x$	$7x$
b) two numbers have a ratio of 7 to 1,	*Ans.* $7x$ and x	$8x$
c) three numbers have a ratio of $4:3:1$,	*Ans.* $4x$, $3x$ and x	$8x$
d) three numbers have a ratio of $2:5:8$,	*Ans.* $2x$, $5x$ and $8x$	$15x$
e) five numbers have a ratio of $8:5:3:2:1$.	*Ans.* $8x$, $5x$, $3x$, $2x$ and x	$19x$

5.7. Ratio in Number Problems

If two numbers in the ratio of 5 : 3 are represented by $5x$ and $3x$, express each statement as an equation; then, find x and the numbers:

	Equations	Value of x	Numbers
a) The sum of the numbers is 88.	*Ans. a*) $8x = 88$	$x = 11$	55 and 33
b) The difference of the numbers is 4.	*Ans. b*) $2x = 4$	$x = 2$	10 and 6
c) Twice the larger added to three times the smaller is 57.	*Ans. c*) $10x + 9x = 57$	$x = 3$	15 and 9
d) Three times the smaller equals the larger increased by 20.	*Ans. d*) $9x = 5x + 20$	$x = 5$	25 and 15

5.8. Ratio in a Triangle Problem

The sides of a triangle are in the ratio of 2 : 3 : 4. If the perimeter of the triangle is 45 in., find each side.

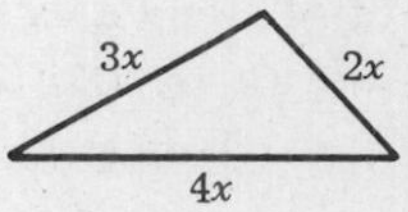

Solution:

Let $2x$, $3x$, and $4x$ represent the sides in inches.

Then the perimeter, $9x = 45$;

$\underline{x = 5}$, $2x = 10$, $3x = 15$, $4x = 20$. *Ans.* 10 in., 15 in., 20 in.

5.9. Ratio in a Money Problem

Will has $3\frac{1}{2}$ times as much money as Lester. If Will gives Lester a quarter, they will then have the same amount. How much did each have?

Solution:

Since the ratio of their money, $3\frac{1}{2}$ to 1, is 7 : 2,

let $7x$ = Will's money in ¢ and $2x$ = Lester's money in ¢.

Then, $7x - 25 = 2x + 25$, and $\underline{x = 10}$, $7x = 70$, $2x = 20$.

Ans. 70¢ and 20¢

5.10. Ratio in a Will Problem

In his will, a man left his wife $20,000 and his son $12,000. Upon his death, his estate amounted to only $16,400. If the court divides the estate in the ratio of the bequests in the will, what should each receive?

Solution:

The ratio of $20,000 to $12,000 is 5 : 3.

Let $5x$ = the wife's share in $ and $3x$ = the son's share in $.

Then, $8x = 16{,}400$, and $\underline{x = 2050}$, $5x = 10{,}250$, $3x = 6150$.

Ans. $10,250 and $6150

5.11. Ratio in a Wage Problem

Henry and George receive the same hourly wage. After Henry worked 4 hr. and George $3\frac{1}{2}$ hr., Henry found he had $1.50 more than George. What did each earn?

Solution:

Since the ratio of 4 to $3\frac{1}{2}$ is 8 : 7,

let $8x$ = Henry's earnings in $ and $7x$ = George's earnings in $.

Then, $8x = 7x + 1.50$, and $\underline{x = 1.50}$, $8x = 12$, $7x = 10.50$. *Ans.* $12 and $10.50

6. ANGLE PROBLEMS

Pairs of Angles:

(1) Adjacent Angles

Adjacent angles are two angles having the same vertex and a common side between them.

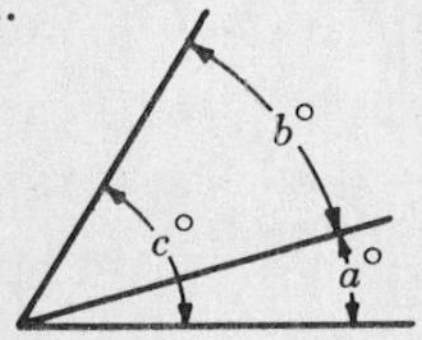

Fig. 1

Rule 1. If an angle of c° consists of two adjacent angles of a° and b°, as in Fig. 1, then

$$a^\circ + b^\circ = c^\circ.$$

Thus, if $a^\circ = 25^\circ$ and $b^\circ = 32^\circ$, then $c^\circ = 57^\circ$.

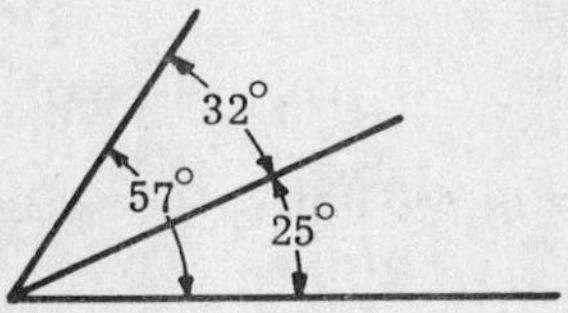

(2) Complementary Angles

Complementary angles are two angles whose sum equals 90° or a right angle.

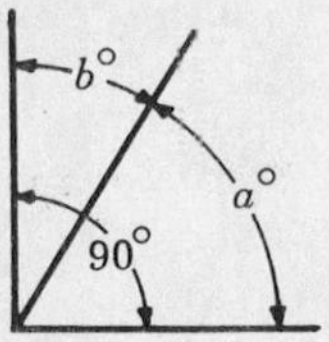

Fig. 2

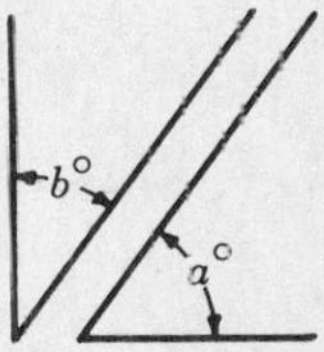

Fig. 3

Rule 2. If two complementary angles contain a° and b°, then

$$a^\circ + b^\circ = 90^\circ \text{ or } b^\circ = 90^\circ - a^\circ.$$

Thus, 70° and 20° are complementary angles.

Complementary angles may be adjacent as in Fig. 2, or non-adjacent as in Fig. 3.

Either of two complementary angles is the complement of the other. If a° and b° are complementary,

$$b^\circ = \text{complement of } a^\circ = 90^\circ - a^\circ$$
$$a^\circ = \text{complement of } b^\circ = 90^\circ - b^\circ.$$

(3) Supplementary Angles

Supplementary angles are two angles whose sum equals 180° or a straight angle.

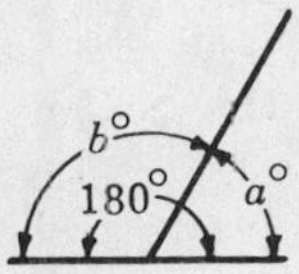

Fig. 4

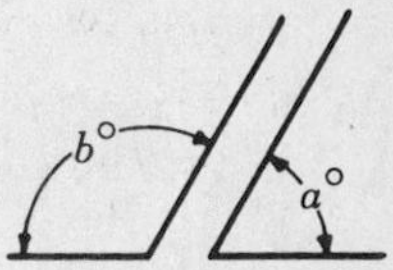

Fig. 5

Rule 3. If two supplementary angles contain a° and b°, then

$$a^\circ + b^\circ = 180^\circ \text{ or } b^\circ = 180^\circ - a^\circ.$$

Thus, 70° and 110° are supplementary angles.

Supplementary angles may be adjacent as in Fig. 4, or non-adjacent as in Fig. 5.

Either of two supplementary angles is the supplement of the other. If a° and b° are supplementary,

$$b^\circ = \text{supplement of } a^\circ = 180^\circ - a^\circ$$
$$a^\circ = \text{supplement of } b^\circ = 180^\circ - b^\circ.$$

(4) Angles of any Triangle

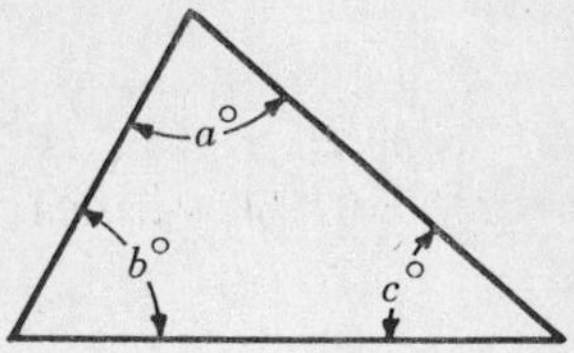

Fig. 6

Rule 4. The sum of the angles of any triangle equals 180°.

Thus, if the angles of a triangle contain a°, b° and c° as in Fig. 6, then

$$a^\circ + b^\circ + c^\circ = 180^\circ$$

(5) Angles of a Right Triangle

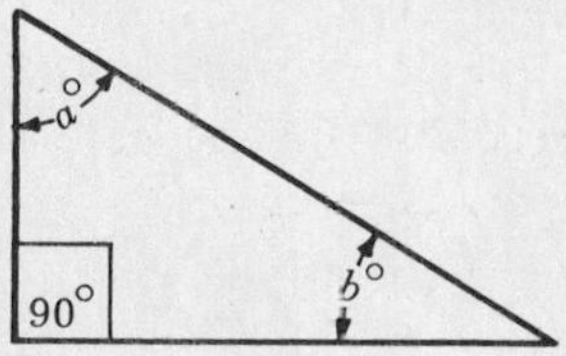

Fig. 7

Rule 5. The sum of the acute angles of a right triangle equals 90°.

Thus, if the acute angles of a right triangle as in Fig. 7, contain a° and b°, then

$$a^\circ + b^\circ = 90^\circ$$

(6) Angles of Isosceles Triangle

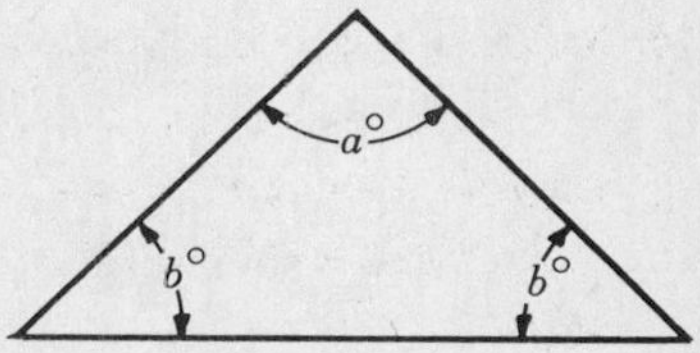

Fig. 8

Rule 6. The angles opposite the equal sides of an isosceles triangle are equal.

Thus, if the equal angles of an isosceles triangle contain b° and the third angle a°, as in Fig. 8, then

$$a^\circ + 2b^\circ = 180^\circ$$

Note. Henceforth, unless otherwise indicated, the word "angle" will mean "the number of degrees in the angle".

6.1. Using One Equation for Two Unknowns in Angle Problems

If two angles in the ratio of 3 to 2 are represented by $3x$ and $2x$, express each statement as an equation; then find x and the angles:

	Equation	x	Angles
a) The angles are adjacent and form an angle of 40°.	*Ans.* $3x+2x = 40$	8	24°, 16°
b) The angles are complementary.	*Ans.* $3x+2x = 90$	18	54°, 36°
c) The angles are supplementary.	*Ans.* $3x+2x = 180$	36	108°, 72°
d) The larger angle is 28° less than twice the smaller.	*Ans.* $3x = 4x - 28$	28	84°, 56°
e) The smaller is 40° more than one-third of the larger.	*Ans.* $2x = \frac{3x}{3} + 40$	40	120°, 80°
f) The angles are two angles of a triangle whose third angle is 70°.	*Ans.* $3x+2x+70 = 180$	22	66°, 44°
g) The angles are the acute angles of a right triangle.	*Ans.* $3x+2x = 90$	18	54°, 36°
h) The first angle is one of the two equal angles of an isosceles triangle and the other is the remaining angle.	*Ans.* $3x+3x+2x = 180$	$22\frac{1}{2}$	$67\frac{1}{2}$°, 45°

6.2. Using Two Equations for Two Unknowns in Angle Problems

If two angles are represented by a and b, obtain two equations for each problem; then find the angles:

	Equations	Angles
a) The angles are adjacent forming an angle of 88°. One is 36° more than the other.	*Ans.* $a+b = 88$, $a = b+36$	62°, 26°
b) The angles are complementary. One is twice as large as the other.	*Ans.* $a+b = 90$, $a = 2b$	60°, 30°
c) The angles are supplementary. One is 60° less than twice the other.	*Ans.* $a+b = 180$, $a = 2b - 60$	100°, 80°
d) The angles are two angles of a triangle whose third angle is 40°. The difference of the angles is 24°	*Ans.* $a+b = 140$, $a-b = 24$	82°, 58°
e) The first angle is one of the two equal angles of an isosceles triangle and the other is the remaining angle. The second angle is three times the first.	*Ans.* $2a+b = 180$, $b = 3a$	36°, 108°

6.3. Three Angles Having a Fixed Ratio

If three angles in the ratio of 4 : 3 : 2 are represented by $4x$, $3x$ and $2x$, express each statement as an equation; then find x and the angles:

	Equation	x	Angles
a) The first and second are adjacent and form an angle of 84°.	*Ans.* $4x+3x = 84$	12	48°, 36°, 24°
b) The second and third are complementary.	*Ans.* $3x+2x = 90$	18	72°, 54°, 36°
c) The first and third are supplementary.	*Ans.* $4x+2x = 180$	30	120°, 90°, 60°
d) The angles are the three angles of a triangle.	*Ans.* $4x+3x+2x = 180$	20	80°, 60°, 40°
e) The sum of the first and second is 27° more than twice the third.	*Ans.* $4x+3x = 4x+27$	9	36°, 27°, 18°
f) The first and third are acute angles of a right triangle.	*Ans.* $4x+2x = 90$	15	60°, 45°, 30°

6.4. Supplementary Angles Problem

The number of degrees in each of two supplementary angles are consecutive odd integers. Find each angle.

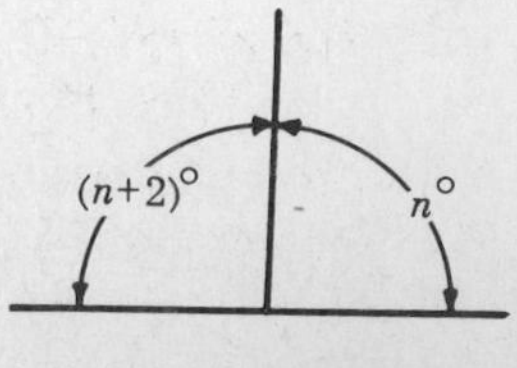

Solution: Let n and $n+2$ = the no. of degrees in each of the angles.
Then $2n+2 = 180$, $\underline{n = 89}$. *Ans.* 89°, 91°

6.5. Sum of Angles of a Triangle Problem

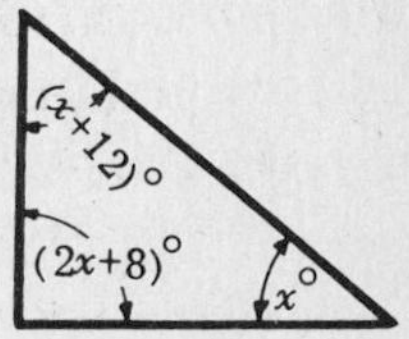

In a triangle, one angle exceeds another by 12°. The third angle is 4° less than the sum of the other two. Find the angles.

Solution: Let x = the no. of degrees in one angle
$x+12$ = the no. of degrees in the second
$2x+8$ = the no. of degrees in the third, since $x+(x+12)-4 = 2x+8$.

Then their sum $4x+20 = 180$, and
$\underline{x = 40}$, $x+12 = 52$, $2x+8 = 88$. *Ans.* 40°, 52°, 88°

7. *PERIMETER PROBLEMS*

Note. Unless otherwise indicated, the word "side" means "the number of linear units in the side".

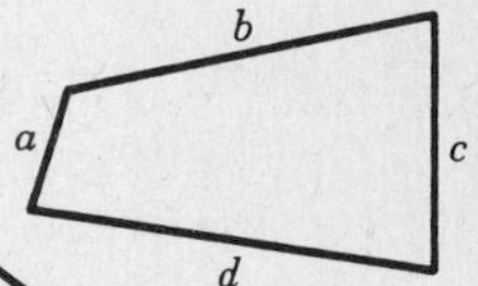

The **perimeter of a polygon** is the sum of its sides.
Thus, for a quadrilateral as shown, if p = the perimeter,

$$p = a+b+c+d$$

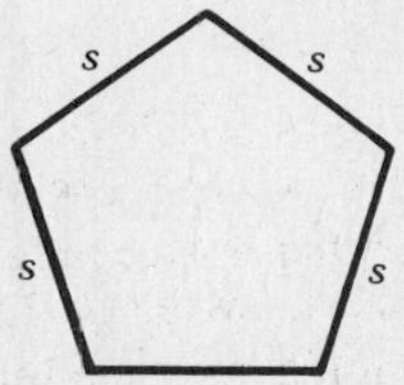

The perimeter of a regular polygon equals the product of one side and the number of sides.
Thus, for a regular pentagon, if p = the perimeter,

$$p = 5s$$

7.1. One Equation Method: Perimeter of Rectangle

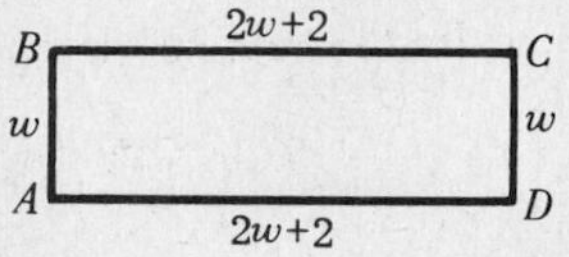

The width and length of a rectangle are represented by w and $2w+2$, respectively. Express each statement as an equation; then find w and the dimensions of the rectangle:

		Equation	w	Width, Length
a) The perimeter is 76 ft.	*Ans.*	$6w+4 = 76$	12	12 ft., 26 ft.
b) The semi-perimeter (half-perimeter) is 56 in.	*Ans.*	$3w+2 = 56$	18	18 in., 38 in.
c) The sum of three sides not including BC is 82 yd.	*Ans.*	$4w+2 = 82$	20	20 yd., 42 yd.
d) A rectangle having each dimension 3 rods less than those of the one shown has a perimeter of 28 rods.	*Ans.*	$2(2w-1)+2(w-3) = 28$	6	6 rd., 14 rd.

7.2. Two Equation Method: Perimeter of Triangle

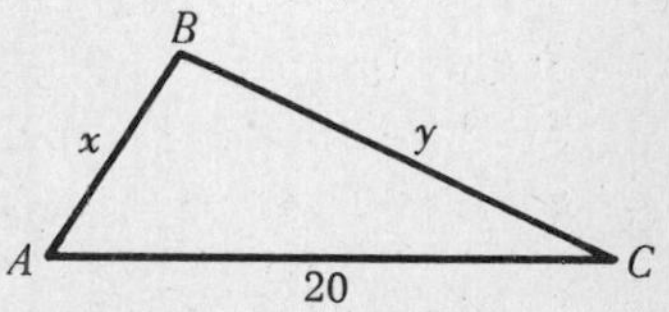

The base of a triangle is 20. If x and y represent the remaining sides, obtain two equations for each problem; then find AB and BC.

		Equations	AB, BC
a) The perimeter is 50. BC is twice AB.	*Ans.*	$x+y+20 = 50$ $y = 2x$	10, 20
b) The sum of AB and BC is 25. BC is 15 less than three times AB.	*Ans.*	$x+y = 25$ $y = 3x-15$	10, 15
c) AB is 4 more than half of BC. The difference of BC and AB is 2.	*Ans.*	$x = \frac{y}{2}+4$ $y-x = 2$	10, 12

7.3. One Equation Method: Perimeter of Trapezoid

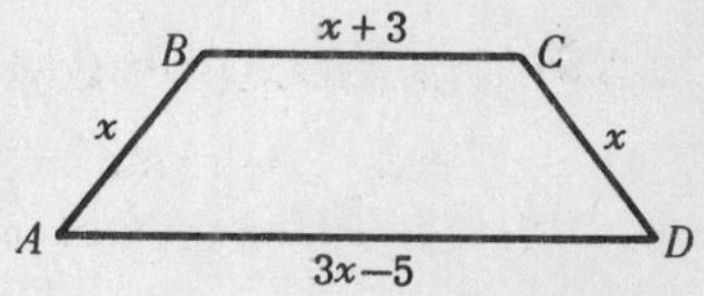

In a trapezoid having equal sides, the bases and sides are represented as shown in the adjoining figure. Express each statement as an equation; then find x and the length of each base:

	Equation	x	BC, AD
a) The perimeter is 70 yd.	*Ans.* $6x - 2 = 70$	12	15 yd., 31 yd.
b) The combined length of the upper and lower bases is 36 in.	*Ans.* $4x - 2 = 36$	$9\frac{1}{2}$	$12\frac{1}{2}$ in., $23\frac{1}{2}$ in.
c) The sum of three sides not including the upper base is 65 ft.	*Ans.* $5x - 5 = 65$	14	17 ft., 37 ft.
d) The perimeter of the trapezoid exceeds the perimeter of a square having BC as a side by 10 rd.	*Ans.* $6x - 2 = 4(x+3) + 10$	12	15 rd., 31 rd.

7.4. Perimeter of a Quadrilateral: Ratio of Sides

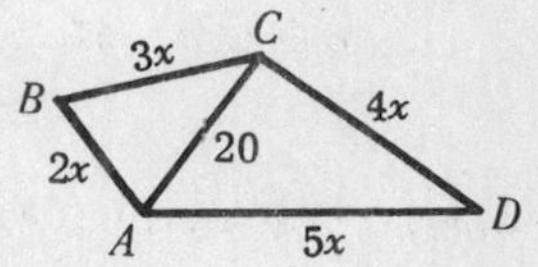

The sides of a quadrilateral in the ratio of 2 : 3 : 4 : 5 are represented by $2x$, $3x$, $4x$ and $5x$, as shown in the adjoining figure. Express each statement as an equation; then find x and AB:

	Equation	x	AB
a) The perimeter is 98 ft.	*Ans.* $14x = 98$	7	14 ft.
b) The largest side is 8 in. less than the sum of the other three sides.	*Ans.* $5x = 9x - 8$	2	4 in.
c) The sum of twice the smaller side and the other sides is 84 mi.	*Ans.* $2(2x) + 12x = 84$	$5\frac{1}{4}$	$10\frac{1}{2}$ mi.
d) The perimeter of the quadrilateral is 22 rd. more than that of a square having a side 4 rd. less than BC.	*Ans.* $14x = 4(3x - 4) + 22$	3	6 rd.
e) The perimeter of triangle ACD exceeds that of triangle ABC by 18 yd.	*Ans.* $9x + 20 = (5x+20) + 18$	$4\frac{1}{2}$	9 yd.

7.5. Perimeter of an Isosceles Triangle Problem

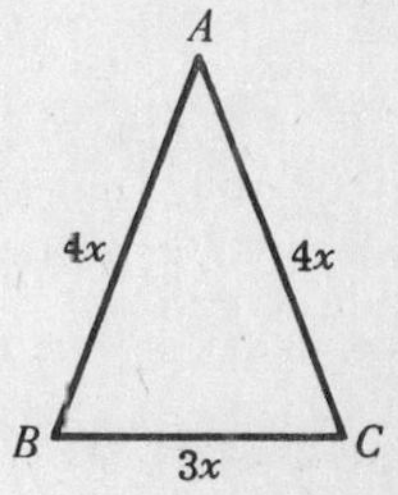

In an isosceles triangle, the ratio of one of the equal sides to the base is 4 : 3. A piece of wire 55 inches long is to be used to make a wire model of it. Find the base and the side.

Solution: Let $4x$ = each equal side in in. and $3x$ = base in in.

Hence, $4x + 4x + 3x = 55$, $11x = 55$,

$\underline{x = 5}$, $3x = 15$, $4x = 20$. *Ans.* 15 in. and 20 in.

7.6. Perimeter of a Square Problem

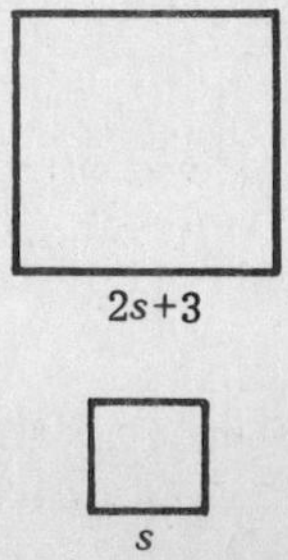

The side of a larger square is 3 feet more than twice the side of a smaller square. The perimeter of the larger square exceeds that of the smaller square by 46 feet. Find the sides of the squares.

Solution: Let s = each side of the smaller square in ft.

and $2s+3$ = each side of the larger square in ft.

Then $4(2s+3) = 4s + 46$, $8s + 12 = 4s + 46$,

$\underline{s = 8\frac{1}{2}}$, $2s + 3 = 20$. *Ans.* $8\frac{1}{2}$ ft. and 20 ft.

8. COIN OR STAMP PROBLEMS

The total value (T) of a number of coins or stamps of the same kind equals the number (N) of the coins or stamps multiplied by the value (V) of one of the coins or stamps.

Formula: $T = NV$

Thus, 8 nickels have a total value = 8(5) = 40¢,
20 3-cent stamps have a total value = 20(3) = 60¢.

Note: In $T = NV$, T and V must have the same unit of money; that is, if V is in dollars, then T must be in dollars.

8.1. Finding Total Values

Find the total value *in cents* of each of the following:

a) 3 nickels and 5 dimes. *Ans. a*) $3(5) + 5(10) = 65$

b) q quarters and 7 nickels. *Ans. b*) $25q + 7(5) = 25q + 35$

c) 8 2-cent and t 3-cent stamps. *Ans. c*) $8(2) + 3t = 3t + 16$

d) x 4-cent and $(x+5)$ 6-cent stamps. *Ans. d*) $4x + 6(x+5) = 10x + 30$

8.2. Representation in Coin or Stamp Problems

Write the expression that represents the total value *in cents* of each; then simplify it:

		Representation	In Simplified Form
a) n dimes and $(n+3)$ quarters.	*Ans. a*)	$10n + 25(n+3)$	$35n + 75$
b) n nickels and $(2n-3)$ half-dollars.	*Ans. b*)	$5n + 50(2n-3)$	$105n - 150$
c) n 2-cent and $2n$ $1\frac{1}{2}$-cent stamps.	*Ans. c*)	$2n + 2n(1\frac{1}{2})$	$5n$
d) n 6-cent and $(3n-10)$ 12-cent stamps.	*Ans. d*)	$6n + 12(3n-10)$	$42n - 120$

8.3. Using One Equation in Coin or Stamp Problems

Express each statement as an equation; then find n and the number of each coin or stamp:

a) The total value of n dimes and $2n$ nickels is \$2.40.

b) The total value of n nickels and $(n-3)$ quarters is \$7.65.

c) The value of $(n+5)$ dimes is three times n nickels.

d) The value of n 2-cent and $(n+3)$ 3-cent stamps is 2¢ more than $(n-1)$ 6¢ stamps.

Ans.

Equations	n	Number of Each Coin or Stamp
a) $10n + 5(2n) = 240$	12	12 dimes and 24 nickels
b) $5n + 25(n-3) = 765$	28	28 nickels and 25 quarters
c) $10(n+5) = 3(5n)$	10	10 nickels and 15 dimes
d) $2n + 3(n+3) = 6(n-1) + 2$	13	13 2-cent, 16 3-cent and 12 6-cent stamps

8.4. Using Two Equations in Coin or Stamp Problems

Using x for the value in ¢ of the first kind of coin and y for the value in ¢ of the second, obtain two equations for each problem; then find the kind of coins.

Equations

a) The value of 5 of the first kind and 6 of the second is 85¢. Each second coin has twice the value of each first.

Ans. a) $5x + 6y = 85$
$y = 2x$
$x = 5, \quad y = 10$
NICKELS and DIMES

b) The value of 4 of the first kind is 85¢ less than 5 of the second. The value of 2 of the first kind and 4 of the second is \$1.20.

Ans. b) $4x = 5y - 85$
$2x + 4y = 120$
$x = 10, \quad y = 25$
DIMES and QUARTERS

8.5. Complete Solution of Coin Problem

In his coin bank, Joe has three times as many quarters as nickels. If the value of the quarters is \$5.60 more than the value of the nickels, how many of each kind does he have?

Solution: Let n = the number of nickels. $NV = T$

	Number of Coins	Value of Each in (¢)	Total Value in ¢
nickels	n	5	$5n$
quarters	$3n$	25	$75n$

The value of the quarters is \$5.60 more than the value of the nickels.

$75n = 5n + 560$, $\underline{n = 8}$, $3n = 24$. *Ans.* 8 nickels, 24 quarters

8.6. Complete Solution of a Change Problem

Is it possible to change a \$10 bill into an equal number of half-dollars, quarters and dimes?

Solution: Let the no. of half-dollars, the no. of quarters and the no. of dimes, each be represented by n. The combined value is \$10.

Then, $10n + 25n + 50n = 1000$, $85n = 1000$, $\underline{n = 11\frac{13}{17}}$

Ans. Impossible, the number of coins must be a whole number.

8.7. Complete Solution of a Fares Problem

On a trip, the fare for each adult was 50¢ and for each child 25¢. The number of passengers was 30 and the total paid was \$12.25. How many adults and children were there?

Solution: Let a = the no. of adults and c = the no. of children.

Then, $50a + 25c = 1225$

$a + c = 30$

Solving, $a = 19$, $c = 11$. *Ans.* 19 adults, 11 children

9. COST AND MIXTURE PROBLEMS

The total value (T) of a number of units of the same kind equals the number (N) of the units multiplied by the value (V) of one of the units.

Formula: $T = NV$

Thus, the total value of 5 books at 15¢ a book = 5(15) or 75¢.

The total value of 6 tickets at 50¢ each and 5 tickets at 25¢ each = 6(50) + 5(25) or 425¢.

Coin or Stamp Problems

The coin or stamp problems previously considered, are special cases of mixture problems. Stamps and coins may be "mixed" as coffees, teas, nuts and other items are mixed.

Cost Problems

When $T = NV$ is applied to a cost problem, the formula $C = NP$ should be used as follows:

$$C = NP \begin{cases} C = \text{total cost of number of units of a kind} \\ N = \text{number of such units} \\ P = \text{price of each unit purchased} \end{cases}$$

Thus, the cost of 5 pencils at 8¢ a pencil = 5(8) or 40¢.

9.1. Representation in Mixture Problems: $T = NV$

Write the expression that represents the total value in *cents*; then simplify it:

	Representation	In Simplified Form
a) *n* lb. of coffee valued at 90¢ a lb. and (*n*+3) lb. of coffee valued at $1.05 a lb.	*Ans. a*) $90n + 105(n+3)$	$195n + 315$
b) 3 lb. of tea valued at $1.50 per lb. and *n* lb. of tea valued at $1.75 per lb.	*Ans. b*) $3(150) + 175n$	$175n + 450$
c) *n* stamps valued at 35¢ each and (20 − *n*) stamps valued at 50¢ each.	*Ans. c*) $35n + 50(20-n)$	$1000 - 15n$
d) *d* dozen pencils at $1.30 a dozen and 3 dozen pencils at 60¢ a dozen.	*Ans. d*) $130d + 3(60)$	$130d + 180$

9.2. Representation in Cost Problems

Write the expression that represents the total cost in *dollars*; then simplify it:

	Representation	In Simplified Form
a) 10 tables priced at $5 apiece and *n* tables at $7.50 apiece.	*Ans. a*) $5(10) + 7.50n$	$7.5n + 50$
b) *n* lb. of candy priced at 65¢ a lb. and (30 − *n*) lb. of candy at $1.45 a lb.	*Ans. b*) $.65n + 1.45(30-n)$	$43.5 - .8n$
c) *q* qt. of cream at 35¢ a qt. and (2*q*+3) qt. of cream at 42¢ a qt.	*Ans. c*) $.35q + .42(2q+3)$	$1.19q + 1.26$

9.3. Translation in Value and Cost Problems

Express each statement as an equation; then find *n* and the number of each kind:

	Equation
a) The cost of *n* lb. of coffee at 95¢ a lb. and (25−*n*) lb. of coffee at $1.10 a lb. is $26.00.	*Ans.* $95n + 110(25-n) = 2600$, $n = 10$ 10 lb. at 95¢ and 15 lb. at $1.10
b) The value of 40 tickets at 75¢ each and *n* tickets at $1.35 each is $40.80.	*Ans.* $40(75) + 135n = 4080$, $n = 8$ 8 tickets at $1.35
c) The cost of *n* lb. of cookies at 80¢ a lb. and (2*n*−3) lb. of cookies at $1.60 a lb. is $15.20.	*Ans.* $80n + 160(2n-3) = 1520$, $n = 5$ 5 lb. at 80¢ and 7 lb. at $1.60
d) The value of *n* dollar bills, (*n*+10) five-dollar bills and (3*n*−2) ten-dollar bills is $174.	*Ans.* In dollars: $n + 5(n+10) + 10(3n-2) = 174$, $n = 4$ 4 $1 bills, 14 $5 bills, 10 $10 bills

9.4. Complete Solution: Blending Coffee Problem

A coffee merchant blended coffee worth $.93 a lb. with coffee worth $1.20 a lb. The mixture of 30 lb. was valued by him at $1.02 a lb. How many lb. of each grade of coffee did he use?

Solutions:

Method 1, using one letter.

	Value per lb. (¢)	No. of lb.	Total Value (¢)
cheaper	93	x	$93x$
better	120	$30-x$	$120(30-x)$
mixture	102	30	3060

$$93x + 120(30-x) = 3060$$
$$93x + 3600 - 120x = 3060$$
$$x = 20,\quad 30 - x = 10$$

Ans. 20 lb. of $.93¢ coffee and 10 lb. of $1.20 coffee

Method 2, using two letters

	Value per lb. (¢)	No. of lb.	Total Value (¢)
cheaper	93	x	$93x$
better	120	y	$120y$
mixture	102	30	3060

$93x + 120y = 3060$ $\qquad$ $93x + 120y = 3060$

M_{93} $\quad x + y = 30$ $\qquad$ $93x + 93y = 2790$

Subtract: $\qquad 27y = 270$

$y = 10,\ x = 20$

9.5. Complete Solution: Selling Tickets Problem

At a game, tickets were bought at 30¢, 50¢ and 75¢ each. The number of 50¢ tickets was three times the number of 30¢ tickets and ten less than the number at 75¢. The receipts amounted to $88.50. How many of each were sold?

Solution:

Let n = the number of tickets at 30¢ each
$3n$ = the number of tickets at 50¢ each
$3n+10$ = the number of tickets at 75¢ each.

Hence, $30n + 50(3n) + 75(3n+10) = 8850$
$30n + 150n + 225n + 750 = 8850$ or $405n = 8100$
$n = 20$, $3n = 60$, $3n+10 = 70$ *Ans.* 20 at 30¢, 60 at 50¢, 70 at 75¢

10. INVESTMENT OR INTEREST PROBLEMS

Annual interest (I) equals the principal (P) multiplied by the rate of interest (R) per year.

Formula: $I = PR$

Thus, the annual interest from $200 at 6% per year is (200)(.06) or $12.

Note. Unless otherwise stated, the rate of interest is the rate per yr.; that is, "5%" means "5% per yr.".

10.1. Representation in Interest Problems: $I = PR$

Write the expression that represents the annual interest in dollars earned by each principal; then simplify it:

	Representation	In Simplified Form
a) $2000 at 4% and $($P$+200) at 3%.	*Ans.* $.04(2000)+.03(P+200)$	$.03P+86$
b) $$P$ at 2% and 2P$ at $5\frac{1}{2}$%.	*Ans.* $.02P+.05\frac{1}{2}(2P)$	$.13P$
c) $$P$ at 3% and $(2$P$−400) at 6%.	*Ans.* $.03P+.06(2P-400)$	$.15P-24$
d) $$P$ at 7%, 2P$ at 5% and 3P$ at 3%.	*Ans.* $.07P+.05(2P)+.03(3P)$	$.26P$

10.2. Translation in an Interest Problem

Express each statement as an equation; then solve and state each principal:

	Equation
a) The total annual income from $500 at 4% and $$P$ at 5% is $55.	*Ans. a*) $.04(500)+.05P = 55$, $P = 700$ $700 at 5%
b) The total annual income from $$P$ at 3% and $(5000−$P$) at 2% is $120.	*Ans. b*) $.03P+.02(5000-P) = 120$, $P = 2000$ $2000 at 3% and $3000 at 2%
c) The annual interest from $$P$ at 3% equals that from $($P$−2000) at 5%.	*Ans. c*) $.03P = .05(P-2000)$, $P = 5000$ $5000 at 3% and $3000 at 5%
d) The annual interest from 2P$ at 5% exceeds that from 3P$ at 3% by $25.	*Ans. d*) $.05(2P) = .03(3P)+25$, $P = 2500$ $5000 at 5% and $7500 at 3%
e) The total annual interest from $$P$ at 6%, 2P$ at 5% and $(2$P$−300) at 4% is $180.	*Ans. e*) $.06P+.05(2P)+.04(2P-300) = 180$, $P = 800$ $800 at 6%, $1600 at 5%, $1300 at 4%

10.3. Complete Solution: Ratio in an Interest Problem

Mr. White invested two sums of money in the ratio of 5 : 3. The first sum was invested at 4% and the second at 2%. The annual interest from the first exceeds that of the second by $56. How much was each investment?

Solution: Since the ratio is 5 : 3, let $5x$ = first investment in $
and $3x$ = second investment in $.

	($) Principal	Rate of Interest	Annual Interest ($)
First Investment	$5x$	.04	$.04(5x)$
Second Investment	$3x$	.02	$.02(3x)$

$$.04(5x) = .02(3x) + 56$$
$$.20x = .06x + 56$$
$$\underline{x = 400},\ 5x = 2000,\ 3x = 1200$$

Ans. $2000 at 4%, $1200 at 2%

10.4. Alternate Methods in an Investment Problem

Mr. White invested $8000, part at 5% and the rest at 2%. The yearly income from the 5% investment exceeded that from the 2% investment by $85. Find the investment at each rate.

Solutions:

Method 1, using one letter

	($) Principal	Rate of Interest	Annual Interest ($)
1st	x	.05	$.05x$
2nd	$(8000-x)$	.02	$.02(8000-x)$

$$M_{100}\quad .05x - .02(8000-x) = 85$$
$$5x - 2(8000-x) = 8500$$
$$5x - 16{,}000 + 2x = 8500$$
$$7x = 24{,}500$$
$$\underline{x = 3500},\ 8000 - x = 4500$$

Ans. $3500 at 5%, $4500 at 2%

Method 2, using two letters

	($) Principal	Rate of Interest	Annual Interest ($)
1st	x	.05	$.05x$
2nd	y	.02	$.02y$

$$M_2\quad x + y = 8000$$
$$M_{100}\quad .05x - .02y = 85$$
$$2x + 2y = 16{,}000$$
$$\text{Add:}\quad 5x - 2y = 8{,}500$$
$$7x = 24{,}500$$
$$\underline{x = 3500},\ y = 4500$$

10.5. Alternate Methods in a Profit and Loss Investment Problem

Mr. Brown invested a total of $4000. On part of this he earned 4%. On the remainder he lost 3%. Combining his earnings and losses, he found his annual income to be $55. Find the amounts at each rate.

Solutions:

Method 1, using one letter

	($) Principal	Rate of Interest	Annual Interest ($)
1st	x	.04	$.04x$
2nd	$4000-x$	$-.03$	$-.03(4000-x)$

$$M_{100}\quad .04x - .03(4000-x) = 55$$
$$4x - 3(4000-x) = 5500$$
$$7x - 12{,}000 = 5500$$
$$\underline{x = 2500},\ 4000 - x = 1500$$

Method 2, using two letters

	($) Principal	Rate of Interest	Annual Interest ($)
1st	x	.04	$.04x$
2nd	y	$-.03$	$-.03y$

$$M_3\quad x + y = 4000 \rightarrow 3x + 3y = 12{,}000$$
$$M_{100}\quad .04x - .03y = 55 \rightarrow 4x - 3y = 5{,}500$$
$$7x = 17{,}500$$
$$\underline{x = 2500},\ y = 1500$$

Ans. Earnings on $2500 at 4%, losses on $1500 at 3%.

10.6. Complete Solution: Adding a Third Investment

Mr. Black has \$3000 invested at 3% and \$1000 at 4%. How much must he invest at 6% so that his annual income will be 5% of the entire investment?

Solution:

Let x = the principal to be added at 6%; then
$4000 + x$ = the entire principal at 5%.

$$.03(3000) + .04(1000) + .06x = .05(4000+x)$$
$$90 + 40 + .06x = 200 + .05x$$
$$\underline{x = 7000}$$ *Ans.* \$7000 at 6%

11. MOTION PROBLEMS

The distance (D) traveled equals the rate of speed (R) multiplied by the time spent traveling (T).

Formula: $D = RT$ **Other Forms:** $R = \frac{D}{T}$, $T = \frac{D}{R}$

Thus, the distance traveled in 5 hr. at a rate of 30 mph is 150 mi., and the distance traveled in 5 sec. at a rate of 30 ft. per sec. is 150 ft.

Note. In using $D = RT$, units for rate, time and distance must be in agreement.
Thus, if the rate is in miles per hour (mph), use miles for distance and hours for time.

Uniform and Average Rates of Speed

Unless otherwise stated "rate of speed" or simply "rate" may be taken to mean (1) uniform rate of speed or (2) average rate of speed.

(1) **A uniform rate of speed** is an unchanging or fixed rate of speed for each unit of time.
Thus, a uniform rate of 40 mph for 3 hr. means that 40 mi. was covered during each of the 3 hours.

(2) **An average rate of speed** is a rate that is the average per unit of time for changing rates of speed.
Thus, an average rate of 40 mph for 3 hr. may mean the average rate in a situation where 30 mi. was covered during the first hr., 40 mi. during the second hr. and 50 mi. during the third hr.

11.1. Representation of Distance:

Write the expression that represents the distance in miles traveled; then simplify it:

		Representation	In Simplified Form
a) In 5 hr. at 20 mph and in 6 hr. more at R mph.	*Ans.*	$5(20)+6R$	$6R + 100$
b) In 6 hr. at 45 mph and in T hr. more at 50 mph.	*Ans.*	$6(45) + 50T$	$50T + 270$
c) In T hr. at 40 mph and in $(10-T)$ hr. more at 30 mph.	*Ans.*	$40T+30(10-T)$	$10T + 300$
d) In 2 hr. at R mph and in 30 min. more at 20 mph.	*Ans.*	$2R + \frac{1}{2}(20)$	$2R + 10$
e) In 4 hr. at R mph and in 8 hr. more at $(R+10)$ mph.	*Ans.*	$4R + 8(R+10)$	$12R + 80$
f) In 5 min. at R mi. per min. and in 1 hr. more at 3 mi. per min.	*Ans.*	$5R + 60(3)$	$5R + 180$

11.2. Representation of Time: $T = \frac{D}{R}$

Write the expression that represents the time *in hr.* needed to travel; then simplify it:

		Representation	In Simplified Form
a) 100 mi. at 20 mph and 80 mi. farther at R mph.	*Ans. a*)	$\frac{100}{20} + \frac{80}{R}$	$5 + \frac{80}{R}$
b) 120 mi. at 30 mph and D mi. farther at 20 mph.	*Ans. b*)	$\frac{120}{30} + \frac{D}{20}$	$4 + \frac{D}{20}$
c) 60 mi. at R mph and 80 mi. farther at 2 mi. per min.	*Ans. c*)	$\frac{60}{R} + \frac{80}{120}$	$\frac{60}{R} + \frac{2}{3}$

11.3. Separation Situation:

Two travelers start from the same place at the same time and travel **in opposite directions.**

Two planes leave the same airport at the same time and fly in opposite directions. The speed of the faster plane is 100 mph faster than the slower one. At the end of 5 hr., they are 2000 miles apart. Find the rate of each plane.

Solutions:

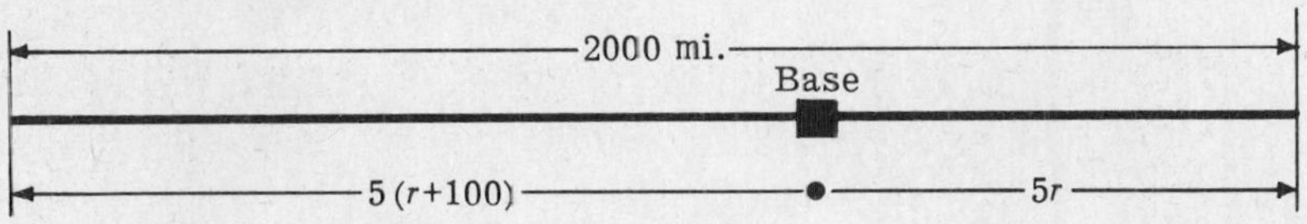

ALTERNATE METHODS OF SOLUTION

Method 1

	(mph) Rate	(hr.) Time	D (mi.) Distance
faster plane	$r+100$	5	$5(r+100)$
slower plane	r	5	$5r$

The sum of the distances equals 2000 mi.

$$5r + 5(r+100) = 2000$$
$$5r + 5r + 500 = 2000$$
$$10r = 1500$$

$r = 150$, $r + 100 = 250$.

Method 2

(mph) Rate of Separation	(hr.) Time of Separation	(mi.) Distance of Separation
R	5	2000

Here, R is the rate at which the two planes are separating from each other!

$$5R = 2000, \quad R = 400$$

Hence, the rate of separation is 400 mph.

Since 400 mph is the sum of both rates,

$$r + (r+100) = 400$$

$r = 150$, $r + 100 = 250$

Ans. The rates are 150 mph and 250 mph.

11.4. Separation Situations:

In each of the following, *two travelers start from the same place at the same time and travel in* **opposite directions.** Using the letter indicated, express each statement as an equation; then solve and find each quantity represented:

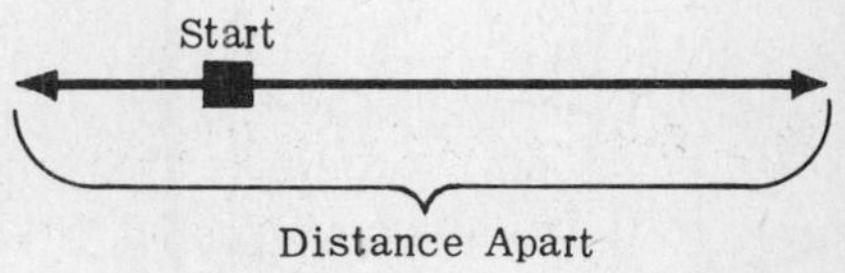

a) After 5 hr., both are 300 mi. apart, one going 40 mph faster. Find the rate of the slower, using R in mph.

Ans. a) $5R + 5(R+40) = 300$, $R = 10$
Slower rate is 10 mph.

b) At speeds of 40 mph and 20 mph, both travel until they are 420 mi. apart. Find time of each, using T in hr.

Ans. b) $40T + 20T = 420$, $T = 7$
Time for each is 7 hr.

c) At speeds in the ratio of 7 : 3, both are 360 mi. apart at the end of 3 hr. Find their repective rates in mph, using $7x$ and $3x$ for these.

Ans. c) $3(7x) + 3(3x) = 360$, $x = 12$
Faster rate is 84 mph.
Slower rate is 36 mph.

11.5. Closure Situation:

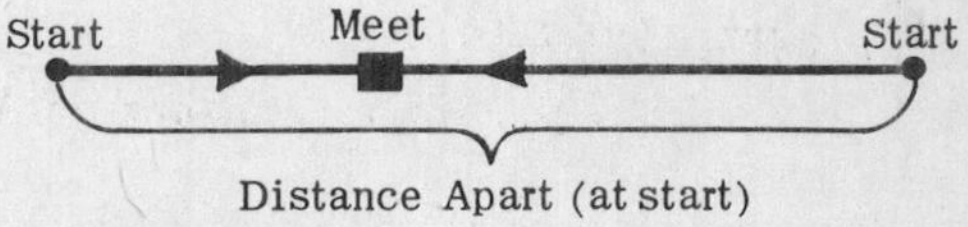

Two travelers start from distant points at the same time and travel toward each other until they meet.

A car leaves Albany en route to New York at the same time that another car leaves New York for Albany. The car from Albany travels at an average rate of 40 mph, while the other averages 20 mph. If Albany and New York are 150 mi. apart, how soon will the cars meet and how far will each have traveled?

Solutions:

ALTERNATE METHODS OF SOLUTION

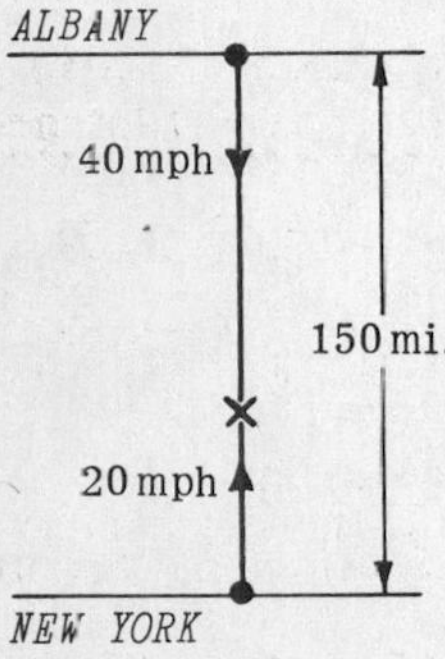

Method 1

	(mph) Rate	(hr.) Time	(mi.) Distance
from ALBANY	40	T	$40T$
from N.Y.	20	T	$20T$

The sum of the distances is 150 mi.

$$40T + 20T = 150$$
$$60T = 150$$
$$T = 2\tfrac{1}{2}$$

Method 2

(mph) Rate of Closure	(hr.) Time of Closure	(mi.) Distance of Closure
(40+20) or 60	T	150

Here, the cars are coming together at a rate of closure of 60 mph!

Hence, $60T = 150$

$$T = 2\tfrac{1}{2}$$

Ans. The cars will meet in $2\frac{1}{2}$ hr. The Albany car will have traveled 100 miles and the New York car 50 miles.

11.6. Closure Situations:

In each of the following, *two travelers start from distant places at the same time and travel toward each other until they meet.*

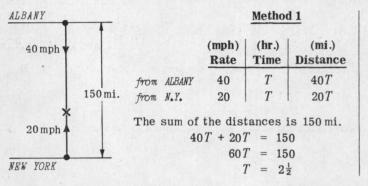

Using the letter indicated, express each statement in an equation; then solve and find each quantity represented:

a) Starting 600 mi. apart, they meet in 10 hr., one traveling 20 mph faster. Find the rate of the slower in mph, *using R for this.*

Ans. a) $10R + 10(R+20) = 600$, $R = 20$
Slower rate is 20 mph.

b) Starting 420 mi. apart, they travel at rates of 27 mph and 33 mph. Find the time of each in hr., *using T for this.*

Ans. b) $27T + 33T = 420$, $T = 7$
Time for each is 7 hr.

c) Starting 560 mi. apart, they travel for 4 hr., one rate being four times the other. Find the rate of the slower in mph, *using R for this.*

Ans. c) $4R + 4(4R) = 560$, $R = 28$
Slower rate is 28 mph.

d) Starting 480 mi. apart, they travel at rates in the ratio of 4 : 3. They meet in 5 hr. The slower is delayed for 1 hr. along the way. Find the rates of speed in mph, *using 4x and 3x for these.*

Ans. d) $4(3x) + 5(4x) = 480$, $x = 15$
Slower rate is 45 mph.
Faster rate is 60 mph.

11.7. Round Trip Situation:

A traveler travels out and back to the starting place along the same road.

Henry drove from his home to Boston and back again along the same road in a total of 10 hr. His average rate going was 20 mph while his average rate on the return trip was 30 mph. How long did he take in each direction and what distance did he cover each way?

Solutions:

ALTERNATE METHODS OF SOLUTION

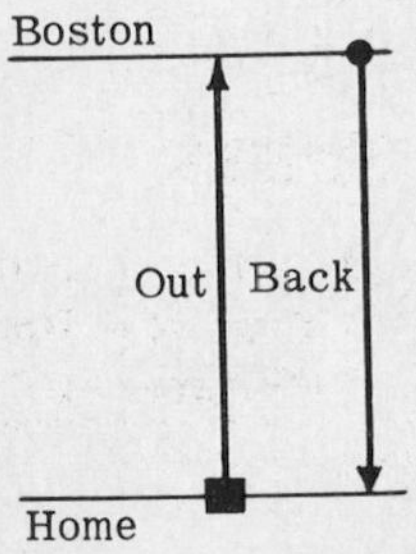

Method 1, using $D = RT$

	(mph) Rate	(hr.) Time	(mi.) Distance
going	20	T	$20T$
return	30	$(10-T)$	$30(10-T)$

The distance out equals the distance back.

$$20T = 30(10-T)$$
$$20T = 300 - 30T$$
$$50T = 300$$
$$T = 6, \quad 10-T = 4$$

Method 2, using $T = \frac{D}{R}$

	(mph) Rate	(hr.) Time	(mi.) Distance
going	20	$\frac{D}{20}$	D
return	30	$\frac{D}{30}$	D

The total time is 10 hr.

$$M_{60} \quad \frac{D}{20} + \frac{D}{30} = 10 \qquad \textbf{L.C.D} = 60$$
$$3D + 2D = 600, \quad 5D = 600$$
$$D = 120$$

Ans. Henry took 6 hr. going and 4 hr. back in going 120 mi. each way.

11.8. Round Trip Situations:

In each situation, *a traveler travels out and back to the starting place along the same road.*

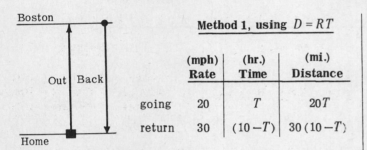

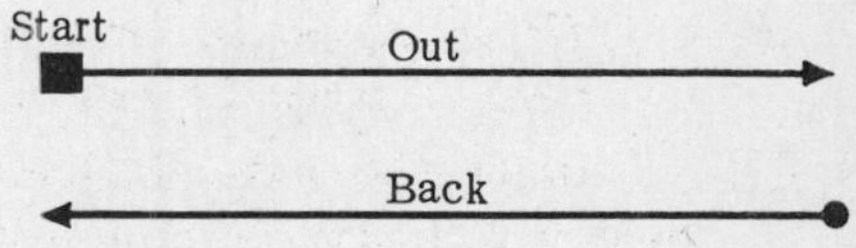

Using the letter indicated, express each statement as an equation; then solve and find each quantity represented.

a) A traveler required a total of 12 hr. on a round trip, averaging 20 mph out and 30 mph back. Find the time in hr. going, *using T for this.*

Ans. a) $20T = 30(12-T)$, $T = 7.2$
Time out was 7.2 hr.

b) A traveler took 2 hr. more to travel back than to go out, averaging 50 mph out and 45 mph back. Find the time in hr. going, *using T for this.*

Ans. b) $50T = 45(T+2)$, $T = 18$
Time out was 18 hr.

c) A traveler traveled to his destination at an average rate of 25 mph. By traveling at 5 mph faster, he took 30 minutes less to return. Find his time in hr. going out, *using T for this.*

Ans. c) $25T = 30(T-\frac{1}{2})$, $T = 3$
Time out was 3 hr.

d) After taking 5 hr. going out, a traveler came back in 3 hr. at a rate that was 20 mph faster. Find the average rate going in mph, *using R for this.*

Ans. d) $5R = 3(R+20)$, $R = 30$
Rate out was 30 mph.

e) After taking 5 hr. going out, a traveler came back in 3 hr. at a rate that was 20 mph faster. Find the distance either way in mi., *using D for this.*

Ans. e) $\frac{D}{3} = \frac{D}{5} + 20$, $D = 150$
Each distance was 150 mi.

11.9. Gain or Overtake Situation:

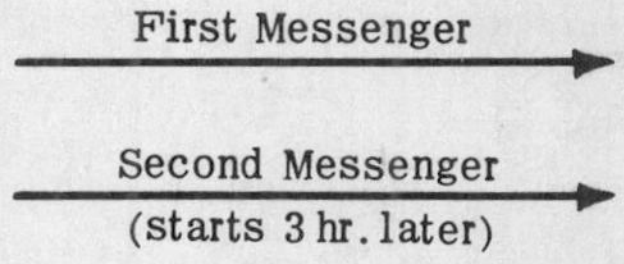

After a traveler has begun his trip, a second one starts from the same place and, going in the same direction, overtakes the first.

A messenger going at 30 mph has been gone for 3 hr. Another messenger, sent to overtake him, travels at 50 mph. How long will it take the second messenger to overtake the first and what distance will he cover?

Solutions:

ALTERNATE METHODS OF SOLUTION

Method 1

	(mph) Rate	(hr.) Time	(mi.) Distance
First	30	$T+3$	$30(T+3)$
Second	50	T	$50T$

The distances are equal.

$$30(T+3) = 50T$$
$$30T + 90 = 50T, \quad 90 = 20T$$
$$4\tfrac{1}{2} = T$$

Method 2

(mph) Rate of Gain	(hr.) Time of Gain	(mi.) Distance to Gain
$(50-30)$ or 20	T	$3(30)$ or 90

Here the second messenger is gaining at the rate of 20 mph. The first has a head start of 90 mi. since he started 3 hr. earlier and his rate was 30 mph.

$$20T = 90$$
$$T = 4\tfrac{1}{2}$$

Ans. To overtake the first, the second messenger must take $4\frac{1}{2}$ hr. and cover 225 mi.

11.10. Gain or Overtake Situations:

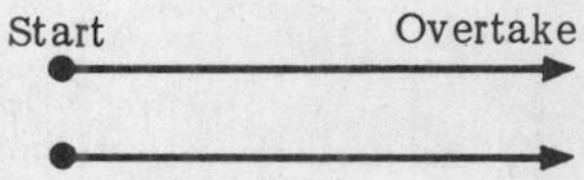

In each of the following, *after a traveler has begun his trip, a second traveler starts from the same place and, going in the same direction, overtakes the first.*

Using the letter indicated, express each statement as an equation; then solve and find each quantity represented:

a) Their speeds are 20 mph and 40 mph. The second starts 3 hr. later. Find the time of the second in hr., *using T for this.*

Ans. a) $20(T+3) = 40T, \quad T = 3$
Time of second is 3 hr.

b) The second starts 3 hr. later and travels 6 mi. an hr. faster to overtake the first in 8 hr. Find the rate of the first in mph, *using R for this.*

Ans. b) $11R = 8(R+6), \quad R = 16$
Rate of first is 16 mph.

c) The second travels 9 mph slower than twice the speed of the first and starts out $2\frac{1}{2}$ hr. later, overtaking in 4 hr. Find the rate of the first in mph, *using R for this.*

Ans. c) $6\frac{1}{2}R = 4(2R-9), \quad R = 24$
Rate of first is 24 mph.

11.11. Motion Problem: A Trip in Two Stages

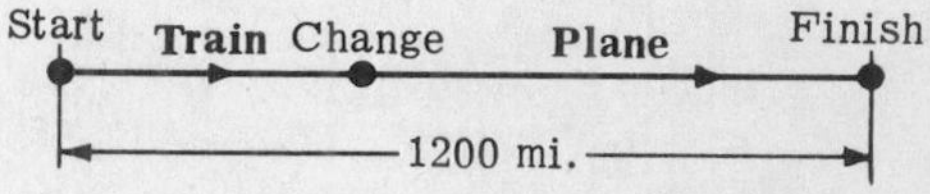

In going 1200 mi., William used a train first and later a plane. The train going at 30 mph took 2 hr. longer than the plane traveling at 150 mph. How long did the trip take?

Solution:

	Rate (mph)	Time (hr.)	Distance (mi.)
Plane trip	150	T	$150T$
Train trip	30	$(T+2)$	$30T + 60$

The total distance is 1200 mi.

$$30T + 60 + 150T = 1200$$
$$180T = 1140, \quad T = 6\tfrac{1}{3}\text{ hr.}$$

Ans. The trip took $14\frac{2}{3}$ hr., $6\frac{1}{3}$ by plane and $8\frac{1}{3}$ by train.

12. COMBINATION PROBLEMS

Combination problems such as those included here are solved best by means of simultaneous equations. In each of these problems, it is especially important to keep in mind that a letter may represent a number only.

Thus, s may stand for the *number* of shirts, the *number* of dollars in the price of a shirt, the *number* of cents in the cost of laundering a shirt; but s may not stand for "shirts".

12.1. Cost Problems: Combining Prices or Numbers

a) Mr. Jones bought 5 shirts and 4 hats for \$39. Later, at the same prices, he bought 4 shirts and 2 hats for \$24. Find the price of each.

Solution:

a) Let s = price of one shirt in \$
h = price of one hat in \$.

Then,
$$5s + 4h = 39$$
$$4s + 2h = 24.$$

Solving, $s = 3$ and $h = 6$.

Ans. The prices were \$3 per shirt and \$6 per hat.

b) Mr. Dodge paid \$65 for a number of shirts at \$5 apiece and a number of hats at \$10 each. Had the shirts been \$10 each and the hats \$5 apiece, he would have paid \$55. Find the number of shirts and the number of hats purchased.

Solution:

b) Let s = no. of shirts
h = no. of hats.

Then,
$$5s + 10h = 65$$
$$10s + 5h = 55.$$

Solving, $s = 3$ and $h = 5$.

Ans. The number of shirts was 3 and the number of hats was 5.

12.2. Motion Problems: Combining Rates or Times

a) Mr. Ford covered 310 mi. by traveling for 4 hr. at one speed and then for 5 hr. at another. Had he gone 5 hr. at the first speed and 4 hr. at the second, he would have gone 320 mi. Find the two rates.

Solution:

a) Let r = first rate in mph
R = second rate in mph.

$$4r + 5R = 310$$
$$5r + 4R = 320.$$

Solving, $r = 40$ and $R = 30$.

Ans. The rates were 40 mph and 30 mph.

b) By going 20 mph for one period of time and then 30 mph for another, Mr. Nash traveled 280 mi. Had he gone 10 mph faster throughout, he would have covered 390 miles. How many hours did he travel at each speed?

Solution:

b) Let t = first time in hr.
T = second time in hr.

$$20t + 30T = 280$$
$$30t + 40T = 390.$$

Solving, $t = 5$ and $T = 6$.

Ans. The times were 5 hr. at one speed and 6 hr. at the other.

12.3. Earnings Problems: Combining Rates of Pay or Times Worked

a) A plumber and his helper received \$30 for their work when the plumber worked 3 hr. and his helper 5 hr. Had each worked 2 hr. longer, they would have earned \$46. Find their hourly rates of pay.

Solution:

a) Let p = hourly rate of plumber in \$
h = hourly rate of helper in \$.

Then, $3p + 5h = 30$
$5p + 7h = 46.$

Solving, $p = 5,\ h = 3.$

Ans. Rates of pay were \$5 per hr. for the plumber and \$3 per hr. for the helper.

b) The rate of pay for a plumber is \$7 per hr. and for his helper, \$3 an hr. They received \$53 for a certain job, the plumber working for one length of time, the helper for another. Had each received \$1 per hr. less, they would have earned \$42. Find the time worked by each.

Solution:

b) Let p = time worked by plumber in hr.
h = time worked by helper in hr.

Then, $7p + 3h = 53$
$6p + 2h = 42.$

Solving, $p = 5,\ h = 6.$

Ans. Plumber worked 5 hr. and helper worked 6 hr.

SUPPLEMENTARY PROBLEMS

Number Problems Having One Unknown

1. Express each statement as an equation; then find the unknown: **(1.1)**

		Equation	Unknown
a) The sum of n and 12 is 21.	*Ans. a*)	$n + 12 = 21$	$n = 9$
b) The result of adding m and 15 is $4m$.	*Ans. b*)	$m + 15 = 4m$	$m = 5$
c) 27 is the difference obtained when p is subtracted from 40.	*Ans. c*)	$27 = 40 - p$	$p = 13$
d) 20 increased by twice a is 32.	*Ans. d*)	$20 + 2a = 32$	$a = 6$
e) 7 less than three times b is 23.	*Ans. e*)	$3b - 7 = 23$	$b = 10$
f) Four times d exceeds 15 by d.	*Ans. f*)	$4d - 15 = d$	$d = 5$
g) The product of 3 and $n+6$ equals 20.	*Ans. g*)	$3(n+6) = 20$	$n = \frac{2}{3}$
h) 5 more than half of x equals 29.	*Ans. h*)	$\frac{x}{2} + 5 = 29$	$x = 48$
i) The difference between 15 and half of y is 6.	*Ans. i*)	$15 - \frac{y}{2} = 6$	$y = 18$

2. Check each statement as indicated: **(1.2)**

a) Is 5 the correct number for "the result of adding a number and 15 is four times the number"?
Ans. a) $5 + 15 \stackrel{?}{=} 4(5)$
$20 = 20$ Yes.

b) Is 11 the correct number for "7 less than three times the number is 23"?
Ans. b) $3(11) - 7 \stackrel{?}{=} 23$
$26 \neq 23$ No.

c) Is 2 the correct number for "the product of 3 and 6 more than a number equals 20"?
Ans. c) $3(6+2) \stackrel{?}{=} 20$
$24 \neq 20$ No.

d) Is 48 the correct number for "5 more than half a number equals 29"?
Ans. d) $\frac{48}{2} + 5 \stackrel{?}{=} 29$
$29 = 29$ Yes.

3. Find the number in each of the following: (1.3)

a) Four times the number increased by 2 equals 30. *Ans. a*) 7

b) Seven times the number less 6 equals 29. *Ans. b*) 5

c) The result of combining five times the number and the number is 24. *Ans. c*) 4

d) If the number is added to 42 the sum is eight times the number. *Ans. d*) 6

e) If twice the number is subtracted from 13, the remainder is 8. *Ans. e*) $2\frac{1}{2}$

f) If twice the number is increased by 6, the result is the same as decreasing four times the number by 14. *Ans. f*) 10

g) Five times the number minus 12 equals three times the number increased by 10. *Ans. g*) 11

h) The sum of one-half the number and one-third the number is 15. *Ans. h*) 18

Number Problems Having Two Unknowns

4. If n represents a number, represent another number that is (2.1)

a) 4 more than three times n.

b) 5 less than twice n.

c) one-half of n increased by 3.

d) two-thirds of n decreased by 6.

e) the sum of three times n and 20.

f) 40 reduced by twice n.

g) the product of $(n+4)$ and 7.

h) the quotient of $(n-2)$ and 4.

Ans. a) $3n+4$, *b*) $2n-5$, *c*) $\frac{n}{2}+3$, *d*) $\frac{2}{3}n-6$, *e*) $3n+20$, *f*) $40-2n$, *g*) $7(n+4)$, *h*) $\frac{n-2}{4}$

5. Express each statement as an equation; then find the numbers: (2.2)

a) Two numbers are represented by n and $(n-8)$.
The first is 16 less than three times the second. *Ans. a*) $n = 3(n-8)-16$, 20 and 12

b) Two numbers are represented by n and $(30-n)$.
The first is six more than twice the second. *Ans. b*) $n = 2(30-n)+6$, 22 and 8

c) Two numbers are represented by n and $8n$.
Half the second exceeds three times the first by 15. *Ans. c*) $4n-3n = 15$, 15 and 120

6. Using s for the smaller number and l for the larger, obtain two equations for each problem; then find l and s: (2.3)

a) The sum of two numbers is 20. Their difference is 11.

b) The sum of two numbers is 12. Twice the larger plus the smaller equals 21.

c) The difference of two numbers is 7. The larger is one less than twice the smaller.

d) Separate 50 into two parts such that the larger is nine times the smaller.

e) Separate 42 into two parts such that the smaller is 3 less than one-half the larger.

Ans.

a)	*b*)	*c*)	*d*)	*e*)
$l+s = 20$	$l+s = 12$	$l-s = 7$	$l+s = 50$	$l+s = 42$
$l-s = 11$	$2l+s = 21$	$l = 2s-1$	$l = 9s$	$s = \frac{l}{2}-3$
$l = 15\frac{1}{2}$, $s = 4\frac{1}{2}$	$l = 9$, $s = 3$	$l = 15$, $s = 8$	$l = 45$, $s = 5$	$l = 30$, $s = 12$

7. Can you find two integers (2.4)

a) whose sum is 15 and whose difference is 5?

b) whose sum is 20 and whose difference is 9?

c) whose sum is 40 and the larger is 10 more than the smaller?

Ans. a) Yes, 10 and 5. *b*) No, the numbers $14\frac{1}{2}$ and $5\frac{1}{2}$ are not integers. *c*) Yes, 25 and 15.

Consecutive Integer Problems

8. Using n for the first integer, represent (3.1)

a) two consecutive odd integers and their sum. *Ans.* n, $n+2$; sum $= 2n+2$

b) four consecutive even integers and their sum. *Ans.* n, $n+2$, $n+4$, $n+6$; sum $= 4n+12$

c) five consecutive integers and their sum. *Ans.* n, $n+1$, $n+2$, $n+3$, $n+4$; sum $= 5n+10$

d) six consecutive odd integers and their sum. *Ans.* n, $n+2$, $n+4$, $n+6$, $n+8$, $n+10$; sum $= 6n+30$

9. Using n for the middle integer, represent (3.2)

a) three consecutive odd integers and their sum. *Ans.* $n-2$, n, $n+2$; sum $= 3n$

b) five consecutive integers and their sum. *Ans.* $n-2$, $n-1$, n, $n+1$, $n+2$; sum $= 5n$

10. Using n, $n+1$ and $n+2$ for three consecutive integers, express each statement as an equation; then find the integers: (3.3)

a) Their sum is 3 less than five times the first.

b) The sum of the first and second is six more than the third.

c) The sum of the first and third is 11 less than three times the second.

d) Twice the sum of the first and second is 16 more than twice the third.

e) Twice the third is 21 more than three times the sum of the first and second.

f) The product of the second and third is 8 more than the square of the first.

Ans. *a*) $3n+3 = 5n-3$; 3, 4, 5
b) $2n+1 = (n+2)+6$; 7, 8, 9
c) $2n+2 = 3(n+1)-11$; 10, 11, 12
d) $2(2n+1) = 2(n+2)+16$; 9, 10, 11
e) $2(n+2) = 3(2n+1)+21$; -5, -4, -3
f) $(n+1)(n+2) = n^2+8$; 2, 3, 4

11. Using n, $n+2$ and $n+4$ for three consecutive odd integers, express each statement as an equation; then find the integers: (3.4)

a) Their sum is 45.

b) The sum of the first and second is 19 less than three times the third.

c) Five times the sum of the first and third equals 18 more than eight times the second.

d) Twenty times the sum of the second and third equals 3 more than the first.

e) Twice the sum of the first and second equals four times the sum of the second and third.

Ans. *a*) $3n+6 = 45$; 13, 15, 17
b) $2n+2 = 3(n+4)-19$; 9, 11, 13
c) $5(2n+4) = 8(n+2)+18$; 7, 9, 11
d) $20(2n+6) = n+3$; -3, -1, 1
e) $2(2n+2) = 4(2n+6)$; -5, -3, -1

12. Can you find (3.5)

a) three consecutive integers whose sum is 36?

b) three consecutive odd integers whose sum is 33?

c) five consecutive even integers whose sum is 60?

d) five consecutive odd integers whose sum is 50?

Ans. *a*) Yes; 11, 12, 13 *b*) Yes; 9, 11, 13 *c*) Yes; 8, 10, 12, 14, 16
d) No. If n is used for the middle odd integer, $5n = 50$, $n = \underline{10}$ which is not odd.

Age Problems

13. Represent the age of a person (4.1, 4.2)

a) in y yr. if his present age is 50 yr. *Ans.* *a*) $50+y$

b) in 5 yr. if his present age is F yr. *Ans.* *b*) $F+5$

c) in 5 yr. if his age 10 yr. ago was S yr. *Ans.* *c*) $S+10+5$ or $S+15$

d) 10 yr. ago if his present age is J yr. *Ans.* *d*) $J-10$

e) y yr. ago if his present age is 40 yr. *Ans.* *e*) $40-y$

14. Using $5S$ and S to represent the present ages of a father and son, express each statement as an equation; then find their present ages: (4.3)

a) In 7 yr., the father will be three times as old as his son will be then.

b) 2 yr. ago, the father was seven times as old as his son was then.

c) In 3 yr., the father will be 1 yr. less than four times as old as his son will be then.

d) 9 yr. ago, the father was three times as old as his son will be 3 yr. hence.

Ans. *a*) $5S+7 = 3(S+7)$; 35, 7
b) $5S-2 = 7(S-2)$; 30, 6
c) $5S+3 = 4(S+3)-1$; 40, 8
d) $5S-9 = 3(S+3)$; 45, 9

15. Obtain two equations for each problem using J and C to represent the present ages of John and Charles respectively; then find their present ages: **(4.4)**

a) The sum of their present ages is 45.
John is 5 yr. older than Charles.

b) John is 10 yr. older than Charles.
12 yr. ago, John was three times as old as Charles was then.

c) In 5 yr., John will be twice as old as Charles will be then.
4 yr. ago, John was three times as old as Charles was then.

d) At present John is six times as old as Charles.
2 yr. hence, John will be ten times as old as Charles was 3 yr. ago.

e) In 5 yr., John will be 25 yr. older than Charles will be then.
2 yr. ago, John was seven times as old as Charles was last year.

Ans. *a*) $J+C=45$, $J=C+5$; 25, 20
b) $J=C+10$, $J-12=3(C-12)$; 27, 17
c) $J+5=2(C+5)$, $J-4=3(C-4)$; 31, 13
d) $J=6C$, $J+2=10(C-3)$; 48, 8
e) $J+5=(C+5)+25$, $J-2=7(C-1)$; 30, 5

16. George is now 8 yr. older than Harry. Find their present ages if **(4.5)**

a) 14 yr, ago, George was twice as old as Harry was then. *Ans.* *a*) 30, 22
b) 2 yr. hence, George will be three times as old as Harry will be then. *Ans.* *b*) 10, 2
c) 5 yr. hence, George will be twice as old as Harry was 2 yr. ago. *Ans.* *c*) 25, 17

17. The sum of the present ages of Will and Sam is 60 yr. In 8 yr., Will will be 5 yr. older than three times as old as Sam was 3 yr. ago. **(4.5)**
Ans. 42 and 18 $\{W+S=60$ and $W+8=3(S-3)+5\}$

Ratio Problems

18. Express each ratio in lowest terms: **(5.1)**

a) 20¢ to 5¢
b) 5 dimes to 15 dimes
c) 30 lb. to 25 lb.
d) 2 oz. to 14 oz.
e) 27 min. to 21 min.
f) 50% to 25%
g) 15% to 75%
h) 33% to 77%
i) $2.20 to $3.30
j) $.84 to $.96
k) $\frac{1}{2}$ lb. to $\frac{1}{4}$ lb.
l) $2\frac{1}{2}$ da. to $3\frac{1}{2}$ da.
m) 5 ft. to $\frac{1}{4}$ ft.
n) $\frac{1}{2}$ yd. to $1\frac{1}{2}$ yd.
o) $16\frac{1}{2}$ ft. to $5\frac{1}{2}$ ft.

Ans. *a*) 4 *c*) $\frac{6}{5}$ *e*) $\frac{9}{7}$ *f*) 2 *h*) $\frac{3}{7}$ *j*) $\frac{7}{8}$ *k*) 2 *m*) 20 *o*) 3
b) $\frac{1}{3}$ *d*) $\frac{1}{7}$ *g*) $\frac{1}{5}$ *i*) $\frac{2}{3}$ *l*) $\frac{5}{7}$ *n*) $\frac{1}{3}$

19. Express each ratio in lowest terms: **(5.2)**

a) 1 yr. to 2 mo.
b) 2 wk. to 5 da.
c) 3 da. to 3 wk.
d) $\frac{1}{2}$ hr. to 20 min.
e) 2 yd. to 2 ft.
f) $2\frac{1}{3}$ yd. to 2 ft.
g) $1\frac{1}{2}$ ft. to 9 in.
h) 2 lb. to 8 oz.
i) 100 lb. to 1 ton
j) $2 to 25¢
k) 2 quarters to 3 dimes
l) 1 sq. yd. to 2 sq. ft.

Ans. *a*) 6 *c*) $\frac{1}{7}$ *e*) 3 *g*) 2 *i*) $\frac{1}{20}$ *k*) $\frac{5}{3}$
b) $\frac{14}{5}$ *d*) $\frac{3}{2}$ *f*) $\frac{7}{2}$ *h*) 4 *j*) 8 *l*) $\frac{9}{2}$

20. Express each ratio in lowest terms: **(5.3)**

a) 20¢ to 30¢ to $1
b) $3 to $1.50 to 25¢
c) 1 quarter to 1 dime to 1 nickel
d) 1 da. to 4 da. to 1 wk.
e) $\frac{1}{2}$ da. to 9 hr. to 3 hr.
f) 2 hr. to $\frac{1}{2}$ hr. to 15 min.
g) 1 ton to 200 lb. to 40 lb.
h) 3 lb. to 1 lb. to 8 oz.
i) 1 gal. to 1 qt. to 1 pt.

Ans. *a*) 2 : 3 : 10 *c*) 5 : 2 : 1 *e*) 4 : 3 : 1 *g*) 50 : 5 : 1 *i*) 8 : 2 : 1
b) 12 : 6 : 1 *d*) 1 : 4 : 7 *f*) 8 : 2 : 1 *h*) 6 : 2 : 1

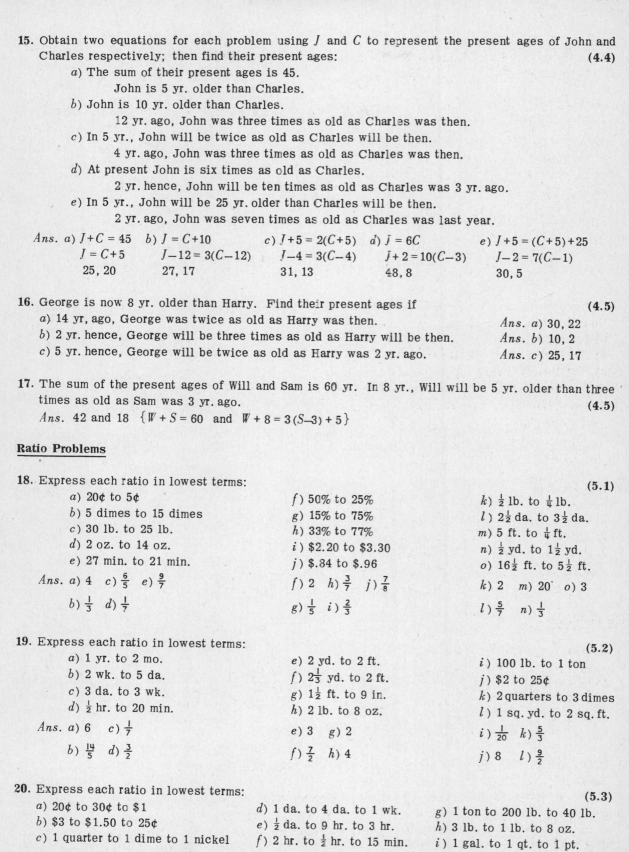

21. Express each ratio in lowest terms: (5.4)

a) 60 to 70 *Ans.* $\frac{6}{7}$
b) 84 to 7 *Ans.* 12
c) 65 to 15 *Ans.* $\frac{13}{3}$
d) 125 to 500 *Ans.* $\frac{1}{4}$
e) 630 to 105 *Ans.* 6
f) 1760 to 990 *Ans.* $\frac{16}{9}$
g) .7 to 2.1 *Ans.* $\frac{1}{3}$
h) .36 to .24 *Ans.* $\frac{3}{2}$
i) .002 to .007 *Ans.* $\frac{2}{7}$
j) .055 to .005 *Ans.* 11
k) 6.4 to 8 *Ans.* .8 or $\frac{4}{5}$
l) 144 to 2.4 *Ans.* 60
m) $7\frac{1}{2}$ to $2\frac{1}{2}$ *Ans.* 3
n) $1\frac{1}{2}$ to 12 *Ans.* $\frac{1}{8}$
o) 5 to $\frac{1}{3}$ *Ans.* 15
p) $\frac{1}{3}$ to $3\frac{1}{3}$ *Ans.* $\frac{1}{10}$
q) $\frac{5}{6}$ to $1\frac{2}{3}$ *Ans.* $\frac{1}{2}$
r) $\frac{7}{4}$ to $\frac{1}{8}$ *Ans.* 14

22. Express each ratio in lowest terms: (5.5)

a) x to $8x$
b) $15c$ to 5
c) $11d$ to 22
d) $2\pi r$ to πD
e) πab to πa^2
f) $4S$ to S^2
g) S^3 to $6S^2$
h) $9r^2t$ to $6rt^2$
i) x to $4x$ to $10x$
j) $15y$ to $10y$ to $5y$
k) x^3 to x^2 to x
l) $12w$ to $10w$ to $8w$ to $2w$

Ans. *a*) $\frac{1}{8}$, *b*) $3c$, *c*) $\frac{d}{2}$, *d*) $\frac{2r}{D}$, *e*) $\frac{b}{a}$, *f*) $\frac{4}{S}$, *g*) $\frac{S}{6}$, *h*) $\frac{3r}{2t}$,
i) 1:4:10, *j*) 3:2:1, *k*) $x^2:x:1$, *l*) 6:5:4:1.

23. Using x as their common factor, represent the numbers and their sum if (5.6)

a) two numbers have a ratio of 5:4. *Ans. a*) $5x$ and $4x$; sum $= 9x$
b) two numbers have a ratio of 9 to 1. *Ans. b*) $9x$ and x; sum $= 10x$
c) three numbers have a ratio of 2:5:11. *Ans. c*) $2x$, $5x$ and $11x$; sum $= 18x$
d) five numbers have a ratio of 1:2:2:3:7. *Ans. d*) x, $2x$, $2x$, $3x$ and $7x$; sum $= 15x$

24. If two numbers in the ratio of 7 to 4 are represented by $7x$ and $4x$, express each statement as an equation; then find x and the numbers: (5.7)

a) The sum of the numbers is 99. *Ans. a*) $11x = 99$, $x = 9$, 63 and 36
b) The difference of the numbers is 39. *Ans. b*) $3x = 39$, $x = 13$, 91 and 52
c) Twice the smaller is 2 more than the larger. *Ans. c*) $8x = 7x + 2$, $x = 2$, 14 and 8
d) The sum of the larger and one-half the smaller is 36. *Ans. d*) $7x + 2x = 36$, $x = 4$, 28 and 16

25. The perimeter of a triangle is 60 in. Find each side if (5.8)

a) the sides are in the ratio of 5:4:3.
b) the sides are in the ratio of 2:6:7.
c) two sides are in the ratio of 3 to 2 and the third side is 25 in.

Ans. a) Using $5x$, $4x$ and $3x$ for sides: $12x = 60$, $x = 5$. Sides are 25in., 20 in., 15 in.
b) Using $2x$, $6x$ and $7x$ for sides: $15x = 60$, $x = 4$. Sides are 8 in., 24 in., 28 in.
c) Using $3x$ and $2x$ for sides; $5x = 35$, $x = 7$. Sides are 21 in., 14 in., 25 in.

26. The ratio of Fred's money to Lee's money is 5:2. How much does each have according to the following: (5.9)

a) They will have equal amount if Fred gives Lee 24¢.
b) Fred will have twice as much as Lee if Fred gives Lee 30¢.

Ans. a) Using $5x$ and $2x$; $5x - 24 = 2x + 24$, $x = 16$. Amounts are 80¢ and 32¢.
b) Using $5x$ and $2x$; $5x - 30 = 2(2x + 30)$, $x = 90$. Amounts are \$4.50 and \$1.80.

27. An estate of \$8800 is to be divided among three heirs according to the conditions of a will. Find the amounts to be received if (5.10)

a) the estate is to be divided in the ratio of 5:2:1.
b) one heir is to get \$4000 and the others are to get the rest in the ratio of 7 to 5.

Ans. a) Using $5x$, $2x$ and x; $8x = 8800$, $x = 1100$. Amounts are \$5500, \$2200, \$1100.
b) Using $7x$ and $5x$; $12x = 8800 - 4000$, $x = 400$. Amounts are \$2800, \$2000, \$4000.

28. The hourly wages of James and Stanley are in the ratio of 8:7. Find their hourly wages (5.11)
a) if in 3 hours James earns 60¢ more than Stanley.
b) if in 5 hours their combined earnings are \$22.50.

Ans. *a*) Using $8x$ and $7x$; $3(8x) = 3(7x) + 60$, $x = 20$. Wages are \$1.60 and \$1.40 per hr.
b) Using $8x$ and $7x$; $5(8x+7x) = 2250$, $x = 30$. Wages are \$2.40 and \$2.10 per hr.

Angle Problems

29. If two angles in the ratio of 5:4 are represented by $5x$ and $4x$, express each statement as an equation; then find x and the angles: (6.1)
a) The angles are adjacent and form an angle of 45°.
b) The angles are complementary.
c) The angles are supplementary.
d) The larger angle is 30° more than one-half the smaller.
e) The smaller angle is 25° more than three-fifths the larger.
f) The angles are the acute angles of a right triangle.
g) The angles are two angles of a triangle whose third angle is their difference.
h) The first angle is one of two equal angles of an isosceles triangle. The remaining angle of the triangle is half of the other angle.

Ans. *a*) $5x + 4x = 45$, $x = 5$, 25° and 20°.
b) $5x + 4x = 90$, $x = 10$, 50° and 40°.
c) $5x + 4x = 180$, $x = 20$, 100° and 80°.
d) $5x = 2x + 30$, $x = 10$, 50° and 40°.
e) $4x = 3x + 25$, $x = 25$, 125° and 100°.
f) $5x + 4x = 90$, $x = 10$, 50° and 40°.
g) $5x + 4x + x = 180$, $x = 18$, 90° and 72°.
h) $5x + 5x + 2x = 180$, $x = 15$, 75° and 60°.

30. If two angles are represented by a and b, obtain two equations for each problem; then find the angles: (6.2)
a) The angles are adjacent forming an angle of 75°. Their difference is 21°.
b) The angles are complementary. One is 10° less than three times the other.
c) The angles are supplementary. One is 20° more than four times the other.
d) The angles are two angles of a triangle whose third angle is 50°. Twice the first added to three times the second equals 300°.

Ans.

a)	*b*)	*c*)	*d*)
$a + b = 75$	$a + b = 90$	$a + b = 180$	$a + b = 130$
$a - b = 21$	$a = 3b - 10$	$a = 4b + 20$	$2a + 3b = 300$
48°, 27°	65°, 25°	148°, 32°	90°, 40°

31. If three angles in the ratio of 7:6:5 are represented by $7x$, $6x$ and $5x$, express each statement as an equation; then find x and the angles: (6.3)
a) The first and second are adjacent and form an angle of 91°.
b) The first and third are supplementary.
c) The first and one-half the second are complementary.
d) The angles are the three angles of a triangle.
e) The sum of the second and third is 20° more than the first.
f) The second is 12° more than one-third the sum of the first and third.

Ans. *a*) $7x+6x = 91$, $x = 7$, 49°, 42° and 35°.
b) $7x+5x = 180$, $x = 15$, 105°, 90° and 75°.
c) $7x+3x = 90$, $x = 9$, 63°, 54° and 45°.
d) $7x+6x+5x = 180$, $x = 10$, 70°, 60° and 50°.
e) $6x+5x = 7x+20$, $x = 5$, 35°, 30° and 25°.
f) $6x = 4x + 12$, $x = 6$, 42°, 36° and 30°.

32. *a*) One of two complementary angles is 5° less than four times the other. Find the angles. (6.4)
Ans. 19°, 71° {If x is the other angle, $x + (4x-5) = 90$}

b) One of two supplementary angles is 10° more than two-thirds of the other. Find the angles.
Ans. 78°, 102° {If x is the other angle, $x + (\frac{2}{3}x+10) = 180$}

33. In a triangle, one angle is 9° less than twice another. The third angle is 18° more than twice their sum. Find the angles. (6.5)
Ans. 33°, 21°, 126° {If x is the second angle, $x+(2x-9)+6x = 180$}

Perimeter Problems

34. The length and width of a rectangle are represented by l and $2l-20$ respectively. Express each statement as an equation; then find l and the dimensions of the rectangle. (7.1)

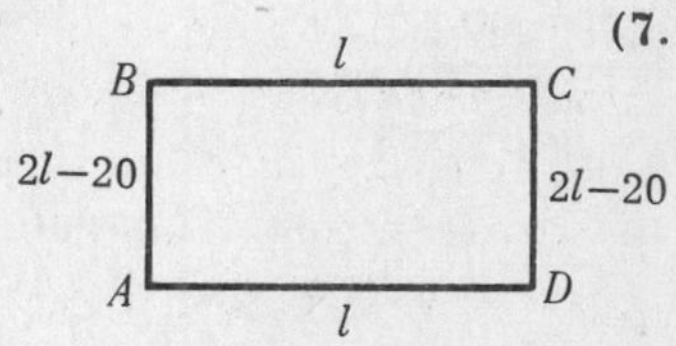

a) The perimeter is 50 yd.
b) The semi-perimeter is 22 in.
c) A fence around the rectangle but not including BC is 25 ft.
d) If each dimension is increased 10 rd., the perimeter of the rectangle will be 72 rd.
e) If the width is unchanged and the length is doubled, the perimeter will be 128 in.
f) The rectangle is a square.
g) The perimeter of a square having the width as a side exceeds the perimeter of an equilateral triangle having the length as a side by 10 ft.

Ans. a) $6l-40 = 50$, $l=15$, 15 yd. and 10 yd.
b) $3l-20 = 22$, $l=14$, 14 in. and 8 in.
c) $5l-40 = 25$, $l=13$, 13 ft. and 6 ft.
d) $2(l+10)+2(2l-10) = 72$, $l=12$, 12 rd. and 4 rd.
e) $2(2l)+2(2l-20) = 128$, $l=21$, 21 in. and 22 in.
f) $l = 2l-20$, $l=20$, 20 and 20.
g) $4(2l-20) = 3l+10$, $l=18$, 18 ft. and 16 ft.

35. Using b for the base and a for each of the equal sides of an isosceles triangle, obtain two equations for each problem; then find AB and AC. (7.2)

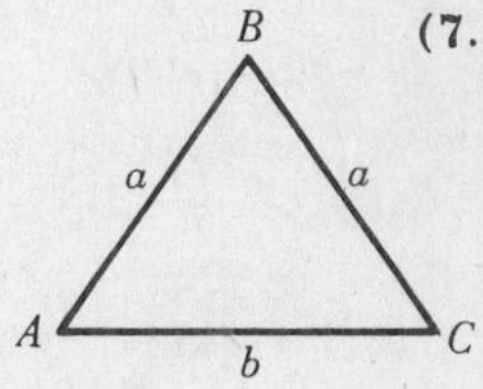

a) The perimeter is 32. AB is 4 more than AC.
b) The perimeter is 42. AB is 9 less tan twice AC.
c) The perimeter is 28. The sum of twice AB and one-half of AC is 24.
d) If each of the equal sides is doubled, the new perimeter will be 43.
If each of the equal sides is increased by 10, the new perimeter will be 45.

Ans. a) $2a+b = 32$, $a = b+4$, $AB=12$, $AC=8$
b) $2a+b = 42$, $a = 2b-9$, $AB=15$, $AC=12$
c) $2a+b = 28$, $2a+\frac{b}{2} = 24$, $AB=10$, $AC=8$
d) $4a+b = 43$, $2(a+10)+b = 45$, $AB=9$, $AC=7$

36. In a trapezoid having equal sides, the bases and sides are represented as shown in the adjoining diagram. Express each statement as an equation; then find x, AB and AD: (7.3)

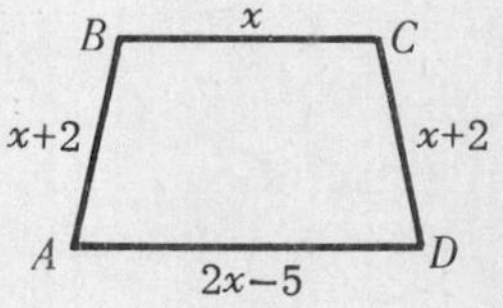

a) The perimeter is 49 in.
b) The combined length of the upper and lower bases is 28 ft.
c) A fence around the trapezoid not including CD is 19 yd.

Ans. a) $5x-1 = 49$, $x=10$, $AB=12$ in., $AD=15$ in., b) $3x-5 = 28$, $x=11$, $AB=13$ ft., $AD=17$ ft.
c) $4x-3=19$, $x=5\frac{1}{2}$, $AB=7\frac{1}{2}$ yd., $AD=6$ yd.

37. Three sides of a quadrilateral in the ratio of 2 : 3 : 5 are represented by $2x$, $3x$ and $5x$, respectively. If the base is 20, express each statement as an equation; then find x and AB. (7.4)

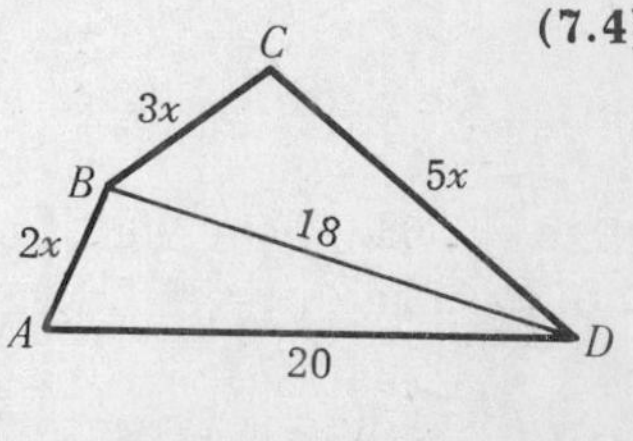

a) The perimeter is 95.
b) AD is 40 less than the combined length of the other three sides.
c) BC is 12 less than the combined length of AD and AB.
d) The perimeter of triangle BCD exceeds that of triangle ABD by 22.

Ans. a) $10x+20 = 95$, $x=7\frac{1}{2}$, $AB=15$
b) $20 = 10x-40$, $x=6$, $AB=12$
c) $3x = (2x+20)-12$, $x=8$, $AB=16$
d) $8x+18 = (2x+38)+22$, $x=7$, $AB=14$

38. A piece of wire 72 inches long is to be used to make a wire model of an isosceles triangle. How long should the sides be if **(7.5)**

a) the ratio of one of the equal sides to the base is 3 : 2?

b) the base is to be 6 inches less than one of the equal sides?

Ans. *a*) 27, 27 and 18 in. *b*) 26, 26 and 20 in.

39. The side of one square is 7 ft. longer than the side of another. **(7.6)**
Find the sides of the squares if

a) the sum of the two perimeters is 60 ft.

b) the perimeter of the larger square is twice that of the smaller.

Ans. *a*) 11 ft. and 4 ft. *b*) 14 ft. and 7 ft.

Coin or Stamp Problems

40. Write the expression that represents the total value *in cents* of each; then simplify it: **(8.1, 8.2)**

a) n nickels and $(n-5)$ dimes. *Ans.* *a*) $5n+10(n-5)$; $15n-50$

b) n dimes and $(3n-10)$ quarters. *Ans.* *b*) $10n+25(3n-10)$; $85n-250$

c) n 3-cent and $(2n+5)$ 2-cent stamps. *Ans.* *c*) $3n+2(2n+5)$; $7n+10$

41. Express each statement as an equation; then find n and the number of each coin or stamp: **(8.3)**

a) The total value of n nickels and $2n$ dimes is $1.25.

b) The total value of n 2-cent and $(n-3)$ 3-cent stamps is 91¢.

c) The value of n quarters is $5.75 less than the value of $(n+10)$ half-dollars.

d) The value of n 6-cent stamps equals that of $(2n-7)$ 10-cent stamps.

Ans. *a*) $5n+10(2n)=125$, $n=5$, 5 nickels and 10 dimes

b) $2n+3(n-3)=91$, $n=20$, 20 2-cent and 17 3-cent stamps

c) $25n=50(n+10)-575$, $n=3$, 3 quarters and 13 half-dollars

d) $6n=10(2n-7)$, $n=5$, 5 6-cent and 3 10-cent stamps.

42. Using x for the number of the first coin and y for the number of the second coin, obtain two equations for each problem; then find the number of each: **(8.4)**

	Equations	No. of Coins
a) John has 25 coins, nickels and half-dollars. Their value is $6.65.	*Ans.* $x+y=25$ $5x+50y=665$	13 nickels 12 half-dollars
b) The value of a number of dimes and quarters is $2.05. If the quarters were replaced by nickels, the value would be $1.05.	*Ans.* $10x+25y=205$ $10x+5y=105$	8 dimes 5 quarters
c) The value of a number of nickels and dimes is $1.40. If the nickels were replaced by dimes and the dimes by nickels, the value would be $1.75.	*Ans.* $5x+10y=140$ $10x+5y=175$	14 nickels 7 dimes
d) The value of a number of nickels and quarters is $3.25. If the number of nickels were increased by 3 and the number of quarters were doubled, the value would be $5.90.	*Ans.* $5x+25y=325$ $5(x+3)+25(2y)=590$	15 nickels 10 quarters

43. Jack has a number of coins worth $4.05. The collection consisted of 5 more nickels than dimes, and the number of quarters was 3 less than twice the number of dimes. How many of each had he? **(8.5)**

Ans. Using d for the number of dimes: $5(d+5)+10d+25(2d-3)=405$, $d=7$.
12 nickels, 7 dimes, 11 quarters.

44. Is it possible to change a \$20 bill into an equal number of nickels, dimes and quarters? If so, how many of each would be needed? **(8.6)**

Ans. Yes. Using n for the number of nickels: $5n + 10n + 25n = 2000$, $n = 50$.
50 nickels, 50 dimes, 50 quarters.

45. At a show, the price of admission for a child was 10¢ and for an adult 25¢. Find the number of each if **(8.7)**

a) 40 were admitted and the total paid was \$6.40.
b) twice as many children as adults were admitted and the total paid was \$4.05.
c) 10 more children than adults were admitted and the total paid was \$6.25.
d) 15 more children than adults were admitted and the totals paid for each were equal.

Ans. *a*) 24 children and 16 adults. $\{10c + 25(40-c) = 640,\ c = 24\}$
b) 18 children and 9 adults. $\{10(2a) + 25a = 405,\ a = 9\}$
c) 25 children and 15 adults. $\{10(a+10) + 25a = 625,\ a = 15\}$
d) 25 children and 10 adults. $\{10(a+15) = 25a,\ a = 10\}$

Cost and Mixture Problems

46. Write the expression that represents the total value *in cents*; then simplify it: **(9.1)**

a) n lb. of tea valued at \$1.25 per lb. and 5 lb. of tea valued at \$1.40 per lb.
b) n stamps valued at 50¢ each and $(n-8)$ stamps valued at 75¢ each.
c) $2n$ notebooks at 10¢ each and $(3n-5)$ notebooks at 15¢ each.
d) d dozen pens at \$3.25 a dozen and $(2d+4)$ doz. pens at \$2.80 a dozen.

Ans. *a*) $125n + 5(140)$; $125n + 700$
b) $50n + 75(n-8)$; $125n - 600$
c) $10(2n) + 15(3n-5)$; $65n - 75$
d) $325d + 280(2d+4)$; $885d + 1120$

47. Write the expression that represents the total cost *in dollars*; then simplify it: **(9.2)**

a) n souvenirs priced at \$3 apiece and $(n-2)$ souvenirs priced at \$1.50 each.
b) $2n$ tickets at \$3.30 each, $\frac{n}{2}$ tickets at \$4.40 each and n tickets at \$1.60 each.
c) 12 chairs at \$7 each, n chairs at \$8.50 each and $(2n+1)$ chairs at \$10 each.
d) n boxes of cards at \$2.50 a box and $(5n-4)$ boxes at \$3.75 a box.

Ans. *a*) $3n + 1.50(n-2)$; $4.5n - 3$
b) $3.30(2n) + 4.40(\frac{n}{2}) + 1.60n$; $10.4n$
c) $7(12) + 8.50n + 10(2n+1)$; $28.5n + 94$
d) $2.50n + 3.75(5n-4)$; $21.25n - 15$

48. Express each statement as an equation; then find n and the number of each kind: **(9.3)**

a) The cost of n lb. of tea at 85¢ a lb. and $(10-n)$ lb. of tea at 90¢ a lb. is \$8.80.
b) The value of n tickets at 90¢ each and $(3n+2)$ tickets at \$1.20 each is \$29.40.
c) John's earnings for n hr. at \$1.70 per hr. and for $(n-4)$ hr. at \$2.40 per hr. is \$31.40.
d) The cost of $(n+2)$ gifts at \$8 each, $2n$ gifts at \$10 each and n gifts at \$15 each is \$231.

Ans. *a*) $85n + 90(10-n) = 880$, $n = 4$, 4 lb. at 85¢ and 6 lb. at 90¢.
b) $90n + 120(3n+2) = 2940$, $n = 6$, 6 tickets at 90¢ and 20 tickets at \$1.20.
c) $170n + 240(n-4) = 3140$, $n = 10$, 10 hr. at \$1.70 and 6 hr. at \$2.40.
d) $8(n+2) + 10(2n) + 15n = 231$, $n = 5$, 7 gifts at \$8, 10 gifts at \$10 and 5 gifts at \$15.

49. A coffee merchant blended coffee worth 75¢ a lb. with coffee worth 95¢ a lb. How many lb. of each grade did he use to make **(9.4)**

		No. of Pounds 75¢ grade	95¢ grade
a) a mixture of 30 lb. valued at 85¢ a lb?	*Ans. a*)	15 lb.,	15 lb.
b) a mixture of 12 lb. valued at 80¢ a lb?	*Ans. b*)	9 lb.,	3 lb.
c) a mixture of 24 lb. valued at 90¢ a lb?	*Ans. c*)	6 lb.,	18 lb.

50. At a game, tickets were sold at 25¢, 50¢ and 80¢ each. The number sold at 50¢ was 10 more than the number sold at 80¢. The total receipts were $16.50. **(9.5)**

Find the number sold at each price

a) if there were 20 sold at 25¢.

b) if the number sold at 25¢ equalled the sum of the other tickets.

c) if the number sold at 25¢ was 5 more than that at 50¢.

Ans. *a*) 15 at 50¢, 5 at 80¢, 20 at 25¢ $\{20(25) + 50(x+10) + 80x = 1650,\ x = 5\}$

b) 15 at 50¢, 5 at 80¢, 20 at 25¢ $\{25(2x+10) + 50(x+10) + 80x = 1650,\ x = 5\}$

c) 15 at 50¢, 5 at 80¢, 20 at 25¢ $\{25(x+15) + 50(x+10) + 80x = 1650,\ x = 5\}$

Investment and Interest Problems

51. Write the expression that represents the annual interest earned by each principal; then simplify it: **(10.1)**

a) \$1000 at 5%, \$2000 at 3% and \$$P$ at 4%. *Ans.* *a*) $.05(1000) + .03(2000) + .04P$; $.04P + 110$

b) \$$P$ at 2%, \$$2P$ at 5% and \$7500 at 6%. *Ans.* *b*) $.02P + .05(2P) + .06(7500)$; $.12P + 450$

c) \$$P$ at 3% and \$$(2P - 600)$ at 7%. *Ans.* *c*) $.03P + .07(2P - 600)$; $.17P - 42$

d) \$$P$ at 8%, \$$3P$ at 3% and \$$(P + 2500)$ at 4%. *Ans.* *d*) $.08P + .03(3P) + .04(P + 2500)$; $.21P + 100$

52. Express each statement as an equation; then solve and state each principal: **(10.2)**

a) The total annual income from \$800 at 3% and \$$P$ at 4% is \$56.

b) The total annual interest from \$$P$ at 2% and \$$(P + 2000)$ at 5% is \$380.

c) The annual interest from \$$P$ at 6% equals that from \$$(2P - 3000)$ at 5%.

d) The annual interest from \$$2P$ at $3\frac{1}{2}$% exceeds that from \$$P$ at 5% by \$40.

Ans. *a*) $.03(800) + .04P = 56,\ P = 800.$ \$800 at 4%

b) $.02P + .05(P + 2000) = 380,\ P = 4000.$ \$4000 at 2%, \$6000 at 5%

c) $.06P = .05(2P - 3000),\ P = 3750.$ \$3750 at 6%, \$4500 at 5%

d) $.03\frac{1}{2}(2P) = .05P + 40,\ P = 2000.$ \$4000 at $3\frac{1}{2}$%, \$2000 at 5%

53. Mr. Baker invested two sums of money, the first at 5% and the second at 6%. State the amount invested at each rate if the ratio of the investments was **(10.3)**

a) 3 : 1 and the interest from the 5% investment exceeded that from the 6% investment by \$180.

b) 4 : 3 and the total interest was \$342.

Ans. *a*) Using $3x$ and x, $.05(3x) - .06x = 180,\ x = 2000,$ \$6000 at 5% and \$2000 at 6%.

b) Using $4x$ and $3x$, $.05(4x) + .06(3x) = 342,\ x = 900,$ \$3600 at 5% and \$2700 at 6%.

54. A total of \$1200 is invested, partly at 3% and the rest at 5%. Find the amounts invested at each rate **(10.4)**

a) if the total interest is \$54. *Ans.* *a*) \$300 at 3%, \$900 at 5%

b) if the annual interests are both equal. *Ans.* *b*) \$750 at 3%, \$450 at 5%

c) if the annual interest from the 3% investment exceeds that from the 5% investment by \$4. *Ans.* *c*) \$800 at 3%, \$400 at 5%

55. On a total investment of \$5400, Mr. Adams lost 4% on one part and earned 5% on the other. How large was each investment **(10.5)**

a) if his losses equalled his earnings? *Ans.* \$3000 at (−4%) and \$2400 at 5%

b) if his net income was \$144? *Ans.* \$1400 at (−4%) and \$4000 at 5%

56. Mr. Howard has \$400 invested at 2% and \$600 at 3%. State the amount he must invest at 6% so that his annual income will be *a*) 4% of the entire investment, *b*) 5% of the entire investment.

Ans. Using x for the added investment: **(10.6)**

a) $.02(400) + .03(600) + .06x = .04(1000 + x),$ \$700 to be added at 6%

b) $.02(400) + .03(600) + .06x = .05(1000 + x),$ \$2400 to be added at 6%

Motion Problems

57. Write the expression, in simplified form, that represents the distance *in miles* traveled (11.1)

a) in 7 hr. at 20 mph and in T hr. more at 42 mph. *Ans. a*) $7(20) + 42T$; $42T + 140$

b) in 10 hr. at R mph and in 12 hr. more at 10 mph. *Ans. b*) $10R + 12(10)$; $10R + 120$

c) in T hr. at 25 mph and in $(T+8)$ hr. more at 15 mph. *Ans. c*) $25T + 15(T+8)$; $40T + 120$

d) in 5 hr. at R mph and in $3\frac{1}{2}$ hr. more at $(2R+8)$ mph. *Ans. d*) $5R + 3\frac{1}{2}(2R+8)$; $12R + 28$

58. Write the expression, in simplified form, that represents the time *in hours* needed to travel (11.2)

a) 240 mi. at 40 mph and D mi. farther at 27 mph. *Ans. a*) $\frac{240}{40} + \frac{D}{27}$; $\frac{D}{27} + 6$

b) 100 mi. at R mph and 75 mi. farther at 15 mph. *Ans. b*) $\frac{100}{R} + \frac{75}{15}$; $\frac{100}{R} + 5$

c) D mi. at 15 mph and $3D$ mi. farther at 45 mph. *Ans. c*) $\frac{D}{15} + \frac{3D}{45}$; $\frac{2D}{15}$

d) 40 mi. at 4 mi. per min. and 50 mi. farther at R mph. *Ans. d*) $\frac{40}{240} + \frac{50}{R}$; $\frac{50}{R} + \frac{1}{6}$

59. Two trains leave the same terminal at the same time and travel in opposite directions. After 8 hr. they are 360 mi. apart. The speed of the faster train is 3 mph less than twice that of the slower train. Find the rate of each train. (11.3)

Ans. Using R for speed of slower train in mph, $8R + 8(2R-3) = 360$, rates are 16 mph and 29 mph.

60. In each situation, the travelers start from the same place at the same time and travel in *opposite directions*. Using the letter indicated, express each statement in an equation; then solve and find each quantity represented: (11.4)

a) At speeds of 25 mph and 15 mph, they travel until they are 210 mi. apart. Find the time of each using T in hr.

b) After 7 hr., both are 707 mi. apart, one going 39 mph slower. Find the faster rate, using R in mph for this.

c) At speeds in the ratio of 6 : 5, both are 308 mi. apart in 14 hr. Find their respective rates, using $6x$ and $5x$ in mph. for these.

Ans. a) $25T + 15T = 210$, time is $5\frac{1}{4}$ hr. *b*) $7R + 7(R-39) = 707$, faster rate is 70 mph.
c) $14(6x) + 14(5x) = 308$, $x = 2$, rates are 12 mph and 10 mph.

61. Two planes leave from points 1925 mi. apart at the same time and fly toward each other. Their average speeds are 225 mph and 325 mph. How soon will the planes meet and how far will each have traveled? (11.5)

Ans. Using T for the time of travel, $225T + 325T = 1925$. Planes will meet in $3\frac{1}{2}$ hr.
The distances traveled will be $787\frac{1}{2}$ mi. and $1137\frac{1}{2}$ mi.

62. In each situation, the travelers start from distant places and travel toward each other until they meet. Using the letter indicated, express each statement in an equation; then solve and find each quantity represented: (11.6)

a) Starting 297 mi. apart, they travel at 38 mph and 28 mph. Find the time of each in hr., using T.

b) Starting 630 mi. apart, they meet in 5 hr., one traveling 46 mph faster than the other. Find the rate of the faster in mph, using R for this.

c) Starting 480 mi. apart, they meet in 6 hr., one traveling five times as fast as the other. Find the rate of the slower in mph, using R for this.

Ans. a) $38T + 28T = 297$, time is $4\frac{1}{2}$ hr.
b) $5R + 5(R-46) = 630$, rate of faster is 86 mph.
c) $6R + 6(5R) = 480$, rate of slower is $13\frac{1}{3}$ mph.

63. A plane traveled from its base to a distant point and back again along the same route in a total of 8 hr. Its average rate going was 180 mph and its average rate returning was 300 mph, how long did it take in each direction and what was the distance covered each way? **(11.7)**

Ans. Using T for the time in hr. going, $180T = 300(8-T)$, time going was 5 hr.
The distance covered each way was 900 mi.

64. In each situation, a traveler traveled out and back to the starting place along the same road. Using the letter indicated, express each statement as an equation, then solve and find each quantity represented: **(11.8)**

a) A traveler traveled out for 3 hr. at an average rate of 44 mph and returned in $5\frac{1}{2}$ hr. Find the average rate returning in mph, using R.

b) A traveler took 3 hr. less to travel back than to go out. If he averaged 45 mph out and 54 mph back, find his time in hr. returning, using T for this.

c) A traveler required a total of 8 hr. for a round trip. If he averaged 24 mph out and 36 mph back, find his time in hr. returning, using T for this.

d) After taking 3 hr. going out, a traveler came back in 5 hr., averaging 28 mph slower on the way back. Find the average rate going in mph, using R for this.

Ans. *a*) $3(44) = 5\frac{1}{2}R$, rate returning 24 mph. *c*) $24(8-T) = 36T$, time back 3.2 hr.
b) $45(T+3) = 54T$, time back 15 hr. *d*) $3R = 5(R-28)$, rate going 70 mph.

65. A boat travels at 24 mph. A patrol boat starts 3 hr. later from the same place and travels at 32 mph in the same direction. How long will it need to overtake the first and what distance will it cover?

Ans. Using T for the time of the patrol boat in hr., $32T = 24(T+3)$. **(11.9)**
The time needed to overtake is 9 hr. and the distance to cover is 288 mi.

66. In each situation, after a traveler has begun his trip, a second traveler starts from the same place along the same road and, going in the same direction, overtakes the first. **(11.10)**

a) The second starts $2\frac{1}{2}$ hr. later and travels for 6 hr. at 34 mph. Find the rate of the first in mph, using R for this.

b) Their speeds are 15 mph and 24 mph. The second starts 2 hr. after the first. Find the time of the first in hr., using T for this.

c) The first travels 18 mph slower than twice the speed of the second. If the second starts out 2 hr. later and overtakes the first in 5 hr., find the rate of the second in mph, using R for this.

d) The ratio of their rates is 4 : 5. The first was delayed 2 hr. along the way and also made a detour of 10 extra mi. The second overtook the first in 6 hr. after starting 4 hr. later. Find their rates in mph, using $4x$ and $5x$ for these.

Ans. *a*) $8\frac{1}{2}R = 6(34)$; rate of first is 24 mph.
b) $15T = 24(T-2)$; time of first is $5\frac{1}{3}$ hr.
c) $7(2R-18) = 5R$; rate of second is 14 mph.
d) $8(4x) - 10 = 6(5x)$, $x = 5$; rates are 20 mph and 25 mph.

67. On a trip, Phillip used a boat first and later a train. The boat trip took 3 hr. longer than the train ride. How long did the trip take if the speeds of the boat and train were **(11.11)**

a) 20 mph and 30 mph respectively and the trip was 210 mi.?
b) 10 mph and 40 mph respectively and the trip was 255 mi.?

Ans. Using T for the time of the train:
a) $20(T+3) + 30T = 210$, $T = 3$, total time 9 hr.
b) $10(T+3) + 40T = 255$, $T = 4\frac{1}{2}$, total time 12 hr.

Combination Problems

68. Find the price of a shirt and the price of a hat in each problem: **(12.1*a*)**
 a) 6 shirts and 8 hats cost \$64. At the same price, 4 shirts and 1 hat cost \$21.
 b) 3 shirts and 5 hats cost \$65. If each price is increased by \$1, 2 shirts and 6 hats would cost \$78.

 Ans. Using s for the price of 1 shirt in \$ and h for the price of 1 hat in \$:
 a) $6s + 8h = 64$ and $4s + h = 21$. Prices are \$4 per shirt and \$5 per hat.
 b) $3s + 5h = 65$ and $2(s+1) + 6(h+1) = 78$. Prices are \$5 per shirt and \$10 per hat.

69. Find the number of lb. of each grade of tea in each problem: **(12.1*b*)**
 a) A mixture of tea worth \$1.20 a lb. and tea worth \$1.80 a lb. is valued at \$18.
 If the price per lb. of each grade is increased 30¢, the new value would be \$21.30.
 b) A mixture of tea worth \$1 a lb. and tea worth \$1.40 a lb. is valued at \$17.
 Had there been twice as many lb. of the second tea, the value would have been \$24.

 Ans. Using n for the no. of lb. of the cheaper tea and N for the no. of lb. of the dearer tea:
 a) $120n + 180N = 1800$ and $150n + 210N = 2130$. 3 lb. at \$1.20 and 8 lb. at \$1.80.
 b) $100n + 140N = 1700$ and $100n + 140(2N) = 2400$. 10 lb. at \$1 and 5 lb. at \$1.40.

70. Find the two rates in each problem: **(12.2*a*)**
 a) By traveling for 5 hr. at one speed and then for 3 hr. at another, Mr. Hart covered 250 mi. Had he traveled for two hours longer at each speed, he would have covered 370 mi.
 b) By traveling for 2 hr. at one speed and then 7 hr. at another, Mr. Hayes completed a trip of 258 mi. Had the first rate been 10 mph faster and the second rate twice as fast, he would have gone 488 mi.

 Ans. Using r for the first rate in mph and R for the second rate in mph:
 a) $5r + 3R = 250$ and $7r + 5R = 370$. Rates were 35 mph and 25 mph.
 b) $2r + 7R = 258$ and $2(r+10) + 7(2R) = 488$. Rates were 24 mph and 30 mph.

71. Find the number of hours traveled at each speed in each problem: **(12.2*b*)**
 a) Mr. Ford completed 282 mi. by going 22 mph for one period of time and then 27 mph for another. Had he increased each speed 5 mph, he would have covered 337 mi.
 b) Mr. Plymouth completed 420 mi. by going 15 mph for one period of time and then 36 mph for another. Had he gone twice as fast during the first time and one-half as fast during the second, he would have gone 300 mi.

 Ans. Using t for the first time in hr. and T for the second time in hr.:
 a) $22t + 27T = 282$ and $27t + 32T = 337$. Times are 3 hr. and 8 hr.
 b) $15t + 36T = 420$ and $30t + 18T = 300$. Times are 4 hr. and 10 hr.

72. *a*) Joseph and Richard earned \$25.30 together when Joseph worked 4 hr. and Richard 5 hr. **(12.3)**
 Had each worked 1 hr. longer, they would have been paid \$31. Find their hourly rates of pay.
 b) Will and Robert earned \$46.90 together for a certain job. Will was paid at the rate of \$3.50 an hr. and Robert at the rate of \$4.20 an hr. Had Will worked as long as Robert and Robert as long as Will, they would have earned \$45.50. Find the time worked by each.

 Ans. *a*) Using r for Joseph's hourly rate and R for Richard's hourly rate, in ¢,
 $4r + 5R = 2530$ and $5r + 6R = 3100$. Rates were \$3.20 and \$2.50 per hr.
 b) Using t for Will's time of work and T for Richard's time of work, in hr.,
 $3.50t + 4.20T = 46.90$ and $4.20t + 3.50T = 45.50$. Times were 5 hr. and 7 hr.

Chapter 10

SPECIAL PRODUCTS and FACTORING

1. UNDERSTANDING FACTORS AND PRODUCTS

A Product and Its Factors

A product is the result obtained by multiplying two or more numbers.

The factors of the product are the numbers being multiplied.
Thus, 2, x and y are the factors of the product $2xy$.

To find the product of a monomial and a polynomial, multiply the monomial by every term of the polynomial.
Thus, $3(4x+2y) = 12x + 6y$.

Two factors of any number are 1 and the number itself.
Thus, 1 and 31 are factors of 31, while 1 and x are factors of x.

A prime number is a whole number which has no whole number factors except 1 and itself.
Thus, 2, 5, 17 and 31 are prime numbers.

Note how products may be divided into factors or into prime factors:

Product		Factors		Prime Factors
36	=	$4 \cdot 9$	=	$2 \cdot 2 \cdot 3 \cdot 3$
36	=	$6 \cdot 6$	=	$2 \cdot 3 \cdot 2 \cdot 3$
$10a^2$	=	$10 \cdot a^2$	=	$2 \cdot 5 \cdot a \cdot a$
$2ax + 6a$	=	$2(ax + 3a)$	=	$2a(x + 3)$

To factor a number or expression is to find its factors, not including 1 and itself.
Thus, $2ax+6a$ may be factored into $2(ax+3a)$ or further into $2a(x+3)$.

To factor a polynomial (2 or more terms) completely, continue factoring until the polynomial factors cannot be factored further.
Thus, $2ax+6a$ may be factored completely into $2a(x+3)$.

Note. Monomial factors need not be factored further.
Thus, $6ax^2 + 12x^2$ may be factored completely into $6x^2(a+2)$; 6 and x^2 need not be further factored.

1.1. Finding the Product of Monomial Factors

Find each product:

a) $5 \cdot 7 \cdot x$ *Ans. a*) $35x$
b) $-3xxx$ *Ans. b*) $-3x^3$
c) $8x^2 \cdot x^3 \cdot x^4$ *Ans. c*) $8x^9$
d) $4(3a)(10b)$ *Ans. d*) $120ab$
e) $(5x^2)(-11y)$ *Ans. e*) $-55x^2y$
f) $(2ab)(3ac)(4ad)$ *Ans. f*) $24a^3bcd$

1.2. Finding the Product of Two Factors, a Monomial and a Polynomial

Find each product:

a) $4(a+b)$	Ans. $4a+4b$	d) $-7x(x-2)$	Ans. $-7x^2+14x$	g) $\pi r(r+h)$	Ans. $\pi r^2+\pi rh$
b) $y(w+z)$	Ans. $wy+yz$	e) $3a^2(3a-5)$	Ans. $9a^3-15a^2$	h) $a(x-y+1)$	Ans. $ax-ay+a$
c) $3(4y-1)$	Ans. $12y-3$	f) $P(1+rn)$	Ans. $P+Prn$	i) $-3(a+b-2)$	Ans. $-3a-3b+6$

1.3. Finding Products Involving Fractions and Decimals

Find each product:

a) $6(5+\frac{1}{2})$	Ans. a) $30+3=\underline{33}$	f) $24(\frac{x}{2}+\frac{x}{3})$	Ans. f) $12x+8x=\underline{20x}$
b) $12(\frac{1}{3}+\frac{3}{4})$	Ans. b) $4+9=\underline{13}$	g) $20(\frac{x}{5}-\frac{x}{4})$	Ans. g) $4x-5x=\underline{-x}$
c) $14(10-\frac{1}{7})$	Ans. c) $140-2=\underline{138}$	h) $9(\frac{x}{3}-\frac{2x}{9})$	Ans. h) $3x-2x=\underline{x}$
d) $.02\frac{1}{2}(1000-40)$	Ans. d) $25-1=\underline{24}$	i) $.05(2000-x)$	Ans. i) $100-.05x$
e) $.03(2000+250)$	Ans. e) $60+7.50=\underline{67.50}$	j) $.03\frac{1}{2}(4000-2x)$	Ans. j) $140-.07x$

2. FACTORING A POLYNOMIAL HAVING A COMMON MONOMIAL FACTOR

A common monomial factor of a polynomial is a monomial that is a factor of each term of the polynomial.

Thus, 7 and a^2 are common monomial factors of $7a^2x+7a^2y-7a^2z$.

The highest common monomial factor of a polynomial (H. C. F.) is the product of all its common monomial factors.

Thus, $7ab$ is the highest common monomial factor of $7abc+7abd$.

To Factor a Polynomial Having a Common Monomial Factor

Procedure:	Factor: $7ax^2+14bx^2-21cx^2$ Solution:
1. Use the highest common monomial factor (H. C. F.) as one factor:	1. H. C. F. is $7x^2$.
2. Find the other factor by dividing each term of the polynomial by the H. C. F.:	2. Divide each term by H. C. F.: $\frac{7ax^2+14bx^2-21cx^2}{7x^2}=a+2b-3c$ Hence, by factoring: $7ax^2+14bx^2-21cx^2=7x^2(a+2b-3c)$

Note. A fraction may be used as a common factor if it is the numerical coefficient of each term of the original polynomial.

Thus, by factoring, $\frac{1}{2}x+\frac{1}{2}y=\frac{1}{2}(x+y)$.

2.1. Factoring Polynomials Having a Common Monomial Factor

Factor, removing the highest common factor:

a) $5a-5b$	Ans. $5(a-b)$	d) $9x^2-9x$	Ans. $9x(x-1)$	g) $\pi R^2-\pi r^2$	Ans. $\pi(R^2-r^2)$
b) $\frac{1}{2}h+\frac{1}{2}k$	Ans. $\frac{1}{2}(h+k)$	e) x^5+3x^2	Ans. $x^2(x^3+3)$	h) $5bx+10by-15b$	Ans. $5b(x+2y-3)$
c) $2ay-2by$	Ans. $2y(a-b)$	f) $S-Snd$	Ans. $S(1-nd)$	i) $4x^3+8x^2-24x$	Ans. $4x(x^2+2x-6)$

2.2. Factoring Numerical Polynomials

Evaluate each, using factoring:

a) $7(14) - 6\frac{1}{2}(14)$	Ans. a) $14(7-6\frac{1}{2}) = 14(\frac{1}{2}) = 7$
b) $6(8\frac{1}{2}) + 4(8\frac{1}{2})$	Ans. b) $8\frac{1}{2}(6+4) = 8\frac{1}{2}(10) = 85$
c) $17\frac{1}{4}(10) - 15\frac{1}{4}(10)$	Ans. c) $10(17\frac{1}{4} - 15\frac{1}{4}) = 10(2) = 20$
d) $.03(800) + .03(750) - .03(550)$	Ans. d) $.03(800+750-550) = .03(1000) = 30$
e) $.02\frac{1}{2}(8500) - .02\frac{1}{2}(7500) - .02\frac{1}{2}(1000)$	Ans. e) $.02\frac{1}{2}(8500-7500-1000) = .02\frac{1}{2}(0) = 0$
f) $8.4(3^3) + 5.3(3^3) - 3.7(3^3)$	Ans. f) $3^3(8.4+5.3-3.7) = 3^3(10) = 270$

3. SQUARING A MONOMIAL

The square of a number is the product of the number multiplied by itself. The number is used twice as a factor.

Thus, the square of 7 or $7^2 = (7)(7) = 49$.
Also, the square of (−7) or $(-7)^2 = (-7)(-7) = 49$.

Opposites have the same square.

Thus, both (+7) and (−7) have the same square, 49; that is, $(+7)^2 = (-7)^2$.

To square a fraction, square both its numerator and its denominator.

Thus, the square of $\frac{2}{3}$ or $(\frac{2}{3})^2 = (\frac{2}{3})(\frac{2}{3}) = \frac{2^2}{3^2} = \frac{4}{9}$.

In general, $(\frac{a}{b})^2 = \frac{a^2}{b^2}$.

To square a monomial, square its numerical coefficient, keep each base and double the exponent of each base.

Thus, $(5ab^3)^2 = (5ab^3)(5ab^3)$ or $25a^2b^6$.

3.1. Squaring Numbers

Find each square:

a) $(+8)^2$	Ans. 64	f) $(\frac{1}{5})^2$	Ans. f) $\frac{1}{25}$	k) $(.002)^2$	Ans. k) .000004
b) $(-8)^2$	Ans. b) 64	g) $(-\frac{1}{5})^2$	Ans. g) $\frac{1}{25}$	l) $(-1.2)^2$	Ans. l) 1.44
c) $(.3)^2$	Ans. c) .09	h) $(\frac{10}{11})^2$	Ans. h) $\frac{100}{121}$	m) $(1.25)^2$	Ans. m) 1.5625
d) $(-.3)^2$	Ans. d) .09	i) $(\frac{5}{3})^2$ or $(1\frac{2}{3})^2$	Ans. i) $\frac{25}{9}$ or $2\frac{7}{9}$	n) $(-.125)^2$	Ans. n) .015625
e) $(-.07)^2$	Ans. e) .0049	j) $(-1\frac{2}{3})^2$	Ans. j) $\frac{25}{9}$ or $2\frac{7}{9}$	o) $(-.101)^2$	Ans. o) .010201

3.2. Squaring Monomials

Find each square:

a) $(r^3)^2$	Ans. a) r^6	e) $(2ab)^2$	Ans. e) $4a^2b^2$	i) $(-8a^2b^3c)^2$	Ans. i) $64a^4b^6c^2$
b) $(x^4)^2$	Ans. b) x^8	f) $(-4x^4)^2$	Ans. f) $16x^8$	j) $(.1x^{100})^2$	Ans. j) $.01x^{200}$
c) $(\frac{a^5}{b^3})^2$	Ans. c) $\frac{a^{10}}{b^6}$	g) $(-\frac{ab^2}{c^7})^2$	Ans. g) $\frac{a^2b^4}{c^{14}}$	k) $(\frac{rst}{uv})^2$	Ans. k) $\frac{r^2s^2t^2}{u^2v^2}$
d) $(\frac{x^{11}}{y})^2$	Ans. d) $\frac{x^{22}}{y^2}$	h) $(-\frac{10}{3}w^{20})^2$	Ans. h) $\frac{100}{9}w^{40}$	l) $(-\frac{7}{10}x^7y^{10})^2$	Ans. l) $\frac{49}{100}x^{14}y^{20}$

3.3. Finding Areas of Squares: $A = s^2$

Find the area of a square whose side is

a) 3 ft.	*Ans. a*) 9 sq. ft.	*d*) $8x$	*Ans. d*) $64x^2$
b) $\frac{3}{4}$ yd.	*Ans. b*) $\frac{9}{16}$ sq. yd.	*e*) $1.5y^2$	*Ans. e*) $2.25y^4$
c) 1.3 mi.	*Ans. c*) 1.69 sq. mi.	*f*) $\frac{2}{3}m^3$	*Ans. f*) $\frac{4}{9}m^6$

4. FINDING THE SQUARE ROOT OF A MONOMIAL

The square root of a number is one of its two equal factors.

Thus, the square root of 49 is either +7 or −7, since $49 = (+7)(+7) = (-7)(-7)$.

A positive number has two square roots which are opposites of each other; that is, they have the same absolute value but differ in sign.

Thus, $\frac{25}{36}$ has two square roots, either $+\frac{5}{6}$ or $-\frac{5}{6}$.

The principal square root of a number is its positive square root.

Thus, the principal square root of 81 is +9 .

The symbol "$\sqrt{\ }$" is used to indicate the principal or positive square root of a number.

Thus, $\sqrt{81} = 9$, $\sqrt{\frac{49}{64}} = \frac{7}{8}$, $\sqrt{.09} = .3$.

To find the principal square root of a monomial, find the principal square root of its numerical coefficient, keep each base and use half the exponent of each base.

Thus, $\sqrt{16y^{16}} = 4y^8$.

The principal square root of a fraction is the principal square root of its numerator divided by the principal square root of its denominator.

Thus, $\sqrt{\frac{100}{121}} = \frac{10}{11}$, $\sqrt{\frac{x^8}{y^6}} = \frac{x^4}{y^3}$.

4.1. Finding Principal Square Roots

Find each principal square root:

a) $\sqrt{100}$	*Ans. a*) 10	*e*) $\sqrt{a^6}$	*Ans. e*) a^3	*i*) $\sqrt{900a^2b^2}$	*Ans. i*) $30ab$
b) $\sqrt{2500}$	*Ans. b*) 50	*f*) $\sqrt{a^8b^{12}}$	*Ans. f*) a^4b^6	*j*) $\sqrt{.09c^{20}}$	*Ans. j*) $.3c^{10}$
c) $\sqrt{.0001}$	*Ans. c*) .01	*g*) $\sqrt{100s^{100}}$	*Ans. g*) $10s^{50}$	*k*) $\sqrt{36r^{36}}$	*Ans. k*) $6r^{18}$
d) $\sqrt{\frac{81}{400}}$	*Ans. d*) $\frac{9}{20}$	*h*) $\sqrt{\frac{x^{18}}{y^2}}$	*Ans. h*) $\frac{x^9}{y}$	*l*) $\sqrt{\frac{169}{x^{10}}}$	*Ans. l*) $\frac{13}{x^5}$

4.2. Finding Sides of Squares: $s = \sqrt{A}$

Find the side of a square whose area is

a) 49 sq. in.	*Ans. a*) 7 in.	*c*) $\frac{25}{81}$ sq. yd.	*Ans. c*) $\frac{5}{9}$ yd.	*e*) $1.21b^4$	*Ans. e*) $1.1b^2$
b) .0004 sq. ft.	*Ans. b*) .02 ft.	*d*) $169a^2$	*Ans. d*) $13a$	*f*) $\frac{100}{169x^2y^2}$	*Ans. f*) $\frac{10}{13xy}$

5. FINDING THE PRODUCT OF THE SUM AND DIFFERENCE OF TWO NUMBERS

$$(x+y)(x-y) = x^2 - y^2$$

If the sum of two numbers is multiplied by their difference, the product is the square of the first minus the square of the second.

Thus, $(x+5)(x-5) = x^2 - 25$.

Note below how the middle term drops out in each case.

(1) Multiply $(x+y)$ by $(x-y)$:	(2) Multiply (a^3+8) by (a^3-8):	(3) Multiply 103×97:
$x+y$	a^3+8	$103 = 100 + 3$
$x-y$	a^3-8	$97 = 100 - 3$
$x^2 + xy$	$a^6 + 8a^3$	$10{,}000 + 300$
$-xy - y^2$	$-8a^3 - 64$	$-300 - 9$
Ans. $x^2 \quad -y^2$	Ans. $a^6 \quad -64$	Ans. $10{,}000 \quad -9 = 9991$

Note, in (3) above, a new method for multiplying 103×97. This process of multiplying the sum of two numbers by their difference is a valuable shortcut in arithmetic in such cases as the following:

(a) $33 \times 27 = (30+3)(30-3) = 900 - 9$ or 891
(b) $2\frac{1}{2} \times 1\frac{1}{2} = (2+\frac{1}{2})(2-\frac{1}{2}) = 4 - \frac{1}{4}$ or $3\frac{3}{4}$
(c) $9.8 \times 10.2 = (10-.2)(10+.2) = 100 - .04$ or 99.96

5.1. Multiplying the Sum of Two Numbers by their Difference

Find each product:

(First + Second)(First − Second)	1. (First)^2	2. (Second)^2	3. Product (Ans.)
a) $(m + 7)(m - 7)$	m^2	49	$m^2 - 49$
b) $(8 + 3x)(8 - 3x)$	64	$9x^2$	$64 - 9x^2$
c) $(11a + 5b)(11a - 5b)$	$121a^2$	$25b^2$	$121a^2 - 25b^2$
d) $(x^2 + y^3)(x^2 - y^3)$	x^4	y^6	$x^4 - y^6$
e) $(1 + y^4z^5)(1 - y^4z^5)$	1	y^8z^{10}	$1 - y^8z^{10}$
f) $(ab + \frac{2}{3})(ab - \frac{2}{3})$	a^2b^2	$\frac{4}{9}$	$a^2b^2 - \frac{4}{9}$
g) $(.2 + \frac{3}{5}p)(.2 - \frac{3}{5}p)$	$.04$	$\frac{9}{25}p^2$	$.04 - \frac{9}{25}p^2$

5.2. Multiplying Two Numbers by the Sum-Product Method

Multiply, using the sum and difference of two numbers:

a) 18×22	Ans. a) $18 \times 22 = (20-2)(20+2) = 400 - 4 = 396$
b) 25×35	Ans. b) $25 \times 35 = (30-5)(30+5) = 900 - 25 = 875$
c) $.96 \times 1.04$	Ans. c) $.96 \times 1.04 = (1-.04)(1+.04) = 1 - .0016 = .9984$
d) 7.1×6.9	Ans. d) $7.1 \times 6.9 = (7+.1)(7-.1) = 49 - .01 = 48.99$
e) $2\frac{1}{4} \times 1\frac{3}{4}$	Ans. e) $2\frac{1}{4} \times 1\frac{3}{4} = (2+\frac{1}{4})(2-\frac{1}{4}) = 4 - \frac{1}{16} = 3\frac{15}{16}$

5.3. Multiplying a Monomial by Sum and Difference Factors

Find each product: (*First, multiply the sum and difference factors.*)

a) $2(x+5)(x-5)$	b) $x^2(1-3y)(1+3y)$	c) $t^2(t+u)(t-u)$	d) $24(m^2+\frac{1}{2})(m^2-\frac{1}{2})$
Solutions:			
$2(x^2-25)$	$x^2(1-9y^2)$	$t^2(t^2-u^2)$	$24(m^4-\frac{1}{4})$
Ans. $2x^2 - 50$	Ans. $x^2 - 9x^2y^2$	Ans. $t^4 - t^2u^2$	Ans. $24m^4 - 6$

6. FACTORING THE DIFFERENCE OF TWO SQUARES

$$x^2 - y^2 = (x-y)(x+y)$$

The expression x^2-y^2 is the difference of two squares, x^2 and y^2.

The factor, $x+y$, is the sum of the principal square roots of x^2 and y^2, while the other factor, $x-y$, is their difference.

To Factor the Difference of Two Squares

Procedure:	Factor: $c^2 - 49$ **Solution:**
1. Obtain principal square root of each square:	**1. Principal square roots:** $\sqrt{c^2} = c$, $\sqrt{49} = 7$
2. One factor is the sum of the principal square roots. The other factor is their difference:	**2. Factors:** **Ans.** $(c+7)(c-7)$

6.1. Procedure for Factoring the Difference of Two Squares

Factor:

	First Square − Second Square	1. $\sqrt{\text{First Square}}$	2. $\sqrt{\text{Second Square}}$	3. Factors (Ans.)
a)	$36 - b^2$	6	b	$(6+b)(6-b)$
b)	$1 - 25y^2$	1	$5y$	$(1+5y)(1-5y)$
c)	$a^2b^2 - 100$	ab	10	$(ab+10)(ab-10)$
d)	$x^6y^8 - z^{10}$	x^3y^4	z^5	$(x^3y^4+z^5)(x^3y^4-z^5)$
e)	$16 - x^{16}$	4	x^8	$(4+x^8)(4-x^8)$
f)	$u^4 - \frac{9}{25}$	u^2	$\frac{3}{5}$	$(u^2+\frac{3}{5})(u^2-\frac{3}{5})$
g)	$(a+b)^2 - c^2$	$a+b$	c	$(a+b+c)(a+b-c)$

6.2. Complete Factoring Involving the Difference of Two Squares

Factor completely: *(Hint. Remove the highest common monomial factor first.)*

a) $10 - 40x^2$ *b*) $75y^3 - 27y$ *c*) $\pi R^2 - \pi r^2$ *d*) $5abc^4 - 80ab$

Solutions:

a) 10 is **H.C.F.**	*b*) $3y$ is **H.C.F.**	*c*) π is **H.C.F.**	*d*) $5ab$ is **H.C.F.**
$10(1-4x^2)$	$3y(25y^2-9)$	$\pi(R^2-r^2)$	$5ab(c^4-16)$
Ans. $10(1+2x)(1-2x)$	*Ans.* $3y(5y+3)(5y-3)$	*Ans.* $\pi(R+r)(R-r)$	*Ans.* $5ab(c^2+4)(c+2)(c-2)$

7. FINDING THE PRODUCT OF TWO BINOMIALS WITH LIKE TERMS

Two Methods of Multiplying $3x+5$ by $2x+4$

Method 1. (Usual Method)

$$\begin{array}{rrr} 3x & + & 5 \\ 2x & + & 4 \\ \hline 6x^2 + 10x & & \\ + 12x & + & 20 \\ \hline 6x^2 + 22x & + & 20 \end{array}$$

Ans. $6x^2 + 22x + 20$

Using **Method 1**, each separate operation is shown. The middle term of the answer, $22x$, is obtained by adding the cross-products, $12x$ and $10x$. The arrows indicate the cross-products.

Method 2. (Method by Inspection)

$12x$ (outer product)

$(3x+5)\ (2x+4)$

$10x$ (inner product)

Ans. $6x^2 + 22x + 20$

Using **Method 2**, the answer is written **by inspection**. The middle term of the answer, $22x$, is obtained **mentally** by adding the product of the outer terms to the product of the inner terms. The arrows indicate the outer and inner product.

To Multiply Two Binomials By Inspection

Procedure:

1. To obtain the first term of the product, **multiply the first terms:**
   Key: *First*

2. To obtain the middle term of the product, **add the product of the outer terms to the product of the inner terms:**
   Key: *Outer + Inner*

3. To obtain the last term of the product, **multiply the last terms:**
   Key: *Last*

4. **Combine** the results to obtain answer:

Multiply: $(3x+5)(2x+4)$

Solution:

1. Multiply **first** terms:
   $(3x)(2x) = 6x^2$

2. Add **outer** and **inner** products:

$$\underset{10x}{\overset{12x}{(3x+5)\ (2x+4)}}$$

$$12x + 10x = 22x$$

3. Multiply **last** terms:
   $(+5)(+4) = +20$

4. Combine: $6x^2 + 22x + 20$ **Ans.**

Since a **mixed number** is the sum of an integer and a fraction, it may be expressed as a binomial. Thus, $3\frac{1}{2}$ and $4\frac{1}{3}$ are mixed numbers.

Two mixed numbers may be multiplied in the same way as two binomials. Thus,

$$3\tfrac{1}{2} \times 4\tfrac{1}{3} = \underset{2}{\overset{1}{(3+\tfrac{1}{2})\ (4+\tfrac{1}{3})}} = 12 + 3 + \tfrac{1}{6} = 15\tfrac{1}{6} \quad \textit{Ans.}$$

7.1. Products of Binomials by Steps

Multiply, showing each separate product:

	Binomial × Binomial	*(First)*	*(Outer + Inner)*	*(Last)*	Product (Ans.)
a)	$(x+7)\ (x+3)$	x^2	$(+3x) + (+7x)$	$+21$	$x^2 + 10x + 21$
b)	$(x-7)\ (x-3)$	x^2	$(-3x) + (-7x)$	$+21$	$x^2 - 10x + 21$
c)	$(x+7)\ (x-3)$	x^2	$(-3x) + (+7x)$	-21	$x^2 + 4x - 21$
d)	$(3w-5)\ (4w+7)$	$12w^2$	$(+21w) + (-20w)$	-35	$12w^2 + w - 35$
e)	$(a+3b)\ (a-8b)$	a^2	$(-8ab) + (+3ab)$	$-24b^2$	$a^2 - 5ab - 24b^2$

7.2. Products of Two Binomials Mentally

Multiply mentally:

a) $(x+7)(x+7)$ *Ans.* $x^2+14x+49$

b) $(x-3)(x-3)$ *Ans.* x^2-6x+9

c) $(3y+1)(2y+1)$ *Ans.* $6y^2+5y+1$

d) $(2c-b)(5c-b)$ *Ans.* $10c^2-7bc+b^2$

e) $(3-7d)(1-d)$ *Ans.* $3-10d+7d^2$

f) $(2r+3s)(3r+2s)$ *Ans.* $6r^2+13rs+6s^2$

g) $(ab+4)(ab-3)$ *Ans.* $a^2b^2+ab-12$

h) $(p-qr)(p-3qr)$ *Ans.* $p^2-4pqr+3q^2r^2$

i) $(pqr-7)(pqr-10)$ *Ans.* $p^2q^2r^2-17pqr+70$

j) $(t^2+5)(t^2+9)$ *Ans.* t^4+14t^2+45

k) $(10-v^2)(6-v^2)$ *Ans.* $60-16v^2+v^4$

l) $(c^2d-5g)(c^2d+3g)$ *Ans.* $c^4d^2-2c^2dg-15g^2$

7.3. Representing Areas of Rectangles: $A = LW$

Represent the area of a rectangle whose dimensions are:

a) $l+8$ and $l-2$ *Ans.* $A = (l+8)(l-2) = l^2+6l-16$

b) $2l-1$ and $l+5$ *Ans.* $A = (2l-1)(l+5) = 2l^2+9l-5$

c) $w+3$ and $w+7$ *Ans.* $A = (w+3)(w+7) = w^2+10w+21$

d) $2w-1$ and $3w+1$ *Ans.* $A = (2w-1)(3w+1) = 6w^2-w-1$

e) $3s-5$ and $s+8$ *Ans.* $A = (3s-5)(s+8) = 3s^2+19s-40$

8. FACTORING TRINOMIALS IN FORM OF $x^2 + bx + c$

A trinomial in the form of $x^2 + bx + c$ may or may not be factorable into binomial factors. If factoring is possible, use the following procedure.

To Factor a Trinomial in Form of $x^2 + bx + c$

Factor: *a*) $x^2 + 6x + 5$ *b*) $x^4 - 6x^2 + 8$

Procedure:	**Solutions:** *a*)	*b*)
1. **Obtain the factors x and x of x^2.** Use each as the first term of each binomial:	1. **Factor x^2:** (x, x)	1. **Factor x^4:** (x^2, x^2)
2. **Select from the factors of the last term, c,** those factors whose sum = b, the coefficient of x. Use each as the second term of each binomial:	2. **Factor +5:** Select (+5, +1) since sum = +6. Discard (−5, −1)	2. **Factor +8:** Select (−4, −2) since sum = −6. Discard (+4, +2), (−8, −1) and (+8, +1)
3. **Form binomial factors** from factors obtained in steps **1** and **2**:	3. $(x+5)(x+1)$ *Ans.*	3. $(x^2-4)(x^2-2)$ *Ans.*

8.1. Factoring Trinomials in Form of $x^2 + bx + c$

Factor each trinomial:

Trinomial Form: $x^2 + bx + c$	**1. Factors of x^2, each** = $\sqrt{x^2}$	**2. Factors of c whose sum** = b	**3. Binomial Factors (Ans.) (Combine 1 and 2)**
a) $x^2 + 4x + 3$	x, x	+3, +1	$(x+3)(x+1)$
b) $x^2 - 4x + 3$	x, x	−3, −1	$(x-3)(x-1)$
c) $y^2 + 4y - 12$	y, y	+6, −2	$(y+6)(y-2)$
d) $w^2 - w - 12$	w, w	−4, +3	$(w-4)(w+3)$
e) $r^2 + 6rs + 5s^2$	r, r	$+5s, +s$	$(r+5s)(r+s)$
f) $a^2b^2 - 12ab + 20$	ab, ab	−10, −2	$(ab-10)(ab-2)$
g) $x^4 - 5x^2 - 14$	x^2, x^2	−7, +2	$(x^2-7)(x^2+2)$

9. FACTORING A TRINOMIAL IN FORM OF $ax^2 + bx + c$

A trinomial in the form of $ax^2 + bx + c$ may or may not be factorable into binomial factors. If this is possible, use the following procedure.

To Factor a Trinomial in Form of $ax^2 + bx + c$

Factor: *a*) $2x^2 - 11x + 5$ *b*) $3a^2 + 10ab + 7b^2$

Procedure:	**Solutions:** *a*)	*b*)
1. *Factor ax^2, **the first term.** Use each as the first term of each binomial factor.	1. **Factor $2x^2$:** $(2x, x)$	1. **Factor $3a^2$:** $(3a, a)$
2. **Select from the factors of c, the last term,** those factors to be used as the second term of each binomial such that the middle term, bx, results.	2. **Factor** + 5: Select (−1, −5) to obtain middle term, $-11x$. Discard (+1, +5).	2. **Factor** + $7b^2$: Select $(+7b, +b)$ to obtain middle term, $10ab$. Discard $(-7b, -b)$.
3. **Form binomial factors** from factors obtained in steps **1** and **2** and test for middle term, bx.	3. $(2x-1)\ (x-5)$, with $-x$ and $-10x$ marked: Middle term, $-11x$, results. **Ans.** $(2x-1)(x-5)$	3. $(3a+7b)\ (a+b)$, with $7ab$ and $3ab$ marked: Middle term, $10ab$, results. **Ans.** $(3a+7b)(a+b)$

* When the first term ax^2 is positive, use positive factors. Thus, in $2x^2 - 11x + 5$, do not use $-2x$ and $-x$ as factors of $2x^2$.

9.1. Factoring Trinomials in Form of $ax^2 + bx + c$.

Factor each trinomial:

Trinomial Form: $ax^2 + bx + c$	1. **Factors of** ax^2	2. **Factors of** c **to Obtain Middle Term,** bx	3. **Binomial Factors (Ans.) (Test Middle Term)**
a) $5x^2 + 11x + 2$	$5x,\ x$	$+1,\ +2$	x $(5x+1)\ (x+2)$ $10x$
b) $4w^2 + 7w + 3$	$4w,\ w$ Discard $(2w, 2w)$	$+3,\ +1$	$+3w$ $(4w+3)\ (w+1)$ $+4w$
c) $4y^2 - 8w + 3$	$2y,\ 2y$ Discard $(4y, y)$	$-3,\ -1$	$-6y$ $(2y-3)\ (2y-1)$ $-2y$
d) $4w^2 + 13wx + 3x^2$	$4w,\ w$ Discard $(2w, 2w)$	$+x,\ +3x$	wx $(4w+x)\ (w+3x)$ $+12wx$
e) $8 + 15h - 2h^2$	$8,\ 1$ Discard $(4, 2)$	$-h,\ +2h$	$-h$ $(8-h)\ (1+2h)$ $+16h$

10. SQUARING A BINOMIAL

$$(x+y)^2 = x^2 + 2xy + y^2$$

The square of a binomial is a perfect square trinomial.

Thus, the square of $x+y$ or $(x+y)^2$ is the perfect square trinomial, $x^2 + 2xy + y^2$.

To Square a Binomial

Procedure:	Square: $3x + 5$ **Solution:**
1. **Square the first term** to obtain the first term of the trinomial:	1. **Square** $3x$: $(3x)^2 = 9x^2$
2. **Double the product of both terms** to obtain the middle term of the trinomial:	2. **Double** $(3x)(+5)$: $2(3x)(+5) = 30x$
3. **Square the last term** to obtain the last term of the trinomial:	3. **Square** $+5$: $5^2 = 25$
4. **Combine results** into answer:	4. **Combine:** $9x^2 + 30x + 25$ **Ans.**

Squaring Numbers Expressed as a Binomial

Numbers may be squared by expressing them first as a binomial.

Thus to square 35, express 35 as $(30+5)$.

$$35^2 = (30+5)^2 = 30^2 + 2(5)(30) + 5^2$$
$$= 900 + 300 + 25 = 1225$$

10.1. Squaring a Binomial

Find each square:

(Binomial)^2	1. Square First Term	2. Double Product	3. Square Last Term	4. Combine Into Square (Ans.)
a) $(b-6)^2$	b^2	$-12b$	$+36$	$b^2 - 12b + 36$
b) $(b+6)^2$	b^2	$+12b$	$+36$	$b^2 + 12b + 36$
c) $(1-3b)^2$	1	$-6b$	$+9b^2$	$1 - 6b + 9b^2$
d) $(c^2+3d)^2$	c^4	$+6c^2d$	$+9d^2$	$c^4 + 6c^2d + 9d^2$
e) $(5x+7y)^2$	$25x^2$	$+70xy$	$+49y^2$	$25x^2 + 70xy + 49y^2$
f) $(ab-11)^2$	a^2b^2	$-22ab$	$+121$	$a^2b^2 - 22ab + 121$
g) $(x^3+y^3)^2$	x^6	$+2x^3y^3$	$+y^6$	$x^6 + 2x^3y^3 + y^6$
h) $(t+\frac{1}{2})^2$	t^2	$+t$	$+\frac{1}{4}$	$t^2 + t + \frac{1}{4}$
i) $(t^2-1.3)^2$	t^4	$-2.6t$	$+1.69$	$t^4 - 2.6t^2 + 1.69$

10.2. Squaring a Number

Find each square by expressing each number as a binomial:

(Number)^2	1. As (Binomial)^2	2. Square First Term		3. Double Product		4. Square Last Term		5. Combine Into Square (Ans.)
a) 25^2	$(20+5)^2$	400	+	200	+	25	=	625
b) $(2\frac{1}{2})^2$	$(2+\frac{1}{2})^2$	4	+	2	+	$\frac{1}{4}$	=	$6\frac{1}{4}$
c) $(12\frac{1}{4})^2$	$(12+\frac{1}{4})^2$	144	+	6	+	$\frac{1}{16}$	=	$150\frac{1}{16}$
d) 10.4^2	$(10+.4)^2$	100	+	8	+	.16	=	108.16
e) 9.5^2	$(10-.5)^2$	100	−	10	+	.25	=	90.25
f) $(29\frac{2}{3})^2$	$(30-\frac{1}{3})^2$	900	−	20	+	$\frac{1}{9}$	=	$880\frac{1}{9}$

10.3. Representing the Area of a Square: $A = s^2$

Represent the area of a square whose side is

a) $(l+10)$ in. b) $(w-8)$ ft. c) $(2s+5)$ yd. d) $9-5s$

Solutions:

a)	b)	c)	d)
$A = (l+10)^2$	$A = (w-8)^2$	$A = (2s+5)^2$	$A = (9-5s)^2$
$A = l^2 + 20l + 100$	$A = w^2 - 16w + 64$	$A = 4s^2 + 20s + 25$	$A = 81 - 90s + 25s^2$
Ans. $(l^2+20l+100)$ sq. in.	*Ans.* $(w^2-16w+64)$ sq. ft.	*Ans.* $(4s^2+20s+25)$ sq. yd.	*Ans.* $81-90s+25s^2$

11. FACTORING A PERFECT SQUARE TRINOMIAL

$$x^2 + 2xy + y^2 = (x+y)(x+y) = (x+y)^2$$
$$x^2 - 2xy + y^2 = (x-y)(x-y) = (x-y)^2$$

The factors of a perfect square trinomial are two equal binomials.

Thus, the factors of the perfect square trinomial $x^2 + 2xy + y^2$ are $(x+y)$ and $(x+y)$.

A perfect square trinomial has

(1) two terms which are positive perfect squares,

(2) a remaining term which is double the product of the square roots of the other two terms. This term may be positive or negative.

Thus, $x^2+14x+49$ and $x^2-14x+49$ are perfect square trinomials.

The last term of each binomial factor has the same sign as the middle term of the perfect square trinomial.

Thus, $x^2-14x+49 = (x-7)^2$ and $x^2+14x+49 = (x+7)^2$.

To Factor a Perfect Square Trinomial

Procedure:	Factor: $4x^2 - 20x + 25$ Solution:
1. **Find the principal square root of the first term.** This becomes the first term of each binomial:	1. Find $\sqrt{4x^2}$: $\sqrt{4x^2} = 2x$
2. **Find the principal square root of the last term and prefix the sign of the middle term.** This becomes the last term of each binomial:	2. Find $\sqrt{25}$: $\sqrt{25} = 5$. Prefix $-$ before 5 since middle term is negative.
3. **Form binomial** from results in steps 1 and 2. The answer is the binomial squared:	3. Form $2x - 5$. **Ans.** $4x^2 - 20x + 25 = (2x-5)^2$.

11.1. Factoring a Perfect Square Trinomial

Factor each perfect square trinomial:

Perfect Square Trinomial	1. $\sqrt{\text{First Term}}$	2. Sign of Middle Term	3. $\sqrt{\text{Last Term}}$	(Binomial)^2 Ans.
a) $25y^2 - 10y + 1$	$5y$	$-$	1	$(5y-1)^2$
b) $25y^2 + 10y + 1$	$5y$	$+$	1	$(5y+1)^2$
c) $9a^2 + 42a + 49$	$3a$	$+$	7	$(3a+7)^2$
d) $9a^2 - 42a + 49$	$3a$	$-$	7	$(3a-7)^2$
e) $49 + 14ab + a^2b^2$	7	$+$	ab	$(7+ab)^2$
f) $16x^6 - 24x^3 + 9$	$4x^3$	$-$	3	$(4x^3-3)^2$
g) $16t^2 + 4t + \frac{1}{4}$	$4t$	$+$	$\frac{1}{2}$	$(4t+\frac{1}{2})^2$
h) $a^4b^8 - 2a^2b^4c^5 + c^{10}$	a^2b^4	$-$	c^5	$(a^2b^4-c^5)^2$

11.2. Representing the Side of a Square: $s = \sqrt{A}$

Represent the side of a square whose area is

a) $(l^2 - 14w + 49)$ sq. in. b) $(w^2 + 20w + 100)$ sq. ft. c) $4s^2 + 4s + 1$ d) $9s^2 - 30s + 25$

Ans. a) $(l-7)$ in. b) $(w+10)$ ft. c) $2s+1$ d) $3s-5$

12. COMPLETELY FACTORING POLYNOMIALS

To factor an expression completely, continue factoring until the polynomial factors cannot be factored further. Thus, to factor $5x^2 - 5$ completely, first factor it into $5(x^2-1)$. Then factor further into $5(x-1)(x+1)$.

If an expression has a common monomial factor:

1. **Remove its highest common factor (H.C.F.).**
2. **Continue factoring its polynomial factors** until no further factors remain.

Procedure to Completely Factor Expressions Having Common Monomial Factor

Factor completely:	a) $8a^2 - 50$	b) $3y^3 - 60y^2 + 300y$	c) $10y^2 - 15xy^2 + 5x^2y^2$
Procedure:	**Solutions:**		
1. **Remove highest common factor (H.C.F.):**	1. **H.C.F.** = 2 $2(4a^2-25)$	1. **H.C.F.** = $3y$ $3y(y^2-20y+100)$	1. **H.C.F.** = $5y^2$ $5y^2(2-3x+x^2)$
2. **Continue factoring polynomial factors:**	2. **Factor** $4a^2-25$: $2(2a+5)(2a-5)$	2. **Factor** $y^2-20y+100$: $3y(y-10)(y-10)$	2. **Factor** $2-3x+x^2$: $5y^2(2-x)(1-x)$
	Ans. $2(2a+5)(2a-5)$	**Ans.** $3y(y-10)^2$	**Ans.** $5y^2(2-x)(1-x)$

Procedure to Completely Factor Expressions Having No Common Monomial Factor

	Factor completely: *a*) $x^4 - 1$	*b*) $16a^4 - 81$
Procedure:	**Solutions:**	
1. Factor into polynomials:	**1.** $(x^2+1)(x^2-1)$	**1.** $(4a^2+9)(4a^2-9)$
2. Continue factoring:	**2.** $(x^2+1)(x+1)(x-1)$ **Ans.**	**2.** $(4a^2+9)(2a+3)(2a-3)$ **Ans.**

12.1. Factoring Completely Expressions Having Common Monomial Factor

Factor completely:

Polynomial	1. Remove H.C.F.	2. Continue Factoring
a) $3b^2 - 27$	$3(b^2-9)$	$3(b+3)(b-3)$ *Ans.*
b) $a^5 - 16a^3$	$a^3(a^2-16)$	$a^3(a+4)(a-4)$ *Ans.*
c) $a^5 - 16a$	$a(a^4-16)$	$a(a^2+4)(a^2-4)$ $a(a^2+4)(a+2)(a-2)$ *Ans.*
d) $5x^2+10x+5$	$5(x^2+2x+1)$	$5(x+1)^2$ *Ans.*
e) $x^5 - 6x^4y + 9x^3y^2$	$x^3(x^2-6xy+9y^2)$	$x^3(x-3y)^2$ *Ans.*
f) $y^5 - 4y^3 + 3y$	$y(y^4-4y^2+3)$	$y(y^2-3)(y^2-1) = y(y^2-3)(y+1)(y-1)$ *Ans.*

12.2. Factoring Completely Expressions Having No Common Monomial Factor

Factor completely:

Polynomial	1. Factor Into Polynomials	2 2. Continue Factoring
a) $1 - a^4$	$(1+a^2)(1-a^2)$	$(1+a^2)(1+a)(1-a)$ *Ans.*
b) $x^4 - 16y^4$	$(x^2+4y^2)(x^2-4y^2)$	$(x^2+4y^2)(x+2y)(x-2y)$ *Ans.*
c) $p^8 - q^{12}$	$(p^4+q^6)(p^4-q^6)$	$(p^4+q^6)(p^2+q^3)(p^2-q^3)$ *Ans.*

SUPPLEMENTARY PROBLEMS

1. Find each product: **(1.1)**

a) $3 \cdot x \cdot 4 \cdot x$	*c*) $(-2)5xy^3x^4$	*e*) $(-5ab^2)(2a^2b)$	*g*) $-3 \cdot \frac{1}{x} \cdot \frac{1}{y}$
b) $(-3)4xyy$	*d*) $(-3x)(-7x^2)y$	*f*) $(3a^4b)(10ac^4)$	*h*) $5 \cdot \frac{1}{a^2} \cdot \frac{1}{b^2} \cdot c$

Ans. *a*) $12x^2$	*c*) $-10x^5y^3$	*e*) $-10a^3b^3$	*g*) $-\frac{3}{xy}$
b) $-12xy^2$	*d*) $21x^3y$	*f*) $30a^5bc^4$	*h*) $\frac{5c}{a^2b^2}$

2. Find each product: **(1.2)**

a) $3(x-3)$	*d*) $a(b+c)$	*g*) $x^2(x-3)$	*j*) $a(b-c+d)$
b) $-5(a+4)$	*e*) $-3c(d-g)$	*h*) $-3x(x^2-5x)$	*k*) $-3x(x^2-2x+5)$
c) $-8(r-7)$	*f*) $2gh(3k-5)$	*i*) $7a^3(5a^2-8a)$	*l*) $2b^2c(b-3c-10c^2)$

Ans. *a*) $3x-9$	*d*) $ab+ac$	*g*) x^3-3x^2	*j*) $ab-ac+ad$
b) $-5a-20$	*e*) $-3cd+3cg$	*h*) $-3x^3+15x^2$	*k*) $-3x^3+6x^2-15x$
c) $-8r+56$	*f*) $6ghk-10gh$	*i*) $35a^5-56a^4$	*l*) $2b^3c-6b^2c^2-20b^2c^3$

3. Find each product: (1.3)

a) $4(8-\frac{1}{2})$ *d*) $50(\frac{2x}{5}-\frac{3x}{10})$ *g*) $.06(2000+500+25)$

b) $20(5-.7)$ *e*) $12(\frac{x}{4}+\frac{x}{3}-\frac{x}{2})$ *h*) $.03(7000-x)$

c) $36(1\frac{1}{4}+\frac{2}{9})$ *f*) $70(2-\frac{5r}{14})$ *i*) $.05\frac{1}{2}(3000+2x)$

Ans. *a*) $32-2 = 30$ *d*) $20x-15x = 5x$ *g*) $120+30+1.50 = 151.50$

b) $100-14 = 86$ *e*) $3x+4x-6x = x$ *h*) $210-.03x$

c) $45+8 = 53$ *f*) $140-25r$ *i*) $165+.11x$

4. Factor, removing highest common factor: (2.1)

a) $3a-21b$ *e*) $p+prt$ *i*) $\pi rh+\pi r^2$ *m*) $8x^2-16x+32$

b) $-5c-15d$ *f*) $7hm-7h$ *j*) $\pi R^2+\pi r^2$ *n*) $10x^3+20x^2-55x$

c) $xy+2xz$ *g*) $6V^2+3V$ *k*) $\frac{1}{2}bh+\frac{1}{2}b'h$ *o*) $a^2bc+ab^2c-abc^2$

d) $5RS-10RT$ *h*) y^5+y^4 *l*) $\frac{1}{4}mnr-\frac{1}{4}mn$ *p*) $x^2y^3-x^2y^2+x^2y$

Ans. *a*) $3(a-7b)$ *e*) $p(1+rt)$ *i*) $\pi r(h+r)$ *m*) $8(x^2-2x+4)$

b) $-5(c+3d)$ *f*) $7h(m-1)$ *j*) $\pi(R^2+r^2)$ *n*) $5x(2x^2+4x-11)$

c) $x(y+2z)$ *g*) $3V(2V+1)$ *k*) $\frac{1}{2}h(b+b')$ *o*) $abc(a+b-c)$

d) $5R(S-2T)$ *h*) $y^4(y+1)$ *l*) $\frac{1}{4}mn(r-1)$ *p*) $x^2y(y^2-y+1)$

5. Evaluate each, using factoring: (2.2)

a) $6(11)+4(11)$ *d*) $2.7(.3)-.7(.3)$ *g*) $10(8^2)-7(8^2)$

b) $8(11\frac{1}{2})+12(11\frac{1}{2})$ *e*) $18(7^2)+2(7^2)$ *h*) $\frac{1}{2}(13)(6)+\frac{1}{2}(7)(6)$

c) $21(2\frac{1}{3})-12(2\frac{1}{3})$ *f*) $\frac{3}{4}(11)-\frac{3}{4}(3)$ *i*) $5^3(28)-5^3(25)$

Ans. *a*) $11(6+4) = 11(10) = 110$ *d*) $.3(2.7-.7) = .3(2) = .6$ *g*) $8^2(10-7) = 8^2(3) = 192$

b) $11\frac{1}{2}(8+12) = 11\frac{1}{2}(20) = 230$ *e*) $7^2(18+2) = 7^2(20) = 980$ *h*) $\frac{1}{2}(6)(13+7) = \frac{1}{2}(6)(20) = 60$

c) $2\frac{1}{3}(21-12) = 2\frac{1}{3}(9) = 21$ *f*) $\frac{3}{4}(11-3) = \frac{3}{4}(8) = 6$ *i*) $5^3(28-25) = 5^3(3) = 375$

6. Find each square: (3.1)

a) 6^2 *d*) 600^2 *g*) $(-\frac{3}{7})^2$ *j*) 35^2

b) $(-6)^2$ *e*) $(\frac{1}{6})^2$ *h*) $(-\frac{9}{2})^2$ *k*) 3.5^2

c) $(.6)^2$ *f*) $(\frac{2}{5})^2$ *i*) $(1\frac{1}{3})^2$ *l*) $(-.35)^2$

Ans. *a*) 36 *d*) 360,000 *g*) $\frac{9}{49}$ *j*) 1225

b) 36 *e*) $\frac{1}{36}$ *h*) $\frac{81}{4}$ *k*) 12.25

c) .36 *f*) $\frac{4}{25}$ *i*) $\frac{16}{9}$ or $1\frac{7}{9}$ *l*) .1225

7. Find each square: (3.2)

a) $(b^2)^2$ *d*) $(5a^5)^2$ *g*) $(rs^2t^3)^2$

b) $(w^5)^2$ *e*) $(-10b^{10})^2$ *h*) $(-.1r^4s^9)^2$

c) $(\frac{1}{V^7})^2$ *f*) $(\frac{.3}{C^3})^2$ *i*) $(-\frac{5r^5}{3t^3})^2$

Ans. *a*) b^4 *d*) $25a^{10}$ *g*) $r^2s^4t^6$

b) w^{10} *e*) $100b^{20}$ *h*) $.01r^8s^{18}$

c) $\frac{1}{V^{14}}$ *f*) $\frac{.09}{C^6}$ *i*) $\frac{25r^{10}}{9t^6}$

8. Find the area of a square whose side is (3.3)

a) 5 yd. *c*) 2.5 mi. *e*) $.7y^2$

b) $\frac{7}{8}$ rd. *d*) $\frac{x}{7}$ in. *f*) $\frac{5m^2}{3p^3}$

Ans. *a*) 25 sq. yd. *c*) 6.25 sq. mi. *e*) $.49y^4$

b) $\frac{49}{64}$ sq. rd. *d*) $\frac{x^2}{49}$ sq. in. *f*) $\frac{25m^4}{9p^6}$

9. Find each principal square root: **(4.1)**

a) $\sqrt{144}$ d) $\sqrt{c^6}$ g) $\sqrt{36x^{36}}$ j) $\sqrt{.0001h^{50}}$

b) $\sqrt{1.44}$ e) $\sqrt{p^8q^{10}}$ h) $\sqrt{16a^4b^{16}}$ k) $\sqrt{.0144m^{14}}$

c) $\sqrt{\frac{900}{49}}$ f) $\sqrt{\frac{r^{14}}{s^{20}}}$ i) $\sqrt{\frac{.01}{d^{100}}}$ l) $\sqrt{\frac{25h^{10}}{64k^{16}}}$

Ans. a) 12 d) c^3 g) $6x^{18}$ j) $.01h^{25}$

b) 1.2 e) p^4q^5 h) $4a^2b^8$ k) $.12m^7$

c) $\frac{30}{7}$ f) $\frac{r^7}{s^{10}}$ i) $\frac{.1}{d^{50}}$ l) $\frac{5h^5}{8k^8}$

10. Find the side of a square whose area is **(4.2)**

a) 100 sq. ft. *Ans.* 10 ft. d) $9x^2$ sq. in. *Ans.* $3x$ in. g) $100x^{20}$ *Ans.* $10x^{10}$

b) .25 sq. mi. *Ans.* .5 mi. e) $2.25b^8$ sq. yd. *Ans.* $1.5b^4$ yd. h) $169(a+b)^2$ *Ans.* $13(a+b)$

c) $\frac{16}{81}$ sq. rd. *Ans.* $\frac{4}{9}$ rd. f) $2500c^{10}$ sq. cm. *Ans.* $50c^5$ cm. i) $\frac{4x^2y^2}{9}$ *Ans.* $\frac{2xy}{3}$

11. Find each product: **(5.1)**

a) $(s+4)(s-4)$ *Ans.* s^2-16 g) $(a^2+b^2)(a^2-b^2)$ *Ans.* a^4-b^4

b) $(10-t)(10+t)$ *Ans.* $100-t^2$ h) $(ab+c^2)(ab-c^2)$ *Ans.* $a^2b^2-c^4$

c) $(2x+1)(2x-1)$ *Ans.* $4x^2-1$ i) $(v+\frac{1}{5})(v-\frac{1}{5})$ *Ans.* $v^2-\frac{1}{25}$

d) $(3y-7z)(3y+7z)$ *Ans.* $9y^2-49z^2$ j) $(d-1.2)(d+1.2)$ *Ans.* $d^2-1.44$

e) $(rs+1)(rs-1)$ *Ans.* r^2s^2-1 k) $(3x^2-\frac{2}{y})(3x^2+\frac{2}{y})$ *Ans.* $9x^4-\frac{4}{y^2}$

f) $(1-8x^2)(1+8x^2)$ *Ans.* $1-64x^4$ l) $(3c^3+.1)(3c^3-.1)$ *Ans.* $9c^6-.01$

12. Multiply, using the sum and difference of two numbers: **(5.2)**

a) 21×19 c) $10\frac{1}{3}\times 9\frac{2}{3}$ e) 89×91 g) $3\frac{1}{2}\times 2\frac{1}{2}$

b) 3.4×2.6 d) 17×23 f) 10.3×9.7 h) 7.5×8.5

Ans. a) $20^2-1^2 = 399$ c) $10^2-(\frac{1}{3})^2 = 99\frac{8}{9}$ e) $90^2-1^2 = 8099$ g) $3^2-(\frac{1}{2})^2 = 8\frac{3}{4} = 8.75$

b) $3^2-.4^2 = 8.84$ d) $20^2-3^2 = 391$ f) $10^2-.3^2 = 99.91$ h) $8^2-(\frac{1}{2})^2 = 63\frac{3}{4} = 63.75$

13. Find each product: **(5.3)**

a) $3(x+2)(x-2)$ c) $5(1-d)(1+d)$ e) $q^2(10-q)(10+q)$ g) $36(w^2-\frac{1}{3})(w^2+\frac{1}{3})$

b) $a^2(b+c)(b-c)$ d) $x(x-3)(x+3)$ f) $160(1-\frac{q}{4})(10+\frac{q}{4})$ h) $ab(c^3+1)(c^3-1)$

Ans. a) $3x^2-12$ c) $5-5d^2$ e) $100q^2-q^4$ g) $36w^4-4$

b) $a^2b^2-a^2c^2$ d) x^3-9x f) $160-10q^2$ h) abc^6-ab

14. Factor: **(6.1)**

a) r^2-25 e) $t^2-\frac{4}{9}$ i) $d^2-.01e^2$ m) $\frac{9}{v^2}-.25$

b) $64-u^2$ f) $100t^2-225s^2$ j) $.09A^2-49$ n) $\frac{x^4}{81}-\frac{25}{49}$

c) $81-c^2d^2$ g) x^4-9 k) $B^2-.0001$ o) $\frac{p^2}{q^2}-\frac{r^2}{16}$

d) $9x^2-1600$ h) $25-16y^{16}$ l) $R^2S^2-1.21$ p) k^6-25m^{10}

Ans. a) $(r+5)(r-5)$ e) $(t+\frac{2}{3})(t-\frac{2}{3})$ i) $(d+.1e)(d-.1e)$ m) $(\frac{3}{v}+.5)(\frac{3}{v}-.5)$

b) $(8+u)(8-u)$ f) $(10t+15s)(10t-15s)$ j) $(.3A+7)(.3A-7)$ n) $(\frac{x^2}{9}+\frac{5}{7})(\frac{x^2}{9}-\frac{5}{7})$

c) $(9+cd)(9-cd)$ g) $(x^2+3)(x^2-3)$ k) $(B+.01)(B-.01)$ o) $(\frac{p}{q}+\frac{r}{4})(\frac{p}{q}-\frac{r}{4})$

d) $(3x+40)(3x-40)$ h) $(5-4y^8)(5+4y^8)$ l) $(RS+1.1)(RS-1.1)$ p) $(k^3+5m^5)(k^3-5m^5)$

15. Factor completely: **(6.2)**

a) $3x^2 - 3$ *d*) $y^5 - y^3$ *g*) $12x^4 - 12$

b) $5x^3 - 45x$ *e*) $\pi R^3 - 25\pi R$ *h*) $15x^4 - 240$

c) $175 - 7y^2$ *f*) $\frac{1}{3}\pi R^2 h - \frac{1}{3}\pi r^2 h$ *i*) $x^7 - 81x^3$

Ans. *a*) $3(x+1)(x-1)$ *d*) $y^3(y+1)(y-1)$ *g*) $12(x^2+1)(x+1)(x-1)$

b) $5x(x+3)(x-3)$ *e*) $\pi R(R+5)(R-5)$ *h*) $15(x^2+4)(x+2)(x-2)$

c) $7(5+y)(5-y)$ *f*) $\frac{\pi h}{3}(R+r)(R-r)$ *i*) $x^3(x^2+9)(x+3)(x-3)$

16. Multiply, showing each separate product: **(7.1)**

	First	Outer	+	Inner	Last	Combine (Ans.)
a) $(x+5)(x+1)$	x^2	(x)	+	$(5x)$	$+5$	x^2+6x+5
b) $(x+8)(x-2)$	x^2	$(-2x)$	+	$(+8x)$	-16	$x^2+6x-16$
c) $(x-7)(x-6)$	x^2	$(-6x)$	+	$(-7x)$	$+42$	$x^2-13x+42$
d) $(x-10)(x+9)$	x^2	$(+9x)$	+	$(-10x)$	-90	x^2-x-90
e) $(3a-1)(4a+1)$	$12a^2$	$(+3a)$	+	$(-4a)$	-1	$12a^2-a-1$
f) $(5b+2)(5b-2)$	$25b^2$	$(-10b)$	+	$(+10b)$	-4	$25b^2-4$
g) $(c-3)(2c+8)$	$2c^2$	$(+8c)$	+	$(-6c)$	-24	$2c^2+2c-24$
h) $(r-4s)(r-11s)$	r^2	$(-11rs)$	+	$(-4rs)$	$+44s^2$	$r^2-15rs+44s^2$
i) $(3s+2t)(3s-2t)$	$9s^2$	$(-6st)$	+	$(+6st)$	$-4t^2$	$9s^2-4t^2$
j) $(x^2+5)(x^2+8)$	x^4	$(+8x^2)$	+	$(+5x^2)$	$+40$	x^4+13x^2+40
k) $(3w^2+2x)(w^2-x)$	$3w^4$	$(-3w^2x)$	+	$(+2w^2x)$	$-2x^2$	$3w^4-w^2x-2x^2$
l) $(2w^3-3)(6w^3+9)$	$12w^6$	$(+18w^3)$	+	$(-18w^3)$	-27	$12w^6-27$

17. Multiply mentally: **(7.2)**

a) $(3c+1)(4c+5)$ *d*) $(pq-8)(pq+11)$ *g*) $(d^2+6)(d^2+1)$

b) $(2+7c)(1-c)$ *e*) $(2AB+7)(2AB-7)$ *h*) $(8-g^2)(3-2g^2)$

c) $(c+3d)(c+12d)$ *f*) $(3x-2y)(4x-9y)$ *i*) $(4c^3+1)(c^3-2)$

Ans. *a*) $12c^2+19c+5$ *d*) $p^2q^2+3pq-88$ *g*) d^4+7d^2+6

b) $2+5c-7c^2$ *e*) $4A^2B^2-49$ *h*) $24-19g^2+2g^4$

c) $c^2+15cd+36d^2$ *f*) $12x^2-35xy+18y^2$ *i*) $4c^6-7c^3-2$

18. Represent the area of a rectangle whose dimensions are **(7.3)**

a) $l-3$ and $l-8$ *c*) $w+10$ and $w-12$ *e*) $13-2s$ and $2-s$

b) $2l+5$ and $3l-1$ *d*) $6-w$ and $8-w$ *f*) $7s+1$ and $9s+2$

Ans. *a*) $l^2-11l+24$ *c*) $w^2-2w-120$ *e*) $26-17s+2s^2$

b) $6l^2+13l-5$ *d*) $48-14w+w^2$ *f*) $63s^2+23s+2$

19. Factor each trinomial: **(8.1)**

a) $a^2+7a+10$ *g*) x^2-x-2 *m*) $a^2b^2+8ab+15$

b) $b^2+8b+15$ *h*) y^2-3y-4 *n*) $d^2e^2-15def+36f^2$

c) $r^2-12r+27$ *i*) w^2+2w-8 *o*) x^4+5x^2+4

d) $s^2-14s+33$ *j*) $w^2+7w-18$ *p*) y^4-6y^2-7

e) $h^2-27h+50$ *k*) $x^2+14xy+24y^2$ *q*) $x^4+8x^2y^2+16y^4$

f) $m^2+19m+48$ *l*) $c^2-17cd+30d^2$ *r*) $x^4y^4-10x^2y^2+25$

Ans. *a*) $(a+5)(a+2)$ *g*) $(x-2)(x+1)$ *m*) $(ab+5)(ab+3)$

b) $(b+5)(b+3)$ *h*) $(y-4)(y+1)$ *n*) $(de-12f)(de-3f)$

c) $(r-9)(r-3)$ *i*) $(w+4)(w-2)$ *o*) $(x^2+4)(x^2+1)$

d) $(s-11)(s-3)$ *j*) $(w+9)(w-2)$ *p*) $(y^2-7)(y^2+1)$

e) $(h-25)(h-2)$ *k*) $(x+12y)(x+2y)$ *q*) $(x^2+4y^2)(x^2+4y^2)$

f) $(m+16)(m+3)$ *l*) $(c-15d)(c-2d)$ *r*) $(x^2y^2-5)(x^2y^2-5)$

20. Factor each trinomial: **(9.1)**

a) $5x^2 + 6x + 1$ *Ans.* $(5x+1)(x+1)$
b) $5x^2 - 6x + 1$ *Ans.* $(5x-1)(x-1)$
c) $5x^2 + 4x - 1$ *Ans.* $(5x-1)(x+1)$
d) $5x^2 - 4x - 1$ *Ans.* $(5x+1)(x-1)$
e) $5 + 8x + 3x^2$ *Ans.* $(5+3x)(1+x)$
f) $5 + 14x - 3x^2$ *Ans.* $(5-x)(1+3x)$
g) $7h^2 + 10h + 3$ *Ans.* $(7h+3)(h+1)$
h) $7h^2 - 10h + 3$ *Ans.* $(7h-3)(h-1)$
i) $7h^2 - 20h - 3$ *Ans.* $(7h+1)(h-3)$
j) $7h^2 - 22h + 3$ *Ans.* $(7h-1)(h-3)$
k) $7h^4 - 15h^2 + 2$ *Ans.* $(7h^2-1)(h^2-2)$
l) $3 + 4h^2 - 7h^4$ *Ans.* $(3+7h^2)(1-h^2)$

21. Factor each trinomial: **(9.1)**

a) $4a^2 + 5a + 1$
b) $2a^2 + 3a + 1$
c) $3a^2 + 4a + 1$
d) $5 + 7b + 2b^2$
e) $4 + 16b + 15b^2$
f) $7x^2 - 15x + 2$
g) $7x^2 + 13x - 2$
h) $3 - 10y + 7y^2$
i) $3 + 4y - 7y^2$
j) $3x^2 + 10xy + 7y^2$
k) $5c^2 + 11cd + 2d^2$
l) $5c^2d^2 + 7cd + 2$
m) $2x^4 - 9x^2 + 7$
n) $35 - 17r^2 + 2r^4$
o) $3t^6 + 8t^3 - 3$

Ans. a) $(4a+1)(a+1)$
b) $(2a+1)(a+1)$
c) $(3a+1)(a+1)$
d) $(5+2b)(1+b)$
e) $(2+3b)(2+5b)$
f) $(7x-1)(x-2)$
g) $(7x-1)(x+2)$
h) $(3-7y)(1-y)$
i) $(3+7y)(1-y)$
j) $(3x+7y)(x+y)$
k) $(5c+d)(c+2d)$
l) $(5cd+2)(cd+1)$
m) $(2x^2-7)(x^2-1)$
n) $(5-r^2)(7-2r^2)$
o) $(t^3+3)(3t^3-1)$

22. Find each square: **(10.1)**

a) $(b+4)^2$
b) $(b-4)^2$
c) $(5-c)^2$
d) $(5-2c)^2$
e) $(3d+2)^2$
f) $(2d-3)^2$
g) $(x-4y)^2$
h) $(5x+4y)^2$
i) $(7xy+2)^2$
j) $(2xy-7)^2$
k) $(6ab-c)^2$
l) $(8-3abc)^2$
m) $(x^2+2y^2)^2$
n) $(x^3-8)^2$
o) $(1-5x^5)^2$

Ans. a) $b^2 + 8b + 16$
b) $b^2 - 8b + 16$
c) $25 - 10c + c^2$
d) $25 - 20c + 4c^2$
e) $9d^2 + 12d + 4$
f) $4d^2 - 12d + 9$
g) $x^2 - 8xy + 16y^2$
h) $25x^2 + 40xy + 16y^2$
i) $49x^2y^2 + 28xy + 4$
j) $4x^2y^2 - 28xy + 49$
k) $36a^2b^2 - 12abc + c^2$
l) $64 - 48abc + 9a^2b^2c^2$
m) $x^4 + 4x^2y^2 + 4y^4$
n) $x^6 - 16x^3 + 64$
o) $1 - 10x^5 + 25x^{10}$

23. Find each square by expressing each number as a binomial: **(10.2)**

a) 31^2 *Ans.* $(30+1)^2 = 961$
b) 42^2 *Ans.* $(40+2)^2 = 1764$
c) 29^2 *Ans.* $(30-1)^2 = 841$
d) 55^2 *Ans.* $(50+5)^2 = 3025$
e) 5.5^2 *Ans.* $(5+.5)^2 = 30.25$
f) $(6.5)^2$ *Ans.* $(6+.5)^2 = 42.25$
g) $(8\frac{1}{4})^2$ *Ans.* $(8+\frac{1}{4})^2 = 68\frac{1}{16}$
h) $(19\frac{3}{4})^2$ *Ans.* $(20-\frac{1}{4})^2 = 390\frac{1}{16}$
i) 9.8^2 *Ans.* $(10-.2)^2 = 96.04$

24. Represent the area of a square whose side is **(10.3)**

a) $(l+7)$ ft.
b) $(l-2)$ in.
c) $(2l+9)$ rd.
d) $(2w-3)$ cm.
e) $(10-w)$ mi.
f) $7 - 5w$
g) $a + b$
h) $c - 2d$
i) $15 - S^2$

Ans. a) $(l^2+14l+49)$ sq. ft.
b) (l^2-4l+4) sq. in.
c) $(4l^2+36l+81)$ sq. rd.
d) $(4w^2-12w+9)$ sq. cm.
e) $(100-20w+w^2)$ sq. mi.
f) $49 - 70w + 25w^2$
g) $a^2 + 2ab + b^2$
h) $c^2 - 4cd + 4d^2$
i) $225 - 30S^2 + S^4$

25. Factor: (*In (g) to (i), rearrange before factoring.*) **(11.1)**

a) $d^2 + 10d + 25$
b) $9 - 6h + h^2$
c) $144 - 24r + r^2$
d) $a^2b^2 - 40ab + 400$
e) $25x^2 - 30x + 9$
f) $49x^2 + 28xy + 4y^2$
g) $x^2 + 100y^2 + 20xy$
h) $16ab + a^2b^2 + 64$
i) $x^4 + 4y^4 - 4x^2y^2$

Ans. a) $(d+5)^2$
b) $(3-h)^2$
c) $(12-r)^2$
d) $(ab-20)^2$
e) $(5x-3)^2$
f) $(7x+2y)^2$
g) $x^2 + 20xy + 100y^2 = (x+10y)^2$
h) $a^2b^2 + 16ab + 64 = (ab+8)^2$
i) $x^4 - 4x^2y^2 + 4y^4 = (x^2-2y^2)^2$

26. Represent the side of a square whose area is **(11.2)**

a) $(81l^2 + 18l + 1)$ sq. in.
b) $(9 - 30l + 25l^2)$ sq. ft.
c) $(4w^2 - 36w + 81)$ sq. mi.
d) $(9 + 42s + 49s^2)$ sq. cm.
e) $s^2 - 12s + 36$
f) $s^4 - 24s^2 + 144$

Ans. *a*) $(9l+1)$ in.
b) $(3-5l)$ ft.
c) $(2w-9)$ mi.
d) $(3+7s)$ cm.
e) $s - 6$
f) $s^2 - 12$

27. Factor completely: **(12.1)**

a) $b^3 - 16b$
b) $3b^2 - 75$
c) $3abc^2 - 3abd^2$
d) $2x^3 + 8x^2 + 8x$
e) $x^4 - 12x^3 + 36x^2$
f) $100 - 40x + 4x^2$
g) $r^3 - 2r^2 - 15r$
h) $24 - 2r - 2r^2$
i) $x^2y^2z - 16xyz + 64z$

Ans. *a*) $b(b+4)(b-4)$
b) $3(b+5)(b-5)$
c) $3ab(c+d)(c-d)$
d) $2x(x+2)^2$
e) $x^2(x-6)^2$
f) $4(5-x)^2$
g) $r(r-5)(r+3)$
h) $2(3-r)(4+r)$
i) $z(xy-8)^2$

28. Factor completely: **(12.2)**

a) $a^4 - b^4$
b) $1 - c^{12}$
c) $81 - a^4$
d) $16 - 81b^4$
e) $a^8 - 1$
f) $1 - a^{16}$

Ans. *a*) $(a^2+b^2)(a+b)(a-b)$
b) $(1+c^6)(1+c^3)(1-c^3)$
c) $(9+a^2)(3+a)(3-a)$
d) $(4+9b^2)(2+3b)(2-3b)$
e) $(a^4+1)(a^2+1)(a+1)(a-1)$
f) $(1+a^8)(1+a^4)(1+a^2)(1+a)(1-a)$

Chapter 11

FRACTIONS

1. UNDERSTANDING FRACTIONS

The terms of a fraction are its numerator and its denominator.

The Meanings of a Fraction

Meaning 1. A fraction may mean **division.**

Thus, $\frac{3}{4}$ may mean 3 divided by 4 or $3 \div 4$.

When a fraction means division, its numerator is the **dividend** and its denominator is the **divisor.**

Thus, if $\frac{14}{5}$ means $14 \div 5$, then 14 is the dividend and 5 is the divisor.

Meaning 2. A fraction may mean **ratio.**

Thus, $\frac{3}{4}$ may mean the ratio of 3 to 4 or 3:4.

When a fraction means the ratio of two quantities, the quantities must have a common unit.

Thus, the ratio of 3 days to 2 weeks equals 3:14 or $\frac{3}{14}$. This is found by changing 2 weeks to 14 days and eliminating the common unit.

Meaning 3. A fraction may mean **a part of a whole thing** or **a part of a group of things.**

Thus, $\frac{3}{4}$ may mean three-fourths of a dollar, or 3 out of 4 dollars.

Zero Numerators or Zero Denominators

(1) **When the numerator of a fraction is zero**, the value of the fraction is zero provided the denominator is not zero also.

Thus, $\frac{0}{3} = 0$. Also, $\frac{x}{3} = 0$ if $x = 0$.

In $\frac{x-5}{3}$, if $x = 5$, then the fraction equals zero. However, $\frac{0}{0}$ is meaningless.

(2) **Since division by zero is impossible**, a fraction with a zero denominator has no meaning.

Thus, $3 \div 0$ is impossible. Hence, $\frac{3}{0}$ is meaningless.

Also, if $x = 0$, $5 \div x$ is impossible and $\frac{5}{x}$ is meaningless.

1.1. Meaning 1: Fractions Meaning Division

Express each in fractional form:

a) 10 divided by 17 *Ans.* $\frac{10}{17}$

b) 5 divided by a *Ans.* $\frac{5}{a}$

c) $(x+5) \div 3$ *Ans.* $\frac{x+5}{3}$

d) the quotient of x and $(x-2)$. *Ans.* $\frac{x}{x-2}$

1.2. Meaning 2: Fractions Meaning Ratio

Express each in fractional form:

[*In (d) to (f), a common unit must be used.*]

a) the ratio of 5 to 8 *Ans.* $\frac{5}{8}$

b) the ratio of 15 to t *Ans.* $\frac{15}{t}$

c) the ratio of a ft. to b ft. *Ans.* $\frac{a}{b}$

d) the ratio of a ft. to 5 yd. *Ans.* $\frac{a}{15}$

e) the ratio of 7 min. to 1 hr. *Ans.* $\frac{7}{60}$

f) the ratio of 3¢ to one quarter. *Ans.* $\frac{3}{25}$

1.3. Meaning 3: Fractions Meaning Parts of a Whole or of a Group

Express each in fractional form:

a) 3 of 32 equal parts of an inch *Ans.* $\frac{3}{32}$

b) 3 of n equal parts of a circle *Ans.* $\frac{3}{n}$

c) a out of b dollars *Ans.* $\frac{a}{b}$

d) 5 members out of a team of 21 *Ans.* $\frac{5}{21}$

e) n students out of a group of 30 *Ans.* $\frac{n}{30}$

f) s squads out of t squads. *Ans.* $\frac{s}{t}$

1.4. Fractions Having Zero Numerators

State the value of

a) $\frac{0}{10}$ *Ans.* 0

b) $\frac{x}{10}$ when $x=0$ *Ans.* 0

c) $\frac{x-2}{10}$ when $x=2$ *Ans.* 0

d) x when $\frac{x}{5}=0$ *Ans.* 0

e) x when $\frac{x-7}{10}=0$ *Ans.* 7

f) x when $\frac{3x+6}{12}=0.$ *Ans.* -2

1.5. Fractions Having Zero Denominators

For what value of x is the fraction meaningless?

a) $\frac{3}{x}$ *Ans.* when $x=0$

b) $\frac{10}{3x}$ *Ans.* when $3x=0$ or $x=0$

c) $\frac{10}{x-3}$ *Ans.* when $x-3=0$ or $x=3$

d) $\frac{x+4}{4x-16}$ *Ans.* when $4x-16=0$ or $x=4$

e) $\frac{8x}{5x+5}$ *Ans.* when $5x+5=0$ or $x=-1$

f) $\frac{2x-12}{2x+12}$ *Ans.* when $2x+12=0$ or $x=-6$

2. CHANGING FRACTIONS TO EQUIVALENT FRACTIONS

Equivalent fractions are fractions having the same value although they have different numerators and denominators.

Thus, $\frac{2}{3}$, $\frac{20}{30}$ and $\frac{20x}{30x}$ are equivalent fractions since $\frac{2}{3}=\frac{20}{30}=\frac{20x}{30x}$.

To obtain equivalent fractions, use one of the following fraction rules:

Rule 1. The value of a fraction is not changed if its numerator and denominator are **multiplied** by the same number except zero.

Thus, $\frac{3}{4}=\frac{30}{40}$. Here, both 3 and 4 are multiplied by 10.

Also, $\frac{5}{7}=\frac{5x}{7x}$. Here, both 5 and 7 are multiplied by x.

Rule 2. The value of a fraction is not changed if its numerator and denominator are **divided** by the same number except zero.

Thus, $\frac{400}{500}=\frac{4}{5}$. Here, both 40 and 50 are divided by 100.

Also, $\frac{7a^2}{9a^2}=\frac{7}{9}$. Here, both $7a^2$ and $9a^2$ are divided by a^2.

2.1. Using Multiplication to Obtain Equivalent Fractions

Change each to equivalent fractions by multiplying its numerator and denominator (terms) by 2, 5, x, $4x$, $x-3$, x^2 and x^2+x-3:

		Multiply Both Numerator and Denominator by						
		2	5	x	$4x$	$x-3$	x^2	x^2+x-3
a) $\frac{2}{3}$	*Ans.* $\frac{2}{3}$	$=\frac{4}{6}$	$=\frac{10}{15}$	$=\frac{2x}{3x}$	$=\frac{8x}{12x}$	$=\frac{2x-6}{3x-9}$	$=\frac{2x^2}{3x^2}$	$=\frac{2x^2+2x-6}{3x^2+3x-9}$
b) $\frac{a}{7}$	*Ans.* $\frac{a}{7}$	$=\frac{2a}{14}$	$=\frac{5a}{35}$	$=\frac{ax}{7x}$	$=\frac{4ax}{28x}$	$=\frac{ax-3a}{7x-21}$	$=\frac{ax^2}{7x^2}$	$=\frac{ax^2+ax-3a}{7x^2+7x-21}$
c) $\frac{3}{x}$	*Ans.* $\frac{3}{x}$	$=\frac{6}{2x}$	$=\frac{15}{5x}$	$=\frac{3x}{x^2}$	$=\frac{12x}{4x^2}$	$=\frac{3x-9}{x^2-3x}$	$=\frac{3x^2}{x^3}$	$=\frac{3x^2+3x-9}{x^3+x^2-3x}$

2.2. Using Division to Obtain Equivalent Fractions

Change each to equivalent fractions by dividing its numerator and denominator (terms) by 2, 5, x, $4x$, x^2, and $20x^2$.

		Divide Both Numerator and Denominator by					
		2	5	x	$4x$	x^2	$20x^2$
a) $\frac{20x^2}{80x^2}$	*Ans.* $\frac{20x^2}{80x^2}$	$=\frac{10x^2}{40x^2}$	$=\frac{4x^2}{16x^2}$	$=\frac{20x}{80x}$	$=\frac{5x}{20x}$	$=\frac{20}{80}$	$=\frac{1}{4}$
b) $\frac{40x^3}{60x^2}$	*Ans.* $\frac{40x^3}{60x^2}$	$=\frac{20x^3}{30x^2}$	$=\frac{8x^3}{12x^2}$	$=\frac{40x^2}{60x}$	$=\frac{10x^2}{15x}$	$=\frac{40x}{60}$	$=\frac{2x}{3}$

2.3. Obtaining Missing Terms

Show how to obtain each missing term:

a) $\frac{3}{7}=\frac{30}{(?)}$ — *a*) To get 30, multiply 3 by 10. Hence, multiply 7 by 10 to get 70. — *a*) $\frac{3}{7}=\frac{30}{(70)}$ *Ans.* 70

b) $\frac{27}{33}=\frac{9}{(?)}$ — *b*) To get 9, divide 27 by 3. Hence, divide 33 by 3 to get 11. — *b*) $\frac{27}{33}=\frac{9}{(11)}$ *Ans.* 11

c) $\frac{3a}{b}=\frac{3ab}{(?)}$ — *c*) To get $3ab$, multiply $3a$ by b. Hence, multiply b by b to get b^2. — *c*) $\frac{3a}{b}=\frac{3ab}{(b^2)}$ *Ans.* b^2

d) $\frac{6ac}{3c^2}=\frac{(?)}{c}$ — *d*) To get c, divide $3c^2$ by $3c$. Hence, divide $6ac$ by $3c$ to get $2a$. — *d*) $\frac{6ac}{3c^2}=\frac{(2a)}{c}$ *Ans.* $2a$

e) $\frac{5}{x-5}=\frac{25}{(?)}$ — *e*) To get 25, multiply 5 by 5. Hence, multiply $x-5$ by 5 to get $5x-25$. — *e*) $\frac{5}{x-5}=\frac{25}{(5x-25)}$ *Ans.* $5x-25$

3. RECIPROCALS AND THEIR USES

Reciprocal of a Number

The reciprocal of a number is 1 divided by the number. Thus, the reciprocal of 5 is $\frac{1}{5}$ and the reciprocal of a is $\frac{1}{a}$.

Also, the reciprocal of $\frac{2}{3}$ is $\frac{3}{2}$ since $\frac{3}{2}=1\div\frac{2}{3}$.

Rules of Reciprocals

Rule 1. The fractions $\frac{a}{b}$ and $\frac{b}{a}$ are reciprocals of each other; that is, the reciprocal of a fraction is the fraction inverted.

Rule 2. The product of two reciprocals is 1.

Thus, $\frac{2}{3} \cdot \frac{3}{2} = 1$, $\frac{a}{b} \cdot \frac{b}{a} = 1$.

Rule 3. To divide by a number or a fraction, multiply by its reciprocal.

Thus, $8 \div \frac{2}{3} = 8 \cdot \frac{3}{2}$ or 12, $7 \div 5 = 7 \cdot \frac{1}{5}$ or $\frac{7}{5}$.

Rule 4. To solve an equation for an unknown having a fractional coefficient, multiply both members by the reciprocal fraction.

Thus, if $\frac{2}{3}x = 10$, multiply both sides by $\frac{3}{2}$. Then $\frac{3}{2} \cdot \frac{2}{3}x = \frac{3}{2} \cdot 10$ and $x = 15$.

3.1. Rule 1: $\frac{a}{b}$ and $\frac{b}{a}$ are Reciprocals of Each Other

State the reciprocal of each:

a) $\frac{1}{15}$ *Ans.* $\frac{15}{1}$ or 15

b) $\frac{1}{x}$ *Ans.* $\frac{x}{1}$ or x

c) 20 *Ans.* $\frac{1}{20}$

d) y *Ans.* $\frac{1}{y}$

e) $\frac{27}{31}$ *Ans.* $\frac{31}{27}$

f) $\frac{3}{5}$ *Ans.* $\frac{5}{3}$ or $1\frac{2}{3}$

g) $\frac{3x}{10y}$ *Ans.* $\frac{10y}{3x}$

h) $\frac{x+3}{y-10}$ *Ans.* $\frac{y-10}{x+3}$

i) .5 *Ans.* $\frac{10}{5}$ or 2

j) .04 *Ans.* $\frac{100}{4}$ or 25

k) $4\frac{2}{3}$ *Ans.* $\frac{3}{14}$

l) $.001x$ *Ans.* $\frac{1000}{x}$

3.2. Rule 2: The Product of Two Reciprocals is 1

Supply the missing entry:

a) $(\frac{2}{7})(\frac{7}{2}) = (?)$ *Ans.* 1

b) $\frac{3}{8}(?) = 1$ *Ans.* $\frac{8}{3}$

c) $10(\frac{5}{9})(\frac{9}{5}) = ?$ *Ans.* 10

d) $(\frac{x}{17})(\frac{17}{x})(\frac{2}{5}) = (?)$ *Ans.* $\frac{2}{5}$

e) $(\frac{8}{7})(\frac{7}{8}) + ? = 5$ *Ans.* 4

f) $3(\frac{3}{5})(\frac{5}{3}) + (\frac{8}{15})(?) = 4$ *Ans.* $\frac{15}{8}$

3.3. Rule 3: Using Reciprocals to Change Division to Multiplication

Show how to change each division to multiplication by using the reciprocal of a number:

a) $8 \div \frac{4}{5}$ *b)* $\frac{2}{x} \div y$ *c)* $\frac{1}{r} \div \frac{s}{t}$ *d)* $\frac{3}{x} \div \frac{y}{7}$

Ans. a) $8 \cdot \frac{5}{4} = 10$ *b)* $\frac{2}{x} \cdot \frac{1}{y} = \frac{2}{xy}$ *c)* $\frac{1}{r} \cdot \frac{t}{s} = \frac{t}{rs}$ *d)* $\frac{3}{x} \cdot \frac{7}{y} = \frac{21}{xy}$

3.4. Rule 4: Using Reciprocals to Solve Equations

Solve:

a) $\frac{3}{4}x = 21$ *b)* $2\frac{1}{3}y = 70$ *c)* 30% of $P = 24$ *d)* $\frac{ax}{3} = 5a$

Solutions:

a)
$\frac{3}{4}x = 21$
$M_{\frac{4}{3}}$ $\frac{4}{3} \cdot \frac{3}{4}x = \frac{4}{3} \cdot 21$
Ans. $x = 28$

b)
$2\frac{1}{3}y = 70$
$M_{\frac{3}{7}}$ $\frac{3}{7} \cdot \frac{7}{3}y = \frac{3}{7} \cdot 70$
Ans. $y = 30$

c)
30% of $P = 24$
$M_{\frac{10}{3}}$ $\frac{10}{3} \cdot \frac{3}{10} P = \frac{10}{3} \cdot 24$
Ans. $P = 80$

d)
$\frac{ax}{3} = 5a$
$M_{\frac{3}{a}}$ $\frac{3}{a} \cdot \frac{ax}{3} = \frac{3}{a} \cdot 5a$
Ans. $x = 15$

4. REDUCING FRACTIONS TO LOWEST TERMS

A fraction is reduced to lowest terms when its numerator and denominator have no common factor except 1.

Thus, $\frac{3x}{7x}$ is not in lowest terms because x is a common factor of the numerator and denominator. After x has been eliminated by division, the resulting fraction, $\frac{3}{7}$, will be in lowest terms.

To reduce a fraction to lowest terms, use the following rule:

The value of a fraction is not changed if its numerator and denominator are divided by the same number except zero.

To Reduce a Fraction to Lowest Terms

	Reduce: a) $\frac{3ab^2c}{3ab^2d}$	b) $\frac{8a+8b}{12a+12b}$	c) $\frac{2a^2-2b^2}{5a+5b}$
Procedure:	**Solutions:**		
1. Factor its terms: (*Numerator and Denominator*)	$\frac{3ab^2c}{3ab^2d}$	$\frac{8(a+b)}{12(a+b)}$	$\frac{2(a+b)(a-b)}{5(a+b)}$
2. Divide both terms by every common factor: (*This is commonly called "cancelling".*)	$\frac{\overset{1}{\cancel{3ab^2}}c}{\underset{1}{\cancel{3ab^2}}d}$	$\frac{\overset{2}{\cancel{8}}\overset{(1)}{\cancel{(a+b)}}}{\underset{3}{\cancel{12}}\underset{(1)}{\cancel{(a+b)}}}$	$\frac{2\overset{1}{\cancel{(a+b)}}(a-b)}{5\underset{1}{\cancel{(a+b)}}}$
	Ans. $\frac{c}{d}$	*Ans.* $\frac{2}{3}$	*Ans.** $\frac{2(a-b)}{5}$

***Note.** In answers, algebraic factors which remain need not be multiplied out.

Rule 1. If two expressions are exactly alike or have the same value, their quotient is 1.

Thus, $\frac{5abc}{5abc} = 1$, $\frac{a+b}{b+a} = 1$, $\frac{\overset{4}{\cancel{8}}\overset{(1)}{\cancel{(x^2+x-5)}}}{\cancel{2}\cancel{(x^2+x-5)}} = 4$.

Binomials Which are Negatives of Each Other.

Binomials such as $x-y$ and $y-x$ are negatives of each other if x and y have different values.

Thus, if $x=5$ and $y=2$, $x-y=+3$ and $y-x=-3$.

Hence, either $(x-y) = -(y-x)$ or $(y-x) = -(x-y)$.

Rule 2. If two binomials are negatives of each other, their quotient is -1.

Thus, $\frac{x-y}{y-x} = -1$, $\frac{\overset{1}{\cancel{(5+x)}}\overset{(-1)}{\cancel{(5-x)}}}{\cancel{(x+5)}\cancel{(x-5)}} = -1$, $\frac{\overset{(-1)}{\cancel{(a-b)}}\overset{(-1)}{\cancel{(7-c)}}}{\cancel{(b-a)}\cancel{(c-7)}} = 1$.

Warning! How NOT to Reduce a Fraction to Lowest Terms:

1. **Do NOT Subtract** the same number from the numerator and denominator.

   Thus, $\frac{5}{6}$ does **NOT** equal $\frac{5-4}{6-4}$ or $\frac{1}{2}$. Also, $\frac{n+1}{n+2}$ does **NOT** equal $\frac{\cancel{n}+1}{\cancel{n}+2}$ or $\frac{1}{2}$.

2. **Do NOT Add** the same number to both numerator and denominator.

   Thus, $\frac{1}{2}$ does **NOT** equal $\frac{1+3}{2+3}$ or $\frac{4}{5}$. Also, $\frac{x-3}{y-3}$ does **NOT** equal $\frac{x\cancel{-3}}{y\cancel{-3}}$ or $\frac{x}{y}$.

4.1. Reducing Fractions Whose Terms Have Common Monomial Factors

Reduce to lowest terms:

a) $\frac{39rs}{52rs}$ b) $\frac{32a^3b^3}{64a^2b}$ c) $\frac{5x-35}{15x}$ d) $\frac{21a^2}{14a^2-7ab}$

Solutions:

$\frac{\overset{3}{\cancel{39}}\,\overset{(1)}{\cancel{rs}}}{\underset{4}{\cancel{52}}\,\underset{(1)}{\cancel{rs}}}$ $\qquad \frac{\overset{1}{\cancel{32}}\,\overset{ab^2}{\cancel{a^3b^3}}}{\underset{2}{\cancel{64}}\,\underset{(1)}{\cancel{a^2b}}}$ $\qquad \frac{\overset{1}{\cancel{5}}(x-7)}{\underset{3}{\cancel{15}}x}$ $\qquad \frac{\overset{3a}{\cancel{21a^2}}}{\underset{1}{\cancel{7a}}(2a-b)}$

Ans. $\frac{3}{4}$ *Ans.* $\frac{ab^2}{2}$ *Ans.* $\frac{x-7}{3x}$ *Ans.* $\frac{3a}{2a-b}$

4.2. Rule 1. Reducing Fractions Whose Terms Have a Common Binomial Factor

Reduce to lowest terms:

a) $\frac{2x+6}{3ax+9a}$ b) $\frac{x^2+x}{2+2x}$ c) $\frac{3x+3y}{3y^2-3x^2}$ d) $\frac{(b+c)^2}{-acx-abx}$

Solutions:

$\frac{2(x+3)}{3a(x+3)}$ $\qquad \frac{x(x+1)}{2(1+x)}$ $\qquad \frac{3(x+y)}{3(y+x)(y-x)}$ $\qquad \frac{(b+c)(b+c)}{-ax(c+b)}$

$\frac{2\,\overset{1}{\cancel{(x+3)}}}{3a\cancel{(x+3)}}$ $\qquad \frac{x\,\overset{1}{\cancel{(x+1)}}}{2\cancel{(1+x)}}$ $\qquad \frac{\overset{1}{\cancel{3}}\,\overset{(1)}{\cancel{(x+y)}}}{\cancel{3}\cancel{(y+x)}(y-x)}$ $\qquad \frac{(b+c)\overset{1}{\cancel{(b+c)}}}{-ax\cancel{(c+b)}}$

Ans. $\frac{2}{3a}$ *Ans.* $\frac{x}{2}$ *Ans.* $\frac{1}{y-x}$ *Ans.* $-\frac{b+c}{ax}$

4.3. Rule 2. Reducing Fractions Having Binomial Factors Which Are Negatives of Each Other

Reduce to lowest terms:

a) $\frac{4-y}{3y-12}$ b) $\frac{d^2-49}{14-2d}$ c) $\frac{5-5r}{10rt-10t}$ d) $\frac{(w-x)^2}{x^2-w^2}$

Solutions:

$\frac{\overset{-1}{\cancel{(4-y)}}}{3\cancel{(y-4)}}$ $\qquad \frac{\overset{(-1)}{\cancel{(d-7)}}(d+7)}{2\cancel{(7-d)}}$ $\qquad \frac{\overset{1}{\cancel{5}}\,\overset{(-1)}{\cancel{(1-r)}}}{\underset{2}{\cancel{10}}t\cancel{(r-1)}}$ $\qquad \frac{\overset{-1}{\cancel{(w-x)}}(w-x)}{\cancel{(x-w)}(x+w)}$

Ans. $-\frac{1}{3}$ *Ans.* $-\frac{d+7}{2}$ *Ans.* $-\frac{1}{2t}$ *Ans.* $-\frac{w-x}{x+w}$ or $\frac{x-w}{x+w}$

4.4. Fractions Having at Least One Trinomial Term

Reduce to lowest terms:

a) $\frac{b^2+3b}{b^2+10b+21}$ b) $\frac{x^2-9x+20}{4x-x^2}$ c) $\frac{y^2+2y-15}{2y^2-12y+18}$

Solutions:

$\frac{b\,\overset{1}{\cancel{(b+3)}}}{\cancel{(b+3)}(b+7)}$ $\qquad \frac{(x-5)\overset{-1}{\cancel{(x-4)}}}{x\cancel{(4-x)}}$ $\qquad \frac{(y+5)\overset{1}{\cancel{(y-3)}}}{2(y-3)\cancel{(y-3)}}$

Ans. $\frac{b}{b+7}$ *Ans.* $-\frac{x-5}{x}$ or $\frac{5-x}{x}$ *Ans.* $\frac{y+5}{2(y-3)}$

5. MULTIPLYING FRACTIONS

To Multiply Fractions Having No Cancellable Common Factor

Procedure:	Multiply: a) $\frac{3}{5}\cdot\frac{7}{11}$	b) $\frac{x}{3}\cdot\frac{5}{r}$	c) $\frac{a}{4}\cdot\frac{9}{2}\cdot\frac{c}{d}\cdot\frac{7}{a+c}$
	Solutions:		
1. Multiply numerators to obtain numerator of answer:	**1.** $3(7)$	**1.** $x(5)$	**1.** $a(9)(c)(7)$
2. Multiply denominators to obtain denominator of answer:	**2.** $5(11)$	**2.** $3(r)$	**2.** $4(2)(d)(a+c)$
	Ans. $\frac{21}{55}$	**Ans.** $\frac{5x}{3r}$	**Ans.** $\frac{63ac}{8d(a+c)}$

To Multiply Fractions Having a Cancellable Common Factor

Procedure:	Multiply: a) $\frac{3}{5}\cdot\frac{10}{7}\cdot\frac{77}{6}$	b) $\frac{7x}{a}\cdot\frac{2a}{7+7x}$	c) $\frac{5x}{a^2-b^2}\cdot\frac{3a+3b}{x}$
	Solutions:		
1. Factor those numerators and denominators which are polynomials:	$\frac{3}{5}\cdot\frac{10}{7}\cdot\frac{77}{6}$	$\frac{7x}{a}\cdot\frac{2a}{7(1+x)}$	$\frac{5x}{(a+b)(a-b)}\cdot\frac{3(a+b)}{x}$
2. Divide out factors common to any numerator and any denominator:	$\frac{\overset{(1)}{\cancel{3}}}{\underset{(1)}{\cancel{5}}}\cdot\frac{\overset{(2)}{\cancel{10}}}{\underset{(1)}{\cancel{7}}}\cdot\frac{\overset{(11)}{\cancel{77}}}{\underset{(2)}{\cancel{6}}}$	$\frac{\overset{1}{\cancel{7}}x}{\underset{(1)}{\cancel{a}}}\cdot\frac{2\overset{(1)}{\cancel{a}}}{\underset{(1)}{\cancel{7}}(1+x)}$	$\frac{5\overset{1}{\cancel{x}}}{\underset{(1)}{\cancel{(a+b)}}(a-b)}\cdot\frac{3\overset{1}{\cancel{(a+b)}}}{\underset{(1)}{\cancel{x}}}$
3. Multiply remaining factors:	**Ans.** 11	**Ans.** $\frac{2x}{1+x}$	**Ans.** $\frac{15}{a-b}$

5.1. Multiplying Fractions Having No Cancellable Common Factor

Multiply:

a) $\frac{1}{5}\cdot\frac{2}{5}\cdot\frac{4}{7}$ b) $\frac{2}{d}\cdot\frac{c}{5}\cdot\frac{x}{y}$ c) $5xy\cdot\frac{3c}{ab}$ d) $\frac{a}{4}\cdot\frac{3}{r}\cdot\frac{r+5}{a-2}$

Solutions:

a) $\frac{1(2)(4)}{5(5)(7)}$ *Ans.* $\frac{8}{175}$

b) $\frac{2(c)(x)}{d(5)(y)}$ *Ans.* $\frac{2cx}{5dy}$

c) $\frac{5xy\cdot 3c}{ab}$ *Ans.* $\frac{15cxy}{ab}$

d) $\frac{a(3)(r+5)}{4(r)(a-2)}$ *Ans.* $\frac{3a(r+5)}{4r(a-2)}$

5.2. Multiplying Fractions Having a Cancellable Common Factor

Multiply:

a) $\frac{9}{5}\cdot\frac{7}{3}\cdot\frac{15}{14}$ b) $\frac{12c}{d}\cdot\frac{h}{3}\cdot\frac{d^2}{h^2}$ c) $\frac{c}{a+b}\cdot\frac{5a+5b}{2c^2+2c}$ d) $\frac{9x}{3x-15}\cdot\frac{(x-5)^2}{2(5-x)}$

Solutions:

a) $\frac{\overset{3}{\cancel{9}}}{\underset{(1)}{\cancel{5}}}\cdot\frac{\overset{(1)}{\cancel{7}}}{\underset{(1)}{\cancel{3}}}\cdot\frac{\overset{3}{\cancel{15}}}{\underset{(2)}{\cancel{14}}}$ *Ans.* $\frac{9}{2}$

b) $\frac{\overset{4}{\cancel{12}}c}{\underset{(1)}{\cancel{d}}}\cdot\frac{\overset{(1)}{\cancel{h}}}{\underset{(1)}{\cancel{3}}}\cdot\frac{\overset{d}{\cancel{d^2}}}{\underset{h}{\cancel{h^2}}}$ *Ans.* $\frac{4cd}{h}$

c) $\frac{\overset{(1)}{\cancel{c}}}{\cancel{(a+b)}}\cdot\frac{5\overset{1}{\cancel{(a+b)}}}{2\cancel{c}(c+1)}$ *Ans.* $\frac{5}{2(c+1)}$

d) $\frac{\overset{3}{\cancel{9}}x}{\cancel{3}\cancel{(x-5)}}\cdot\frac{\overset{(1)}{\cancel{(x-5)}}\overset{(-1)}{\cancel{(x-5)}}}{2\cancel{(5-x)}}$ *Ans.* $-\frac{3x}{2}$

5.3. More Difficult Multiplication of Fractions

Multiply:

a) $\frac{y^2+6y+5}{7y^2-63}\cdot\frac{7y+21}{(5+y)^2}$ b) $\frac{a+4}{4a}\cdot\frac{2a-8}{4+a}\cdot\frac{a^2-4}{24-12a}\cdot\frac{4a^2}{4-a}$

Solutions:

a) $\frac{\overset{(1)}{\cancel{(y+5)}}(y+1)}{\cancel{7}(y-3)\cancel{(y+3)}}\cdot\frac{\overset{(1)}{\cancel{7}}\overset{(1)}{\cancel{(y+3)}}}{\cancel{(5+y)}(5+y)}$

Ans. $\frac{y+1}{(y-3)(5+y)}$ or $\frac{y+1}{(y-3)(y+5)}$

b) $\frac{\overset{(1)}{\cancel{a+4}}}{\cancel{4a}}\cdot\frac{\overset{(1)}{\cancel{2}}\overset{(-1)}{\cancel{(a-4)}}}{\cancel{(4+a)}}\cdot\frac{(a+2)\overset{(-1)}{\cancel{(a-2)}}}{\underset{6}{\cancel{12}}\cancel{(2-a)}}\cdot\frac{\overset{a}{\cancel{4a^2}}}{\cancel{4-a}}$

Ans. $\frac{a(a+2)}{6}$

6. DIVIDING FRACTIONS

Rule. To divide by a fraction, invert the fraction and multiply.

To Divide Fractions

	Divide: $a)\ \frac{2}{3} \div 5$	$b)\ \frac{9}{4} \div \frac{a}{3}$	$c)\ \frac{14}{x} \div 2\frac{1}{3}$	$d)\ \frac{a^2}{b^2} \div \frac{a}{b}$
Procedure:	**Solutions:**			
1. Invert fraction which is divisor:	Invert $\frac{5}{1}$	Invert $\frac{a}{3}$	$2\frac{1}{3} = \frac{7}{3}$. Invert $\frac{7}{3}$	Invert $\frac{a}{b}$
2. Multiply resulting fractions:	$\frac{2}{3} \cdot \frac{1}{5}$	$\frac{9}{4} \cdot \frac{3}{a}$	$\frac{\overset{2}{\cancel{14}}}{x} \cdot \frac{3}{\underset{1}{\cancel{7}}}$	$\frac{\overset{a}{\cancel{a^2}}}{\underset{b}{\cancel{b^2}}} \cdot \frac{\overset{1}{\cancel{b}}}{\underset{1}{\cancel{a}}}$
	Ans. $\frac{2}{15}$	*Ans.* $\frac{27}{4a}$	*Ans.* $\frac{6}{x}$	*Ans.* $\frac{a}{b}$

6.1. Dividing Fractions Having Monomial Terms

Divide:

$a)\ 2\frac{3}{4} \div 22$ $\qquad b)\ \frac{8}{x^3} \div \frac{12}{x^2}$ $\qquad c)\ b^2 \div \frac{7}{b^3}$ $\qquad d)\ \frac{5y}{7} \cdot \frac{2x}{y} \div \frac{x^5}{42}$

Solutions:

$a)\ \frac{\overset{1}{\cancel{11}}}{4} \cdot \frac{1}{\underset{2}{\cancel{22}}}$ $\qquad b)\ \frac{\overset{2}{\cancel{8}}}{\underset{x}{\cancel{x^3}}} \cdot \frac{\overset{(1)}{\cancel{x^2}}}{\underset{3}{\cancel{12}}}$ $\qquad c)\ \frac{b^2}{1} \cdot \frac{b^3}{7}$ $\qquad d)\ \frac{\overset{1}{5\cancel{y}}}{\underset{1}{\cancel{7}}} \cdot \frac{\overset{(1)}{2\cancel{x}}}{\underset{1}{\cancel{y}}} \cdot \frac{\overset{6}{\cancel{42}}}{\underset{x^4}{\cancel{x^5}}}$

Ans. $\frac{1}{8}$ $\qquad$ *Ans.* $\frac{2}{3x}$ $\qquad$ *Ans.* $\frac{b^5}{7}$ $\qquad$ *Ans.* $\frac{60}{x^4}$

6.2. Dividing Fractions Having Polynomial Terms

Divide:

$a)\ \frac{a^2-100}{8} \div \frac{2a+20}{20}$ $\qquad b)\ \frac{5a^2}{b^2-36} \div \frac{25ab-25a}{b^2-7b+6}$ $\qquad c)\ \frac{4x^2-1}{9x-3x^2} \div \frac{2x^2-7x-4}{x^2-7x+12}$

Solutions:

$a)\ \frac{a^2-100}{8} \cdot \frac{20}{2a+20}$

$\frac{\overset{1}{\cancel{(a+10)}}(a-10)}{\underset{(2)}{\cancel{8}}} \cdot \frac{\overset{5}{\cancel{20}}}{2\cancel{(a+10)}}$

Ans. $\frac{5(a-10)}{4}$

$b)\ \frac{5a^2}{b^2-36} \cdot \frac{b^2-7b+6}{25ab-25a}$

$\frac{\overset{(1)\,a}{\cancel{5}a^2}}{(b+6)\cancel{(b-6)}} \cdot \frac{\overset{1}{\cancel{(b-6)}}\overset{1}{\cancel{(b-1)}}}{\underset{5(1)}{25\cancel{a}}\cancel{(b-1)}}$

Ans. $\frac{a}{5(b+6)}$

$c)\ \frac{4x^2-1}{9x-3x^2} \cdot \frac{x^2-7x+12}{2x^2-7x-4}$

$\frac{\overset{1}{\cancel{(2x+1)}}(2x-1)}{3x\cancel{(3-x)}} \cdot \frac{\overset{(-1)}{\cancel{(x-3)}}\overset{1}{\cancel{(x-4)}}}{\cancel{(2x+1)}\cancel{(x-4)}}$

Ans. $-\frac{2x-1}{3x}$ or $\frac{1-2x}{3x}$

7. ADDING OR SUBTRACTING FRACTIONS HAVING THE SAME DENOMINATOR

To Combine (Add or Subtract) Fractions Having the Same Denominator

	Combine: $a)\ \frac{2a}{15} + \frac{7a}{15} - \frac{4a}{15}$	$b)\ \frac{5a}{3} - \frac{2a-9}{3}$	$c)\ \frac{7}{x-2} - \frac{5+x}{x-2}$
Procedure:	**Solutions:**		
1. * Keep denominator and combine numerators:	$\frac{2a + 7a - 4a}{15}$	$\frac{5a - (2a-9)}{3}$	$\frac{7 - (5+x)}{x-2}$
2. Reduce resulting fraction:	$\frac{\overset{(1)}{\cancel{5}a}}{\underset{3}{\cancel{15}}}$	$\frac{3a+9}{3} = \frac{\overset{(1)}{\cancel{3}}(a+3)}{\underset{(1)}{\cancel{3}}}$	$\frac{\overset{-1}{\cancel{2-x}}}{\cancel{x-2}}$
	Ans. $\frac{a}{3}$	*Ans.* $a+3$	*Ans.* -1

*** Note.** In combining numerators, enclose each polynomial numerator in parentheses preceded by the sign of its fraction.

7.1. Combining Fractions Having Same Monomial Denominator

Combine:

a) $2\frac{4}{5}+\frac{4}{5}-1\frac{3}{5}$ b) $\frac{8}{3c}-\frac{1}{3c}+\frac{11}{3c}$ c) $\frac{5x}{8}-\frac{x-4}{8}$ d) $\frac{x+5}{3x}-\frac{1-x}{3x}-\frac{7x+4}{3x}$

Solutions:

a) $\frac{14+4-8}{5}$ b) $\frac{8-1+11}{3c}$ c) $\frac{5x-(x-4)}{8}$ d) $\frac{(x+5)-(1-x)-(7x+4)}{3x}$

a) $\frac{\overset{2}{\cancel{10}}}{\underset{1}{\cancel{5}}}$ b) $\frac{\overset{6}{\cancel{18}}}{\underset{(1)}{\cancel{3}c}}$ c) $\frac{4x+4}{8}=\frac{\overset{1}{\cancel{4}}(x+1)}{\underset{2}{\cancel{8}}}$ d) $\frac{x+5-1+x-7x-4}{3x}=\frac{-5\overset{(1)}{\cancel{x}}}{3\underset{(1)}{\cancel{x}}}$

Ans. 2 *Ans.* $\frac{6}{c}$ *Ans.* $\frac{x+1}{2}$ *Ans.* $-\frac{5}{3}$

7.2. Combining Fractions Having Same Polynomial Denominator

Combine:

a) $\frac{10x}{2x-6}-\frac{9x+3}{2x-6}$ b) $\frac{2b}{a-b}-\frac{2a}{a-b}$ c) $\frac{5}{x^2+3x-4}+\frac{7x-8}{x^2+3x-4}-\frac{3x+1}{x^2+3x-4}$

Solutions:

a) $\frac{10x-(9x+3)}{2x-6}$ b) $\frac{2b-2a}{a-b}$ c) $\frac{5+(7x-8)-(3x+1)}{x^2+3x-4}$

a) $\frac{\overset{1}{\cancel{x-3}}}{2(\cancel{x-3})}$ b) $\frac{2\overset{(-1)}{(\cancel{b-a})}}{\cancel{a-b}}$ c) $\frac{4x-4}{x^2+3x-4}$ or $\frac{4\overset{1}{(\cancel{x-1})}}{(\cancel{x-1})(x+4)}$

Ans. $\frac{1}{2}$ *Ans.* -2 *Ans.* $\frac{4}{x+4}$

8. ADDING OR SUBTRACTING FRACTIONS HAVING DIFFERENT DENOMINATORS

The Lowest Common Denominator (L.C.D.)

The lowest common denominator of two or more fractions is the smallest number divisible without remainder by their denominators.

Thus, 12 is the **L.C.D.** of $\frac{1}{3}$ and $\frac{1}{4}$. Of the common denominators, 12, 24, 36, etc., the lowest or smallest is 12.

To Find the L.C.D.

Rule 1. If no two denominators have a common factor, find the **L.C.D.** by multiplying all the denominators.

Thus, $3ax$ is the **L.C.D.** of $\frac{1}{3}$, $\frac{1}{a}$ and $\frac{1}{x}$.

Rule 2. If two of the denominators have a common factor, find the **L.C.D.** by multiplying the common factor by the remaining factors.

Thus, for $\frac{1}{3xy}$ and $\frac{1}{5xy}$, the **L.C.D.**, $15xy$, is obtained by multiplying the common factor, xy, by the remaining factors 3 and 5.

Rule 3. If there is a common literal factor with more than one exponent, use its highest exponent in the **L.C.D.**

Thus, $3y^5$ is the **L.C.D.** of $\frac{1}{3y}$, $\frac{1}{y^2}$ and $\frac{1}{y^5}$.

To Combine (Add or Subtract) Fractions Having Different Denominators

Combine: a) $\frac{1}{3}+\frac{3}{4}-\frac{1}{12}$ b) $\frac{5}{x}+\frac{3}{y}$

Procedure:	Solutions: a)	b)
1. **Find the L.C.D.:**	1. **L.C.D.** = 12	1. **L.C.D.** = xy
2. **Change each fraction to an equivalent fraction whose denominator is L.C.D.:** *(Note how each fraction is multiplied by 1.)*	2. $\frac{1}{3}\cdot\frac{4}{4}+\frac{3}{4}\cdot\frac{3}{3}-\frac{1}{12}$ $\frac{4}{12}+\frac{9}{12}-\frac{1}{12}$ $\frac{12}{12}$	2. $\frac{5}{x}\cdot\frac{y}{y}+\frac{3}{y}\cdot\frac{x}{x}$ $\frac{5y}{xy}+\frac{3x}{xy}$
3. Combine fractions having the same denominator and reduce, if necessary:	3. 1 *Ans.*	3. $\frac{5y+3x}{xy}$ *Ans.*

8.1. Rule 1. Combining Fractions whose Denominators Have No Common Factor

Combine: a) $\frac{2}{5}+\frac{3}{4}$ b) $\frac{3}{x}-\frac{3}{x+1}$ c) $r+2-\frac{4r-1}{2r}$

Procedure:	Solutions: a)	b)	c)
1. **Find L.C.D. (Rule 1):**	**L.C.D.** = (5)(4) = 20	**L.C.D.** = $x(x+1)$	**L.C.D.** = $2r$
2. Change to equivalent fractions having same **L.C.D.:**	$\frac{2}{5}\cdot\frac{4}{4}+\frac{3}{4}\cdot\frac{5}{5}$ $\frac{8}{20}+\frac{15}{20}$	$\frac{3}{x}\cdot\frac{(x+1)}{(x+1)}-\frac{3}{(x+1)}\cdot\frac{x}{x}$ $\frac{3x+3}{x(x+1)}-\frac{3x}{x(x+1)}$	$\frac{(r+2)}{1}\cdot\frac{2r}{2r}-\frac{4r-1}{2r}$ $\frac{2r^2+4r}{2r}-\frac{4r-1}{2r}$
3. Combine fractions:	$\frac{23}{20}$ or $1\frac{3}{20}$ *Ans.*	$\frac{3}{x(x+1)}$ *Ans.*	$\frac{2r^2+1}{2r}$ *Ans.*

8.2. Rule 2. Combining Fractions whose Denominators Have a Common Factor

Combine: a) $\frac{5}{6}-\frac{1}{12}+\frac{3}{2}$ b) $\frac{2}{5a}+\frac{7}{5a+5}$ c) $\frac{a+4}{3a}-\frac{2-4a}{6a}$

Procedure:	Solutions: a)	b)	c)
1. **Find L.C.D. (Rule 2):**	**L.C.D.** = 12	**L.C.D.** = $5a(a+1)$	**L.C.D.** = $6a$
2. Change to equivalent fractions having same **L.C.D.:**	$\frac{5}{6}\cdot\frac{2}{2}-\frac{1}{12}+\frac{3}{2}\cdot\frac{6}{6}$ $\frac{10}{12}-\frac{1}{12}+\frac{18}{12}$	$\frac{2}{5a}\cdot\frac{(a+1)}{(a+1)}+\frac{7}{5(a+1)}\cdot\frac{a}{a}$ $\frac{2a+2}{5a(a+1)}+\frac{7a}{5a(a+1)}$	$\frac{(a+4)}{3a}\cdot\frac{2}{2}-\frac{2-4a}{6a}$ $\frac{2a+8}{6a}-\frac{2-4a}{6a}$
3. Combine fractions:	$\frac{27}{12}=\frac{9}{4}$ *Ans.*	$\frac{9a+2}{5a(a+1)}$ *Ans.*	$\frac{6a+6}{6a}=\frac{a+1}{a}$ *Ans.*

8.3. Rule 3. Combining Fractions whose L.C.D. Includes Base with Highest Exponent

Combine:

a) $\frac{5}{x}+\frac{7}{x^2}$	b) $\frac{3}{x^3}-\frac{2}{x^7}+\frac{1}{x^5}$	c) $\frac{s^2}{9r^2}-\frac{s^3}{12r^3}$	d) $\frac{2}{a^2b}+\frac{3}{ab^2}$
Solutions:			
L.C.D. = x^2	**L.C.D.** = x^7	**L.C.D.** = $36r^3$	**L.C.D.** = a^2b^2
$\frac{5}{x}\cdot\frac{x}{x}+\frac{7}{x^2}$	$\frac{3}{x^3}\cdot\frac{x^4}{x^4}-\frac{2}{x^7}+\frac{1}{x^5}\cdot\frac{x^2}{x^2}$	$\frac{s^2}{9r^2}\cdot\frac{4r}{4r}-\frac{s^3}{12r^3}\cdot\frac{3}{3}$	$\frac{2}{a^2b}\cdot\frac{b}{b}+\frac{3}{ab^2}\cdot\frac{a}{a}$
$\frac{5x}{x^2}+\frac{7}{x^2}$	$\frac{3x^4}{x^7}-\frac{2}{x^7}+\frac{x^2}{x^7}$	$\frac{4rs^2}{36r^3}-\frac{3s^3}{36r^3}$	$\frac{2b}{a^2b^2}+\frac{3a}{a^2b^2}$
$\frac{5x+7}{x^2}$ *Ans.*	$\frac{3x^4+x^2-2}{x^7}$ *Ans.*	$\frac{4rs^2-3s^3}{36r^3}$ *Ans.*	$\frac{2b+3a}{a^2b^2}$ *Ans.*

8.4. Combining Fractions Having Binomial Denominators

Combine:

$a)\ \dfrac{3x}{x-2}+\dfrac{5x}{x+2}$ $\quad b)\ \dfrac{3}{2y+4}-\dfrac{5}{3y+6}$ $\quad c)\ \dfrac{2a-3}{a^2-25}-\dfrac{7}{5a-25}$

Solutions: (*Factor denominators first.*)

a)

L.C.D. $=(x-2)(x+2)$

$$\frac{3x}{(x-2)}\cdot\frac{(x+2)}{(x+2)}+\frac{5x}{(x+2)}\cdot\frac{(x-2)}{(x-2)}$$

$$\frac{3x^2+6x}{(x-2)(x+2)}+\frac{5x^2-10x}{(x+2)(x-2)}$$

Ans. $\dfrac{8x^2-4x}{(x-2)(x+2)}$

b) $\dfrac{3}{2(y+2)}-\dfrac{5}{3(y+2)}$

L.C.D. $=6(y+2)$

$$\frac{3}{2(y+2)}\cdot\frac{3}{3}-\frac{5}{3(y+2)}\cdot\frac{2}{2}$$

$$\frac{9}{6(y+2)}-\frac{10}{6(y+2)}$$

Ans. $-\dfrac{1}{6(y+2)}$

c) $\dfrac{2a-3}{(a+5)(a-5)}-\dfrac{7}{5(a-5)}$

L.C.D. $=5(a+5)(a-5)$

$$\frac{(2a-3)}{(a+5)(a-5)}\cdot\frac{5}{5}-\frac{7}{5(a-5)}\cdot\frac{(a+5)}{(a+5)}$$

$$\frac{10a-15}{5(a+5)(a-5)}-\frac{(7a+35)}{5(a+5)(a-5)}$$

Ans. $\dfrac{3a-50}{5(a+5)(a-5)}$

9. SIMPLIFYING COMPLEX FRACTIONS

A complex fraction is a fraction containing at least one other fraction within it. Thus, $\dfrac{\frac{3}{4}}{2}$, $\dfrac{5}{\frac{2}{3}}$ and $\dfrac{x+\frac{1}{2}}{x-\frac{1}{4}}$ are complex fractions.

To Simplify a Complex Fraction: L.C.D.–Multiplication Method

Simplify: $a)\ \dfrac{\frac{2}{3}}{\frac{3}{4}}$ $\quad b)\ \dfrac{\frac{1}{2}-\frac{1}{3}}{5}$ $\quad c)\ \dfrac{x-\frac{1}{3}}{\frac{3}{5}+\frac{7}{10}}$

Procedure:	**Solutions:** a)	b)	c)
1. **Find L.C.D. of fractions in complex fraction:**	**L.C.D.** = 12	**L.C.D.** = 6	**L.C.D.** = 30
2. **Multiply both numerator and denominator by L.C.D.** and reduce, if necessary:	$\dfrac{\frac{2}{3}}{\frac{3}{4}}\cdot\dfrac{12}{12}$	$\dfrac{(\frac{1}{2}-\frac{1}{3})}{5}\cdot\dfrac{6}{6}$	$\dfrac{(x-\frac{1}{3})}{(\frac{3}{5}+\frac{7}{10})}\cdot\dfrac{30}{30}$
	$\dfrac{8}{9}$ *Ans.*	$\dfrac{3-2}{30}=\dfrac{1}{30}$ *Ans.*	$\dfrac{30x-10}{18+21}=\dfrac{30x-10}{39}$ *Ans.*

To Simplify a Complex Fraction: Combining–Division Method

Simplify: $a)\ \dfrac{\frac{1}{2}-\frac{1}{3}}{\frac{1}{2}+\frac{1}{3}}$ $\quad b)\ \dfrac{x-\frac{1}{3}}{x+\frac{1}{3}}$ $\quad c)\ \dfrac{\frac{x}{2}+\frac{x}{5}}{2x-\frac{3x}{10}}$ $\quad d)\ \dfrac{1+\frac{2}{y}}{1-\frac{4}{y^2}}$

Procedure:	**Solutions:** a)	b)	c)	d)
1. **Combine terms of numerator:**	$\frac{1}{2}-\frac{1}{3}=\frac{1}{6}$	$x-\frac{1}{3}=\frac{3x-1}{3}$	$\frac{x}{2}+\frac{x}{5}=\frac{7x}{10}$	$1+\frac{2}{y}=\frac{y+2}{y}$
2. **Combine terms of denominator:**	$\frac{1}{2}+\frac{1}{3}=\frac{5}{6}$	$x+\frac{1}{3}=\frac{3x+1}{3}$	$2x-\frac{3x}{10}=\frac{17x}{10}$	$1-\frac{4}{y^2}=\frac{y^2-4}{y^2}$
3. **Divide new numerator by new denominator:**	$\frac{1}{6}\div\frac{5}{6}$	$\frac{3x-1}{3}\div\frac{3x+1}{3}$	$\frac{7x}{10}\div\frac{17x}{10}$	$\frac{y+2}{y}\div\frac{y^2-4}{y^2}$
	$\frac{1}{6}\cdot\frac{6}{5}$	$\frac{3x-1}{3}\cdot\frac{3}{3x+1}$	$\frac{7x}{10}\cdot\frac{10}{17x}$	$\frac{y+2}{y}\cdot\frac{y^2}{y^2-4}$
	$\frac{1}{5}$ *Ans.*	$\frac{3x-1}{3x+1}$ *Ans.*	$\frac{7}{17}$ *Ans.*	$\frac{y}{y-2}$ *Ans.*

9.1. Simplifying Numerical Complex Fractions

Simplify:

a) $\dfrac{4-\frac{1}{3}}{5}$

L.C.D. = 3

$\dfrac{(4-\frac{1}{3})}{5}\cdot\dfrac{3}{3}$

$\dfrac{12-1}{15}$

$\dfrac{11}{15}$ *Ans.*

b) $\dfrac{5+\frac{2}{5}}{7-\frac{1}{10}}$

Division Method

$\dfrac{27}{5}\div\dfrac{69}{10}$

$\dfrac{\overset{9}{\cancel{27}}}{\underset{(1)}{\cancel{5}}}\cdot\dfrac{\overset{(2)}{\cancel{10}}}{\underset{23}{\cancel{69}}}$

$\dfrac{18}{23}$ *Ans.*

c) $\dfrac{\frac{1}{2}-\frac{1}{4}}{\frac{3}{8}+\frac{1}{16}}$

L.C.D. = 16

$\dfrac{(\frac{1}{2}-\frac{1}{4})}{(\frac{3}{8}+\frac{1}{16})}\cdot\dfrac{16}{16}$

$\dfrac{8-4}{6+1}$

$\dfrac{4}{7}$ *Ans.*

d) $\dfrac{1-\frac{1}{6}+\frac{2}{3}}{\frac{2}{9}+3-\frac{1}{2}}$

L.C.D. = 18

$\dfrac{(1-\frac{1}{6}+\frac{2}{3})}{(\frac{2}{9}+3-\frac{1}{2})}\cdot\dfrac{18}{18}$

$\dfrac{18-3+12}{4+54-9}$

$\dfrac{27}{49}$ *Ans.*

9.2. Simplifying Complex Fractions

Simplify:

a) $\dfrac{x-\frac{1}{2}}{4}$

L.C.D. = 2

$\dfrac{(x-\frac{1}{2})}{4}\cdot\dfrac{2}{2}$

$\dfrac{2x-1}{8}$ *Ans.*

b) $\dfrac{\frac{x}{3}-\frac{x}{5}}{\frac{1}{2}}$

Division Method

$\dfrac{2x}{15}\div\dfrac{1}{2}$

$\dfrac{2x}{15}\cdot\dfrac{2}{1}=\dfrac{4x}{15}$ *Ans.*

c) $\dfrac{\frac{1}{x}-\frac{1}{x^2}}{\frac{4}{x^3}}$

L.C.D. = x^3

$\dfrac{(\frac{1}{x}+\frac{1}{x^2})}{(\frac{4}{x^3})}\cdot\dfrac{x^3}{x^3}$

$\dfrac{x^2+x}{4}$ *Ans.*

d) $\dfrac{\frac{1}{2x}-\frac{4}{y}}{\frac{1}{x}+\frac{2}{3y}}$

L.C.D. = $6xy$

$\dfrac{(\frac{1}{2x}-\frac{4}{y})}{(\frac{1}{x}+\frac{2}{3y})}\cdot\dfrac{6xy}{6xy}$

$\dfrac{3y-24x}{6y+4x}$ *Ans.*

SUPPLEMENTARY PROBLEMS

1. Express each in fractional form: **(1.1 to 1.3)**

a) $10 \div 19$
b) $n \div (p+3)$
c) 34 divided by x
d) the ratio of 25 to 11
e) the ratio of 7¢ to a dime
f) the quotient of $(r-7)$ and 30
g) 3 of 8 equal portions of a pie
h) 3 out of 8 equal pies
i) the ratio of 1 sec. to 1 hr.

Ans. *a)* $\dfrac{10}{19}$, *b)* $\dfrac{n}{p+3}$, *c)* $\dfrac{34}{x}$, *d)* $\dfrac{25}{11}$, *e)* $\dfrac{7}{10}$, *f)* $\dfrac{r-7}{30}$, *g)* $\dfrac{3}{8}$, *h)* $\dfrac{3}{8}$, *i)* $\dfrac{1}{3600}$

2. State the value of x when **(1.4, 1.5)**

a) $\dfrac{x}{10}=0$ *Ans.* 0
b) $\dfrac{x-3}{5}=0$ *Ans.* 3
c) $\dfrac{2x+5}{7x}=0$ *Ans.* $-2\frac{1}{2}$
d) $\dfrac{3x-12}{3x+12}=0$ *Ans.* 4
e) $\dfrac{3x+12}{3x-12}=0$ *Ans.* -4
f) $\dfrac{1}{3x}$ is meaningless *Ans.* 0
g) $\dfrac{5}{x-10}$ is meaningless *Ans.* 10
h) $\dfrac{x}{5x+20}$ is meaningless *Ans.* -4
i) $\dfrac{2x+1}{2x-1}$ is meaningless *Ans.* $\frac{1}{2}$

3. Change each to equivalent fractions by performing the operations indicated on its numerator and denominator (terms): **(2.1, 2.2)**

a) Multiply terms of $\frac{1}{2}$ by 5.

b) Multiply terms of $\frac{2}{3}$ by $5x$.

c) Multiply terms of $\frac{a}{7}$ by $(a+2)$.

d) Divide terms of $\frac{10}{15}$ by 5.

e) Divide terms of $\frac{3x}{6x^2}$ by $3x$.

f) Divide terms of $\frac{5(a-2)^2}{7(a-2)}$ by $(a-2)$.

Ans. *a*) $\frac{5}{10}$, *b*) $\frac{10x}{15x}$, *c*) $\frac{a(a+2)}{7(a+2)} = \frac{a^2+2a}{7a+14}$, *d*) $\frac{2}{3}$, *e*) $\frac{1}{2x}$, *f*) $\frac{5(a-2)}{7}$

4. Change $\frac{x}{5}$ to equivalent fractions by multiplying its terms by **(2.1)**

a) 6, *b*) y, *c*) x^2, *d*) $3x-2$, *e*) $a+b$.

Ans. *a*) $\frac{6x}{30}$, *b*) $\frac{xy}{5y}$, *c*) $\frac{x^3}{5x^2}$, *d*) $\frac{3x^2-2x}{15x-10}$, *e*) $\frac{ax+bx}{5a+5b}$

5. Change $\frac{12a^2}{36a^3}$ to equivalent fractions by dividing its terms by **(2.2)**

a) 2, *b*) 12, *c*) $4a$, *d*) a^2, *e*) $12a^2$

Ans. *a*) $\frac{6a^2}{18a^3}$, *b*) $\frac{a^2}{3a^3}$, *c*) $\frac{3a}{9a^2}$, *d*) $\frac{12}{36a}$, *e*) $\frac{1}{3a}$

6. Obtain each missing term: **(2.3)**

a) $\frac{3}{7} = \frac{(?)}{28}$ *Ans.* 12

b) $\frac{5}{9} = \frac{20}{(?)}$ *Ans.* 36

c) $\frac{7a}{8} = \frac{(?)}{8a}$ *Ans.* $7a^2$

d) $\frac{10}{x} = \frac{10ab}{(?)}$ *Ans.* abx

e) $\frac{2a}{3} = \frac{(?)}{30bc}$ *Ans.* $20abc$

f) $\frac{x+2}{3} = \frac{(?)}{3(x-2)}$ *Ans.* x^2-4

g) $\frac{7ac}{14cd} = \frac{a}{(?)}$ *Ans.* $2d$

h) $\frac{a^2-b^2}{2a+2b} = \frac{(?)}{2}$ *Ans.* $a-b$

i) $\frac{24(a+2)}{36(a+2)^2} = \frac{(?)}{3(a+2)}$ *Ans.* 2

7. State the reciprocal of each: **(3.1)**

a) $\frac{1}{100}$ *Ans.* 100

b) 100 *Ans.* $\frac{1}{100}$

c) $\frac{1}{3x}$ *Ans.* $3x$

d) $\frac{3}{x}$ *Ans.* $\frac{x}{3}$

e) $\frac{2}{7}$ *Ans.* $\frac{7}{2}$ or $3\frac{1}{2}$

f) $1\frac{1}{3}$ *Ans.* $\frac{3}{4}$

g) $\frac{3x}{2}$ *Ans.* $\frac{2}{3x}$

h) $\frac{2}{3}x$ *Ans.* $\frac{3}{2x}$

i) .3 *Ans.* $\frac{10}{3}$

j) $10\% x$ *Ans.* $\frac{10}{x}$

k) 25% *Ans.* 4 or 400%

l) 100% *Ans.* 1 or 100%

8. Supply the missing entry: **(3.2)**

a) $(\frac{3}{4})(\frac{4}{3}) = (?)$

b) $(\frac{8}{5})(?) = 1$

c) $(\frac{3}{7})(\frac{7}{3})(?)(\frac{2}{5}) = 1$

d) $(\frac{2}{x})(?) = 1$

e) $(\frac{7}{4})(\frac{4}{7}y) = (?)$

f) $(?)(\frac{6}{5}x) = x$

Ans. *a*) 1 *b*) $\frac{5}{8}$ *c*) $\frac{5}{2}$ *d*) $\frac{x}{2}$ *e*) y *f*) $\frac{5}{6}$

9. Show how to change each division to multiplication by using the reciprocal of a number: **(3.3)**

a) $12 \div \frac{6}{7}$

b) $\frac{6}{7} \div 12$

c) $20y \div \frac{y}{x}$

d) $3x^2 \div \frac{x}{y}$

e) $(3a-6b) \div \frac{3}{2}$

f) $(x^3+8x^2-14x) \div \frac{x}{10}$

Ans. *a*) $12 \cdot \frac{7}{6} = 14$, *b*) $\frac{6}{7} \cdot \frac{1}{12} = \frac{1}{14}$, *c*) $20y(\frac{x}{y}) = 20x$, *d*) $3x^2(\frac{y}{x}) = 3xy$,

e) $\frac{2}{3}(3a-6b) = 2a-4b$, *f*) $\frac{10}{x}(x^3+8x^2-14x) = 10x^2+80x-140$

10. Solve for x, indicating each multiplier of both sides: **(3.4)**

a) $\frac{5}{9}x = 35$ *c)* $1\frac{1}{3}x = 20$ *e)* $\frac{bx}{2} = 7b$

b) $\frac{8}{3}x = 20$ *d)* $.3x = \frac{9}{5}$ *f)* $\frac{3x}{c} = 6c$

Ans. *a)* $M_{\frac{9}{5}}$, $x = 63$ *c)* $M_{\frac{3}{4}}$, $x = 15$ *e)* $M_{\frac{2}{b}}$, $x = 14$

b) $M_{\frac{3}{8}}$, $x = 7\frac{1}{2}$ *d)* $M_{\frac{10}{3}}$, $x = 6$ *f)* $M_{\frac{c}{3}}$, $x = 2c^2$

11. Reduce to lowest terms: **(4.1)**

a) $\frac{21}{35}$ *b)* $\frac{42}{24}$ *c)* $\frac{3ab}{9ac}$ *d)* $\frac{15d^2}{5d^3}$ *e)* $\frac{8g - 4h}{12g}$ *f)* $\frac{10rs}{5s^2 + 20s}$

Ans. *a)* $\frac{3}{5}$ *b)* $\frac{7}{4}$ *c)* $\frac{b}{3c}$ *d)* $\frac{3}{d}$ *e)* $\frac{2g - h}{3g}$ *f)* $\frac{2r}{s + 4}$

12. Reduce to lowest terms: **(4.2)**

a) $\frac{5(x + 5)}{11(x + 5)}$ *c)* $\frac{3a - 3b}{6a - 6b}$ *e)* $\frac{(l - 5)^2}{8l - 40}$ *g)* $\frac{km + kn}{n^2 + nm}$

b) $\frac{8ab(2c - 3)}{4ac(2c - 3)}$ *d)* $\frac{a^2 - 9}{7a + 21}$ *f)* $\frac{c^2 - d^2}{(c - d)^2}$ *h)* $\frac{cd + 5d}{5x + cx}$

Ans. *a)* $\frac{5}{11}$, *b)* $\frac{2b}{c}$, *c)* $\frac{1}{2}$, *d)* $\frac{a - 3}{7}$, *e)* $\frac{l - 5}{8}$, *f)* $\frac{c + d}{c - d}$, *g)* $\frac{k}{n}$, *h)* $\frac{d}{x}$

13. Reduce to lowest terms: **(4.3)**

a) $\frac{6 - x}{5x - 30}$ *c)* $\frac{y^2 - 4}{2a - ay}$ *e)* $\frac{1 - 2p}{2p^2 - p}$ *g)* $\frac{ac^2 - ad^2}{d^2 - cd}$

b) $\frac{x^2 - 3x}{21 - 7x}$ *d)* $\frac{2b - 3ab}{9a^2 - 4}$ *f)* $\frac{3w - 3x}{2x^2 - 2w^2}$ *h)* $\frac{(r - s)^2}{s^2 - r^2}$

Ans. *a)* $-\frac{1}{5}$ *c)* $-\frac{y + 2}{a}$ *e)* $-\frac{1}{p}$ *g)* $-\frac{a(c + d)}{d}$

b) $-\frac{x}{7}$ *d)* $-\frac{b}{3a + 2}$ *f)* $-\frac{3}{2(x + w)}$ *h)* $-\frac{r - s}{r + s}$ or $\frac{s - r}{s + r}$

14. Reduce to lowest terms: **(4.4)**

a) $\frac{c^2 + 5c}{c^2 + 12c + 35}$ *c)* $\frac{4a^2 - b^2}{4a^2 - 4ab + b^2}$ *e)* $\frac{2x^2 - 2}{x^2 - 4x - 5}$ *g)* $\frac{2w^2 - 14w + 24}{12w^2 - 32w - 12}$

b) $\frac{b^2 + 6b - 7}{b^2 - 49}$ *d)* $\frac{9 - 3s}{s^2 - 5s + 6}$ *f)* $\frac{y^2 + 8y + 12}{y^2 - 3y - 10}$ *h)* $\frac{2a^2 - 9a - 5}{6a^2 + 7a + 2}$

Ans. *a)* $\frac{c}{c + 7}$ *c)* $\frac{2a + b}{2a - b}$ *e)* $\frac{2(x - 1)}{x - 5}$ *g)* $\frac{w - 4}{2(3w + 1)}$

b) $\frac{b - 1}{b - 7}$ *d)* $-\frac{3}{s - 2}$ *f)* $\frac{y + 6}{y - 5}$ *h)* $\frac{a - 5}{3a + 2}$

15. Multiply: **(5.1, 5.2)**

a) $\frac{3}{4} \times 5$ *c)* $\frac{3}{4} \times \frac{28}{5}$ *e)* $\frac{3}{5} \times \frac{5}{8} \times \frac{8}{3}$

b) $\frac{2}{5} \times \frac{3}{7} \times 11$ *d)* $\frac{20}{6} \times \frac{9}{4} \times \frac{12}{13}$ *f)* $\frac{2}{9} \times \frac{3}{4} \times \frac{3}{8} \times \frac{16}{17}$

Ans. *a)* $\frac{15}{4}$, *b)* $\frac{66}{35}$, *c)* $\frac{21}{5}$, *d)* $\frac{90}{13}$, *e)* 1, *f)* $\frac{1}{17}$

16. Multiply: **(5.1, 5.2)**

a) $\frac{3b}{c} \cdot \frac{b^2}{5}$ e) $\frac{a+b}{7} \cdot \frac{a-b}{2}$ i) $\frac{15b+15c}{16x^2-9} \cdot \frac{8x+6}{5b+5c}$

b) $7ab \cdot \frac{a}{2} \cdot \frac{3}{c}$ f) $\frac{a+b}{5} \cdot \frac{15}{11a+11b}$ j) $\frac{(v+w)^2}{bv+bw} \cdot \frac{b^2}{3v+3w}$

c) $\frac{x^2}{y} \cdot \frac{3}{y^2} \cdot x^3$ g) $\frac{x^2}{y} \cdot \frac{y^2}{3} \cdot \frac{1}{x^3}$ k) $\frac{r^4-s^4}{5r+5s} \cdot \frac{25}{2r^2+2s^2}$

d) $\frac{m+n}{3} \cdot \frac{m+n}{mn}$ h) $\frac{6r+12s}{2rs-4r} \cdot \frac{7s-14}{3r+6s}$ l) $\frac{5ax+10a}{ab-10b} \cdot \frac{200-2a^2}{ax+2a}$

Ans. a) $\frac{3b^3}{5c}$ c) $\frac{3x^5}{y^3}$ e) $\frac{a^2-b^2}{14}$ g) $\frac{y}{3x}$ i) $\frac{6}{4x-3}$ k) $\frac{5(r-s)}{2}$

b) $\frac{21a^2b}{2c}$ d) $\frac{(m+n)^2}{3mn}$ f) $\frac{3}{11}$ h) $\frac{7}{r}$ j) $\frac{b}{3}$ l) $-\frac{10(10+a)}{b}$

17. Multiply: **(5.3)**

a) $\frac{a^2-4}{(a-2)^2} \cdot \frac{a^2-9a+14}{a^3+2a^2}$ c) $\frac{2c+b}{b^2} \cdot \frac{3b^2-3bc}{bc+2c^2} \cdot \frac{5bc-10c^2}{b^2-3bc+2c^2}$

b) $\frac{(x+3)^2}{x^2-7x-30} \cdot \frac{7x^2-700}{x^3+3x^2}$ d) $\frac{r^2-s^2}{3-r} \cdot \frac{r^2+3rs}{4s-4r} \cdot \frac{rs-3s}{r^2+4rs+3s^2}$

Ans. a) $\frac{a-7}{a^2}$ b) $\frac{7(x+10)}{x^2}$ c) $\frac{15}{b}$ d) $\frac{rs}{4}$

18. Divide: **(6.1)**

a) $10 \div \frac{2}{3}$ c) $3x \div \frac{12x}{7}$ e) $\frac{ab}{4} \div \frac{b^2}{6}$ g) $\frac{r^2s}{16t} \div \frac{rs^2}{8t^2}$

b) $\frac{3}{4} \div 36$ d) $\frac{5}{y} \div \frac{7}{y}$ f) $\frac{x}{2} \div \frac{ax}{6}$ h) $\frac{2x}{y} \div \frac{10y}{3x}$

Ans. a) 15 b) $\frac{1}{48}$ c) $\frac{7}{4}$ d) $\frac{5}{7}$ e) $\frac{3a}{2b}$ f) $\frac{3}{a}$ g) $\frac{rt}{2s}$ h) $\frac{3x^2}{5y^2}$

19. Divide: **(6.2)**

a) $\frac{3x}{x+2} \div \frac{5x}{x+2}$ e) $\frac{d^2-1}{3d+9} \div \frac{d^2h-h}{dh+3h}$ i) $\frac{a^3-64a}{2a^2+16a} \div \frac{a^2-9a+8}{a^2+4a-5}$

b) $\frac{2(y+3)}{7} \div \frac{8(y+3)}{63}$ f) $\frac{c^2-cd}{3c^2+3cd} \div \frac{7c-7d}{c^2-d^2}$ j) $\frac{a^3-ab^2}{6a+6b} \div \frac{a^2-4ab+3b^2}{12a-36b}$

c) $\frac{2a-10}{3b-21} \div \frac{a-5}{4b-28}$ g) $\frac{2x^2+2y^2}{x^2-4y^2} \div \frac{x^2+y^2}{x+2y}$ k) $\frac{3p+p^2}{12+3p} \div \frac{9p-p^3}{12-p-p^2}$

d) $\frac{3(a^2+5)}{b(a^2-3)} \div \frac{18(a^2+5)}{7(a^2-3)}$ h) $\frac{s^2+st}{rs-st} \div \frac{s^2-t^2}{r^2-t^2}$ l) $\frac{c^2+16c+64}{2c^2-128} \div \frac{3c^2+30c+48}{c^2-6c-16}$

Ans. a) $\frac{3}{5}$ c) $\frac{8}{3}$ e) $\frac{1}{3}$ g) $\frac{2}{x-2y}$ i) $\frac{a+5}{2}$ k) $\frac{1}{3}$

b) $\frac{9}{4}$ d) $\frac{7}{6b}$ f) $\frac{c-d}{21}$ h) $\frac{r+t}{s-t}$ j) $2a$ l) $\frac{1}{6}$

20. Combine: **(7.1)**

a) $\frac{4}{9} + \frac{2}{9}$ d) $\frac{3}{5} + \frac{4}{5} - \frac{1}{5}$ g) $3\frac{5}{12} + 2\frac{3}{12}$

b) $\frac{10}{13} - \frac{7}{13}$ e) $\frac{6}{15} - \frac{2}{15} - \frac{2}{15}$ h) $10\frac{2}{7} - 8\frac{1}{7}$

c) $\frac{3}{11} - \frac{8}{11}$ f) $\frac{3}{8} - \frac{1}{8} + \frac{7}{8}$ i) $\frac{5}{16} + \frac{15}{16} - \frac{7}{16} + \frac{3}{16}$

Ans. a) $\frac{6}{9} = \frac{2}{3}$ b) $\frac{3}{13}$ c) $-\frac{5}{11}$ d) $\frac{6}{5}$ e) $\frac{2}{15}$ f) $\frac{9}{8}$ g) $5\frac{2}{3}$ h) $2\frac{1}{7}$ i) 1

21. Combine: **(7.1)**

a) $\frac{x}{3} + \frac{5x}{3}$ *c)* $\frac{x^2}{6} + \frac{x^2}{6}$ *e)* $\frac{17}{3s} - \frac{14}{3s}$ *g)* $\frac{5x}{y} + \frac{x}{y} - \frac{3}{y}$

b) $\frac{7y}{5} - \frac{3y}{5}$ *d)* $\frac{10}{r} + \frac{3}{r}$ *f)* $\frac{5a}{b^2} - \frac{3a}{b^2}$ *h)* $\frac{4p}{3q} - \frac{p}{3q} + \frac{8}{3q} - \frac{2}{3q}$

Ans. *a)* $\frac{6x}{3} = 2x$ *b)* $\frac{4y}{5}$ *c)* $\frac{x^2}{3}$ *d)* $\frac{13}{r}$ *e)* $\frac{1}{s}$ *f)* $\frac{2a}{b^2}$ *g)* $\frac{6x-3}{y}$ *h)* $\frac{p+2}{q}$

22. Combine: **(7.1)**

a) $\frac{a+2}{8} + \frac{7a-2}{8}$ *c)* $\frac{2y-7}{5} - \frac{2y-10}{5}$ *e)* $\frac{b+8}{5a} + \frac{4b+7}{5a}$

b) $\frac{5x}{7} - \frac{5x-35}{7}$ *d)* $\frac{8}{x} + \frac{2}{x} - \frac{3x+10}{x}$ *f)* $\frac{x+3}{2x^2} + \frac{3x-3}{2x^2}$

Ans. *a)* $\frac{8a}{8} = a$ *b)* $\frac{35}{7} = 5$ *c)* $\frac{3}{5}$ *d)* $-\frac{3x}{x} = -3$ *e)* $\frac{5b+15}{5a} = \frac{b+3}{a}$ *f)* $\frac{4x}{2x^2} = \frac{2}{x}$

23. Combine: **(7.2)**

a) $\frac{5}{x+3} + \frac{x-2}{x+3}$ *d)* $\frac{2}{c+3} - \frac{c+5}{c+3}$ *g)* $\frac{3a}{2(a+b)} + \frac{a+4b}{2(a+b)}$

b) $\frac{2x}{x+y} + \frac{2y}{x+y}$ *e)* $\frac{7}{d^2+7} + \frac{d^2}{d^2+7}$ *h)* $\frac{7a-2}{a(b+c)} - \frac{a-2}{a(b+c)}$

c) $\frac{5p}{p-q} - \frac{5q}{p-q}$ *f)* $\frac{6}{2x-3} - \frac{4x}{2x-3}$ *i)* $\frac{3x}{x^2-4} - \frac{x+4}{x^2-4}$

Ans. *a)* $\frac{x+3}{x+3} = 1$ *c)* $\frac{5p-5q}{p-q} = 5$ *e)* 1 *g)* $\frac{4a+4b}{2(a+b)} = 2$ *i)* $\frac{2x-4}{x^2-4} = \frac{2}{x+2}$

b) $\frac{2x+2y}{x+y} = 2$ *d)* $\frac{-c-3}{c+3} = -1$ *f)* $\frac{6-4x}{2x-3} = -2$ *h)* $\frac{6a}{a(b+c)} = \frac{6}{b+c}$

24. Combine: **(8.1)**

a) $\frac{2}{3} + \frac{1}{2}$ *b)* $\frac{3}{4} - \frac{1}{5}$ *c)* $\frac{a}{7} - \frac{b}{2}$ *d)* $\frac{4}{3x} + \frac{5}{2y}$ *e)* $\frac{a}{b} + \frac{b}{a}$

Ans. *a)* $\frac{7}{6}$ *b)* $\frac{11}{20}$ *c)* $\frac{2a-7b}{14}$ *d)* $\frac{8y+15x}{6xy}$ *e)* $\frac{a^2+b^2}{ab}$

25. Combine: **(8.1)**

a) $\frac{4}{5} + \frac{2}{x+3}$ *c)* $\frac{8c}{3c+1} - \frac{1}{2}$ *e)* $\frac{x}{2-x} - \frac{x}{2+x}$

b) $\frac{3}{a} + \frac{2}{5-a}$ *d)* $\frac{2}{y+1} + \frac{3}{y-1}$ *f)* $\frac{4}{a+5} + \frac{2}{a+1}$

Ans. *a)* $\frac{4x+22}{5(x+3)}$ *b)* $\frac{15-a}{a(5-a)}$ *c)* $\frac{13c-1}{2(3c+1)}$ *d)* $\frac{5y+1}{y^2-1}$ *e)* $\frac{2x^2}{4-x^2}$ *f)* $\frac{6a+14}{(a+5)(a+1)}$

26. Combine: **(8.2)**

a) $\frac{1}{3} + \frac{3}{4} + \frac{1}{6}$ *c)* $\frac{11}{10} - \frac{2}{5} + \frac{1}{2}$ *e)* $\frac{5x}{8} - \frac{x}{12}$ *g)* $\frac{p}{8} + \frac{q}{3} - \frac{r}{6}$

b) $\frac{7}{8} - \frac{1}{2} - \frac{1}{4}$ *d)* $\frac{a}{3} - \frac{a}{12}$ *f)* $\frac{c}{6} + \frac{2c}{9}$ *h)* $\frac{4}{3x} + \frac{3}{2x} + \frac{1}{6x}$

Ans. *a)* $\frac{5}{4}$ *b)* $\frac{1}{8}$ *c)* $\frac{6}{5}$ *d)* $\frac{a}{4}$ *e)* $\frac{13x}{24}$ *f)* $\frac{7c}{18}$ *g)* $\frac{3p+8q-4r}{24}$ *h)* $\frac{3}{x}$

27. Combine: (8.3)

a) $\frac{3}{a} + \frac{4}{a^2}$ *d*) $\frac{1}{3} + \frac{2}{x} + \frac{1}{x^2}$ *g*) $\frac{8}{r^2s} + \frac{4}{rs^2}$

b) $\frac{5}{b^2} - \frac{2}{b}$ *e*) $\frac{2}{x} - \frac{5}{x^2} + \frac{3}{x^3}$ *h*) $\frac{10}{a^2b^2} - \frac{4}{ab^2} + \frac{5}{a^2b}$

c) $\frac{7}{2b} + \frac{2}{3b^2}$ *f*) $\frac{1}{2x^3} + \frac{3}{x^2} - \frac{7}{5x}$ *i*) $\frac{1}{a^2b^2} + \frac{3}{a^2} + \frac{2}{b^2}$

Ans. *a*) $\frac{3a+4}{a^2}$ *b*) $\frac{5-2b}{b^2}$ *c*) $\frac{21b+4}{6b^2}$ *d*) $\frac{x^2+6x+3}{3x^2}$ *e*) $\frac{2x^2-5x+3}{x^3}$ *f*) $\frac{5+30x-14x^2}{10x^3}$

g) $\frac{8s+4r}{r^2s^2}$ *h*) $\frac{10-4a+5b}{a^2b^2}$ *i*) $\frac{1+3b^2+2a^2}{a^2b^2}$

28. Combine: (8.4)

a) $\frac{7}{2} - \frac{8}{h+3}$ *d*) $\frac{2}{p+2} - \frac{2}{p+3}$ *g*) $\frac{4x}{x^2-36} - \frac{4}{x+6}$

b) $\frac{4a}{3b+6} - \frac{a}{b+2}$ *e*) $\frac{3a}{a+2} + \frac{a}{a-2}$ *h*) $\frac{8}{7-y} - \frac{8y}{49-y^2}$

c) $\frac{10}{x^2+x} + \frac{2}{3x^2+3x}$ *f*) $\frac{2}{3x+3y} - \frac{3}{5x+5y}$ *i*) $\frac{3t^2}{9t^2-16s^2} - \frac{t}{3t+4s}$

Ans. *a*) $\frac{7h+5}{2(h+3)}$ *b*) $\frac{a}{3b+6}$ *c*) $\frac{32}{3(x^2+x)}$ *d*) $\frac{2}{(p+2)(p+3)}$ *e*) $\frac{4a^2-4a}{a^2-4}$ *f*) $\frac{1}{15(x+y)}$

g) $\frac{24}{x^2-36}$ *h*) $\frac{56}{49-y^2}$ *i*) $\frac{4st}{9t^2-16s^2}$

29. Simplify: (9.1)

a) $\frac{5+\frac{2}{3}}{2}$ *c*) $\frac{\frac{1}{4}+\frac{1}{3}}{2}$ *e*) $\frac{\frac{3}{14}-\frac{2}{7}}{5}$ *g*) $\frac{2+\frac{1}{5}+\frac{3}{4}}{10}$

b) $\frac{4+\frac{1}{3}}{4-\frac{1}{3}}$ *d*) $\frac{5}{\frac{1}{2}-\frac{1}{4}}$ *f*) $\frac{\frac{1}{8}+\frac{5}{16}}{\frac{7}{12}}$ *h*) $\frac{\frac{4}{5}-\frac{1}{6}}{\frac{4}{5}+\frac{1}{3}}$

Ans. *a*) $\frac{17}{6}$ *b*) $\frac{13}{11}$ *c*) $\frac{7}{24}$ *d*) 20 *e*) $-\frac{1}{70}$ *f*) $\frac{3}{4}$ *g*) $\frac{59}{200}$ *h*) $\frac{19}{34}$

30. Simplify: (9.2)

a) $\frac{y+\frac{1}{3}}{2}$ *c*) $\frac{5p+2}{\frac{2}{3}}$ *e*) $\frac{5h}{\frac{1}{2}-\frac{1}{5}}$ *g*) $\frac{\frac{3}{x}+\frac{1}{6x}}{\frac{7}{3x}}$

b) $\frac{2a-1}{\frac{1}{2}}$ *d*) $\frac{\frac{x}{4}-\frac{2x}{9}}{3}$ *f*) $\frac{\frac{r}{4}}{\frac{7}{8}-\frac{r}{2}}$ *h*) $\frac{\frac{a}{x}+\frac{x}{a}}{\frac{a}{x}-\frac{x}{a}}$

Ans. *a*) $\frac{3y+1}{6}$ *b*) $4a-2$ *c*) $\frac{15p+6}{2}$ *d*) $\frac{x}{108}$ *e*) $\frac{50h}{3}$ *f*) $\frac{2r}{7-4r}$ *g*) $\frac{19}{14}$ *h*) $\frac{a^2+x^2}{a^2-x^2}$

Chapter 12

ROOTS and RADICALS

1. UNDERSTANDING ROOTS AND RADICALS

The square root of a number is one of its two equal factors.

Thus, +5 is a square root of 25 since $(+5)(+5) = 25$.

Also, −5 is another square root of 25 since $(-5)(-5) = 25$.

Rule. A positive number has two square roots which are **opposites** of each other (same absolute value but unlike signs).

Thus, +10 and −10 are the two square roots of 100.

To indicate both square roots, the symbol ±, which combines + and −, may be used.

Thus, the square roots of 49, +7 and −7, may be written together as ±7.

Read "±7" as "plus or minus 7".

The principal square root of a number is its positive square root.

Thus, the principal square root of 36 is +6.

The symbol, $\sqrt{\ }$, indicates the principal square root of a number. This symbol is a modified form of the letter, *r*, the initial of the Latin word *radix*, meaning root.

Thus, $\sqrt{9}$ = principal square root of 9 = +3.

To indicate the negative square root of a number, place the negative sign before $\sqrt{\ }$.

Thus, $-\sqrt{16} = -4$.

Note. Unless otherwise stated, whenever a square root of a number is to be found it is to be understood that the principal or positive square root is required.

Radical, Radical Sign, Radicand, Index

(1) **A radical** is an indicated root of a number or expression.

Thus, $\sqrt{5}$, $\sqrt[3]{8x}$ and $\sqrt[4]{7x^3}$ are radicals.

(2) The symbols $\sqrt{\ }$, $\sqrt[3]{\ }$ and $\sqrt[4]{\ }$ are **radical signs**.

(3) **The radicand** is the number or expression under the radical sign.

Thus, 80 is the radicand of $\sqrt[3]{80}$ or $\sqrt[4]{80}$.

(4) **The index of a root** is the small number written above and to the left of the radical sign, $\sqrt{\ }$. The index indicates which root is to be taken. In square roots, the index, 2, is not indicated but understood.

Thus, $\sqrt[3]{8}$ indicates the 3rd root or cube root of 8.

The cube root of a number is one of its three equal factors.

Thus, 3 is a cube root of 27 since $(3)(3)(3) = 27$.

This is written, $3 = \sqrt[3]{27}$ and read, "3 is a cube root of 27".

Also, −3 is a cube root of −27 since $(-3)(-3)(-3) = -27$.

Square and Cube Roots of Negative Numbers: Imaginary Numbers

The cube root of a negative number is a negative number. For example, $\sqrt[3]{-8} = -2$. However, the square root of a negative number is not a negative number. For example, the square root of -16 is **not** -4. Numbers such as $\sqrt{-16}$ or $\sqrt{-100}$ are called **imaginary numbers.** This name was given to them before it was realized their great importance in science, engineering and higher mathematics.

In general, the *n*th root of a number is one of its *n* equal factors.

Thus, 2 is the fifth root of 32 since it is one of the five equal factors of 32.

1.1. Opposite Square Roots of a Positive Number

State the square roots of

a) 25 *b)* $25x^2$ *c)* x^4 *d)* $81a^2b^2$ *e)* $\frac{49}{64}$ *f)* $\frac{9a^2}{16b^2}$

Ans. a) ± 5 *b)* $\pm 5x$ *c)* $\pm x^2$ *d)* $\pm 9ab$ *e)* $\pm\frac{7}{8}$ *f)* $\pm\frac{3a}{4b}$

1.2. Principal Square Root of a Number

Find each principal (positive) square root:

a) $\sqrt{36}$ *b)* $\sqrt{.09}$ *c)* $\sqrt{1.44x^2}$ *d)* $\sqrt{\frac{25}{4}}$ *e)* $\sqrt{\frac{x^2}{400}}$ *f)* $\sqrt{\frac{b^6}{c^8}}$

Ans. a) 6 *b)* .3 *c)* $1.2x$ *d)* $\frac{5}{2}$ *e)* $\frac{x}{20}$ *f)* $\frac{b^3}{c^4}$

1.3. Negative Square Root of a Number

Find each negative square root:

a) $-\sqrt{100}$ *b)* $-\sqrt{.16}$ *c)* $-\sqrt{400c^2}$ *d)* $-\sqrt{\frac{9}{4}}$ *e)* $-\sqrt{\frac{81}{r^2}}$ *f)* $-\sqrt{\frac{p^8}{q^{12}}}$

Ans. a) -10 *b)* $-.4$ *c)* $-20c$ *d)* $-\frac{3}{2}$ *e)* $-\frac{9}{r}$ *f)* $-\frac{p^4}{q^6}$

1.4. Evaluating Numerical Square Root Expressions

Find the value of

a) $3\sqrt{4} + \sqrt{25} - 2\sqrt{9}$ *b)* $\sqrt{\frac{25}{9}} + \sqrt{\frac{64}{9}}$ *c)* $4\sqrt{\frac{81}{4}}$ *d)* $\frac{1}{3}\sqrt{36} + \frac{2}{5}\sqrt{100}$

Solutions:

a) $3(2) + 5 - 2(3)$ *b)* $\frac{5}{3} + \frac{8}{3}$ *c)* $4(\frac{9}{2})$ *d)* $\frac{1}{3}(6) + \frac{2}{5}(10)$

Ans. 5 *Ans.* $\frac{13}{3}$ *Ans.* 18 *Ans.* 6

2. UNDERSTANDING RATIONAL AND IRRATIONAL NUMBERS

Rational Numbers

A rational number is one that can be expressed as the quotient or **ratio** of two integers. (*Note the word* **ratio** *in* **rational**.)

Thus, $\frac{3}{5}$ and $-\frac{3}{5}$ are rational numbers.

Kinds of Rational Numbers

1. **All integers**, that is, all positive and negative whole numbers and zero.
   Thus, 5 is rational since 5 can be expressed as $\frac{5}{1}$.

2. **Fractions** whose numerator and denominator are integers, after simplification.
   Thus, $\frac{1.5}{2}$ is rational because it equals $\frac{3}{4}$ when simplified. However, $\frac{\sqrt{2}}{3}$ is not rational.

3. **Decimals** which have a limited number of decimal places.
   Thus, 3.14 is rational since it can be expressed as $\frac{314}{100}$.

4. **Decimals** which have an unlimited number of decimal places and the digits continue to repeat themselves. Thus, .6666... is rational since it can be expressed as $\frac{2}{3}$.
   Note. ... is a symbol meaning "continued without end".

5. **Square root expressions** whose radicand is a perfect square, such as $\sqrt{25}$; **cube root expressions** whose radicand is a perfect cube, such as $\sqrt[3]{27}$; etc.

Irrational Numbers

An irrational number is one that cannot be expressed as the ratio of two integers. Thus,

(1) $\sqrt{2}$ is an irrational number since it cannot equal a fraction whose terms are integers. However, we can approximate $\sqrt{2}$ to any number of decimal places. For example, $\sqrt{2} = 1.4142$ to the nearest ten-thousandth. Such approximations are rational numbers.

(2) π is also irrational. An approximation of π has been carried out to over 10,000 places.

(3) Other examples of irrationals are $\frac{\sqrt{5}}{2}$, $\frac{2}{\sqrt{5}}$ and $2+\sqrt{7}$.

Surds

A surd is either an indicated root of a number which can only be approximated, or it is a polynomial involving one or more such roots. A surd is irrational but not imaginary.

Thus, $\sqrt{5}$, $2+\sqrt{5}$ and $10+\sqrt[3]{7}-2\sqrt{3}$ are surds.

However, $\sqrt{-4}$ is not a surd since it is imaginary.

2.1. Classifying Rational and Irrational Numbers

Classify each number in the table. If rational, express it as the ratio of two integers:

	NUMBER												
	-3	0	$\sqrt{100}$	20%	.333...	.333	$\sqrt{.09}$	$\frac{\sqrt{25}}{12}$	$\sqrt{7}$	$\frac{2}{3+\sqrt{4}}$	$-\sqrt{\frac{32}{2}}$	$-\frac{\sqrt{32}}{2}$	$\sqrt[3]{25}$
Positive Integer			√										
Negative Integer	√										√		
Rational Number	√	√	√	√	√	√	√	√		√	√		
Expressed as Ratio of Two Integers	$\frac{-3}{1}$	$\frac{0}{1}$	$\frac{10}{1}$	$\frac{1}{5}$	$\frac{1}{3}$	$\frac{333}{1000}$	$\frac{3}{10}$	$\frac{5}{12}$		$\frac{2}{5}$	$\frac{-4}{1}$		
Irrational Number									√			√	√

2.2. Expressing Rational Numbers as the Ratio of Two Integers

Express each of the following rational numbers as the ratio of two integers:

a) 10	*Ans.* $\frac{10}{1}$	*e*) $\frac{\sqrt{100}}{7}$	*Ans.* $\frac{10}{7}$	*i*) $\frac{\sqrt{16}}{\sqrt{121}}$	*Ans.* $\frac{4}{11}$
b) -7	*Ans.* $\frac{-7}{1}$	*f*) $\frac{8}{\sqrt{49}}$	*Ans.* $\frac{8}{7}$	*j*) $\sqrt{\frac{18}{8}}$	*Ans.* $\frac{3}{2}$
c) 6.3	*Ans.* $\frac{63}{10}$	*g*) $3\sqrt{25}$	*Ans.* $\frac{15}{1}$	*k*) $\sqrt{6\frac{1}{4}}$	*Ans.* $\frac{5}{2}$
d) $-5\frac{1}{2}$	*Ans.* $\frac{-11}{2}$	*h*) $\frac{1}{3}\sqrt{.81}$	*Ans.* $\frac{3}{10}$	*l*) $\sqrt{.0001}$	*Ans.* $\frac{1}{100}$

3. FINDING THE SQUARE ROOT OF A NUMBER BY USING A GRAPH

Approximate square roots of numbers can be obtained by using a graph of $x = \sqrt{y}$.

Table of values used to graph $x = \sqrt{y}$ from $x = 0$ to $x = 8$:

If x =	0	1	$1\frac{1}{2}$	2	$2\frac{1}{2}$	3	$3\frac{1}{2}$	4	$4\frac{1}{2}$	5	$5\frac{1}{2}$	6	$6\frac{1}{2}$	7	$7\frac{1}{2}$	8
then y =	0	1	$2\frac{1}{4}$	4	$6\frac{1}{4}$	9	$12\frac{1}{4}$	16	$20\frac{1}{4}$	25	$30\frac{1}{4}$	36	$42\frac{1}{4}$	49	$56\frac{1}{4}$	64

In the table, each x-value is the principal square root of the corresponding y-value; that is, $x = \sqrt{y}$. On the graph, the x-value of each point is the principal square root of its y-value.

To Find the Square Root of a Number Graphically

Procedure:

1. Find the number on the y-axis:
2. From the number proceed horizontally to the curve:
3. From the curve proceed vertically to the x-axis:
4. Read the approximate square root value on the x-axis:

Find $\sqrt{27}$, graphically.

Solution:

1. Find 27 on the y-axis.
2. From 27 follow the horizontal line to point A on the curve.
3. From A follow the vertical line to the x-axis.
4. Read 5.2 on the x-axis.

Ans. $\sqrt{27} = 5.2$, approximately.

To find the square of a number graphically, proceed in the reverse direction; from the x-axis to the curve, then to the y-axis.

Thus, to square 7.2, follow this path: $x = 7.2$ to curve, then to $y = 52$.

Ans. 7.2 = 52, approximately.

GRAPH of $x = \sqrt{y}$

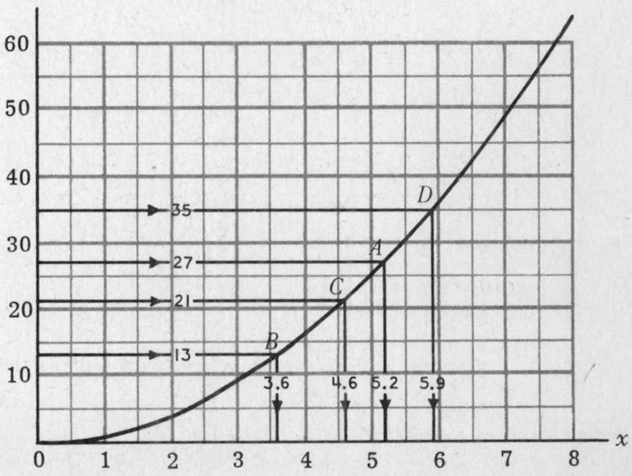

3.1. Finding Square Roots Graphically

From the graph of $x = \sqrt{y}$, find the square root of each number to the nearest tenth, indicating the path followed:

	Number	Sq. Rt.	Path
a)	13	3.6	($y = 13$ to B to $x = 3.6$)
b)	21	4.6	($y = 21$ to C to $x = 4.6$)
c)	35	5.9	($y = 35$ to D to $x = 5.9$)

4. FINDING THE SQUARE ROOT OF A NUMBER BY USING A TABLE

Approximate square roots of numbers can be obtained using the table on page 290. The square root values obtained from such a table are more precise than those from the graph of $x = \sqrt{y}$. The table saves time for the scientist, engineer, machinist or mathematician.

To find the principal square root of a number, look for the number under N. Read the square root of the number under $\sqrt{N}$, immediately to the right of the number.

Thus, $\sqrt{5} = 2.236$, $\sqrt{87} = 9.327$, $\sqrt{200} = 14.14$.

Multiplying or Dividing the Radicand in a Square Root by 100

Rule 1. If the radicand of a square root is multiplied by 100, the square root is multiplied by 10.

Rule 2. If the radicand of a square root is divided by 100, the square root is divided by 10.

Thus, applying **Rule 1**:

$$\sqrt{9} = 3$$
$$\sqrt{9\,\underline{00}} = 30$$
$$\sqrt{9\,\underline{00}\,\underline{00}} = 300$$
$$\sqrt{9\,\underline{00}\,\underline{00}\,\underline{00}} = 3000$$

Thus, applying **Rule 2**:

$$\sqrt{9} = 3$$
$$\sqrt{.\underline{09}} = .3$$
$$\sqrt{.\underline{00}\,\underline{09}} = .03$$
$$\sqrt{.\underline{00}\,\underline{00}\,\underline{09}} = .003$$

Note. Each underlining above indicates a multiplication or division of the radicand by 100. This process of underlining pairs of digits is used later in the process of computing square roots.

4.1. Finding Square Roots, Using a Table

Find the approximate value of each square root, using the table of square roots:

a) $\sqrt{5}$ *Ans.* 2.236
b) $\sqrt{50}$ *Ans.* 7.071
c) $\sqrt{500}$ *Ans.* 22.36
d) $\sqrt{262}$ *Ans.* 16.19
e) $\sqrt{562}$ *Ans.* 23.71
f) $\sqrt{862}$ *Ans.* 29.36

4.2. Evaluating, Using a Square Root Table

Find the value of each to the nearest hundredth, using the table of square roots:

a) $3\sqrt{2}$

b) $2\sqrt{300} - 25$

c) $\dfrac{\sqrt{45}}{3} + \dfrac{1}{\sqrt{2}}$

Solutions:

a) Since $\sqrt{2} = 1.414$,

$3\sqrt{2} = 3(1.414)$
$= 4.242$

Ans. 4.24

b) Since $\sqrt{300} = 17.32$,

$2\sqrt{300} - 25 = 2(17.32) - 25$
$= 9.64$

Ans. 9.64

c) Since $\sqrt{45} = 6.708$ and $\sqrt{2} = 1.414$,

$\dfrac{\sqrt{45}}{3} + \dfrac{1}{\sqrt{2}} = \dfrac{6.708}{3} + \dfrac{1}{1.414}$
$= 2.236 + .707$
$= 2.943$ *Ans.* 2.94

4.3. Rule 1. Multiplying the Radicand of a Square Root by a Multiple of 100

Using the table of square roots and applying **Rule 1**, find the approximate value of each square root:

a) $\sqrt{50000}$ and $\sqrt{5000000}$ *b)* $\sqrt{4000}$ and $\sqrt{400000}$ *c)* $\sqrt{53400}$ and $\sqrt{5340000}$

Solutions:

Since $\sqrt{5} = 2.236$,	Since $\sqrt{40} = 6.325$,	Since $\sqrt{534} = 23.11$,
$\sqrt{5\,00\,00} = 223.6$ *Ans.*	$\sqrt{40\,00} = 63.25$ *Ans.*	$\sqrt{534\,00} = 231.1$ *Ans.*
$\sqrt{5\,00\,00\,00} = 2236.$ *Ans.*	$\sqrt{40\,00\,00} = 632.5$ *Ans.*	$\sqrt{534\,00\,00} = 2311.$ *Ans.*

4.4. Rule 2. Dividing the Radicand of a Square Root by a Multiple of 100

Using the table of square roots and applying **Rule 2**, find the approximate value of each square root:

a) $\sqrt{2.55}$ and $\sqrt{.0255}$ *b)* $\sqrt{.37}$ and $\sqrt{.0037}$ *c)* $\sqrt{.05}$ and $\sqrt{.0005}$

Solutions:

Since $\sqrt{255} = 15.97$,	Since $\sqrt{37} = 6.083$,	Since $\sqrt{5} = 2.236$,
$\sqrt{2\,.55} = 1.597$ *Ans.*	$\sqrt{.37} = .6083$ *Ans.*	$\sqrt{.05} = .2236$ *Ans.*
$\sqrt{.02\,55} = .1597$ *Ans.*	$\sqrt{.00\,37} = .06083$ *Ans.*	$\sqrt{.00\,05} = .02236$ *Ans.*

5. COMPUTING THE SQUARE ROOT OF A NUMBER

Step 1	Step 2	Step 3
Square and Square Root	**Doubling and Dividing**	**Duplicating and Multiplying**

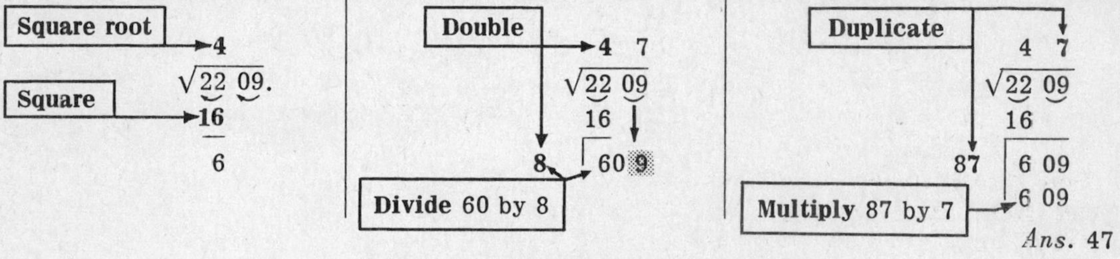

Ans. 47

To Compute the Square Root of a Number

To compute the square root of a number, such as $\sqrt{2209}$, the following major steps and their substeps are needed.

Step 1 (Square and Square Root Step)
   (1) Pair the digits of the radicand 2209, counting from the decimal point.
   (2) Under the first pair 22, place 16 which is the largest perfect **square** less than 22.
   (3) On top, place 4, the **square root** of 16.
   (4) Subtract to obtain the remainder 6.

Step 2 (Doubling and Dividing Step)
   (1) Bring down 09, the next pair of digits.
   (2) **Double** 4, the number on top, and place its double, 8, on the outside as shown.
   (3) Cover up 9, the last digit of 609.
   (4) **Divide** 8 into 60 to obtain 7. Disregard the remainder.
   (5) Make 7 the next digit on top.

Step 3 (Duplicating and Multiplying Step)

(1) **Duplicate** the next digit 7 and place it after 8 as shown.

(2) **Multiply** 87 by 7 (as in long division) to obtain 609.

(3) Subtract to obtain a remainder of zero, showing that 47 is an exact square root. $\sqrt{2209} = 47$ *Ans.*

Step 4 (Repeating Operation Step)

If there is a remainder, **repeat Steps 2 and 3** until the desired number of decimal places are found.

Note. Decimal point of answer should be written directly above decimal point of radicand.

5.1. Computing Exact Square Roots Showing Separate Steps

Find $\sqrt{3806.89}$.

Step 1 Square and Square Root	Step 2 Double and Divide	Step 3 Duplicate and Multiply	Step 4 Repeat Steps 2 and 3
6 $\sqrt{38\ 06.\ 89}$ 36 2	6 1. $\sqrt{38\ 06.\ 89}$ 36 12	20 6 (12 into 20 yields 1)	6 1. $\sqrt{38\ 06.\ 89}$ 36 121	2 06 1 21 ← (121×1) 85	6 1. 7 $\sqrt{38\ 06.\ 89}$ 36 121	2 06 1 21 1227	85 89 85 89 *Ans.* 61.7

5.2. Approximating Square Roots, Showing Separate Steps

Find $\sqrt{12.1}$, to nearest tenth.

Step 1 Square and Square Root	Step 2 Double and Divide	Step 3 Duplicate and Multiply	Step 4 Repeat Steps 2 and 3
3. $\sqrt{12.\ 10\ 00}$ 9 3 **Note.** Extra zeros are needed to obtain nearest tenth.	3. 4 $\sqrt{12.\ 10\ 00}$ 9 6	31 0 **Note.** 6 into 31 yields 5 but this is too large.	3. 4 $\sqrt{12.\ 10\ 00}$ 9 64	3 10 2 56 ← (64×4) 54	3. 4 7 $\sqrt{12.\ 10\ 00}$ 9 64	3 10 2 56 687	54 00 48 09 5 91 *Ans.* 3.5 (*to nearest tenth*)

5.3. Computing Exact Square Roots, Showing Entire Process

Find each square root:

a) $\sqrt{961}$	*b*) $\sqrt{.4225}$	*c*) $\sqrt{219.04}$
Solutions:		
3 1 $\sqrt{9\ 61}$ 9 61	61 61 *Ans.* 31	. 6 5 $\sqrt{.42\ 25}$ 36 125	6 25 6 25 *Ans.* .65	1 4. 8 $\sqrt{2\ 19.\ 04}$ 1 24	1 19 96 288	23 04 23 04 *Ans.* 14.8

5.4. Approximating Square Roots, Showing Entire Process

a) Find $\sqrt{17}$, to nearest tenth.

Solutions:

$$\begin{array}{rl} & 4.\ \ 1\ \ 2 \\ & \sqrt{17.\ 00\ 00} \\ & 16 \\ 81 & 1\ 00 \\ & 81 \\ 822 & 19\ 00 \\ & 16\ 44 \\ & 2\ 56 \end{array}$$

Ans. 4.1

b) Find $\sqrt{.083}$, to nearest hundredth.

$$\begin{array}{rl} & .\ 2\ \ 8\ \ 8 \\ & \sqrt{.08\ 30\ 00} \\ & 4 \\ 48 & 4\ 30 \\ & 3\ 84 \\ 568 & 46\ 00 \\ & 45\ 44 \\ & 56 \end{array}$$

Ans. .29

c) Find $\sqrt{\frac{1}{2}}$, to nearest hundredth.

$\sqrt{\frac{1}{2}} = \sqrt{.5}$

$$\begin{array}{rl} & .\ 7\ \ 0\ \ 7 \\ & \sqrt{.50\ 00\ 00} \\ & 49 \\ 1407 & 1\ 00\ 00 \\ & 98\ 49 \\ & 1\ 51 \end{array}$$

Ans. .71

6. SIMPLIFYING THE SQUARE ROOT OF A PRODUCT

Formulas: $\sqrt{ab} = \sqrt{a}\sqrt{b}$, $\sqrt{abc} = \sqrt{a}\sqrt{b}\sqrt{c}$

Rule 1. The square root of a product of two or more numbers equals the product of the separate square roots of these numbers.

Thus, $\sqrt{3600} = \sqrt{(36)(100)} = \sqrt{36}\sqrt{100} = (6)(10) = 60$

To Simplify the Square Root of a Product

Procedure:	Simplify $\sqrt{72}$. **Solution:**
1. Factor the radicand, choosing perfect square factors:	**1.** $\sqrt{(4)(9)(2)}$
2. Form a separate square root for each factor:	**2.** $\sqrt{4}\sqrt{9}\sqrt{2}$
3. Extract the square roots of each perfect square:	**3.** $(2)(3)\sqrt{2}$
4. Multiply the factors outside the radical:	**4.** $6\sqrt{2}$

Simplifying Square Roots of Powers

Rule 2. To find the square root of a power, keep the base and take one-half of the exponent.

Thus, $\sqrt{x^6} = x^3$ since $x^3 \cdot x^3 = x^6$.

Rule 3. To find the square root of the product of powers, keep each base and take one-half of the exponent.

Thus, $\sqrt{x^2y^4} = xy^2$ since $\sqrt{x^2y^4} = \sqrt{x^2}\sqrt{y^4} = xy^2$.

6.1. Rule 1. Finding Exact Square Root Through Simplification

Find each square root through simplification:

a) $\sqrt{1600}$	*b*) $\sqrt{256}$	*c*) $\sqrt{225}$	*d*) $\sqrt{324}$	*e*) $\frac{1}{2}\sqrt{576}$
Solutions:				
$\sqrt{(16)(100)}$	$\sqrt{(4)(64)}$	$\sqrt{(25)(9)}$	$\sqrt{(9)(36)}$	$\frac{1}{2}\sqrt{(4)(144)}$
$\sqrt{16}\sqrt{100}$	$\sqrt{4}\sqrt{64}$	$\sqrt{25}\sqrt{9}$	$\sqrt{9}\sqrt{36}$	$\frac{1}{2}\sqrt{4}\sqrt{144}$
$(4)(10)$	$(2)(8)$	$(5)(3)$	$(3)(6)$	$\frac{1}{2}(2)(12)$
Ans. 40	*Ans.* 16	*Ans.* 15	*Ans.* 18	*Ans.* 12

6.2. Rule 1. Simplifying Square Roots of Numbers

Simplify:

a) $\sqrt{75}$	b) $\sqrt{90}$	c) $\sqrt{112}$	d) $5\sqrt{288}$	e) $\frac{1}{3}\sqrt{486}$	f) $\frac{1}{2}\sqrt{1200}$
Solutions:					
$\sqrt{(25)(3)}$	$\sqrt{(9)(10)}$	$\sqrt{(16)(7)}$	$5\sqrt{(144)(2)}$	$\frac{1}{3}\sqrt{(81)(6)}$	$\frac{1}{2}\sqrt{(400)(3)}$
$\sqrt{25}\sqrt{3}$	$\sqrt{9}\sqrt{10}$	$\sqrt{16}\sqrt{7}$	$5\sqrt{144}\sqrt{2}$	$\frac{1}{3}\sqrt{81}\sqrt{6}$	$\frac{1}{2}\sqrt{400}\sqrt{3}$
Ans. $5\sqrt{3}$	*Ans.* $3\sqrt{10}$	*Ans.* $4\sqrt{7}$	*Ans.* $60\sqrt{2}$	*Ans.* $3\sqrt{6}$	*Ans.* $10\sqrt{3}$

6.3. Evaluating Simplified Square Roots

Using $\sqrt{2} = 1.414$ or $\sqrt{3} = 1.732$, evaluate to the nearest tenth:

a) $\sqrt{72}$	b) $\frac{1}{3}\sqrt{243}$	c) $3\sqrt{450}$	d) $\frac{3}{4}\sqrt{4800}$	e) $1.2\sqrt{363}$
Solutions:				
$\sqrt{36 \cdot 2}$	$\frac{1}{3}\sqrt{81 \cdot 3}$	$3\sqrt{9 \cdot 25 \cdot 2}$	$\frac{3}{4}\sqrt{16 \cdot 100 \cdot 3}$	$1.2\sqrt{121 \cdot 3}$
$\sqrt{36}\sqrt{2}$	$\frac{1}{3}\sqrt{81}\sqrt{3}$	$3\sqrt{9}\sqrt{25}\sqrt{2}$	$\frac{3}{4}\sqrt{16}\sqrt{100}\sqrt{3}$	$1.2\sqrt{121}\sqrt{3}$
$6\sqrt{2}$	$\frac{1}{3}(9)\sqrt{3}$	$3(3)(5)\sqrt{2}$	$\frac{3}{4}(4)(10)\sqrt{3}$	$1.2(11)\sqrt{3}$
$6(1.414)$	$3(1.732)$	$45(1.414)$	$30(1.732)$	$(13.2)(1.732)$
Ans. 8.5	*Ans.* 5.2	*Ans.* 63.6	*Ans.* 52.0	*Ans.* 22.9

6.4. Rules 2 and 3. Simplifying Square Roots of Powers

Simplify:

a) $\sqrt{49b^8}$	b) $\sqrt{900c^{10}d^{16}}$	c) $\sqrt{98b^6}$	d) $2\sqrt{25t^3}$
Solutions:			
$\sqrt{49}\sqrt{b^8}$	$\sqrt{900}\sqrt{c^{10}}\sqrt{d^{16}}$	$\sqrt{49}\sqrt{2}\sqrt{b^6}$	$2\sqrt{25}\sqrt{t^2}\sqrt{t}$
Ans. $7b^4$	*Ans.* $30c^5d^8$	*Ans.* $7b^3\sqrt{2}$	*Ans.* $10t\sqrt{t}$

7. SIMPLIFYING THE SQUARE ROOT OF A FRACTION

Formula: $\sqrt{\dfrac{a}{b}} = \dfrac{\sqrt{a}}{\sqrt{b}}$

Rule. The square root of a fraction equals the square root of the numerator divided by the square root of the denominator.

Thus, $\sqrt{\dfrac{25}{64}} = \dfrac{\sqrt{25}}{\sqrt{64}} = \dfrac{5}{8}$

To simplify the square root of a fraction whose denominator is not a perfect square, **change the fraction to an equivalent fraction** which has a denominator that is the smallest perfect square. Thus,

$$\sqrt{\frac{1}{8}} = \sqrt{\frac{2}{16}} = \frac{\sqrt{2}}{\sqrt{16}} = \frac{\sqrt{2}}{4} \quad \text{or} \quad \tfrac{1}{4}\sqrt{2}$$

7.1. Finding Exact Square Roots Involving Fractions

Simplify:

a) $\sqrt{2\frac{1}{4}}$	b) $6\sqrt{\dfrac{1600}{9}}$	c) $5\sqrt{\dfrac{144}{625}}$	d) $\dfrac{8}{5}\sqrt{\dfrac{225}{256}}$	e) $\sqrt{\dfrac{9}{49}}\sqrt{\dfrac{49}{400}}$
Solutions:				
$\sqrt{\dfrac{9}{4}}$	$\dfrac{6\sqrt{1600}}{\sqrt{9}}$	$\dfrac{5\sqrt{144}}{\sqrt{625}}$	$\dfrac{8}{5} \cdot \dfrac{\sqrt{225}}{\sqrt{256}}$	$\dfrac{\sqrt{9}}{\sqrt{49}} \cdot \dfrac{\sqrt{49}}{\sqrt{400}}$
$\dfrac{\sqrt{9}}{\sqrt{4}}$	$\dfrac{6(40)}{3}$	$\dfrac{5(12)}{25}$	$\dfrac{8}{5} \cdot \dfrac{(15)}{(16)}$	$\dfrac{3}{7} \cdot \dfrac{7}{20}$
Ans. $\frac{3}{2}$	*Ans.* 80	*Ans.* $\frac{12}{5}$	*Ans.* $\frac{3}{2}$	*Ans.* $\frac{3}{20}$

7.2. Simplifying Square Roots of Fractions Whose Denominators are Perfect Squares

Simplify:

	a) $6\sqrt{\frac{7}{9}}$	*b*) $\frac{2}{3}\sqrt{\frac{11}{100}}$	*c*) $\frac{1}{5}\sqrt{\frac{75}{16}}$	*d*) $x\sqrt{\frac{200}{x^2}}$	*e*) $10\sqrt{\frac{y}{25}}$
Solutions:	$\frac{6\sqrt{7}}{\sqrt{9}}$	$\frac{2\sqrt{11}}{3\sqrt{100}}$	$\frac{\sqrt{25}\sqrt{3}}{5\sqrt{16}}$	$\frac{x\sqrt{100}\sqrt{2}}{\sqrt{x^2}}$	$\frac{10\sqrt{y}}{\sqrt{25}}$
	$\frac{6\sqrt{7}}{3}$	$\frac{2\sqrt{11}}{3(10)}$	$\frac{5\sqrt{3}}{5(4)}$	$\frac{10x\sqrt{2}}{x}$	$\frac{10\sqrt{y}}{5}$
	Ans. $2\sqrt{7}$	*Ans.* $\frac{\sqrt{11}}{15}$	*Ans.* $\frac{\sqrt{3}}{4}$	*Ans.* $10\sqrt{2}$	*Ans.* $2\sqrt{y}$

7.3. Simplifying Square Roots of Fractions Whose Denominators Are Not Perfect Squares

Simplify:

	a) $\sqrt{\frac{1}{5}}$	*b*) $\sqrt{\frac{3}{8}}$	*c*) $\sqrt{\frac{a}{3}}$	*d*) $\sqrt{\frac{x^2}{27}}$	*e*) $\sqrt{\frac{3}{y}}$
Solutions:	$\sqrt{\frac{1}{5}\cdot\frac{5}{5}}$	$\sqrt{\frac{3}{8}\cdot\frac{2}{2}}$	$\sqrt{\frac{a}{3}\cdot\frac{3}{3}}$	$\sqrt{\frac{x^2}{27}\cdot\frac{3}{3}}$	$\sqrt{\frac{3}{y}\cdot\frac{y}{y}}$
	$\sqrt{\frac{5}{25}}$	$\sqrt{\frac{6}{16}}$	$\sqrt{\frac{3a}{9}}$	$\sqrt{\frac{3x^2}{81}}$	$\sqrt{\frac{3y}{y^2}}$
	$\frac{\sqrt{5}}{\sqrt{25}}$	$\frac{\sqrt{6}}{\sqrt{16}}$	$\frac{\sqrt{3a}}{\sqrt{9}}$	$\frac{\sqrt{3x^2}}{\sqrt{81}}$	$\frac{\sqrt{3y}}{\sqrt{y^2}}$
	Ans. $\frac{\sqrt{5}}{5}$	*Ans.* $\frac{\sqrt{6}}{4}$	*Ans.* $\frac{\sqrt{3a}}{3}$	*Ans.* $\frac{x\sqrt{3}}{9}$	*Ans.* $\frac{\sqrt{3y}}{y}$

7.4. Evaluating Square Roots after Simplification

Using $\sqrt{3} = 1.732$ or $\sqrt{5} = 2.236$, evaluate each to the nearest hundredth:

	a) $\sqrt{\frac{3}{4}}$	*b*) $\sqrt{\frac{4}{3}}$	*c*) $30\sqrt{\frac{20}{9}}$	*d*) $10\sqrt{\frac{4}{5}}$	*e*) $\sqrt{\frac{49}{20}}$
Solutions:	$\frac{\sqrt{3}}{\sqrt{4}}$	$\sqrt{\frac{4}{3}\cdot\frac{3}{3}}$	$\frac{30\sqrt{20}}{\sqrt{9}}$	$10\sqrt{\frac{4}{5}\cdot\frac{5}{5}}$	$\sqrt{\frac{49}{20}\cdot\frac{5}{5}}$
	$\frac{\sqrt{3}}{2}$	$\frac{\sqrt{4}\sqrt{3}}{\sqrt{9}}$	$10\sqrt{4}\sqrt{5}$	$\frac{10\sqrt{4}\sqrt{5}}{5}$	$\frac{\sqrt{49}\sqrt{5}}{\sqrt{100}}$
	$\frac{1.732}{2}$	$\frac{2}{3}(1.732)$	$20(2.236)$	$4(2.236)$	$\frac{7}{10}(2.236)$
	Ans. .87	*Ans.* 1.15	*Ans.* 44.7	*Ans.* 8.94	*Ans.* 1.6

8. ADDING AND SUBTRACTING SQUARE ROOTS OF NUMBERS

Like radicals are radicals having the same index and the same radicand.

Thus:	Like Radicals	Unlike Radicals
	$5\sqrt{3}$ and $2\sqrt{3}$	$5\sqrt{3}$ and $2\sqrt{5}$
	$8\sqrt{x}$ and $-3\sqrt{x}$	$8\sqrt{x}$ and $-3\sqrt{y}$
	$7\sqrt[3]{6}$ and $3\sqrt[3]{6}$	$7\sqrt[3]{6}$ and $3\sqrt{6}$

Rule. To combine (add or subtract) like radicals, keep the common radical and combine their coefficients.

Thus, $5\sqrt{3} + 2\sqrt{3} - 4\sqrt{3} = (5+2-4)\sqrt{3} = 3\sqrt{3}$.

Note. Combining like radicals involves the same process as that for combining like terms. Hence, to combine $5\sqrt{3}$ and $2\sqrt{3}$, think of combining $5x$ and $2x$ when $x = \sqrt{3}$.

Unlike radicals may be combined into one radical if like radicals can be obtained by simplifying. Thus, $\sqrt{50} + \sqrt{32} = 5\sqrt{2} + 4\sqrt{2} = 9\sqrt{2}$.

8.1. Combining Like Radicals

Combine: (*Keep radical and combine coefficients.*)

a) $3\sqrt{2} + 2\sqrt{2}$
b) $5\sqrt{3} - 10\sqrt{3}$
c) $\sqrt{6} + 9\sqrt{6}$
d) $\frac{1}{2}\sqrt{5} + 1\frac{1}{2}\sqrt{5}$
e) $7\frac{3}{4}\sqrt{10} - 5\frac{3}{4}\sqrt{10}$
f) $\sqrt{2x} + 3\sqrt{2x}$
g) $2x\sqrt{7} + 3x\sqrt{7}$
h) $20\sqrt{a} - 13\sqrt{a}$
i) $5a\sqrt{y} - 2a\sqrt{y}$

Solutions:

a) $(3+2)\sqrt{2} = 5\sqrt{2}$ *Ans.*
b) $(5-10)\sqrt{3} = -5\sqrt{3}$ *Ans.*
c) $(1+9)\sqrt{6} = 10\sqrt{6}$ *Ans.*
d) $(\frac{1}{2}+1\frac{1}{2})\sqrt{5} = 2\sqrt{5}$ *Ans.*
e) $(7\frac{3}{4}-5\frac{3}{4})\sqrt{10} = 2\sqrt{10}$ *Ans.*
f) $(1+3)\sqrt{2x} = 4\sqrt{2x}$ *Ans.*
g) $(2x+3x)\sqrt{7} = 5x\sqrt{7}$ *Ans.*
h) $(20-13)\sqrt{a} = 7\sqrt{a}$ *Ans.*
i) $(5a-2a)\sqrt{y} = 3a\sqrt{y}$ *Ans.*

8.2. Combining Unlike Numerical Radicals After Simplification

Combine after simplifying:

a) $\sqrt{5} + \sqrt{20} + \sqrt{45}$
b) $\sqrt{27} + 3\sqrt{12}$
c) $\sqrt{8} - 6\sqrt{\frac{1}{2}}$
d) $\sqrt{4a} + \sqrt{25a} + \sqrt{\frac{a}{4}}$

Solutions:

a) $\sqrt{5} + 2\sqrt{5} + 3\sqrt{5}$
Ans. $6\sqrt{5}$

b) $3\sqrt{3} + 3(2\sqrt{3})$
$3\sqrt{3} + 6\sqrt{3}$
Ans. $9\sqrt{3}$

c) $2\sqrt{2} - 6(\frac{1}{2}\sqrt{2})$
$2\sqrt{2} - 3\sqrt{2}$
Ans. $-\sqrt{2}$

d) $2\sqrt{a} + 5\sqrt{a} + \frac{1}{2}\sqrt{a}$
Ans. $7\frac{1}{2}\sqrt{a}$

9. MULTIPLYING SQUARE ROOTS OF NUMBERS

Formulas: $\sqrt{a}\sqrt{b} = \sqrt{ab}$, $\sqrt{a}\sqrt{b}\sqrt{c} = \sqrt{abc}$

Rule 1. The product of the square roots of two or more non-negative numbers equals the square root of their product.

Thus, $\sqrt{2}\sqrt{3}\sqrt{6} = \sqrt{36} = 6$. However $\sqrt{-5}\sqrt{-5} = -5$. $\sqrt{-5}\sqrt{-5} \neq \sqrt{25}$ or 5.

Rule 2. The square of the square root of a number equals the number.

Thus, $(\sqrt{7})^2 = (\sqrt{7})(\sqrt{7}) = 7$. In general, $(\sqrt{x})^2 = x$. Hence, when squaring the square root of a number, merely eliminate the radical sign.

To Multiply Square Root Monomials

Multiply: a) $3\sqrt{2}\cdot 2\sqrt{6}$ b) $\frac{2}{3}\sqrt{2x}\cdot 6\sqrt{2y}$

Procedure:	Solutions: a)	b)
1. **Multiply** coefficients and radicals separately:	1. $3\cdot 2\sqrt{2}\sqrt{6}$	1. $\frac{2}{3}\cdot 6\sqrt{2x}\sqrt{2y}$
2. **Multiply** the resulting products:	2. $6\sqrt{12}$	2. $4\sqrt{4xy}$
3. **Simplify**, if possible:	3. $6(2\sqrt{3})$	3. $4\cdot 2\sqrt{xy}$
	Ans. $12\sqrt{3}$	**Ans.** $8\sqrt{xy}$

9.1. Multiplying Square Root Monomials

Multiply:

a) $\sqrt{2}\sqrt{5}\sqrt{6}\sqrt{15}$

Solutions:

$\sqrt{2\cdot5\cdot6\cdot15}$

$\sqrt{900}$

Ans. 30

b) $(2\sqrt{3})(3\sqrt{5})(5\sqrt{10})$

$2\cdot3\cdot5\sqrt{3\cdot5\cdot10}$

$30\sqrt{150} = 30(5\sqrt{6})$

Ans. $150\sqrt{6}$

c) $\sqrt{\frac{x}{7}}\sqrt{14x}$

$\sqrt{(\frac{x}{7})(14x)}$

$\sqrt{2x^2}$

Ans. $x\sqrt{2}$

d) $(4\sqrt{m})(2\sqrt{25m})$

$4\cdot2\sqrt{(m)(25m)}$

$8\sqrt{25m^2} = 8(5m)$

Ans. $40m$

9.2. Rule 2. Squaring Like Radicals

Multiply:

a) $(2\sqrt{17})(3\sqrt{17})$

Solutions:

$2\cdot3(\sqrt{17})^2$

$6(17)$

Ans. 102

b) $(3\sqrt{\frac{2}{3}})^2$

$3^2(\sqrt{\frac{2}{3}})^2$

$9(\frac{2}{3})$

Ans. 6

c) $(\frac{5}{2}\sqrt{8})^2$

$(\frac{5}{2})^2(\sqrt{8})^2$

$\frac{25}{4}(8)$

Ans. 50

d) $(2a\sqrt{2a})^2$

$(2a)^2(\sqrt{2a})^2$

$4a^2(2a)$

Ans. $8a^3$

e) $\left(\frac{x}{3}\sqrt{\frac{3}{x}}\right)^2$

$(\frac{x}{3})^2(\sqrt{\frac{3}{x}})^2$

$\frac{x^2}{9}(\frac{3}{x})$

Ans. $\frac{x}{3}$

9.3. Multiplying a Monomial by a Polynomial

Multiply:

a) $\sqrt{3}(\sqrt{3}+\sqrt{27})$

Solutions:

$\sqrt{3}\sqrt{3} + \sqrt{3}\sqrt{27}$

$3 + \sqrt{81}$

Ans. 12

b) $3\sqrt{2}(\sqrt{2}+\sqrt{5})$

$3\sqrt{2}\sqrt{2} + 3\sqrt{2}\sqrt{5}$

$3(2) + 3\sqrt{10}$

Ans. $6 + 3\sqrt{10}$

c) $\frac{\sqrt{5}}{2}(4\sqrt{3} - 6\sqrt{20})$

$(\frac{\sqrt{5}}{2})(4\sqrt{3}) - (\frac{\sqrt{5}}{2})(6\sqrt{20})$

$2\sqrt{15} - 3\sqrt{100}$

Ans. $2\sqrt{15} - 30$

9.4. Multiplying Binomials Containing Radicals

Multiply:

Procedure:	*a)* $(5+\sqrt{2})(5-\sqrt{2})$	*b)* $(\sqrt{10}+\sqrt{3})(\sqrt{10}-\sqrt{3})$	*c)* $(\sqrt{7}+\sqrt{2})(\sqrt{7}+\sqrt{2})$
	Solutions:		
1. Multiply first terms:	$5\cdot5 = 25$	$\sqrt{10}\sqrt{10} = 10$	$\sqrt{7}\sqrt{7} = 7$
2. Add products of outer and inner terms:	$5\sqrt{2} - 5\sqrt{2} = 0$	$\sqrt{30} - \sqrt{30} = 0$	$\sqrt{14} + \sqrt{14} = 2\sqrt{14}$
3. Multiply last terms:	$(\sqrt{2})(-\sqrt{2}) = -2$	$(\sqrt{3})(-\sqrt{3}) = -3$	$\sqrt{2}\sqrt{2} = 2$
4. Combine results:	*Ans.* 23	*Ans.* 7	*Ans.* $9 + 2\sqrt{14}$

10. DIVIDING SQUARE ROOTS OF NUMBERS

Formula: $\frac{\sqrt{a}}{\sqrt{b}} = \sqrt{\frac{a}{b}}$

Rule. The square root of a number **divided** by the square root of another number equals the square root of their quotient.

Thus, $\frac{\sqrt{6}}{\sqrt{2}} = \sqrt{\frac{6}{2}} = \sqrt{3}$.

To Divide Square Root Monomials

Procedure:	Divide: a) $\frac{14\sqrt{40}}{2\sqrt{5}}$	b) $\frac{6\sqrt{x^4}}{3\sqrt{x}}$
	Solutions:	
1. **Divide** coefficients and radicals separately:	1. $\frac{14}{2}\sqrt{\frac{40}{5}}$	1. $\frac{6}{3}\sqrt{\frac{x^4}{x}}$
2. **Multiply** the resulting quotients:	2. $7\sqrt{8}$	2. $2\sqrt{x^3}$
3. **Simplify**, if possible:	3. $7(2\sqrt{2})$	3. $2(\sqrt{x^2}\sqrt{x})$
	Ans. $14\sqrt{2}$	**Ans.** $2x\sqrt{x}$

10.1. Dividing Square Root Monomials

Divide:

a) $\frac{8\sqrt{8}}{2\sqrt{2}}$	b) $\frac{6\sqrt{2}}{12\sqrt{8}}$	c) $\frac{10\sqrt{56}}{4\sqrt{7}}$	d) $\frac{9\sqrt{x^3}}{3\sqrt{x}}$	e) $\frac{5a\sqrt{5a}}{10a\sqrt{10a}}$
Solutions:				
$\frac{8}{2}\sqrt{\frac{8}{2}}$	$\frac{6}{12}\sqrt{\frac{2}{8}}$	$\frac{10}{4}\sqrt{\frac{56}{7}}$	$\frac{9}{3}\sqrt{\frac{x^3}{x}}$	$\frac{5a}{10a}\sqrt{\frac{5a}{10a}}$
$4\sqrt{4} = 4(2)$	$\frac{1}{2}\sqrt{\frac{1}{4}} = \frac{1}{2}(\frac{1}{2})$	$\frac{5}{2}\sqrt{8} = \frac{5}{2}(2\sqrt{2})$	$3\sqrt{x^2}$	$\frac{1}{2}\sqrt{\frac{1}{2}} = \frac{1}{2}(\frac{1}{2}\sqrt{2})$
Ans. 8	*Ans.* $\frac{1}{4}$	*Ans.* $5\sqrt{2}$	*Ans.* $3x$	*Ans.* $\frac{1}{4}\sqrt{2}$

10.2. Dividing a Polynomial by a Monomial

Divide:

a) $\frac{\sqrt{50}+\sqrt{98}}{\sqrt{2}}$	b) $\frac{2\sqrt{24}+\sqrt{96}}{\sqrt{3}}$	c) $\frac{18\sqrt{40}-10\sqrt{80}}{2\sqrt{10}}$	d) $\frac{15\sqrt{y^5}-6\sqrt{y^3}}{3\sqrt{y}}$
Solutions:			
$\frac{\sqrt{50}}{\sqrt{2}} + \frac{\sqrt{98}}{\sqrt{2}}$	$\frac{2\sqrt{24}}{\sqrt{3}} + \frac{\sqrt{96}}{\sqrt{3}}$	$\frac{18\sqrt{40}}{2\sqrt{10}} - \frac{10\sqrt{80}}{2\sqrt{10}}$	$\frac{15\sqrt{y^5}}{3\sqrt{y}} - \frac{6\sqrt{y^3}}{3\sqrt{y}}$
$\sqrt{25} + \sqrt{49}$	$2\sqrt{8} + \sqrt{32}$	$9\sqrt{4} - 5\sqrt{8}$	$5\sqrt{\frac{y^5}{y}} - 2\sqrt{\frac{y^3}{y}}$
$5 + 7$	$2(2\sqrt{2}) + 4\sqrt{2}$	$18 - 5(2\sqrt{2})$	$5\sqrt{y^4} - 2\sqrt{y^2}$
Ans. 12	*Ans.* $8\sqrt{2}$	*Ans.* $18 - 10\sqrt{2}$	*Ans.* $5y^2 - 2y$

11. RATIONALIZING THE DENOMINATOR OF A FRACTION

To rationalize the denominator of a fraction is to change the denominator from an irrational number to a rational number. To do this when the denominator is a monomial surd, multiply both terms of the fraction by the smallest surd which will make the denominator rational.

Thus, to rationalize the denominator of $\frac{4}{\sqrt{8}}$, multiply by $\frac{\sqrt{2}}{\sqrt{2}}$:

$$\frac{4}{\sqrt{8}} \cdot \frac{\sqrt{2}}{\sqrt{2}} = \frac{4\sqrt{2}}{\sqrt{16}} = \frac{4\sqrt{2}}{4} = \sqrt{2}$$

Rationalizing the denominator simplifies evaluation.

Thus, $\frac{4}{\sqrt{8}} = \sqrt{2} = 1.414$. Otherwise, $\frac{4}{\sqrt{8}} = \frac{4}{2.828}$, thus leading to a lengthy division.

11.1. Rationalizing the Denominator

Rationalize each denominator:

a) $\frac{12}{\sqrt{6}}$	b) $\frac{\sqrt{3}}{\sqrt{10}}$	c) $\frac{24}{\sqrt{8}}$	d) $\frac{3}{\sqrt{x}}$	e) $\frac{1}{\sqrt{c^3}}$	f) $\frac{3+\sqrt{2}}{\sqrt{2}}$
Solutions:					
$\frac{12}{\sqrt{6}}\cdot\frac{\sqrt{6}}{\sqrt{6}}$	$\frac{\sqrt{3}}{\sqrt{10}}\cdot\frac{\sqrt{10}}{\sqrt{10}}$	$\frac{24}{\sqrt{8}}\cdot\frac{\sqrt{2}}{\sqrt{2}}$	$\frac{3}{\sqrt{x}}\cdot\frac{\sqrt{x}}{\sqrt{x}}$	$\frac{1}{\sqrt{c^3}}\cdot\frac{\sqrt{c}}{\sqrt{c}}$	$\frac{\sqrt{2}}{\sqrt{2}}\cdot\frac{3+\sqrt{2}}{\sqrt{2}}$
$\frac{12\sqrt{6}}{6}$	*Ans.* $\frac{\sqrt{30}}{10}$	$\frac{24\sqrt{2}}{\sqrt{16}}$	*Ans.* $\frac{3\sqrt{x}}{x}$	$\frac{\sqrt{c}}{\sqrt{c^4}}$	$\frac{\sqrt{2}(3+\sqrt{2})}{2}$
Ans. $2\sqrt{6}$		*Ans.* $6\sqrt{2}$		*Ans.* $\frac{\sqrt{c}}{c^2}$	*Ans.* $\frac{3\sqrt{2}+2}{2}$

11.2. Evaluating Fractions With Irrational Denominators

Evaluate to the nearest tenth, after rationalizing the denominator:

a) $\frac{7}{\sqrt{7}}$	b) $\frac{20}{\sqrt{50}}$	c) $\frac{3}{\sqrt{6}}$	d) $\frac{1}{\sqrt{12}}$	e) $\frac{3+\sqrt{5}}{\sqrt{5}}$
Solutions:				
$\frac{7}{\sqrt{7}}\cdot\frac{\sqrt{7}}{\sqrt{7}}$	$\frac{20}{\sqrt{50}}\cdot\frac{\sqrt{2}}{\sqrt{2}}$	$\frac{3}{\sqrt{6}}\cdot\frac{\sqrt{6}}{\sqrt{6}}$	$\frac{1}{\sqrt{12}}\cdot\frac{\sqrt{3}}{\sqrt{3}}$	$\frac{(3+\sqrt{5})}{\sqrt{5}}\cdot\frac{\sqrt{5}}{\sqrt{5}}$
$\frac{7\sqrt{7}}{7}$	$\frac{20\sqrt{2}}{10}$	$\frac{3\sqrt{6}}{6}$	$\frac{\sqrt{3}}{\sqrt{36}}$	$\frac{\sqrt{5}(3+\sqrt{5})}{5}$
$\sqrt{7}$	$2\sqrt{2}$	$\frac{\sqrt{6}}{2}$	$\frac{\sqrt{3}}{6}$	$\frac{3\sqrt{5}+5}{5}$
2.646	2(1.414)	$\frac{2.449}{2}$	$\frac{1.732}{6}$	$\frac{3(2.236)+5}{5}$
Ans. 2.6	*Ans.* 2.8	*Ans.* 1.2	*Ans.* .3	*Ans.* 2.3

12. SOLVING RADICAL EQUATIONS

Radical equations are equations in which the unknown is included in a radicand.

Thus, $2\sqrt{x}+5 = 9$ is a radical equation.

but $2x+\sqrt{5} = 9$ is not a radical equation,

To Solve a Radical Equation

Procedure:	Solve: $\sqrt{2x}+5 = 9$ **Solution:**
1. **Isolate** the term containing the radical:	**1.** $\sqrt{2x}+5 = 9$ $\sqrt{2x} = 4$
2. **Square** both sides:	**2.** By squaring, $2x = 16$
3. **Solve** for the unknown:	**3.** $x = 8$
4. **Check** the roots obtained in the original equation:	**4.** Check for $x = 8$: $\sqrt{2x}+5 \stackrel{?}{=} 9$ $\sqrt{16}+5 \stackrel{?}{=} 9$ $9 = 9$ *Ans.* $x = 8$

Note. In the solution of equations, the symbol "**Sq**" means "square both sides".

An extraneous root of an equation is a value of the unknown which satisfies, **not the original equation**, but a later equation obtained from the original equation. It must be rejected.

Thus, if $\sqrt{2x} = -4$, when both sides are squared, $2x = 16$ or $x = 8$.

However, $x = 8$ does not satisfy $\sqrt{2x} = -4$, the original equation.

Hence, 8 is an *extraneous root* of $\sqrt{2x} = -4$ and must be rejected.

12.1. Solving Simple Radical Equations

Solve and check:

a) $\sqrt{5x} = 10$

Solutions:

By squaring,

$5x = 100$

$x = 20$

Check: $\sqrt{5x} = 10$

$\sqrt{100} \stackrel{?}{=} 10$

$10 = 10$

Ans. $x = 20$

b) $\sqrt{\frac{x}{3}} = 2$

By squaring,

$\frac{x}{3} = 4$

$x = 12$

Check: $\sqrt{\frac{x}{3}} = 2$

$\sqrt{4} \stackrel{?}{=} 2$

$2 = 2$

Ans. $x = 12$

c) $\sqrt{7x-5} = 3$

By squaring,

$7x + 5 = 9$

$x = \frac{4}{7}$

Check: $\sqrt{7x+5} = 3$

$\sqrt{4+5} \stackrel{?}{=} 3$

$3 = 3$

Ans. $x = \frac{4}{7}$

d) $2\sqrt{x-8} = -3$

By squaring,

$4(x-8) = 9$

$x = \frac{41}{4}$

Check: $2\sqrt{x-8} = -3$

$2\sqrt{\frac{9}{4}} \stackrel{?}{=} -3$

$3 \neq -3$

Reject $x = \frac{41}{4}$. (*No root.*)

12.2. Isolating the Radical in Solving Radical Equations

Solve and check

a) $3\sqrt{x} - 6 = 9$ *b*) $5\sqrt{x} = \sqrt{x} - 12$ *c*) $\sqrt{3y+4} - 2 = 3$

Solutions: (*Isolate the radical term first.*)

a)

A_6 $3\sqrt{x} - 6 = 9$

D_3 $3\sqrt{x} = 15$

Sq $\sqrt{x} = 5$

$x = 25$

Check: $3\sqrt{x} - 6 = 9$

$3\sqrt{25} - 6 \stackrel{?}{=} 9$

$9 = 9$

Ans. $x = 25$

b)

$S_{\sqrt{x}}$ $5\sqrt{x} = \sqrt{x} - 12$

D_4 $4\sqrt{x} = -12$

Sq $\sqrt{x} = -3$

$x = 9$

Check: $5\sqrt{x} = \sqrt{x} - 12$

$5\sqrt{9} \stackrel{?}{=} \sqrt{9} - 12$

$15 \neq -9$

Reject $x = 9$. (*No root.*)

c)

A_2 $\sqrt{3y+4} - 2 = 3$

Sq $\sqrt{3y+4} = 5$

$3y + 4 = 25$

$y = 7$

Check: $\sqrt{3y+4} - 2 = 3$

$\sqrt{25} - 2 \stackrel{?}{=} 3$

$3 = 3$

Ans. $y = 7$

12.3. Transforming Radical Equations

Solve for the letter indicated:

a) Solve for x:

$\sqrt{\frac{x}{2}} = y$

Solutions:

By squaring,

$\frac{x}{2} = y^2$

Ans. $x = 2y^2$

b) Solve for y:

$\sqrt{2y-5} = x$

By squaring,

$2y - 5 = x^2$

Ans. $y = \frac{x^2 + 5}{2}$

c) Solve for A:

$r = \sqrt{\frac{A}{\pi}}$

By squaring,

$r^2 = \frac{A}{\pi}$

Ans. $A = \pi r^2$

d) Solve for x:

$\sqrt{x+4} = y + 2$

By squaring,

$x + 4 = y^2 + 4y + 4$

Ans. $x = y^2 + 4y$

SUPPLEMENTARY PROBLEMS

1. State the square roots of **(1.1)**

a) 100 | *b)* $100y^2$ | *c)* $4c^2d^2$ | *d)* $\frac{4}{81}$ | *e)* $\frac{25r^2}{36s^2}$ | *f)* $\frac{x^4z^{16}}{y^6}$

Ans. a) ± 10 | *b)* $\pm 10y$ | *c)* $\pm 2cd$ | *d)* $\pm\frac{2}{9}$ | *e)* $\pm\frac{5r}{6s}$ | *f)* $\pm\frac{x^2z^8}{y^3}$

2. Find each principal square root: **(1.2)**

a) $\sqrt{400}$ | *b)* $\sqrt{.04}$ | *c)* $\sqrt{1600h^2}$ | *d)* $\sqrt{1\frac{9}{16}}$ | *e)* $\sqrt{\frac{16r^2}{s^2t^2}}$ | *f)* $\sqrt{\frac{64m^8}{p^{14}}}$

Ans. a) 20 | *b)* $.2$ | *c)* $40h$ | *d)* $\frac{5}{4}$ | *e)* $\frac{4r}{st}$ | *f)* $\frac{8m^4}{p^7}$

3. Find each negative square root: **(1.3)**

a) $-\sqrt{\frac{1}{49}}$ | *b)* $-\sqrt{a^2b^4}$ | *c)* $-\sqrt{r^{10}t^{20}}$ | *d)* $-\sqrt{\frac{121}{25w^{12}}}$ | *e)* $-\sqrt{\frac{225x^{16}}{y^{36}}}$

Ans. a) $-\frac{1}{7}$ | *b)* $-ab^2$ | *c)* $-r^5t^{10}$ | *d)* $-\frac{11}{5w^6}$ | *e)* $-\frac{15x^8}{y^{18}}$

4. Find the value of **(1.4)**

a) $\sqrt{81} - \sqrt{25}$ *Ans.* 4

b) $2\sqrt{9} + 3\sqrt{4}$ *Ans.* 12

c) $\frac{1}{2}\sqrt{36} - \frac{1}{3}\sqrt{81}$ *Ans.* 0

d) $\sqrt{\frac{4}{9}} + \sqrt{\frac{1}{9}}$ *Ans.* 1

e) $8\sqrt{\frac{1}{16}} + 4\sqrt{\frac{9}{16}}$ *Ans.* 5

f) $\left(-\sqrt{\frac{4}{9}}\right)\left(-\sqrt{\frac{81}{100}}\right)$ *Ans.* $\frac{3}{5}$

g) $4\sqrt{x^2} - x\sqrt{4}$ *Ans.* $2x$

h) $\sqrt{16x^2} + 16\sqrt{x^2}$ *Ans.* $20x$

i) $\sqrt{\frac{16}{x^2}} + \frac{16}{\sqrt{x^2}}$ *Ans.* $\frac{20}{x}$

5. Express each of the following rational numbers as the ratio of two integers: **(2.2)**

a) 5 *b)* -13 *c)* $.07$ *d)* $-11\frac{2}{3}$ *e)* $\frac{\sqrt{25}}{3}$ *f)* $\frac{17}{\sqrt{144}}$ *g)* $\frac{16-\sqrt{16}}{\sqrt{25}}$ *h)* $\sqrt{\frac{50}{32}}$

Ans. a) $\frac{5}{1}$ *b)* $\frac{-13}{1}$ *c)* $\frac{7}{100}$ *d)* $\frac{-35}{3}$ *e)* $\frac{5}{3}$ *f)* $\frac{17}{12}$ *g)* $\frac{12}{5}$ *h)* $\frac{5}{4}$

6. From the graph of $x = \sqrt{y}$, find the square root of each number to the nearest tenth: **(3.1)**

a) $\sqrt{17}$ *b)* $\sqrt{23}$ *c)* $\sqrt{37}$ *d)* $\sqrt{45}$ *e)* $\sqrt{48}$ *f)* $\sqrt{58}$ *g)* $\sqrt{61}$

Ans. a) 4.1 *b)* 4.8 *c)* 6.1 *d)* 6.7 *e)* 6.9 *f)* 7.6 *g)* 7.8

7. Find the approximate value of each square root, using the table of square roots: **(4.1)**

a) $\sqrt{7}$ *d)* $\sqrt{46}$ *g)* $\sqrt{275}$ *j)* $\sqrt{528}$ *m)* $\sqrt{780}$

b) $\sqrt{70}$ *e)* $\sqrt{94}$ *h)* $\sqrt{329}$ *k)* $\sqrt{576}$ *n)* $\sqrt{892}$

c) $\sqrt{700}$ *f)* $\sqrt{138}$ *i)* $\sqrt{486}$ *l)* $\sqrt{747}$ *o)* $\sqrt{917}$

Ans. a) 2.646 *d)* 6.782 *g)* 16.58 *j)* 22.98 *m)* 27.93

b) 8.367 *e)* 9.695 *h)* 18.14 *k)* 24.00 *n)* 29.87

c) 26.46 *f)* 11.75 *i)* 22.05 *l)* 27.33 *o)* 30.28

8. Find the value of each to the nearest tenth, using the table of square roots: **(4.2)**

a) $2\sqrt{3}$ *c)* $\frac{1}{2}\sqrt{140}$ *e)* $100 - 10\sqrt{83}$ *g)* $\frac{\sqrt{3}+\sqrt{2}}{5}$

b) $10\sqrt{50}$ *d)* $\frac{2}{5}\sqrt{280}$ *f)* $\frac{3+\sqrt{2}}{5}$ *h)* $\frac{10-\sqrt{2}}{3}$

Ans. a) 3.5 *b)* 70.7 *c)* 5.9 *d)* 6.7 *e)* 8.9 *f)* .9 *g)* .6 *h)* 2.9

9. Find the approximate value of each, using the table and **Rule 1**: **(4.3)**

a) $\sqrt{3000}$ *c*) $\sqrt{19000}$ *e*) $\sqrt{5000} + \sqrt{500}$, to nearest tenth.

b) $\sqrt{30000}$ *d*) $\sqrt{190000}$ *f*) $\sqrt{8300} - \sqrt{830}$, to nearest tenth.

Ans. *a*) 54.77 *c*) 137.8 *e*) 70.71 + 22.36 = 93.07 or 93.1

b) 173.2 *d*) 435.9 *f*) 91.10 − 28.81 = 62.29 or 62.3

10. Find the approximate value of each, using the table and **Rule 2**: **(4.4)**

a) $\sqrt{.03}$ *d*) $\sqrt{3.85}$ *g*) $\sqrt{55} + \sqrt{.55}$, to nearest hundredth.

b) $\sqrt{.0003}$ *e*) $\sqrt{.0385}$ *h*) $\sqrt{427} - \sqrt{4.27}$, to nearest hundredth.

c) $\sqrt{.30}$ *f*) $\sqrt{.3}$ *i*) $\sqrt{.02} + \sqrt{.0002}$, to nearest hundredth.

Ans. *a*) .1732, *b*) .01732, *c*) .5477, *d*) 1.962, *e*) .1962, *f*) .5477 ($= \sqrt{.30}$),
g) 8.16, *h*) 18.59, *i*) .16

11. Find each square root, using the square root process. **(5.1 to 5.3)**

a) $\sqrt{2304}$ *d*) $\sqrt{29.16}$ *g*) $\sqrt{67.24}$ *j*) $\sqrt{.012544}$

b) $\sqrt{3025}$ *e*) $\sqrt{12.96}$ *h*) $\sqrt{4.1209}$ *k*) $\sqrt{.014641}$

c) $\sqrt{7396}$ *f*) $\sqrt{466.56}$ *i*) $\sqrt{5.0625}$ *l*) $\sqrt{.050625}$

Ans. *a*) 48, *b*) 55, *c*) 86, *d*) 5.4, *e*) 3.6, *f*) 21.6, *g*) 8.2, *h*) 2.03, *i*) 2.25
j) .112, *k*) .121, *l*) .225

12. Find each square root, to the nearest tenth. Use the square root process. **(5.4)**

a) $\sqrt{96.24}$ *Ans.* 9.8 *c*) $\sqrt{463.45}$ *Ans.* 21.5 *e*) $\sqrt{.1750}$ *Ans.* .4

b) $\sqrt{135.06}$ *Ans.* 11.6 *d*) $\sqrt{207.86}$ *Ans.* 14.4 *f*) $\sqrt{.4545}$ *Ans.* .7

13. Verify each value to the nearest hundredth, using the square root process. **(5.4)**

a) $\sqrt{5} = 2.24$ *c*) $\sqrt{28} = 5.29$ *e*) $\sqrt{48} = 6.93$ *g*) $\sqrt{692} = 26.31$

b) $\sqrt{10} = 3.16$ *d*) $\sqrt{34} = 5.83$ *f*) $\sqrt{57} = 7.55$ *h*) $\sqrt{765} = 27.66$

14. Find each square root, to the nearest tenth. Use the square root process. **(5.4)**

a) $\sqrt{7\frac{1}{2}}$ or $\sqrt{7.5}$ *Ans.* 2.7 *b*) $\sqrt{4\frac{1}{4}}$ or $\sqrt{4.25}$ *Ans.* 2.1 *c*) $\sqrt{52\frac{3}{8}}$ or $\sqrt{52.375}$ *Ans.* 7.2

15. Find each square root through simplification: **(6.1)**

a) $\sqrt{2500}$ *Ans.* 50 *c*) $\sqrt{2025}$ *Ans.* 45 *e*) $\sqrt{729}$ *Ans.* 27 *g*) $\sqrt{48400}$ *Ans.* 220

b) $\sqrt{19600}$ *Ans.* 140 *d*) $\sqrt{441}$ *Ans.* 21 *f*) $\sqrt{784}$ *Ans.* 28 *h*) $\sqrt{562500}$ *Ans.* 750

16. Simplify: **(6.2)**

a) $\sqrt{63}$ *d*) $\frac{1}{2}\sqrt{32}$ *g*) $\sqrt{17500}$ *j*) $\frac{3}{7}\sqrt{392}$

b) $\sqrt{96}$ *e*) $\frac{1}{5}\sqrt{300}$ *h*) $\sqrt{4500}$ *k*) $\frac{2}{5}\sqrt{250}$

c) $\sqrt{448}$ *f*) $\frac{7}{8}\sqrt{320}$ *i*) $\sqrt{21600}$ *l*) $\frac{5}{3}\sqrt{999}$

Ans. *a*) $3\sqrt{7}$, *b*) $4\sqrt{6}$, *c*) $8\sqrt{7}$, *d*) $2\sqrt{2}$, *e*) $2\sqrt{3}$, *f*) $7\sqrt{5}$, *g*) $50\sqrt{7}$, *h*) $30\sqrt{5}$
i) $60\sqrt{6}$, *j*) $6\sqrt{2}$, *k*) $2\sqrt{10}$, *l*) $5\sqrt{111}$

17. Using $\sqrt{2} = 1.414$ or $\sqrt{5} = 2.236$, evaluate to the nearest tenth: **(6.3)**

a) $\frac{2}{3}\sqrt{18}$ *Ans.* $2\sqrt{2} = 2.8$ *c*) $10\sqrt{20}$ *Ans.* $20\sqrt{5} = 44.7$ *e*) $\frac{1}{4}\sqrt{8000}$ *Ans.* $10\sqrt{5} = 22.4$

b) $\frac{1}{3}\sqrt{45}$ *Ans.* $\sqrt{5} = 2.2$ *d*) $5\sqrt{98}$ *Ans.* $35\sqrt{2} = 49.5$ *f*) $\frac{1}{2}\sqrt{16200}$ *Ans.* $45\sqrt{2} = 63.6$

18. Simplify: **(6.4)**

$a)$ $\sqrt{49m^2}$ *Ans.* $7m$ | $e)$ $\sqrt{7r^2s^2}$ *Ans.* $rs\sqrt{7}$ | $i)$ $\sqrt{400a^8b^{10}}$ *Ans.* $20a^4b^5$

$b)$ $\sqrt{100a^2b^2}$ *Ans.* $10ab$ | $f)$ $\sqrt{4s^2t^2u}$ *Ans.* $2st\sqrt{u}$ | $j)$ $\sqrt{25x^3}$ *Ans.* $5x\sqrt{x}$

$c)$ $\sqrt{64r^2p^2q^2}$ *Ans.* $8rpq$ | $g)$ $\sqrt{81x^4}$ *Ans.* $9x^2$ | $k)$ $\sqrt{9u^5w^{12}}$ *Ans.* $3u^2w^6\sqrt{u}$

$d)$ $\sqrt{4p^2q}$ *Ans.* $2p\sqrt{q}$ | $h)$ $\sqrt{144y^6}$ *Ans.* $12y^3$ | $l)$ $\sqrt{27t^3x^5}$ *Ans.* $3tx^2\sqrt{3tx}$

19. Simplify: **(7.1)**

$a)$ $\sqrt{6\frac{1}{4}}$ *Ans.* $\frac{5}{2}$ | $c)$ $\sqrt{\frac{2500}{81}}$ *Ans.* $\frac{50}{9}$ | $e)$ $\sqrt{\frac{441}{169}}$ *Ans.* $\frac{21}{13}$ | $g)$ $\sqrt{\frac{4}{9}}\sqrt{\frac{25}{16}}$ *Ans.* $\frac{5}{6}$

$b)$ $\sqrt{7\frac{1}{9}}$ *Ans.* $\frac{8}{3}$ | $d)$ $\sqrt{\frac{9}{6400}}$ *Ans.* $\frac{3}{80}$ | $f)$ $\sqrt{\frac{289}{10000}}$ *Ans.* $\frac{17}{100}$ | $h)$ $\sqrt{\frac{36}{225}}\sqrt{\frac{1}{4}}\sqrt{\frac{25}{16}}$ *Ans.* $\frac{1}{4}$

20. Simplify: **(7.2)**

$a)$ $\sqrt{\frac{5}{16}}$ *Ans.* $\frac{\sqrt{5}}{4}$ | $c)$ $\sqrt{\frac{8}{25}}$ *Ans.* $\frac{2}{5}\sqrt{2}$ | $e)$ $10\sqrt{\frac{x}{25}}$ *Ans.* $2\sqrt{x}$ | $g)$ $\frac{1}{y}\sqrt{\frac{y^3}{900}}$ *Ans.* $\frac{\sqrt{y}}{30}$

$b)$ $9\sqrt{\frac{13}{9}}$ *Ans.* $3\sqrt{13}$ | $d)$ $21\sqrt{\frac{32}{49}}$ *Ans.* $12\sqrt{2}$ | $f)$ $\sqrt{\frac{150}{x^4}}$ *Ans.* $\frac{5}{x^2}\sqrt{6}$ | $h)$ $y^4\sqrt{\frac{45}{y^8}}$ *Ans.* $3y\sqrt{5}$

21. Simplify: **(7.3)**

$a)$ $\sqrt{\frac{1}{7}}$ *Ans.* $\frac{\sqrt{7}}{7}$ | $c)$ $\sqrt{\frac{5}{18}}$ *Ans.* $\frac{\sqrt{10}}{6}$ | $e)$ $\sqrt{\frac{1}{b}}$ *Ans.* $\frac{1}{b}\sqrt{b}$ | $g)$ $\sqrt{\frac{2s}{g}}$ *Ans.* $\frac{1}{g}\sqrt{2sg}$

$b)$ $\sqrt{\frac{3}{11}}$ *Ans.* $\frac{\sqrt{33}}{11}$ | $d)$ $5\sqrt{\frac{7}{20}}$ *Ans.* $\frac{\sqrt{35}}{2}$ | $f)$ $\sqrt{\frac{1}{c^3}}$ *Ans.* $\frac{1}{c^2}\sqrt{c}$ | $h)$ $\sqrt{\frac{x^3}{50}}$ *Ans.* $\frac{x}{10}\sqrt{2x}$

22. Using $\sqrt{2} = 1.414$ or $\sqrt{3} = 1.732$, evaluate each to the nearest hundredth: **(7.4)**

$a)$ $\sqrt{\frac{1}{2}}$ $b)$ $\sqrt{\frac{1}{3}}$ $c)$ $\sqrt{\frac{9}{50}}$ $d)$ $6\sqrt{\frac{1}{12}}$ $e)$ $16\sqrt{\frac{9}{32}}$ $f)$ $45\sqrt{\frac{4}{27}}$ $g)$ $\sqrt{2\frac{1}{12}}$ $h)$ $\sqrt{2\frac{13}{18}}$

Ans. $a)$ $\frac{1}{2}\sqrt{2} = .71$ | $c)$ $\frac{3}{10}\sqrt{2} = .42$ | $e)$ $6\sqrt{2} = 8.48$ | $g)$ $\frac{5}{6}\sqrt{3} = 1.44$

$b)$ $\frac{1}{3}\sqrt{3} = .58$ | $d)$ $\sqrt{3} = 1.73$ | $f)$ $10\sqrt{3} = 17.32$ | $h)$ $\frac{7}{6}\sqrt{2} = 1.65$

23. Combine: **(8.1)**

$a)$ $5\sqrt{3} + \sqrt{3}$ | $c)$ $3\sqrt{2} + 4\sqrt{3} - 2\sqrt{2}$ | $e)$ $x\sqrt{15} + 2x\sqrt{15}$ | $g)$ $12\sqrt{b} - 2\sqrt{a} + 6\sqrt{b}$

$b)$ $7\sqrt{11} - 10\sqrt{11}$ | $d)$ $5\sqrt{5} - 2\sqrt{5} + \sqrt{10}$ | $f)$ $11\sqrt{a} + \sqrt{a} - 3\sqrt{a}$ | $h)$ $5b\sqrt{c} + 3b\sqrt{c} + c\sqrt{b}$

Ans. $a)$ $6\sqrt{3}$ | $c)$ $\sqrt{2} + 4\sqrt{3}$ | $e)$ $3x\sqrt{15}$ | $g)$ $18\sqrt{b} - 2\sqrt{a}$

$b)$ $-3\sqrt{11}$ | $d)$ $3\sqrt{5} + \sqrt{10}$ | $f)$ $9\sqrt{a}$ | $h)$ $8b\sqrt{c} + c\sqrt{b}$

24. Simplify and combine: **(8.2)**

$a)$ $\sqrt{2} + \sqrt{32}$ | $e)$ $2\sqrt{27} - 4\sqrt{12}$ | $i)$ $8\sqrt{3} - 6\sqrt{\frac{1}{3}}$

$b)$ $2\sqrt{5} + 3\sqrt{20}$ | $f)$ $\sqrt{700} - 2\sqrt{63}$ | $j)$ $10\sqrt{18} + 20\sqrt{\frac{1}{2}}$

$c)$ $2\sqrt{28} - 3\sqrt{7}$ | $g)$ $-3\sqrt{90} - 5\sqrt{40}$ | $k)$ $6\sqrt{\frac{5}{12}} - \sqrt{60}$

$d)$ $x\sqrt{3} + x\sqrt{27}$ | $h)$ $\sqrt{9y} + 2\sqrt{25y}$ | $l)$ $\sqrt{\frac{a^2}{2}} - \sqrt{\frac{a^2}{8}}$

Ans. $a)$ $\sqrt{2} + 4\sqrt{2} = 5\sqrt{2}$ | $e)$ $6\sqrt{3} - 8\sqrt{3} = -2\sqrt{3}$ | $i)$ $8\sqrt{3} - 2\sqrt{3} = 6\sqrt{3}$

$b)$ $2\sqrt{5} + 6\sqrt{5} = 8\sqrt{5}$ | $f)$ $10\sqrt{7} - 6\sqrt{7} = 4\sqrt{7}$ | $j)$ $30\sqrt{2} + 10\sqrt{2} = 40\sqrt{2}$

$c)$ $4\sqrt{7} - 3\sqrt{7} = \sqrt{7}$ | $g)$ $-9\sqrt{10} - 10\sqrt{10} = -19\sqrt{10}$ | $k)$ $\sqrt{15} - 2\sqrt{15} = -\sqrt{15}$

$d)$ $x\sqrt{3} + 3x\sqrt{3} = 4x\sqrt{3}$ | $h)$ $3\sqrt{y} + 10\sqrt{y} = 13\sqrt{y}$ | $l)$ $\frac{a}{2}\sqrt{2} - \frac{a}{4}\sqrt{2} = \frac{a}{4}\sqrt{2}$

25. Multiply: **(9.1)**

a) $4\sqrt{2}\cdot5\sqrt{3}$ *d*) $\sqrt{2}\ \sqrt{3}\ \sqrt{12}$ *g*) $\sqrt{21}\ \sqrt{3}$ *j*) $\frac{1}{2}\sqrt{3}\cdot6\sqrt{18}$

b) $7\sqrt{7}\cdot10\sqrt{10}$ *e*) $3\sqrt{6}\cdot5\sqrt{24}$ *h*) $\sqrt{10}\ \sqrt{20}$ *k*) $\frac{2}{3}\sqrt{7}\cdot15\sqrt{14}$

c) $4\sqrt{3}\cdot3\sqrt{2}\cdot6\sqrt{5}$ *f*) $10\sqrt{\frac{1}{3}}\cdot7\sqrt{75}$ *i*) $\sqrt{8}\ \sqrt{12}\ \sqrt{10}$ *l*) $\frac{3}{2}\sqrt{24}\cdot4\sqrt{3}$

Ans. *a*) $20\sqrt{6}$ *d*) $6\sqrt{2}$ *g*) $\sqrt{63}=3\sqrt{7}$ *j*) $3\sqrt{54}=9\sqrt{6}$

b) $70\sqrt{70}$ *e*) $15\sqrt{144}=180$ *h*) $\sqrt{200}=10\sqrt{2}$ *k*) $10\sqrt{98}=70\sqrt{2}$

c) $72\sqrt{30}$ *f*) $70\sqrt{25}=350$ *i*) $\sqrt{960}=8\sqrt{15}$ *l*) $6\sqrt{72}=36\sqrt{2}$

26. Multiply: **(9.1)**

a) $2\sqrt{a}\cdot4\sqrt{b}$ *d*) $\sqrt{pq}\ \sqrt{p}\ \sqrt{q}$ *g*) $\sqrt{5a}\ \sqrt{3a}$ *j*) $\sqrt{\frac{y}{x}}\ \sqrt{\frac{x}{y}}$

b) $c\sqrt{5}\cdot d\sqrt{7}$ *e*) $\sqrt{4m}\ \sqrt{25m}$ *h*) $\sqrt{\frac{1}{2}b}\ \sqrt{40b}$ *k*) $3x\sqrt{x}\cdot5x\sqrt{x}$

c) $r\sqrt{x}\cdot s\sqrt{y}$ *f*) $\sqrt{3n}\ \sqrt{27n^3}$ *i*) $\sqrt{8h^3}\ \sqrt{2k^2}$ *l*) $y^2\sqrt{y}\cdot y\sqrt{y^2}$

Ans. *a*) $8\sqrt{ab}$ *d*) $\sqrt{p^2q^2}=pq$ *g*) $\sqrt{15a^2}=a\sqrt{15}$ *j*) 1

b) $cd\sqrt{35}$ *e*) $\sqrt{100m^2}=10m$ *h*) $\sqrt{20b^2}=2b\sqrt{5}$ *k*) $15x^2\sqrt{x^2}=15x^3$

c) $rs\sqrt{xy}$ *f*) $\sqrt{81n^4}=9n^2$ *i*) $\sqrt{16h^3k^2}=4hk\sqrt{h}$ *l*) $y^3\sqrt{y^3}=y^4\sqrt{y}$

27. Multiply: **(9.2)**

a) $5\sqrt{10}\cdot\sqrt{10}$ *d*) $(\sqrt{14})^2$ *g*) $\sqrt{7x}\cdot\sqrt{7x}$ *j*) $5\sqrt{2b}\cdot4\sqrt{2b}$

b) $\sqrt{12}\cdot\frac{1}{2}\sqrt{12}$ *e*) $(3\sqrt{6})^2$ *h*) $x\sqrt{5}\cdot y\sqrt{5}$ *k*) $(2x\sqrt{2x})^2$

c) $2\sqrt{\frac{2}{3}}\cdot3\sqrt{\frac{3}{2}}$ *f*) $(\frac{1}{2}\sqrt{8})^2$ *i*) $(7\sqrt{2a})^2$ *l*) $(c\sqrt{\frac{5}{c}})^2$

Ans. *a*) $5\cdot10=50$ *d*) 14 *g*) $7x$ *j*) $20(2b)=40b$

b) $\frac{1}{2}\cdot12=6$ *e*) $9\cdot6=54$ *h*) $5xy$ *k*) $4x^2(2x)=8x^3$

c) 6 *f*) $\frac{1}{4}\cdot8=2$ *i*) $49(2a)=98a$ *l*) $c^2(\frac{5}{c})=5c$

28. Multiply: **(9.3)**

a) $\sqrt{2}(5+\sqrt{2})$ *c*) $3\sqrt{3}(\sqrt{27}+7\sqrt{3})$ *e*) $\sqrt{6}(\sqrt{6x}+\sqrt{24x})$

b) $\sqrt{5}(2\sqrt{5}-3\sqrt{20})$ *d*) $\sqrt{x}(\sqrt{16x}-\sqrt{9x})$ *f*) $\sqrt{8}(\sqrt{2x}+\sqrt{18y})$

Ans. *a*) $5\sqrt{2}+2$ *c*) $27+63=90$ *e*) $6\sqrt{x}+12\sqrt{x}=18\sqrt{x}$

b) $10-30=-20$ *d*) $4x-3x=x$ *f*) $4\sqrt{x}+12\sqrt{y}$

29. Multiply: **(9.4)**

a) $(6+\sqrt{3})(6-\sqrt{3})$ *d*) $(3+\sqrt{2})^2$ *g*) $(2-\sqrt{6})(3-\sqrt{6})$

b) $(\sqrt{5}-2)(\sqrt{5}+2)$ *e*) $(\sqrt{11}-3)^2$ *h*) $(\sqrt{7}+3)(\sqrt{7}+2)$

c) $(\sqrt{7}+\sqrt{2})(\sqrt{7}-\sqrt{2})$ *f*) $(\sqrt{3}+\sqrt{5})^2$ *i*) $(4\sqrt{2}-\sqrt{5})(3\sqrt{2}+2\sqrt{5})$

Ans. *a*) $36-3=33$ *d*) $9+6\sqrt{2}+2=11+6\sqrt{2}$ *g*) $6-5\sqrt{6}+6=12-5\sqrt{6}$

b) $5-4=1$ *e*) $11-6\sqrt{11}+9=20-6\sqrt{11}$ *h*) $7+5\sqrt{7}+6=13+5\sqrt{7}$

c) $7-2=5$ *f*) $3+2\sqrt{15}+5=8+2\sqrt{15}$ *i*) $24+5\sqrt{10}-10=14-5\sqrt{10}$

30. Divide: **(10.1)**

a) $\frac{15\sqrt{60}}{5\sqrt{15}}$ *Ans.* $3\sqrt{4}=6$ *d*) $\frac{\sqrt{120}}{2\sqrt{5}}$ *Ans.* $\frac{\sqrt{24}}{2}=\sqrt{6}$ *g*) $\frac{5c\sqrt{5c}}{c\sqrt{c}}$ *Ans.* $5\sqrt{5}$

b) $\frac{24\sqrt{2}}{3\sqrt{32}}$ *Ans.* $8\sqrt{\frac{1}{16}}=2$ *e*) $\frac{3\sqrt{24}}{\sqrt{2}}$ *Ans.* $3\sqrt{12}=6\sqrt{3}$ *h*) $\frac{3x\sqrt{x^5}}{9x\sqrt{x}}$ *Ans.* $\frac{1}{3}\sqrt{x^4}=\frac{x^2}{3}$

c) $\frac{6\sqrt{3}}{3\sqrt{27}}$ *Ans.* $2\sqrt{\frac{1}{9}}=\frac{2}{3}$ *f*) $\frac{14\sqrt{a^3}}{2\sqrt{a}}$ *Ans.* $7\sqrt{a^2}=7a$ *i*) $\frac{a\sqrt{ab}}{ab\sqrt{b}}$ *Ans.* $\frac{1}{b}\sqrt{a}$

31. Divide: **(10.2)**

a) $\frac{\sqrt{200}+\sqrt{50}}{\sqrt{5}}$ *Ans.* $\sqrt{40}+\sqrt{10} = 3\sqrt{10}$

b) $\frac{9\sqrt{200}-12\sqrt{32}}{3\sqrt{2}}$ *Ans.* $3\sqrt{100}-4\sqrt{16} = 14$

c) $\frac{8\sqrt{2}-4\sqrt{8}}{\sqrt{32}}$ *Ans.* $8\sqrt{\frac{1}{16}}-4\sqrt{\frac{1}{4}} = 0$

d) $\frac{\sqrt{20}+\sqrt{80}}{2\sqrt{5}}$ *Ans.* $\frac{\sqrt{4}}{2}+\frac{\sqrt{16}}{2} = 3$

e) $\frac{\sqrt{27a}-2\sqrt{3a}}{\sqrt{a}}$ *Ans.* $\sqrt{27}-2\sqrt{3} = \sqrt{3}$

f) $\frac{\sqrt{y^7}-\sqrt{y^5}}{\sqrt{y^3}}$ *Ans.* $\sqrt{y^4}-\sqrt{y^2} = y^2-y$

32. Rationalize each denominator: **(11.1)**

a) $\frac{2}{\sqrt{5}}$ *Ans.* $\frac{2\sqrt{5}}{5}$

b) $\frac{12}{\sqrt{2}}$ *Ans.* $6\sqrt{2}$

c) $\frac{10}{\sqrt{50}}$ *Ans.* $\sqrt{2}$

d) $\frac{\sqrt{5}}{\sqrt{15}}$ *Ans.* $\frac{\sqrt{3}}{3}$

e) $\frac{12}{\sqrt{6}}$ *Ans.* $2\sqrt{6}$

f) $\frac{\sqrt{2}}{\sqrt{3}}$ *Ans.* $\frac{\sqrt{6}}{3}$

g) $\frac{4+\sqrt{2}}{\sqrt{2}}$ *Ans.* $2\sqrt{2}+1$

h) $\frac{3\sqrt{3}-6}{\sqrt{3}}$ *Ans.* $3-2\sqrt{3}$

i) $\frac{16-2\sqrt{2}}{\sqrt{8}}$ *Ans.* $4\sqrt{2}-1$

j) $\frac{5}{\sqrt{y}}$ *Ans.* $\frac{5\sqrt{y}}{y}$

k) $\frac{2a}{\sqrt{10}}$ *Ans.* $\frac{a\sqrt{10}}{5}$

l) $\frac{3x}{\sqrt{x}}$ *Ans.* $3\sqrt{x}$

33. Evaluate to the nearest tenth, after rationalizing the denominator: **(11.2)**

a) $\frac{2}{\sqrt{2}}$ *Ans.* $\sqrt{2} = 1.4$

b) $\frac{12}{\sqrt{6}}$ *Ans.* $2\sqrt{6} = 4.9$

c) $\frac{10\sqrt{2}}{\sqrt{5}}$ *Ans.* $2\sqrt{10} = 6.3$

d) $\frac{40}{\sqrt{8}}$ *Ans.* $10\sqrt{2} = 14.1$

e) $\frac{30}{\sqrt{45}}$ *Ans.* $2\sqrt{5} = 4.5$

f) $\frac{9\sqrt{5}}{\sqrt{15}}$ *Ans.* $3\sqrt{3} = 5.2$

g) $\frac{12-\sqrt{3}}{\sqrt{3}}$ *Ans.* $4\sqrt{3}-1 = 5.9$

h) $\frac{1+\sqrt{5}}{\sqrt{5}}$ *Ans.* $\frac{\sqrt{5}+5}{5} = 1.4$

i) $\frac{\sqrt{20}+\sqrt{5}}{\sqrt{10}}$ *Ans.* $\frac{3}{2}\sqrt{2} = 2.1$

34. Solve: **(12.1)**

a) $\sqrt{3x} = 6$ *Ans.* $x = 12$

b) $3\sqrt{x} = 15$ *Ans.* $x = 25$

c) $8 = 2\sqrt{y}$ *Ans.* $y = 16$

d) $5\sqrt{a} = 2$ *Ans.* $a = \frac{4}{25}$

e) $\frac{1}{2}\sqrt{b} = 5$ *Ans.* $b = 100$

f) $\frac{2}{3}\sqrt{c} = 8$ *Ans.* $c = 144$

g) $\sqrt{c+3} = 4$ *Ans.* $c = 13$

h) $\sqrt{d-5} = 3$ *Ans.* $d = 14$

i) $\sqrt{3h} = 5$ *Ans.* $h = \frac{25}{3}$

j) $\sqrt{10-x} = 2$ *Ans.* $x = 6$

k) $\sqrt{2r-5} = 3$ *Ans.* $r = 7$

l) $\sqrt{54-3s} = 6$ *Ans.* $s = 6$

35. Solve: **(12.2)**

a) $2\sqrt{a}-3 = 7$

b) $8+3\sqrt{b} = 20$

c) $7-\sqrt{2b} = 3$

d) $8\sqrt{r}-5 = \sqrt{r}+9$

e) $20-3\sqrt{t} = \sqrt{t}-4$

f) $2\sqrt{5x}-3 = 7$

g) $\sqrt{2x-7}+8 = 11$

h) $22 = 17+\sqrt{40-3y}$

i) $30-\sqrt{20-2y} = 26$

Ans. a) $a=25$, b) $b=16$, c) $b=8$, d) $r=4$, e) $t=36$, f) $x=5$, g) $x=8$, h) $y=5$, i) $y=2$

36. Solve and show, by checking, that each value for the unknown is extraneous: **(12.2)**

a) $\sqrt{a} = -2$

b) $-\sqrt{b} = 3$

c) $5+\sqrt{c} = 4$

d) $6-\sqrt{r} = 10$

e) $2\sqrt{t}+5 = \sqrt{t}$

f) $10-\sqrt{3v} = 12$

g) $\sqrt{y+5} = -4$

h) $3-\sqrt{12-x} = 8$

37. Solve for the letter indicated: **(12.3)**

a) $a = \sqrt{2b}$ for b

b) $g = 3\sqrt{h}$ for h

c) $g = \sqrt{3h}$ for h

d) $s = \frac{\sqrt{t}}{3}$ for t

e) $2u = \sqrt{\frac{v}{2}}$ for v

f) $h = \sqrt{3g-4}$ for g

g) $R = \frac{1}{2}\sqrt{\frac{A}{\pi}}$ for A

h) $R = \sqrt{\frac{3V}{\pi H}}$ for V

Ans. a) $b = \frac{a^2}{2}$ b) $h = \frac{g^2}{9}$ c) $h = \frac{g^2}{3}$ d) $t = 9s^2$ e) $v = 8u^2$ f) $g = \frac{h^2+4}{3}$ g) $A = 4\pi R^2$ h) $V = \frac{1}{3}\pi R^2 H$

Chapter 13

QUADRATIC EQUATIONS in ONE UNKNOWN

1. UNDERSTANDING QUADRATIC EQUATIONS IN ONE UNKNOWN

A quadratic equation in one unknown is an equation in which the highest power of the unknown is the second.

Thus, $2x^2+3x-5=0$ is a quadratic equation in x.

Standard Quadratic Equation Form: $ax^2+bx+c=0$

The **standard form of a quadratic equation** in one unknown is $ax^2+bx+c=0$ where a, b and c represent known numbers and x represents the unknown number. The number a cannot equal 0. Thus, $3x^2-5x+6=0$ is in standard form. Here, $a=3$, $b=-5$ and $c=6$.

To Transform a Quadratic Equation into Standard Form, $ax^2+bx+c=0$

(1) **Remove parentheses:** Thus, $x(x+1)-5=0$ becomes $x^2+x-5=0$.

(2) **Clear of fractions:** Thus, $x-4+\frac{3}{x}=0$ becomes $x^2-4x+3=0$.

(3) **Remove radical signs:** Thus, $\sqrt{x^2-3x}=2$ becomes $x^2-3x-4=0$.

(4) **Collect like terms:** Thus, $x^2+7x=2x+6$ becomes $x^2+5x-6=0$.

1.1. Expressing Quadratic Equations in Standard Form

Express each quadratic equation in the standard form, $ax^2+bx+c=0$, so that a has a positive value:

a) $x(x+1)=6$	b) $\frac{3}{x}+x=4$	c) $x^2=25-5x-15$	d) $2x=\sqrt{x+3}$
Solutions:			
Remove ( ):	**Clear fractions:**	**Collect like terms:**	**Remove radical signs:**
$x^2+x=6$	$3+x^2=4x$	$x^2=10-5x$	Squaring: $4x^2=x+3$
Ans. $x^2+x-6=0$	*Ans.* $x^2-4x+3=0$	*Ans.* $x^2+5x-10=0$	*Ans.* $4x^2-x-3=0$

1.2. Values of a, b and c in Standard Quadratic Equation Form

Express each quadratic equation in standard form, so that a has a positive value. Then state the values of a, b and c.

	Standard Form	a	b	c
a) $x^2-9x=10$	*Ans.* $x^2-9x-10=0$	1	-9	-10
b) $5x^2=125$	*Ans.* $5x^2-125=0$	5	0	-125
c) $2x^2=8x$	*Ans.* $2x^2-8x=0$	2	-8	0
d) $2x^2+9x=2x-3$	*Ans.* $2x^2+7x+3=0$	2	$+7$	$+3$
e) $x(x+3)=10$	*Ans.* $x^2+3x-10=0$	1	$+3$	-10
f) $5(x^2+2)=7(x+3)$	*Ans.* $5x^2-7x-11=0$	5	-7	-11
g) $x-5=\frac{7}{x}$	*Ans.* $x^2-5x-7=0$	1	-5	-7
h) $\sqrt{2x^2-1}=x+2$	*Ans.* $x^2-4x-5=0$	1	-4	-5

2. SOLVING QUADRATIC EQUATIONS BY FACTORING

Rule 1. Every quadratic equation has two roots.

Thus, $x^2 = 9$ has two roots, 3 and -3; that is, $x = \pm 3$.

Rule 2. If the product of two factors is zero, then one or the other of the factors or both of them must equal zero.

Thus: **(1)** In $5(x-3) = 0$, the factor $x-3 = 0$.
(2) In $(x-2)(x-3) = 0$, either $x-2 = 0$ or $x-3 = 0$.
(3) In $(x-3)(x-3) = 0$, both factors $x-3 = 0$.
(4) In $(x-3)(y-5) = 0$, either $x-3 = 0$, $y-5 = 0$ or both equal 0.

To Solve a Quadratic Equation by Factoring

Procedure:	Solve: $x(x-4) = 5$ Solution:	
1. Express in form $ax^2 + bx + c = 0$:	**1.** S_5 $\quad x^2 - 4x = 5$ $x^2 - 4x - 5 = 0$	
2. Factor $ax^2 + bx + c$:	**2.** $(x-5)(x+1) = 0$	
3. Let each factor $= 0$:	**3.** $x - 5 = 0$	$x + 1 = 0$
4. Solve each resulting equation:	**4.** $x = 5$	$x = -1$
5. Check each root in original equation:	**5. Check** in $x(x-4) = 5$: If $x=5$, $5(1) \stackrel{?}{=} 5$ $5 = 5$	If $x \stackrel{?}{=} -1$, $(-1)(-5) = 5$ $5 = 5$

Ans. $x = 5$ or -1

2.1. Solving Quadratic Equations by Factoring

Procedure:	Solve: *a)* $x^2 - x = 6$ Solutions:	*b)* $x + \frac{16}{x} = 8$
1. Express as $ax^2 + bx + c = 0$:	**1.** $x^2 - x - 6 = 0$	**1.** $x^2 + 16 = 8x$ $x^2 - 8x + 16 = 0$
2. Factor $ax^2 + bx + c$:	**2.** $(x-3)(x+2) = 0$	**2.** $(x-4)(x-4) = 0$
3. Let each factor $= 0$:	**3.** $x-3 = 0$ | $x+2 = 0$	**3.** $x-4 = 0$ | $x-4 = 0$
4. Solve:	**4.** $x = 3$ | $x = -2$	**4.** $x = 4$ | $x = 4$
5. Check *(in original equation)*:	**5.** *(Check to be done by the student.)*	
	Ans. $x=3$ or -2	**Ans.** $x=4$ (equal roots)

2.2. Solving Quadratic Equations in Standard Form

Solve by factoring:

a) $x^2 + 9x + 20 = 0$	*b)* $a^2 - 49 = 0$	*c)* $3r^2 - 21r = 0$
Solutions:		
$(x+4)(x+5) = 0$	$(a+7)(a-7) = 0$	$3r(r-7) = 0$
$x + 4 = 0$ | $x + 5 = 0$	$a + 7 = 0$ | $a - 7 = 0$	$3r = 0$ | $r - 7 = 0$
$x = -4$ | $x = -5$	$a = -7$ | $a = +7$	$r = 0$ | $r = +7$
Ans. $x = -4$ or -5	*Ans.* $a = -7$ or $+7$	*Ans.* $r = 0$ or $+7$

2.3. More Difficult Solutions by Factoring

Solve by factoring:

a) $2 + \frac{5}{x} = \frac{12}{x^2}$

Solutions:

Clear of fractions:

$2x^2 + 5x = 12$

$2x^2 + 5x - 12 = 0$

$(2x-3)(x+4) = 0$

$2x - 3 = 0$, $x = \frac{3}{2}$ | $x + 4 = 0$, $x = -4$

Ans. $x = \frac{3}{2}$ or -4

b) $(y-8)(2y-3) = 34$

Remove ( ):

$2y^2 - 19y + 24 = 34$

$2y^2 - 19y - 10 = 0$

$(2y+1)(y-10) = 0$

$2y + 1 = 0$, $y = -\frac{1}{2}$ | $y - 10 = 0$, $y = 10$

Ans. $y = -\frac{1}{2}$ or 10

c) $x - 6 = \sqrt{x}$

Remove radical sign:

Sq $x^2 - 12x + 36 = x$

$x^2 - 13x + 36 = 0$

$(x-9)(x-4) = 0$

$x - 9 = 0$, $x = 9$ | $x - 4 = 0$, $x = 4$

Ans. $x = 9$
(Show that 4 is extraneous.)

2.4. Number Problems Involving Quadratic Equations

a) Find a number whose square is 2 less than three times the number.

Solutions:

Let n = the number.

$n^2 = 3n - 2$

$n^2 - 3n + 2 = 0$

$(n-2)(n-1) = 0$

$n = 2$ or 1

Ans. 2 or 1

b) Find two consecutive integers such that the sum of their squares is 25.

Let n and $n+1$ = the consecutive integers.

$n^2 + (n+1)^2 = 25$

$n^2 + n^2 + 2n + 1 = 25$

$2n^2 + 2n - 24 = 0$

$n^2 + n - 12 = 0$

$(n+4)(n-3) = 0$

$n = -4$, $n + 1 = -3$ | $n = 3$, $n + 1 = 4$

Ans. Either -4 and -3 or 3 and 4.

2.5. Area Problems Involving Quadratic Equations

The length of a rectangular lot is 3 yd. more than its width. Find its dimensions if the area

a) equals 40 sq. yd. | *b*) increases 70 sq. yd. when the width is doubled.

Solutions:

In (*a*) and (*b*), let w = width in yd. and $(w+3)$ = length in yd.

(*a*)

Area of lot = 40 sq. yd.

$w(w+3) = 40$

$w^2 + 3w - 40 = 0$

$(w-5)(w+8) = 0$

$w = 5$, $l = 8$ | $w = -8$ Reject since width cannot be negative.

Ans. 8 yd. and 5 yd.

(*b*)

New area = old area + 70 sq. yd.

$2w(w+3) = w(w+3) + 70$

$2w^2 + 6w = w^2 + 3w + 70$

$w^2 + 3w - 70 = 0$

$(w-7)(w+10) = 0$

$w = 7$, $l = 10$ | $w = -10$ Reject.

Ans. 10 yd. and 7 yd.

3. SOLVING INCOMPLETE QUADRATIC EQUATIONS

An **incomplete quadratic equation** in one unknown lacks either

(1) the term containing the first power of the unknown as $x^2 - 4 = 0$, or

(2) the constant term as $x^2 - 4x = 0$.

Rule. If an incomplete quadratic equation lacks the constant term, then one of the roots is zero.

Thus, if $x^2 - 4x = 0$, $x = 0$ or 4.

To Solve an Incomplete Quadratic Equation Lacking the First Power of Unknown

Solve: $2(x^2-8) = 11 - x^2$

Procedure:	**Solution:**
1. **Express in form,** $ax^2 = k$ where k is a constant:	1. **Tr** $2x^2 - 16 = 11 - x^2$ $3x^2 = 27$
2. **Divide** both sides by a, obtaining $x^2 = \frac{k}{a}$:	2. Dividing by 3, $x^2 = 9$ (*See note.*)
3. **Take the square root of both sides**, obtaining $x = \pm\sqrt{\frac{k}{a}}$:	3. Taking a sq. rt. of both sides, $x = \pm 3$
4. **Check** each root in the original equation:	4. (*Check is left to the student.*)

Note. $x^2=9$ may be solved by factoring, as follows: $x^2-9=0$, $(x+3)(x-3)=0$, $x=\pm 3$.

To Solve an Incomplete Quadratic Equation Lacking Constant Term

Solve: $3x^2 = 18x$

Procedure:	**Solution:**
1. **Express in form** $ax^2 + bx = 0$:	1. $3x^2 - 18x = 0$
2. **Factor** $ax^2 + bx$:	2. $3x(x-6) = 0$
3. **Let each factor** $= 0$:	3. $3x = 0$ | $x - 6 = 0$
4. **Solve** each resulting equation:	4. $x = 0$ | $x = 6$
5. **Check** each root in the original equation:	5. Check in $3x^2 = 18x$: If $x = 0$, $3(0^2) \stackrel{?}{=} 18(0)$, $0 = 0$ | If $x = 6$, $3(6^2) \stackrel{?}{=} 18(6)$, $108 = 108$

3.1. Solving Incomplete Quadratic Equations Lacking First Power of Unknown

Solve: (*Leave irrational answers in radical form.*)

a) $4x^2 - 49 = 0$	*b*) $2y^2 = 125 - 3y^2$	*c*) $9x^2 - 2 = 5$	*d*) $\frac{8x}{25} = \frac{16}{x}$
Solutions:			
$\mathbf{D}_4$ $4x^2 = 49$	$\mathbf{D}_5$ $5y^2 = 125$	$\mathbf{D}_9$ $9x^2 = 7$	$\mathbf{D}_8$ $8x^2 = 400$
Sq Rt $x^2 = \frac{49}{4}$	**Sq Rt** $y^2 = 25$	**Sq Rt** $x^2 = \frac{7}{9}$	**Sq Rt** $x^2 = 50$
Ans. $x = \pm\frac{7}{2}$	*Ans.* $y = \pm 5$	*Ans.* $x = \pm\frac{\sqrt{7}}{3}$	*Ans.* $x = \pm 5\sqrt{2}$

3.2. Solving Incomplete Quadratic Equations Lacking Constant Term

Solve:

a) $3x^2 = 21x$	*b*) $5y(y-6) - 8y = 2y$	*c*) $\frac{x^2}{5} = \frac{x}{15}$	*d*) $3(y^2+8) = 24 - 15y$
Solutions:			
	$5y^2 - 30y - 8y = 2y$	$15x^2 = 5x$	$3y^2 + 24 = 24 - 15y$
$3x^2 - 21x = 0$	$5y^2 - 40y = 0$	$15x^2 - 5x = 0$	$3y^2 + 15y = 0$
$3x(x-7) = 0$	$5y(y-8) = 0$	$5x(3x-1) = 0$	$3y(y+5) = 0$
$3x = 0$ | $x - 7 = 0$	$5y = 0$ | $y - 8 = 0$	$5x = 0$ | $3x - 1 = 0$	$3y = 0$ | $y + 5 = 0$
$x = 0$ | $x = 7$	$y = 0$ | $y = 8$	$x = 0$ | $x = \frac{1}{3}$	$y = 0$ | $y = -5$
Ans. $x = 0$ or 7	*Ans.* $y = 0$ or 8	*Ans.* $x = 0$ or $\frac{1}{3}$	*Ans.* $y = 0$ or -5

3.3. Solving Incomplete Quadratics for an Indicated Letter

Solve for letter indicated:

a) Solve for x: $9x^2 - 64y^2 = 0$	*b*) Solve for y: $4y^2 - 95x^2 = 5x^2$	*c*) Solve for a: $3a^2 = b^2 - a^2$	*d*) Solve for a: $a(a-4) = b^2 - 4a$
Solutions:			
$\mathbf{D}_9$ $9x^2 = 64y^2$	$\mathbf{D}_4$ $4y^2 = 100x^2$	$\mathbf{D}_4$ $4a^2 = b^2$	$a^2 - 4a = b^2 - 4a$
Sq Rt $x^2 = \frac{64y^2}{9}$	**Sq Rt** $y^2 = 25x^2$	**Sq Rt** $a^2 = \frac{b^2}{4}$	**Sq Rt** $a^2 = b^2$
Ans. $x = \pm\frac{8y}{3}$	*Ans.* $y = \pm 5x$	*Ans.* $a = \pm\frac{b}{2}$	*Ans.* $a = \pm b$

3.4. Solving Formulas for an Indicated Letter

Solve for *positive* value of letter indicated:

a) Solve for r: $\pi r^2 = A$	*b*) Solve for r: $\pi r^2 h = V$	*c*) Solve for a: $a^2 + b^2 = c^2$	*d*) Solve for s: $\frac{s^2}{4} = A$
Solutions:			
$\mathbf{D}_\pi$ $\frac{\pi r^2}{\pi} = \frac{A}{\pi}$	$\mathbf{D}_{\pi h}$ $\frac{\pi r^2 h}{\pi h} = \frac{V}{\pi h}$	$\mathbf{S}_{b^2}$ $a^2 + b^2 = c^2$	$\mathbf{M}_4$ $4(\frac{s^2}{4}) = 4A$
Sq Rt $r^2 = \frac{A}{\pi}$	**Sq Rt** $r^2 = \frac{V}{\pi h}$	**Sq Rt** $a^2 = c^2 - b^2$	**Sq Rt** $s^2 = 4A$
Ans. $r = \sqrt{\frac{A}{\pi}}$	*Ans.* $r = \sqrt{\frac{V}{\pi h}}$	*Ans.* $a = \sqrt{c^2 - b^2}$	*Ans.* $s = 2\sqrt{A}$

4. SOLVING A QUADRATIC EQUATION BY COMPLETING THE SQUARE

The square of a binomial is a perfect trinomial square.

Thus, $x^2 + 6x + 9$ is the perfect trinomial square of $x+3$.

Rule. If x^2 is the first term of a perfect trinomial square and the term in x is also given, the last term may be found by squaring one-half the coefficient of x.

Thus, if $x^2 + 6x$ is given, 9 is needed to complete the perfect trinomial square, $x^2 + 6x + 9$. This last term, 9, is found by squaring $\frac{1}{2}$ of 6 or 3.

<u>To Solve a Quadratic Equation by Completing the Square</u>

<u>Procedure</u>:	Solve $x^2 + 6x - 7 = 0$ by completing the square. **Solution:**
1. Express the equation in the form of $x^2 + px = q$:	1. Change $x^2 + 6x - 7 = 0$ $x^2 + 6x = 7$
2. Square one-half the coefficient of x and add this to both sides:	2. The square of $\frac{1}{2}(6) = 3^2 = 9$. Add 9 to get $x^2 + 6x + \underline{9} = 7 + \underline{9}$
3. Replace the perfect trinomial square by its binomial squared:	3. $(x+3)^2 = 16$
4. Take a square root of both sides. Set the binomial equal to $\pm$ the square root of the number on the other side:	4. **Sq Rt**: $x + 3 = \pm 4$
5. Solve the two resulting equations:	5. $x + 3 = 4$, $x = 1$ | $x + 3 = -4$, $x = -7$
6. Check both roots in the original equation:	6. (*Check is left to the student.*) **Ans.** $x = 1$ or -7

4.1. Completing the Perfect Trinomial Square

Complete each perfect trinomial square and state its binomial squared:

a) $x^2 + 14x + ?$

Ans. a) The square of $\frac{1}{2}(14) = 7^2 = 49$. Add 49 to get $x^2 + 14x + 49 = (x+7)^2$

b) $x^2 - 20x + ?$

Ans. b) The square of $\frac{1}{2}(-20) = (-10)^2 = 100$. Add 100 to get $x^2 - 20x + 100 = (x-10)^2$

c) $x^2 + 5x + ?$

Ans. c) The square of $\frac{1}{2}(5) = (\frac{5}{2})^2 = \frac{25}{4}$. Add $\frac{25}{4}$ to get $x^2 + 5x + \frac{25}{4} = (x+\frac{5}{2})^2$

d) $y^2 + \frac{4}{3}y + ?$

Ans. d) The square of $\frac{1}{2}(\frac{4}{3}) = (\frac{2}{3})^2 = \frac{4}{9}$. Add $\frac{4}{9}$ to get $y^2 + \frac{4}{3}y + \frac{4}{9} = (y+\frac{2}{3})^2$

4.2. Solving Quadratic Equations by Completing the Square

Solve by completing the square: a) $x^2 + 14x - 32 = 0$ | b) $x(x+18) = 19$

<u>**Procedure:**</u>	**Solutions:** a)	b)
1. Express as $x^2 + px = q$:	1. $x^2 + 14x - 32 = 0$ $x^2 + 14x = 32$	1. $x(x+18) = 19$ $x^2 + 18x = 19$
2. Add the square of half the coefficient of x to both sides:	2. Square of $\frac{1}{2}(14) = 7^2 = 49$ $\mathbf{A}_{49}$ $x^2 + 14x + 49 = 32 + 49$	2. Square of $\frac{1}{2}(18) = 9^2 = 81$ $\mathbf{A}_{81}$ $x^2 + 18x + 81 = 19 + 81$
3. Replace the perfect trinomial square by its binomial squared:	3. $x^2 + 14x + 49 = 81$ $(x+7)^2 = 81$	3. $x^2 + 18x + 81 = 100$ $(x+9)^2 = 100$
4. Take a square root of both sides:	4. **Sq Rt** $x + 7 = \pm 9$	4. **Sq Rt** $x + 9 = \pm 10$
5. Solve each resulting equation:	5. $x+7 = 9$, $x = 2$ | $x+7 = -9$, $x = -16$	5. $x+9 = 10$, $x = 1$ | $x+9 = -10$, $x = -19$
6. Check both roots:	6. *(Check is left to the student.)* *Ans.* $x = 2$ or -16	*Ans.* $x = 1$ or -19

4.3. Equations Requiring Fractions to Complete the Square

Solve by completing the square:

a) $x(x-5) = -4$

Solutions:

$$x^2 - 5x = -4$$

Square of $\frac{1}{2}(-5) = (-\frac{5}{2})^2 = \frac{25}{4}$

$\mathbf{A}_{\frac{25}{4}}$ $\quad x^2 - 5x + \frac{25}{4} = \frac{25}{4} - 4$

$$(x-\frac{5}{2})^2 = \frac{9}{4}$$

Sq Rt $\quad x - \frac{5}{2} = \pm\frac{3}{2}$

$x - \frac{5}{2} = \frac{3}{2}$, $x = 4$ | $x - \frac{5}{2} = -\frac{3}{2}$, $x = 1$

Ans. $x = 4$ or 1

b) $5t^2 - 4t = 33$

$$t^2 - \frac{4t}{5} = \frac{33}{5}$$

Square of $\frac{1}{2}(-\frac{4}{5}) = (-\frac{2}{5})^2 = \frac{4}{25}$

$\mathbf{A}_{\frac{4}{25}}$ $\quad t^2 - \frac{4t}{5} + \frac{4}{25} = \frac{33}{5} + \frac{4}{25}$

$$(t-\frac{2}{5})^2 = \frac{169}{25}$$

Sq Rt $\quad t - \frac{2}{5} = \pm\frac{13}{5}$

$t - \frac{2}{5} = \frac{13}{5}$, $t = 3$ | $t - \frac{2}{5} = -\frac{13}{5}$, $t = -\frac{11}{5}$

Ans. $t = 3$ or $-\frac{11}{5}$

4.4. Solving the Quadratic Equation in Standard Form

If $ax^2 + bx + c = 0$, then $x = \dfrac{-b \pm \sqrt{b^2 - 4ac}}{2a}$. Prove this by completing the square.

Solution:

	$ax^2 + bx + c = 0$
Tr	$ax^2 + bx = -c$
D$_a$	$x^2 + \dfrac{bx}{a} = -\dfrac{c}{a}$
	Square of $\frac{1}{2}(\frac{b}{a}) = (\frac{b}{2a})^2 = \dfrac{b^2}{4a^2}$
A $\frac{b^2}{4a^2}$	$x^2 + \dfrac{bx}{a} + \dfrac{b^2}{4a^2} = \dfrac{b^2}{4a^2} - \dfrac{c}{a}$
	$(x + \dfrac{b}{2a})^2 = \dfrac{b^2 - 4ac}{4a^2}$
Sq Rt	$x + \dfrac{b}{2a} = \pm \dfrac{\sqrt{b^2 - 4ac}}{2a}$
	$x = -\dfrac{b}{2a} \pm \dfrac{\sqrt{b^2 - 4ac}}{2a}$
	By combining fractions,
	$x = \dfrac{-b \pm \sqrt{b^2 - 4ac}}{2a}$ *Ans.*

5. SOLVING A QUADRATIC EQUATION BY QUADRATIC FORMULA

Quadratic Formula: If $ax^2 + bx + c = 0$, then $x = \dfrac{-b \pm \sqrt{b^2 - 4ac}}{2a}$.

(See proof of this in example **4.4** *of the previous unit.)*

To Solve a Quadratic Equation by the Quadratic Formula

Solve $x^2 - 4x = -3$ by quadratic formula.

Procedure:	**Solution:**
1. Express in form of $ax^2 + bx + c = 0$:	1. $x^2 - 4x = -3$ $x^2 - 4x + 3 = 0$
2. State the values of a, b and c:	2. $a = 1$, $b = -4$, $c = 3$
3. Substitute the values of a, b and c in the formula:	3. $x = \dfrac{-b \pm \sqrt{b^2 - 4ac}}{2a}$ $x = \dfrac{-(-4) \pm \sqrt{(-4)^2 - 4(1)(3)}}{2(1)}$
4. Solve for x:	4. $x = \dfrac{+4 \pm \sqrt{16 - 12}}{2}$ $x = \dfrac{+4 \pm \sqrt{4}}{2}$ $x = \dfrac{4 \pm 2}{2}$ $x = \dfrac{4+2}{2}$, $x = 3$ | $x = \dfrac{4-2}{2}$, $x = 1$
5. Check each root in the original equation:	5. *(Check left to student.)* Ans. $x = 3$ or 1

5.1. Finding $b^2 - 4ac$ in the Quadratic Formula

For each equation, state the values of a, b and c. Then find the value of $b^2 - 4ac$.

a) $3x^2 + 4x - 5 = 0$

Solutions:

$a = 3$, $b = 4$, $c = -5$

$b^2 - 4ac$
$4^2 - 4(3)(-5)$
$16 + 60$

Ans. 76

b) $x^2 - 2x - 10 = 0$

$a = 1$, $b = -2$, $c = -10$

$b^2 - 4ac$
$(-2)^2 - 4(1)(-10)$
$4 + 40$

Ans. 44

c) $3x^2 - 5x = 10$

Tr $3x^2 - 5x - 10 = 0$

$a = 3$, $b = -5$, $c = -10$

$b^2 - 4ac$
$(-5)^2 - 4(3)(-10)$
$25 + 120$

Ans. 145

5.2. Expressing Roots in Simplified Radical Form

Express the roots of each equation in simplified radical form, using the value of $b^2 - 4ac$ obtained in example **5.1**:

a) $3x^2 + 4x - 5 = 0$

Solutions:

$$x = \frac{-b \pm \sqrt{b^2 - 4ac}}{2a}$$

$a = 3,\ b = 4,\ c = -5$: $\quad x = \dfrac{-4 \pm \sqrt{76}}{6}$

b) $x^2 - 2x - 10 = 0$

$$x = \frac{-b \pm \sqrt{b^2 - 4ac}}{2a}$$

$a = 1,\ b = -2,\ c = -10$: $\quad x = \dfrac{2 \pm \sqrt{44}}{2}$

c) $3x^2 - 5x = 10$

$$x = \frac{-b \pm \sqrt{b^2 - 4ac}}{2a}$$

$a = 3,\ b = -5,\ c = -10$: $\quad x = \dfrac{5 \pm \sqrt{145}}{6}$

5.3. Evaluating Roots in Radical Form

Solve for x, to the nearest tenth, using the simplified radical form obtained in example **5.2**: (*Find square root to two decimal places.*)

a) $3x^2 + 4x - 5 = 0$

Solutions:

$$x = \frac{-4 \pm \sqrt{76}}{6}$$

```
          8.  7  1
        √76. 00 00
         64
167      12 00
         11 69
1741        31 00
            17 41
            13 59
```

$x = \dfrac{-4 + 8.71}{6}$ | $x = \dfrac{-4 - 8.71}{6}$

$x = \dfrac{4.71}{6}$ | $x = \dfrac{-12.71}{6}$

Ans. $x = .8$ or $x = -2.1$

b) $x^2 - 2x - 10 = 0$

$$x = \frac{2 \pm \sqrt{44}}{2}$$

```
          6.  6  3
        √44. 00 00
         36
126       8 00
          7 56
1323        44 00
            39 69
             4 31
```

$x = \dfrac{2 + 6.63}{2}$ | $x = \dfrac{2 - 6.63}{2}$

$x = \dfrac{8.63}{2}$ | $x = \dfrac{-4.63}{2}$

Ans. $x = 4.3$ or $x = -2.3$

c) $3x^2 - 5x = 10$

$$x = \frac{5 \pm \sqrt{145}}{6}$$

```
          1  2.  0  4
        √1 45. 00 00
         1
22         45
           44
2404        1 00 00
              96 16
               3 84
```

$x = \dfrac{5 + 12.04}{6}$ | $x = \dfrac{5 - 12.04}{6}$

$x = \dfrac{17.04}{6}$ | $x = \dfrac{-7.04}{6}$

Ans. $x = 2.8$ or $x = -1.2$

6. SOLVING QUADRATIC EQUATIONS GRAPHICALLY

To Solve a Quadratic Equation Graphically

Solve graphically, $x^2 - 5x + 4 = 0$.

Procedure:	**Solution:**
1. Express in form of $ax^2 + bx + c = 0$.	1. $x^2 - 5x + 4 = 0$
2. Graph the curve, $y = ax^2 + bx + c$. (*Curve is called a parabola.*)	2. (see graph)
3. Find where $y = 0$ intersects $y = ax^2 + bx + c$. The values of x at the points of intersection are the roots of $ax^2 + bx + c = 0$.	3. $x = 1$ and $x = 4$ *Ans.*

GRAPH of
$y = x^2 - 5x + 4$
(*parabola*)

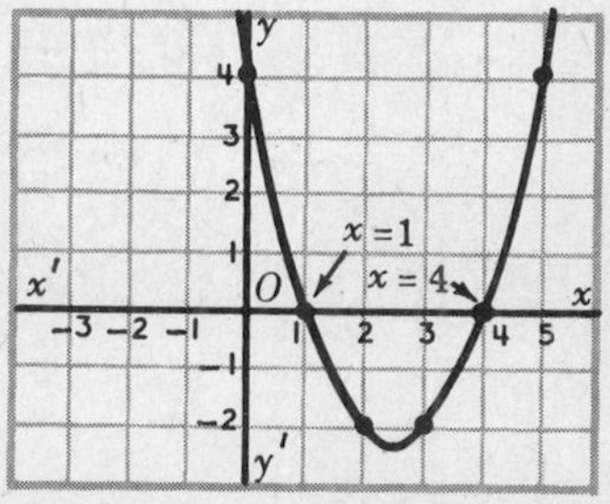

(**Note.** Think of $ax^2 + bx + c = 0$ as the result of combining $y = ax^2 + bx + c$ with $y = 0$.)

The following is the graphic solution of a quadratic equation in greater detail.

To Solve a Quadratic Equation Graphically

Procedure:

1. **Express in form of** $ax^2 + bx + c = 0$.

2. **Graph parabola,** $y = ax^2 + bx + c$.

   a) Obtain a **table of values,** using a suitable sequence of values for x:
   (This may be done by finding the value of $-\frac{b}{2a}$ and choosing values of x greater or smaller than $-\frac{b}{2a}$.)

   b) Join plotted points:
   (Note that $x = -\frac{b}{2a} = 1$ is the folding line or axis of symmetry of the parabola.)

3. **Find roots** where parabola crosses x-axis.

Solve graphically: $x^2 - 2x = 3$

Solution:

1. Tr $x^2 - 2x - 3 = 0$

2. Graph $x^2 - 2x - 3 = y$

x	x^2	−	$2x$	−	3	=	y
4	16	−	8	−	3	=	5
3	9	−	6	−	3	=	0
2	4	−	4	−	3	=	−3
$-\frac{b}{2a} = \frac{2}{2} = 1$	1	−	2	−	3	=	−4
0	0		0	−	3	=	−3
−1	1	+	2	−	3	=	0
−2	4	+	4	−	3	=	5

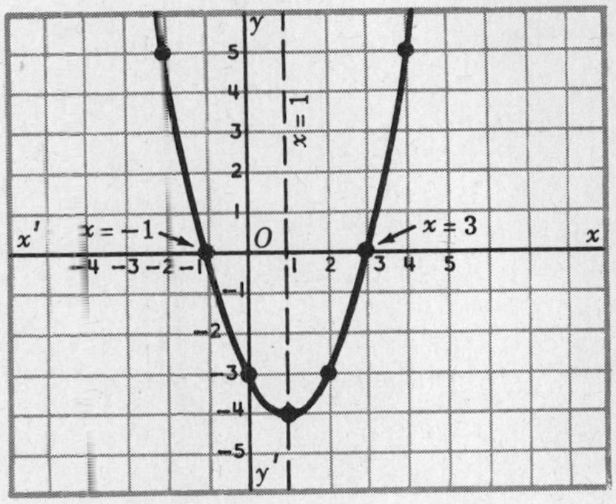

3. $x = -1$ or 3 *Ans.*

6.1. Obtaining Tables of Values for a Parabola

Obtain a table of values for each, using the indicated sequence of values for x:

a) $y = x^2 - 4$ for $x = -3$ to $+3$

b) $y = x^2 - 3x$ for $x = -1$ to 4

c) $y = x^2 - 3x - 4$ for $x = -1$ to 5

Solutions:

a) $x^2 - 4 = y$

x	x^2	−	4	=	y
3	9	−	4	=	5
2	4	−	4	=	0
1	1	−	4	=	−3
→ 0	0	−	4	=	−4
−1	1	−	4	=	−3
−2	4	−	4	=	0
−3	9	−	4	=	5

b) $x^2 - 3x = y$

x	x^2	−	$3x$	=	y
4	16	−	12	=	4
3	9	−	9	=	0
2	4	−	6	=	−2
$x = 1\frac{1}{2}$ →					
1	1	−	3	=	−2
0	0	+	0	=	0
−1	1	+	3	=	4

c) $x^2 - 3x - 4 = y$

x	x^2	−	$3x$	−	4	=	y
5	25	−	15	−	4	=	6
4	16	−	12	−	4	=	0
3	9	−	9	−	4	=	−4
2	4	−	6	−	4	=	−6
$x = 1\frac{1}{2}$ →							
1	1	−	3	−	4	=	−6
0	0	+	0	−	4	=	−4
−1	1	+	3	−	4	=	0

(The arrow in each table indicates $x = -\frac{b}{2a}$, the axis of symmetry of the parabola.)

6.2. Graphing Parabolas $y = ax^2 + bx + c$

Graph each, using the table of values obtained in example **6.1**:

a) $y = x^2 - 4$

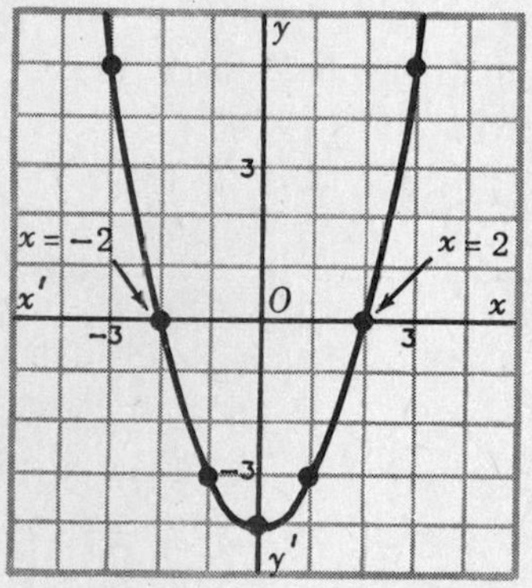

b) $y = x^2 - 3x$

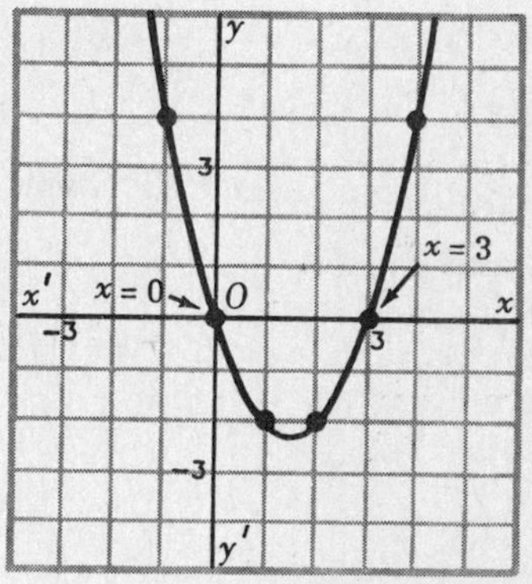

c) $y = x^2 - 3x - 4$

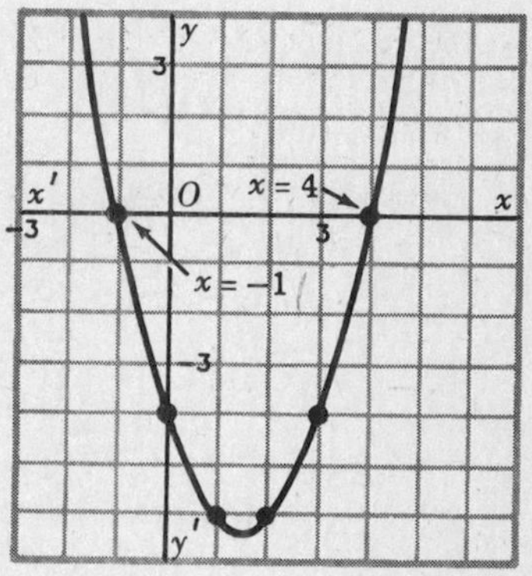

6.3. Finding Roots Graphically

Find the roots of each, using the parabolas obtained in example **6.2**:

a) $0 = x^2 - 4$ *b*) $0 = x^2 - 3x$ *c*) $0 = x^2 - 3x - 4$

Solutions:

(In each case, find the values of x at the points where the parabola crosses the x-axis.)

Ans. $x = -2$ or 2 *Ans.* $x = 0$ or 3 *Ans.* $x = -1$ or 4

SUPPLEMENTARY PROBLEMS

1. Express in standard quadratic equation form, $ax^2 + bx + c = 0$, so that a has a positive value: **(1.1)**

a) $3(x^2 - 5) = 4x$

b) $2x(x+4) = 42$

c) $\frac{7}{x} - 5 = 2x$

d) $\frac{6}{x} + x = \frac{11}{2}$

e) $3x - 9 = x^2 - 7x$

f) $x(2x-7) = 3x^2 - 8$

g) $\sqrt{2x^2 + 3x} = x$

h) $x + 1 = \sqrt{3x + 7}$

Ans. *a*) $3x^2 - 4x - 15 = 0$ *b*) $2x^2 + 8x - 42 = 0$ *c*) $0 = 2x^2 + 5x - 7$ *d*) $2x^2 - 11x + 12 = 0$ *e*) $0 = x^2 - 10x + 9$ *f*) $0 = x^2 + 7x - 8$ *g*) $x^2 + 3x = 0$ *h*) $x^2 - x - 6 = 0$

2. Express in standard quadratic equation form, so that a has a positive value. Then state the values of a, b and c. **(1.2)**

		Standard Form	a	b	c
a) $x^2 = 5x - 4$	*Ans. a*)	$x^2 - 5x + 4 = 0$	1	−5	+4
b) $20 + 6x = 2x^2$	*Ans. b*)	$0 = 2x^2 - 6x - 20$	2	−6	−20
c) $3x^2 = -5x$	*Ans. c*)	$3x^2 + 5x = 0$	3	+5	0
d) $18 = 2x^2$	*Ans. d*)	$0 = 2x^2 - 18$	2	0	−18
e) $x(8 - 2x) = 6$	*Ans. e*)	$0 = 2x^2 - 8x + 6$	2	−8	+6
f) $7(x^2 - 9) = x(x - 5)$	*Ans. f*)	$6x^2 + 5x - 63 = 0$	6	+5	−63
g) $\frac{10}{x} + 1 = 4x$	*Ans. g*)	$0 = 4x^2 - x - 10$	4	−1	−10
h) $\sqrt{x^2 - 5x} = 3x$	*Ans. h*)	$0 = 8x^2 + 5x$	8	+5	0

3. Solve by factoring: (2.1, 2.2)

a) $x^2 - 5x + 6 = 0$ *d*) $4a^2 = 28a$ *g*) $c^2 + 6c = -8$ *j*) $5 = r(2r+3)$

b) $y^2 + y - 20 = 0$ *e*) $4x^2 = 1$ *h*) $d^2 = 5d + 24$ *k*) $t + 8 = \frac{20}{t}$

c) $w^2 - 64 = 0$ *f*) $9b^2 = 3b$ *i*) $p(3p+20) = 7$ *l*) $\frac{7}{x} = 9 - 2x$

Each answer is underlined:

a) $(x-2)(x-3) = 0$, $\underline{x = 2 \text{ or } 3}$ *d*) $4a(a-7) = 0$, $\underline{a = 0 \text{ or } 7}$ *g*) $(c+4)(c+2) = 0$, $\underline{c = -4 \text{ or } -2}$ *j*) $0 = (r-1)(2r+5)$, $\underline{r = 1 \text{ or } -\frac{5}{2}}$

b) $(y+5)(y-4) = 0$, $\underline{y = -5 \text{ or } 4}$ *e*) $(2x+1)(2x-1) = 0$, $\underline{x = -\frac{1}{2} \text{ or } \frac{1}{2}}$ *h*) $(d-8)(d+3) = 0$, $\underline{d = 8 \text{ or } -3}$ *k*) $(t-2)(t+10) = 0$, $\underline{t = 2 \text{ or } -10}$

c) $(w-8)(w+8) = 0$, $\underline{w = 8 \text{ or } -8}$ *f*) $3b(3b-1) = 0$, $\underline{b = 0 \text{ or } \frac{1}{3}}$ *i*) $(3p-1)(p+7) = 0$, $\underline{p = \frac{1}{3} \text{ or } -7}$ *l*) $(2x-7)(x-1) = 0$, $\underline{x = \frac{7}{2} \text{ or } 1}$

4. Solve by factoring: (2.3)

a) $\frac{3x-2}{5} = \frac{8}{x}$ *c*) $(y-2)(3y-1) = 100$ *e*) $x + 1 = \sqrt{5x+1}$

b) $\frac{21}{x-3} = x - 7$ *d*) $(y+3)(y-3) = 2y - 1$ *f*) $\sqrt{x} - 2 = x - 8$

Each answer is underlined:

a) $(x-4)(3x+10) = 0$, $\underline{x = 4 \text{ or } -\frac{10}{3}}$ *c*) $(y-7)(3y+14) = 0$, $\underline{y = 7 \text{ or } -\frac{14}{3}}$ *e*) $x(x-3) = 0$, $\underline{x = 0 \text{ or } 3}$

b) $0 = x(x-10)$, $\underline{x = 0 \text{ or } 10}$ *d*) $(y-4)(y+2) = 0$, $\underline{y = 4 \text{ or } -2}$ *f*) $0 = (x-9)(x-4)$, $\underline{x = 9}$ (4 is an extraneous root.)

5. Find a number such that (2.4)

a) its square is 12 more than the number.
b) its square decreased by three times the number is 18.
c) the product of the number and 4 less than the number is 32.
d) the square of one more than the number is 4 more than four times the number.

Each answer is underlined:

a) $n^2 = n + 12$, $\underline{4 \text{ or } -3}$ *b*) $n^2 - 3n = 18$, $\underline{6 \text{ or } -3}$ *c*) $n(n-4) = 32$, $\underline{8 \text{ or } -4}$ *d*) $(n+1)^2 = 4n + 4$, $\underline{3 \text{ or } -1}$

6. Find two consecutive integers such that (2.4)

a) the sum of their squares is 13.
b) their product is 30.
c) the square of the first added to twice the second is 5.
d) the product of the first and twice the second is 40.

Each answer is underlined:

a) $n^2 + (n+1)^2 = 13$, Either 2, 3 or −3, −2 *c*) $n^2 + 2(n+1) = 5$, Either 1, 2 or −3, −2

b) $n(n+1) = 30$, Either 5, 6 or −6, −5 *d*) $n(2n+2) = 40$, Either 4, 5 or −5, −4

7. The length of a rectangular lot is 2 yd. more than its width. Find the length and the width (2.5)

a) if the area is 80 sq. yd.
b) if the area is 48 sq. yd. when each dimension is made 2 yd. longer.
c) if the area is 30 sq. yd. when the width is doubled.
d) if the area is increased 70 sq. yd. when the width is tripled.
e) if the area is decreased 7 sq. yd. when the length is made 5 yd. shorter and the width doubled.

Each answer is underlined: (*Reject all negative roots.*)

a) $w(w+2) = 80$, 10 yd. and 8 yd. *c*) $2w(w+2) = 30$, 5 yd. and 3 yd. *e*) $w(w+2) - 7 = 2w(w-3)$, 9 yd. and 7 yd.

b) $(w+2)(w+4) = 48$, 6 yd. and 4 yd. *d*) $w(w+2) + 70 = 3w(w+2)$, 7 yd. and 5 yd.

8. Solve: (*Leave irrational answers in radical form.*) **(3.1)**

a) $2x^2 = 8$ *c*) $3a^2 - 4 = 5$ *e*) $\frac{2x}{3} = \frac{6}{x}$ *g*) $p^2 = 4(p^2-10) - 8$

b) $4 = 100y^2$ *d*) $b(b+2) = 2b + 10$ *f*) $\frac{x+2}{5} = \frac{9}{x-2}$ *h*) $r(r-6) = 3(7-2r)$

Ans. *a*) $x = \pm 2$ *c*) $a = \pm\sqrt{3}$ *e*) $x = \pm 3$ *g*) $p = \pm 4$

b) $y = \pm\frac{1}{5}$ *d*) $b = \pm\sqrt{10}$ *f*) $x = \pm 7$ *h*) $r = \pm\sqrt{21}$

9. Solve: **(3.2)**

a) $2x^2 = 8x$ *c*) $3a^2 - 4a = 5a$ *e*) $\frac{x^2}{40} = \frac{x}{5}$ *g*) $(r+3)(r-3) = 3(5r-3)$

b) $8x^2 = 2x$ *d*) $2b^2 + 12b = b^2$ *f*) $\frac{y^2+5}{y+1} = 5$ *h*) $(s+4)(s+7) = 2(14-3s)$

Ans. *a*) $x = 0$ or 4 *c*) $a = 0$ or 3 *e*) $x = 0$ or 8 *g*) $r = 0$ or 15

b) $x = 0$ or $\frac{1}{4}$ *d*) $b = 0$ or -12 *f*) $y = 0$ or 5 *h*) $s = 0$ or -17

10. Solve for the letter indicated: **(3.3)**

a) $2x^2 = 50a^2$ for x *c*) $b^2y^2 = 49$ for y *e*) $aw^2 = a^2w$ for w

b) $3x^2 = 6bx$ for x *d*) $4y^2 = 25c^2$ for y *f*) $6v^2 = h^2 - 3v^2$ for v

Ans. *a*) $x = \pm 5a$ *b*) $x = 0$ or $2b$ *c*) $y = \pm\frac{7}{b}$ *d*) $y = \pm\frac{5c}{2}$ *e*) $w = 0$ or a *f*) $v = \pm\frac{h}{3}$

11. Solve for the *positive* value of the letter indicated: **(3.4)**

a) $S = 16t^2$ for t *c*) $S = 4\pi r^2$ for r *e*) $a^2 + b^2 = c^2$ for b

b) $K = \frac{1}{2}mv^2$ for v *d*) $F = \frac{mv^2}{r}$ for v *f*) $A = \pi(R^2 - r^2)$ for R

Ans. *a*) $t = \frac{\sqrt{S}}{4}$, *b*) $v = \sqrt{\frac{2K}{m}}$, *c*) $r = \frac{1}{2}\sqrt{\frac{S}{\pi}}$, *d*) $v = \sqrt{\frac{Fr}{m}}$, *e*) $b = \sqrt{c^2 - a^2}$, *f*) $R = \sqrt{\frac{A + \pi r^2}{\pi}}$

12. Complete each perfect binomial square and state its binomial squared: **(4.1)**

a) $a^2 + 12a + ?$ *c*) $x^2 + 7x + ?$ *e*) $w^2 + 20w + ?$

b) $c^2 - 18c + ?$ *d*) $y^2 - 11y + ?$ *f*) $r^2 - 30r + ?$

Ans. *a*) $a^2 + 12a + 36$, $(a+6)^2$ *c*) $x^2 + 7x + \frac{49}{4}$, $(x+\frac{7}{2})^2$ *e*) $w^2 + 20w + 100$, $(w+10)^2$

b) $c^2 - 18c + 81$, $(c-9)^2$ *d*) $y^2 - 11y + \frac{121}{4}$, $(y-\frac{11}{2})^2$ *f*) $r^2 - 30r + 225$, $(r-15)^2$

13. Solve by completing the square: (*Use the answers in* **12** *above to help you.*) **(4.2, 4.3)**

a) $a^2 + 12a = 45$ *c*) $x^2 + 7x = 8$ *e*) $w^2 + 20w = -19$

b) $c^2 - 18c = -65$ *d*) $y^2 - 11y = -28$ *f*) $r^2 - 30r = 99$

Each answer is underlined:

a) $\underline{a = 3 \text{ or } -15}$ from $(a+6)^2 = 81$ *c*) $\underline{x = 1 \text{ or } -8}$ from $(x+\frac{7}{2})^2 = \frac{81}{4}$ *e*) $\underline{w = -1 \text{ or } -19}$ from $(w+10)^2 = \pm 9$

b) $\underline{c = 5 \text{ or } 13}$ from $(c-9)^2 = 16$ *d*) $\underline{y = 4 \text{ or } 7}$ from $(y-\frac{11}{2})^2 = \frac{9}{4}$ *f*) $\underline{r = -3 \text{ or } 33}$ from $(r-15)^2 = 324$

14. For each equation, state the values of a, b and c. Then find the value of $b^2 - 4ac$. **(5.1)**

a) $2x^2 + 5x + 1 = 0$ *c*) $3x^2 - 2 = 4x$ *e*) $5x^2 = 10x - 4$ *g*) $x^2 = 10 - 6x$

b) $x^2 + 7x = -5$ *d*) $2x^2 - 5x = 4$ *f*) $6x^2 = 10x - 3$ *h*) $9x^2 = 2 - x$

Ans. *a*) $a = 2$, $b = 5$, $c = 1$; $b^2 - 4ac = 17$ *c*) $a = 3$, $b = -4$, $c = -2$; $b^2 - 4ac = 40$ *e*) $a = 5$, $b = -10$, $c = 4$; $b^2 - 4ac = 20$ *g*) $a = 1$, $b = 6$, $c = -10$; $b^2 - 4ac = 76$

b) $a = 1$, $b = 7$, $c = 5$; $b^2 - 4ac = 29$ *d*) $a = 2$, $b = -5$, $c = -4$; $b^2 - 4ac = 57$ *f*) $a = 6$, $b = -10$, $c = 3$; $b^2 - 4ac = 28$ *h*) $a = 9$, $b = 1$, $c = -2$; $b^2 - 4ac = 73$

15. Express the roots of each equation in simplified radical form, using the value of $b^2 - 4ac$ found in example 14: **(5.2)**

a) $2x^2 + 5x + 1 = 0$ *c*) $3x^2 - 2 = 4x$ *e*) $5x^2 = 10x - 4$ *g*) $x^2 = 10 - 6x$
b) $x^2 + 7x = -5$ *d*) $2x^2 - 5x = 4$ *f*) $6x^2 = 10x - 3$ *h*) $9x^2 = 2 - x$

Ans. *a*) $x = \frac{-5 \pm \sqrt{17}}{4}$ *c*) $x = \frac{4 \pm \sqrt{40}}{6}$ *e*) $x = \frac{10 \pm \sqrt{20}}{10}$ *g*) $x = \frac{-6 \pm \sqrt{76}}{2}$
b) $x = \frac{-7 \pm \sqrt{29}}{2}$ *d*) $x = \frac{5 \pm \sqrt{57}}{4}$ *f*) $x = \frac{10 \pm \sqrt{28}}{12}$ *h*) $x = \frac{-1 \pm \sqrt{73}}{18}$

16. Solve for x, correct to the nearest tenth, using the simplified radical form found in example 15: **(5.3)**

a) $2x^2 + 5x + 1 = 0$ *c*) $3x^2 - 2 = 4x$ *e*) $5x^2 = 10x - 4$ *g*) $x^2 = 10 - 6x$
b) $x^2 + 7x = -5$ *d*) $2x^2 - 5x = 4$ *f*) $6x^2 = 10x - 3$ *h*) $9x^2 = 2 - x$

Ans. *a*) $x = -2.3$ or $-.2$ *c*) $x = 1.7$ or $-.4$ *e*) $x = 1.4$ or $.6$ *g*) $x = 1.4$ or -7.4
b) $x = -6.2$ or $-.8$ *d*) $x = 3.1$ or $-.6$ *f*) $x = 1.3$ or $.4$ *h*) $x = .4$ or $-.5$

17. Solve, correct to the nearest tenth: **(5.3)**

a) $2y^2 = -3y + 1$ *c*) $2w^2 = 3w + 5$ *e*) $x + 5 = \frac{5}{x}$
b) $x(x - 4) = -2$ *d*) $6v^2 + 1 = 5v$ *f*) $\frac{x^2}{2} = 4x - 1$

Ans. *a*) $y = .3$ or -1.8 *c*) $w = 2.5$ or -1 *e*) $x = .9$ or -5.9
b) $x = 3.4$ or $.6$ *d*) $v = .5$ or $.3$ *f*) $x = 7.7$ or $.3$

18. Solve graphically: (*See sketches after example* **19**.) **(6.1 to 6.3)**

a) Graph $y = x^2 - 9$ from $x = -4$ to $x = 4$ and solve $x^2 - 9 = 0$ graphically.
b) Graph $y = x^2 + 3x - 4$ from $x = -5$ to $x = 2$ and solve $x^2 + 3x = 4$ graphically.
c) Graph $y = x^2 - 6x + 9$ from $x = 1$ to $x = 5$ and solve $x^2 = 6x - 9$ graphically.
d) Graph $y = x^2 + 8x + 16$ from $x = -6$ to $x = -2$ and solve $x^2 + 8x = -16$ graphically.
e) Graph $y = x^2 - 2x + 4$ from $x = -1$ to $x = 3$ and show that $x^2 - 2x + 4 = 0$ has no real roots.

Ans. *a*) $x = 3$ or -3 *b*) $x = -4$ or 1
c) $x = 3$ Equal roots since the parabola touches x-axis at one point.
d) $x = -4$ Equal roots since the parabola touches x-axis at one point.
e) Roots are not real (imaginary) since the parabola does not meet x-axis.

19. Solve graphically: (*See sketches following this example*.) **(6.1 to 6.3)**

a) $x^2 - 1 = 0$ *c*) $x^2 - 4x = 0$ *e*) $x^2 - 2x = 8$
b) $4x^2 - 9 = 0$ *d*) $2x^2 + 7x = 0$ *f*) $4x^2 = 12x - 9$

Ans. *a*) $x = -1$ or 1 *c*) $x = 0$ or 4 *e*) $x = 4$ or -2
b) $x = 1\frac{1}{2}$ or $-1\frac{1}{2}$ *d*) $x = 0$ or $-3\frac{1}{2}$ *f*) $x = 1\frac{1}{2}, 1\frac{1}{2}$

Sketches of graphs needed in examples **18** and **19**, showing relationship of parabola and x-axis:

18*a*, 18*b*, 19*a* to 19*e*

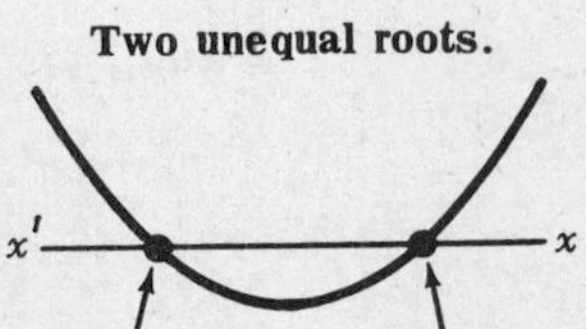

(*Two real values.*)

18*c*, 18*d*, 19*f*

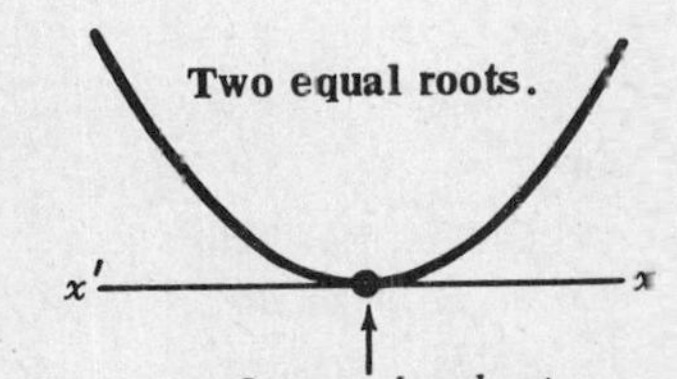

(*One real value.*)

18*e*

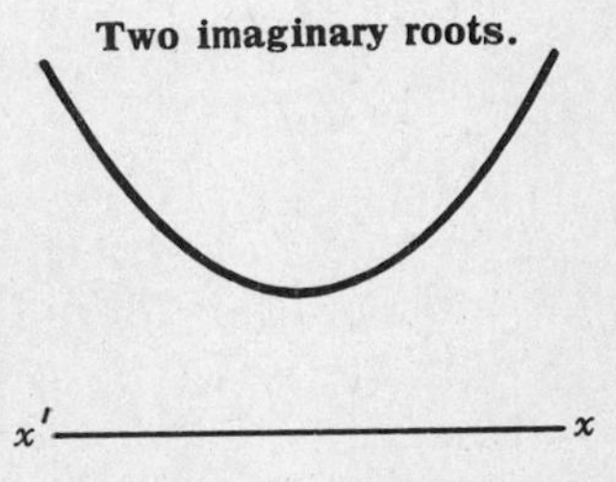

(*No real values.*)

Chapter 14 The VARIABLE: DIRECT, INVERSE, JOINT and POWER VARIATION

1. UNDERSTANDING THE VARIABLE

A **variable** in algebra is a letter which may represent any number of a set of numbers under discussion when the set contains more than one number.

Thus in $y = 2x$, if x represents 1, then y represents 2; if x represents 2, then y represents 4; and if x represents 6, then y represents 12. These three pairs of corresponding values may be tabulated as shown. The set of numbers being discussed is the set of all numbers.

	(1)	*(2)*	*(3)*
y	2	4	12
x	1	2	6

A **constant** is any letter or number which has a fixed value that does not change under discussion.

Thus, 5 and π are constants.

Throughout this chapter, as is customary in mathematics, the letter k shall be used to represent a constant while x, y and z are used to represent variables.

Measuring the Change in a Variable

As a variable x changes in value from one number x_1 to a second number x_2, the change may be measured by finding the difference $x_2 - x_1$, or by finding the ratio $\frac{x_2}{x_1}$. A change in value may be measured by subtracting or dividing.

Thus, the change in the speed of an auto from 20 mph to 60 mph may be expressed as follows:

(*1*) the second speed is $(60 - 20)$ or **40 mph faster** than the first speed.

(*2*) the second speed is $\frac{60}{20}$ or **three times as fast** as the first speed.

Any formula may be considered as a relationship of its variables. Mathematics and science abound in formulas which have exactly the same general structure, $z = xy$. To illustrate, study the following:

	Formula	Rule				
1.	$D = RT$	Distance	=	Rate	×	Time
2.	$A = LW$	Area	=	Length	×	Width
3.	$I = PR$	Interest	=	Price	×	Rate of Interest
4.	$C = NP$	Cost	=	Number	×	Price
5.	$F = PA$	Force	=	Pressure	×	Area
6.	$M = DV$	Mass	=	Density	×	Volume

Note, in each formula which has the form $z = xy$, that there are three variables. Furthermore, one of the variables is the product of the other two. In this chapter, we shall study the relation of these formulas to the three basic types of variation. These types of variation are

1. **Direct Variation**
2. **Inverse Variation**
3. **Joint Variation**

1.1. Using Division to Measure the Change in a Variable

Using division, find the ratio which indicates the change in each variable and express this ratio in a sentence.

a) The price of a suit changes from \$40 to \$60.
b) The speed of an auto changes from 10 mph to 40 mph.
c) A salary changes from \$30 per week to \$40 per week.
d) Length of a line changes from 12 ft. to 4 ft.

Solutions:

a) $\frac{\$60}{\$40} = \frac{3}{2}$. Hence, second price is **three-halves** of first price.

b) $\frac{40\text{ mph}}{10\text{ mph}} = 4$. Hence, second rate is **four times (quadruple)** first rate.

c) $\frac{\$40\text{ per week}}{\$30\text{ per week}} = \frac{4}{3}$. Hence, second salary is **four-thirds** of first salary.

d) $\frac{4\text{ ft.}}{12\text{ ft.}} = \frac{1}{3}$. Hence, second length is **one-third** of first length.

1.2. Multiplying or Dividing Variables

Complete each:

a) If x is doubled, it will change from 13 to ( ) or from ( ) to 90.
b) If y is tripled, it will change from 17 to ( ) or from ( ) to 87.
c) If z is halved, it will change from 7 to ( ) or from ( ) to 110.
d) If s is multiplied by $\frac{5}{4}$, it will change from 16 to ( ) or from ( ) to 55.

Ans. *a*) 26, 45 *b*) 51, 29 *c*) $3\frac{1}{2}$, 220 *d*) 20, 44

2. UNDERSTANDING DIRECT VARIATION: $y = kx$ or $\frac{y}{x} = k$

Direct Variation Formula With Constant Ratio k

$$y = kx \quad \text{or} \quad \frac{y}{x} = k$$

Direct Variation Formula Without Constant

If x varies from x_1 to x_2 while y varies from y_1 to y_2, then

$$\frac{x_2}{x_1} = \frac{y_2}{y_1}$$

Rule 1. If $y = kx$ or $\frac{y}{x} = k$, then:

(*1*) x and y **vary directly** as each other; that is, x varies directly as y and y varies directly as x.

(*2*) x and y are **directly proportional** to each other; that is, $\frac{x_2}{x_1} = \frac{y_2}{y_1}$ as x varies from x_1 to x_2 and y varies from y_1 to y_2.

Thus:

1. If $y = 4x$, y and x vary directly as each other.
2. If $I = .06P$, I and P vary directly as each other.
3. The perimeter of a square equals four times its side; that is, $P = 4S$. For a square, P and S vary directly as each other.

Multiplication and Division in Direct Variation

Rule 2. If x and y vary directly as each other and a value of either x or y is **multiplied** by a number, then the corresponding value of the other is **multiplied** by the same number.

Thus, if $y = 5x$ and x is tripled, then y is tripled, as shown in the table.

$y = 5x$

y is tripled →	y	5	15	45	135
x is tripled →	x	1	3	9	27

Rule 3. If x and y vary directly as each other and a value of either x or y is **divided** by a number, then the corresponding value of the other is **divided** by the same number.

Thus, if $y = 10x$ and x is halved (divided by 2), then y is halved, as shown in the table.

$y = 10x$

y is halved →	y	240	120	60	30
x is halved →	x	24	12	6	3

Direct Variation Applied to a Rectangle

To understand direct variation more fully, note how it applies to a rectangle:

If the length of a rectangle is fixed and the width is tripled, then the area is tripled. As a result, the new area $ABCD$ is three times the area of the old (shaded) rectangle, as shown in the adjoining figure.

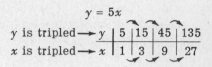

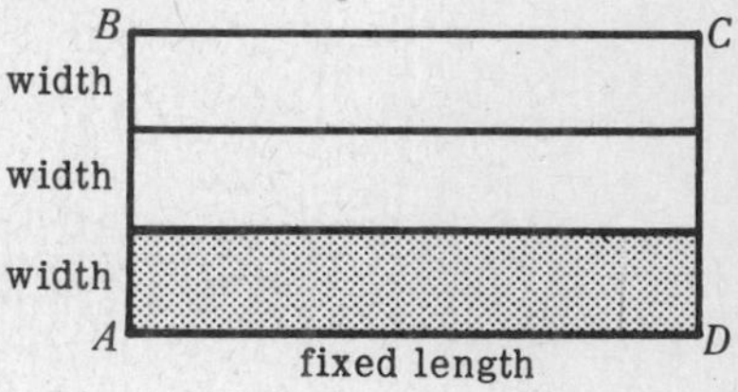

2.1. Rule 1. Direct Variation in Formulas

(*1*) Express each equation in the form $y = kx$.
(*2*) State the variables that vary directly as each other.

a) $120R = P$ *c*) $.05 = \frac{I}{P}$ *e*) $A = \frac{1}{2}bh$ when b is constant

b) $\frac{C}{D} = \pi$ *d*) $f = \frac{n}{d}$ when $f = \frac{2}{3}$ *f*) $V = \pi R^2 H$ when $R = 8$

Ans. *a*) (*1*) $P = 120R$, (*2*) P and R
b) (*1*) $C = \pi D$, (*2*) C and D
c) (*1*) $I = .05P$, (*2*) I and P
d) (*1*) $n = \frac{2}{3}d$, (*2*) n and d
e) (*1*) $A = kh$, (*2*) A and h
f) (*1*) $V = 64\pi H$, (*2*) V and H

2.2. Rules 2 and 3: Multiplication and Division in Direct Variation

Complete each:

a) If $y = 8x$ and x is tripled, then ( ). *Ans.* *a*) y is tripled.
b) If $C = \pi D$ and D is quadrupled, then ( ). *Ans.* *b*) C is quadrupled.
c) If $C = \pi D$ and C is halved, then ( ). *Ans.* *c*) D is halved.
d) If $D = RT$, $T = 12$ and R is divided by 4, then ( ). *Ans.* *d*) D is divided by 4.
e) If $A = LW$, L is constant and W is doubled, then ( ). *Ans.* *e*) A is doubled.

2.3. Applying Direct Variation to Statements

(*1*) Complete each statement and (*2*) state the formula to which direct variation applies.

a) At a uniform speed, doubling time will ( ).
b) If the time of travel is constant, tripling rate will ( ).
c) If a rectangle has a fixed width, to multiply its area by 5 ( ).
d) If the value of a fraction is constant and the numerator is halved, ( ).

Ans. *a*) (*1*) double distance, (*2*) $D = kT$ since $D = RT$ and R is constant.
b) (*1*) triple distance, (*2*) $D = kR$ since $D = RT$ and T is constant.
c) (*1*) multiply its length by 5, (*2*) $A = kL$ since $A = LW$ and W is constant.
d) (*1*) the denominator is halved, (*2*) $n = kd$ since $f = \frac{n}{d}$ and f is constant.

2.4. Finding Values for Directly Varying Variables

a) If y and x vary directly as each other and $y=36$ when $x=9$, find y when $x=27$.

Solution: **Proportion Method**

If y and x vary directly,

	x	y
2nd values	27	y
1st values	9	36

$$\frac{y_2}{y_1} = \frac{x_2}{x_1}$$

$$\frac{y}{36} = \frac{27}{9}$$

$$y = \frac{27}{9}(36) = 108 \qquad \textit{Ans. } 108$$

Ratio Method

(Note the equal ratios in the table.)

	x		y	
2nd values	27	3	y	3
1st values	9	1	36	1

If *2nd x-value* is **three times** *1st x-value*, then *2nd y-value* is **three times** *1st y-value*. Hence, $y = 3(36) = 108$. *Ans.* 108

Equation Method

If y and x vary directly, $kx=y$.
Since $y=36$ when $x=9$, $9k=36$, $k=4$.
Since $k=4$, $y=4x$.
When $x=27$, $y=4(27)=108$. *Ans.* 108

b) If y and x vary directly as each other and $y=25$ when $x=10$, find x when $y=5$.

Solution: **Proportion Method**

If y and x vary directly,

	x	y
2nd values	x	5
1st values	10	25

$$\frac{x_2}{x_1} = \frac{y_2}{y_1}$$

$$\frac{x}{10} = \frac{5}{25}$$

$$x = \frac{5}{25}(10) = 2 \qquad \textit{Ans. } 2$$

Ratio Method

(Note the equal ratios in the table.)

	x		y	
2nd values	x	1	5	1
1st values	10	5	25	5

If *2nd y-value* is **one-fifth** of *1st y-value*, then *2nd x-value* is **one-fifth** of *1st x-value*. Hence, $x = \frac{1}{5}(10) = 2$. *Ans.* 2

Equation Method

If y and x vary directly, $kx=y$.
Since $y=25$ when $x=10$, $10k=25$, $k=2\frac{1}{2}$.
Since $k=2\frac{1}{2}$, $y=2\frac{1}{2}x$.
When $y=5$, $5=2\frac{1}{2}x$, $x=2$. *Ans.* 2

2.5. Applying Direct Variation to a Motion Problem

Henry traveled a distance of 124 miles at 40 mph. If he had taken the same amount of time, how far would he have traveled at 50 mph?

Solution:

	(mph) Rate		(mi.) Distance	
2nd trip	50	5	D	5
1st trip	40	4	124	4

Proportion Method

Since $D = RT$ and the time (T) is constant, then D and R vary directly as each other.

Hence, $\dfrac{D_2}{D_1} = \dfrac{R_2}{R_1}$, $\dfrac{D}{124} = \dfrac{50}{40}$, $\dfrac{D}{124} = \dfrac{5}{4}$,

$D = \frac{5}{4}(124) = 155$ *Ans.* 155 mi.

Ratio Method (*Note use of the equal ratios in the table.*)

If the second rate is **five-fourths** of the first rate, then the second distance is **five-fourths** of the first distance. Hence, $D = \frac{5}{4}(124) = 155$. *Ans.* 155 mi.

2.6. Applying Direct Variation to an Interest Problem

In a bank, the annual interest on \$4500 is \$180. At the same rate, what is the annual interest on \$7500.

Solution:

	(\$) Principal		(\$) Annual Interest	
2nd principal	7500	5	I	5
1st principal	4500	3	180	3

Proportion Method

Since $I=PR$ and the rate (R) is constant, then I and P vary directly as each other.

Hence, $\dfrac{I_2}{I_1} = \dfrac{P_2}{P_1}$, $\dfrac{I}{180} = \dfrac{7500}{4500}$, $\dfrac{I}{180} = \dfrac{5}{3}$,

$I = \frac{5}{3}(180) = 300$ *Ans.* \$300

Ratio Method (*Note use of the equal ratios in the table.*)

If the second principal is **five-thirds** of the first principal, then the second interest is **five-thirds** of the first interest. Hence, $I = \frac{5}{3}(180) = 300$. *Ans.* \$300

3. UNDERSTANDING INVERSE VARIATION: $xy = k$

Inverse Variation Formula With Constant Product k

$$xy = k$$

Inverse Variation Formula Without Constant

If x varies from x_1 to x_2 while y varies from y_1 to y_2, then

$$\frac{x_2}{x_1} = \frac{y_1}{y_2} \quad \text{or} \quad \frac{y_2}{y_1} = \frac{x_1}{x_2}$$

Rule 1. If $xy = k$, then:

(*1*) x and y **vary inversely** as each other; that is, x varies inversely as y and y varies inversely as x.

(*2*) x and y are **inversely proportional** to each other; that is, $\frac{x_2}{x_1} = \frac{y_1}{y_2}$ or $\frac{y_2}{y_1} = \frac{x_1}{x_2}$ as x varies from x_1 to x_2 and y varies from y_1 to y_2.

Thus:

1. If $xy = 12$, y and x vary inversely as each other.

2. If $PR = 150$, P and R vary inversely as each other.

3. For a fixed distance of 120 miles to be traveled, the motion formula $D = RT$ becomes $120 = RT$. In such a case, R and T vary inversely as each other.

Multiplication and Division in Inverse Variation

Rule 2. If x and y vary inversely as each other and a value of either x or y is **multiplied** by a number, then the corresponding value of the other is **divided** by the same number.

Thus, if $xy = 24$ and y is doubled, then x is halved, as shown in the table.

$xy = 24$

y is doubled →	y	2	4	8	16
x is halved →	x	12	6	3	$\frac{3}{2}$

Rule 3. If x and y vary inversely as each other and either x or y is **divided** by a number, then the other is **multiplied** by the same number.

Thus, if $xy = 250$ and y is divided by 5, then x is multiplied by 5, as shown in the table.

$xy = 250$

y is divided by 5 →	y	250	50	10	2
x is multiplied by 5 →	x	1	5	25	125

Inverse Variation Applied to a Rectangle

To understand inverse variation more fully, note how it applies to a rectangle:

If the length of a rectangle is doubled and the width is halved, its area remains constant. As a result, the new area $ABCD$ equals the old (shaded) area.

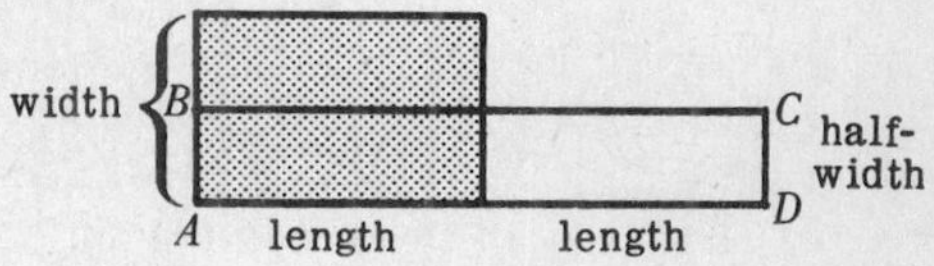

3.1. Inverse Variation in Formulas

(*1*) Express each equation in the form, $xy = k$.
(*2*) State the variables that vary inversely as each other.

a) $5RT = 500$ *Ans. a*) (*1*) $RT = 100$, (*2*) R and T

b) $\frac{1}{2}PV = 7$ *Ans. b*) (*1*) $PV = 14$, (*2*) P and V

c) $f = \frac{n}{d}$ when $n = 3$ *Ans. c*) (*1*) $fd = 3$, (*2*) f and d

d) $A = \frac{1}{2}bh$ when A is constant. *Ans. d*) (*1*) $bh = k$, (*2*) b and h

3.2. Rules 2 and 3: Multiplication and Division in Inverse Variation

Complete each:

a) If $xy = 25$ and x is tripled, then ( ). *Ans. a*) y is divided by 3.

b) If $50 = RT$ and R is divided by 10, then ( ). *Ans. b*) T is multiplied by 10.

c) If $BH = 5$ and B is multiplied by $\frac{3}{2}$, then ( ). *Ans. c*) H is divided by $\frac{3}{2}$.

d) If $A = LW$, A is constant and L is doubled, then ( ). *Ans. d*) W is halved.

e) If $PV = k$ and V is quadrupled, then ( ) *Ans. e*) P is divided by 4.

3.3. Applying Inverse Variation to Statements

(*1*) Complete each statement.

(*2*) State the formula to which inverse variation applies.

a) Over the same distance, doubling the speed ( ).

b) For a fixed area, multiplying the length of a rectangle by 5, ( ).

c) For an enclosed gas at a constant temperature, dividing the pressure by 7 ( ).

d) If the total cost is the same, tripling the price of an article ( ).

Ans. a) (*1*) halves the time (*2*) $RT=k$ since $RT=D$ and D is constant

b) (*1*) divides the width by 5 (*2*) $LW=k$ since $LW=A$ and A is constant

c) (*1*) multiplies the volume by 7 (*2*) $PV=k$ (Boyle's Law)

d) (*1*) divides the number of these articles purchased by 3

(*2*) $NP=k$ since $NP=C$ and C is constant

3.4. Finding Values for Inversely Varying Variables

a) If y and x vary inversely as each other and $y=10$ when $x=6$, find y when $x=15$.

Solution:

Proportion Method

If y and x vary inversely,

	x	y
2nd values	15	y
1st values	6	10

$$\frac{y_2}{y_1} = \frac{x_1}{x_2}$$

$$\frac{y}{10} = \frac{6}{15}$$

$$y = \frac{6}{15}(10) = 4$$

Ans. 4

Ratio Method

(Note the inverse ratios in the table.)

	x		y	
2nd values	15	$\frac{5}{2}$	y	$\frac{2}{5}$
1st values	6		10	

If *2nd x-value* is **five-halves** of *1st x-value*, then *2nd y-value* is **two-fifths** of *1st y-value*.

Hence, $y = \frac{2}{5}(10) = 4$. *Ans.* 4

Equation Method

If x and y vary inversely, $k=xy$.

Since $y=10$ when $x=6$, $k=(6)(10)=60$.

Since $k=60$, $xy=60$.

When $x=15$, $15y=60$, $y=4$. *Ans.* 4

b) If y and x vary inversely as each other and $y=12$ when $x=4$, find x when $y=8$.

Solution:

Proportion Method

If y and x vary inversely,

	x	y
2nd values	x	8
1st values	4	12

$$\frac{x_2}{x_1} = \frac{y_1}{y_2}$$

$$\frac{x}{4} = \frac{12}{8}$$

$$x = \frac{12}{8}(4) = 6$$

Ans. 6

Ratio Method

(Note the inverse ratios in the table.)

	x		y	
2nd values	x	$\frac{3}{2}$	8	$\frac{2}{3}$
1st values	4		12	

If *2nd y-value* is **two-thirds** of *1st y-value*, then *2nd x-value* is **three-halves** of *1st x-value*.

Hence, $x = \frac{3}{2}(4) = 6$. *Ans.* 6

Equation Method

If x and y vary inversely, $k=xy$.

Since $y=12$ when $x=4$, $k=(4)(12)=48$.

Since $k=48$, $xy=48$.

When $y=8$, $8x=48$, $x=6$. *Ans.* 6

3.5. Inverse Variation in a Balanced Lever Problem

Two weights of 35 lb. and 25 lb. are in balance on a lever bar. If the 35 lb. weight is 20 in. from the fulcrum or turning point, how far is the 25 lb. weight from the fulcrum? Consider the weight of the bar as negligible.

Solution:

	(lb.) Weight		(in.) Fulcrum-Distance	
2nd weight	25	$\frac{5}{7}$	D	$\frac{7}{5}$
1st weight	35		20	

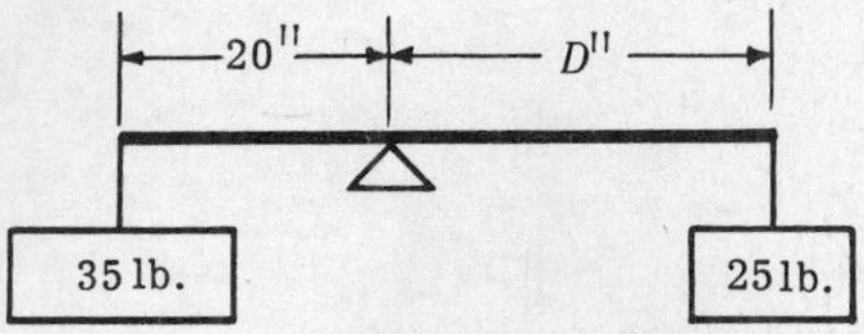

Proportion Method

Since weight and fulcrum-distance on a balanced lever vary inversely,

$$\frac{D_2}{D_1} = \frac{W_1}{W_2}, \quad \frac{D}{20} = \frac{35}{25},$$

$$D = \frac{35}{25}(20) = 28 \qquad \textit{Ans. } 28\text{ in.}$$

Ratio Method (*Note inverse ratios included in table.*)

If the second weight is **five-sevenths** of the first weight, then the second distance is **seven-fifths** of the first distance. Hence, $D = \frac{7}{5}(20) = 28$. *Ans.* 28 in.

3.6. Inverse Variation in a Meshed Gear Problem

A gear having 36 teeth drives another which has 48 teeth. If the first gear makes 100 revolutions, how many revolutions does the second gear make?

Solution:

	No. of Teeth (T)		No. of Revolutions (R)	
2nd gear	48	$\frac{4}{3}$	R	$\frac{3}{4}$
1st gear	36		100	

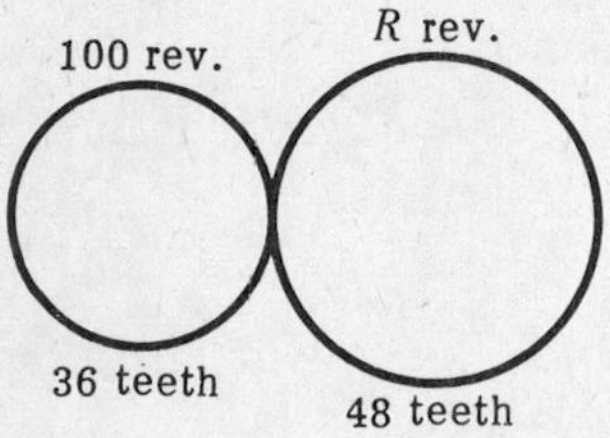

Proportion Method

For meshed gears, the number of teeth and the number of revolutions vary inversely.

$$\text{Hence,} \quad \frac{R_2}{R_1} = \frac{T_1}{T_2}, \quad \frac{R}{100} = \frac{36}{48}$$

$$R = \frac{36}{48}(100) = 75 \qquad \textit{Ans. } 75\text{ rev.}$$

Ratio Method (*Note inverse ratios included in table.*)

If second gear has **four-thirds** as many teeth as first gear, then second gear makes **three-fourths** as many revolutions as first gear.

Hence, $R = \frac{3}{4}(100) = 75$. *Ans.* 75 rev.

3.7. Inverse Variation in Connected Pulleys

Two pulleys connected by the same belt, have diameters of 10 in. and 14 in. If the smaller pulley turns at 560 rpm (revolutions per minute), what is the turning speed of the larger pulley?

Solution:

	(in.) Diameter (D)		(rpm) Revolutions (R)	
2nd pulley	14	$\frac{7}{5}$	R	$\frac{5}{7}$
1st pulley	10		560	

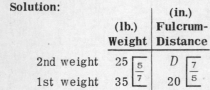

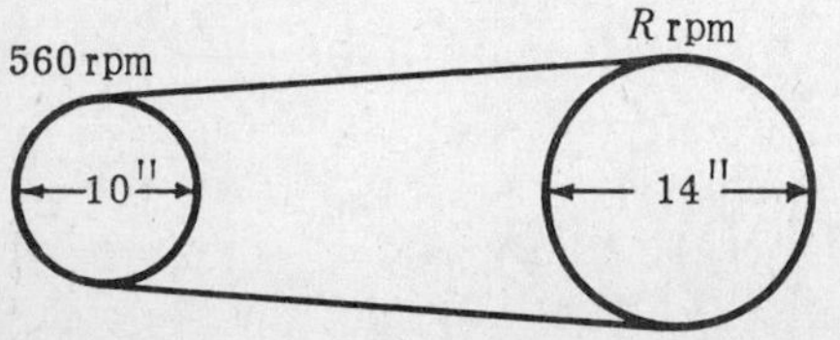

Proportion Method

For pulleys connected by the same belt, diameters and revolutions vary inversely.

$$\text{Hence,} \quad \frac{R_2}{R_1} = \frac{D_1}{D_2}, \quad \frac{R}{560} = \frac{10}{14},$$

$$R = \frac{10}{14}(560) = 400 \qquad \textit{Ans. } 400\text{ rpm}$$

Ratio Method (*Note inverse ratios included in table.*)

If second pulley has **seven-fifths** the diameter of first pulley, then second pulley turns **five-sevenths** as fast as first pulley.

Hence, $R = \frac{5}{7}(560) = 400$. *Ans.* 400 rpm

4. UNDERSTANDING JOINT VARIATION: $z = kxy$

Joint Variation Formula With Constant k

$$z = kxy$$

Note. When $k=1$, we obtain $z=xy$, the formula structure of $D=RT$, $A=LW$, $I=PR$, etc.

Joint Variation Formula Without Constant

$$\frac{z_2}{z_1} = \frac{x_2 y_2}{x_1 y_1}$$

Rule 1. If $z = kxy$, then z **varies jointly as** x and y.

Thus:

(*1*) If $z=5xy$, z varies jointly as x and y.

(*2*) If $D=RT$, D varies jointly as R and T. Hence, in a motion problem, distance varies jointly as rate and time.

Multiplication in Joint Variation

Rule 2. If z varies jointly as x and y and a value of x is multiplied by a while a value of y is multiplied by b, then the corresponding value of z is multiplied by their product ab, as shown in the table.

Thus, if $z = 2xy$ and x is doubled while y is tripled, then z is multiplied by 6, as shown in the table.

$z = 2xy$

x is doubled →	x	1	2	4	8
y is tripled →	y	1	3	9	27
z is multiplied by 6 →	z	2	12	72	432

Applying Joint Variation to a Triangle

To understand joint variation more fully, note how it applies to a triangle:

If the base and altitude of a triangle are tripled, the area becomes nine times as large. As a result, the new triangle ABC is nine times the old (shaded) triangle.

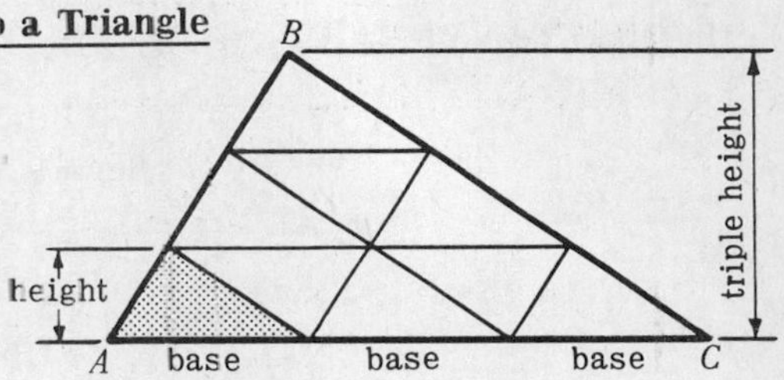

4.1. Rule 1. Joint Variation in Formulas

(*1*) Express each equation in the form $z = kxy$.

(*2*) State how the variables vary jointly.

a) $V = \frac{Bh}{3}$ *Ans.* (*1*) $V = \frac{1}{3}Bh$, (*2*) V varies jointly as B and h.

b) $PV = 5T$ *Ans.* (*1*) $T = \frac{1}{5}PV$, (*2*) T varies jointly as P and V.

c) $V = LWH$ when W is constant. *Ans.* (*1*) $V = kLH$, (*2*) V varies jointly as L and H.

d) $P = \frac{FS}{t}$ when $t=2$. *Ans.* (*1*) $P = \frac{1}{2}FS$, (*2*) P varies jointly as F and S.

4.2. Rule 2. Multiplication in Joint Variation

Complete each:

a) If $z=5xy$ and x and y are each doubled, then (   ).
b) If $A=\frac{1}{2}bh$ and b and h are each halved, then (   ).
c) If $D=RT$, R is tripled while T is doubled, then (   ).
d) If $V=\frac{1}{3}Bh$, B is multiplied by 10 while h is multiplied by $\frac{1}{5}$, then (   ).

Ans. *a*) z is multiplied by $2\cdot 2$ or 4.
b) A is multiplied by $\frac{1}{2}(\frac{1}{2})$ or $\frac{1}{4}$.
c) D is multiplied by $3\cdot 2$ or 6.
d) V is multiplied by $10(\frac{1}{5})$ or 2.

4.3. Finding Values for Jointly Varying Variables

If z varies jointly as x and y, and $z=100$ when $x=4$ and $y=5$, find z when $x=12$ and $y=10$.

Solution:

Proportion Method

	x	y	z
2nd values	12	10	z
1st values	4	5	100

If z varies jointly as x and y,

$$\frac{z_2}{z_1} = \frac{x_2 y_2}{x_1 y_1}, \quad \frac{z}{100} = \frac{(12)(10)}{(4)(5)}$$

$z = 100(6) = 600$ *Ans.* 600

Ratio Method (*Note the ratios included in the table.*)

	x		y		z	
2nd values	12	$\frac{3}{1}$	10	$\frac{2}{1}$	z	$\frac{6}{1}$
1st values	4		5		100	

If x is tripled and y is doubled, then z is multiplied by 6.

Hence, $z = 6(100) = 600$. *Ans.* 600

Equation Method

If z varies jointly as x and y, $kxy = z$.
Since $z=100$ when $x=4$ and $y=5$, $(4)(5)k = 100$, $20k = 100$, $k = 5$.
Since $k = 5$, $z = 5xy$.
When $x=12$ and $y=10$, $z = 5(12)(10) = 600$. *Ans.* 600

5. UNDERSTANDING POWER VARIATION

1. Direct Square Variation: $y = kx^2$, $\frac{y_2}{y_1} = \left(\frac{x_2}{x_1}\right)^2$

Rule 1. If $y=kx^2$, then y varies directly as the square of x. Here, if a value of x is multiplied by a number, then the corresponding value of y is multiplied by the square of that number.

Thus: The surface of a cube equals six times the square of its edge; that is, $S=6e^2$. For a cube, S varies directly as the square of e. If e is doubled, S is quadrupled.

2. Inverse Square Variation: $y = \frac{k}{x^2}$, $\frac{y_2}{y_1} = \left(\frac{x_1}{x_2}\right)^2$

Rule 2. If $y=\frac{k}{x^2}$, then y varies inversely as the square of x. Here, if a value of x is multiplied by a number, then the corresponding value of y is divided by the square of that number.

Thus: If $I=\frac{5}{d^2}$, I varies inversely as the square of d. If d is tripled, I is divided by 9.

3. Direct Cube Variation: $y = kx^3$, $\frac{y_2}{y_1} = \left(\frac{x_2}{x_1}\right)^3$

Rule 3. If $y=kx^3$, then y varies directly as the cube of x. Here, if a value of x is multiplied by a number, then the corresponding value of y is multiplied by the cube of that number.

Thus: The volume of a sphere equals $\frac{4}{3}\pi$ times the cube of its radius; that is, $V = \frac{4}{3}\pi R^3$. For a sphere, the volume varies as the cube of its radius. If R is multiplied by 3, V is multiplied by 27.

5.1. Multiplication and Division in Power Variation

Complete each:

a) If $A = \pi R^2$ and R is multiplied by 5, then ( ).

b) If $V = \frac{4}{3}\pi R^3$ and R is divided by 2, then ( ).

c) If $F = \frac{k}{d^2}$ and d is doubled, then ( ).

Solutions:

a) Using **Rule 1**, A is multiplied by 5^2 or 25.

b) Using **Rule 3**, V is divided by 2^3 or 8.

c) Using **Rule 2**, F is divided by 4 or multiplied by $\frac{1}{4}$.

5.2. Direct Square Variation in Problem Solving

After the brakes have been applied, the distance an automobile goes before stopping varies directly as the square of its speed. For an auto going 30 mph, the stopping distance is 27 ft. What is the stopping distance of an auto going 50 mph?

Solution:

	(ft.) Stopping Distance (D)		(mph) Speed (R)	
2nd stop	D	$\frac{25}{9}$	50	$\frac{5}{3}$
1st stop	27		30	

Proportion Method

Since distance varies directly as the square of speed,

$$\frac{D_2}{D_1} = \left(\frac{R_2}{R_1}\right)^2, \quad \frac{D}{27} = \left(\frac{50}{30}\right)^2 = \frac{25}{9}$$

$$D = \frac{25}{9}(27) = 75 \qquad \textit{Ans. } 75 \text{ ft.}$$

Ratio Method (*Note how the ratios are included in the table.*)

If the second speed is $\frac{5}{3}$ of the first speed, then the second distance is $\frac{25}{9}$ of the first. Hence, $D = \frac{25}{9}(27) = 75$. *Ans.* 75 ft.

5.3. Applying Inverse Square Variation to Illumination

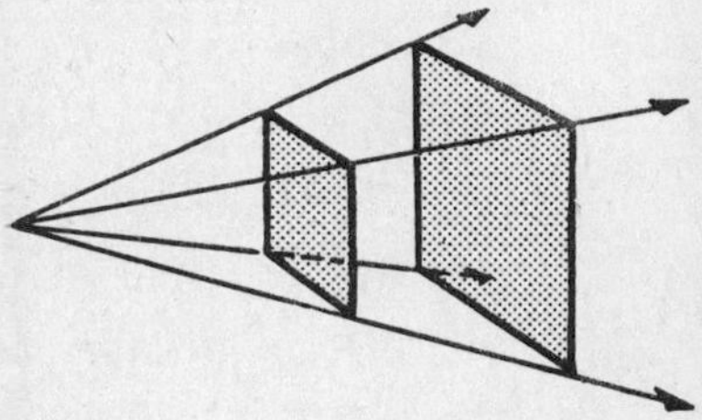

The intensity of light on a surface varies inversely as the square of the distance from a pin-point source. If the distance from the source changes from $2\frac{1}{2}$ ft. to $7\frac{1}{2}$ ft., how many times as bright will the intensity become?

Solution:

	(ft.) Distance (D)		(light units) Intensity (I)	
2nd	$D_2 = 7\frac{1}{2}$	$\frac{3}{1}$	I_2	$\frac{1}{9}$
1st	$D_1 = 2\frac{1}{2}$		I_1	

Since intensity (I) varies inversely as the square of the distance (D), then if D is multiplied by 3, I is divided by 9; that is, the intensity becomes one-ninth as bright.

Otherwise: $\frac{I_2}{I_1} = \left(\frac{D_1}{D_2}\right)^2 = \left(\frac{2\frac{1}{2}}{7\frac{1}{2}}\right)^2 = \frac{1}{9}$ or $I_2 = \frac{1}{9}I_1$

Ans. one-ninth

5.4. Finding Values of Variables in Direct Square Variation

If y varies directly as the square of x and $y = 9$ when $x = 5$, find y when $x = 20$.

Solution:

	x		y	
2nd values	20	$\frac{4}{1}$	y	$\frac{16}{1}$
1st values	5		9	

Proportion Method

Since y varies directly as the square of x, $\frac{y_2}{y_1} = \left(\frac{x_2}{x_1}\right)^2$.

Hence, $\frac{y}{9} = \left(\frac{20}{5}\right)^2$, $\frac{y}{9} = 16$, $y = 144$. *Ans.* 144

Ratio Method (*Note ratios included in the table.*)

If x is multiplied by 4, then y is multiplied by 16.
Hence, $y = 16(9) = 144$. *Ans.* 144

5.5. Finding Values of Variable in Inverse Square Variation

If y varies inversely as the square of x and $x=8$ when $y=48$, find y when $x=32$.

Solution:

	x	ratio	y	ratio
2nd values	32	$\frac{4}{1}$	y	$\frac{1}{16}$
1st values	8		48	

Proportion Method

Since y varies inversely as the square of x, $\frac{y_2}{y_1} = (\frac{x_1}{x_2})^2$.

Hence, $\frac{y}{48} = (\frac{8}{32})^2$, $\frac{y}{48} = \frac{1}{16}$, $y = 3$. *Ans.* 3

Ratio Method (*Note ratios included in the table.*)

If x is multiplied by 4, y is divided by 16. Hence, $y = \frac{48}{16} = 3$. *Ans.* 3

SUPPLEMENTARY PROBLEMS

1. Using division, find the ratio which indicates the change in each variable and express this ratio in a sentence. **(1.1)**

a) The speed of a plane changes from 200 mph to 100 mph.
b) A salary changes from \$50 per week to \$60 per week.
c) The length of a line changes from 2 ft. to 1 ft. 6 in.
d) The price of a radio set changes from \$100 to \$175.
e) John's monthly income remains at \$275 per month.
f) The bus fare changes from 25¢ to 40¢.

Ans. *a*) $\frac{1}{2}$; second speed is **one-half** of first.
b) $\frac{6}{5}$; second salary is **six-fifths** of first.
c) $\frac{3}{4}$; second length is **three-quarters** of first.
d) $\frac{7}{4}$; second price is **seven-fourths** of first.
e) 1; second income **is equal to** first.
f) $\frac{8}{5}$; second fare is **eight-fifths** of first.

2. Complete each: **(1.2)**

a) If x is tripled, it will change from 7 to ( ) or from ( ) to 72. *Ans.* 21, 24
b) If y is multiplied by $\frac{3}{2}$, it will change from 12 to ( ) or from ( ) to 48. *Ans.* 18, 32
c) If z is divided by 5, it will change from 35 to ( ) or from ( ) to 75. *Ans.* 7, 375
d) If k which equals 5.7 remains constant, the new value is ( ). *Ans.* 5.7
e) If r is divided by $\frac{2}{5}$, it will change from 40 to ( ) or from ( ) to 30. *Ans.* 100, 12

3. (*1*) Express each equation in the form, $y=kx$. **(2.1)**
(*2*) State the variables that vary directly as each other.

a) $\frac{C}{R} = 2\pi$
b) $8s = p$
c) $24 = \frac{P}{T}$
d) $f = \frac{n}{d}$ when $d=5$
e) $V = \frac{1}{3}Bh$ when $h=60$
f) $I = PRT$ when R and T are constant

Ans. *a*) (*1*) $C=2\pi R$, (*2*) C and R
b) (*1*) $p=8s$, (*2*) p and s
c) (*1*) $P=24T$, (*2*) P and T
d) (*1*) $f=\frac{1}{5}n$, (*2*) f and n
e) (*1*) $V=20B$, (*2*) V and B
f) (*1*) $I=kP$, (*2*) I and P

4. Complete each: **(2.2)**

a) If $b=5c$ and c is doubled, then ( ). *Ans.* b is doubled.
b) If $\frac{c}{d}=1.5$ and d is halved, then ( ). *Ans.* c is halved.
c) If $h=\frac{7}{5}p$ and p is divided by 5, then ( ). *Ans.* h is divided by 5.
d) If $LW=A$, L is constant and W is multiplied by 7, then ( ). *Ans.* A is multiplied by 7.
e) If $NP=C$, $P=150$ and C is multiplied by $3\frac{1}{2}$, then ( ). *Ans.* N is multiplied by $3\frac{1}{2}$.

5. (*1*) Complete each statement. (*2*) State the formula to which direct variation applies. (2.3)
a) If the time of travel remains fixed, halving rate will ( ).
b) Doubling the length of a rectangle which has a constant width will ( ).
c) If one-third as many articles are purchased at the same price per article, then the cost is ( ).
d) At the same rate of interest, to obtain three times as much annual interest, ( ).
e) If the circumference of a circle is quadrupled, then its radius ( ).

Ans. *a*) (*1*) halve distance. (*2*) $D=kR$ since $D=RT$ and T is constant.
b) (*1*) double area. (*2*) $A=kL$ since $A=LW$ and W is constant.
c) (*1*) one-third as much. (*2*) $C=kN$ since $C=NP$ and P is constant.
d) (*1*) triple principal. (*2*) $I=kP$ since $I=PR$ and R is constant.
e) (*1*) is quadrupled. (*2*) $C=2\pi R$.

6. Find each missing value: (2.4)
a) If y varies directly as x and $y=10$ when $x=5$, find y when $x=15$. *Ans.* 30
b) If r varies directly as s and $r=80$ when $s=8$, find r when $s=6$. *Ans.* 60
c) If L varies directly as A and $L=6$ when $A=21$, find L when $A=28$. *Ans.* 8
d) If D varies directly as T and $D=100$ when $T=2$, find T when $D=300$. *Ans.* 6
e) If N varies directly as C and $C=25$ when $N=10$, find C when $N=16$. *Ans.* 40

7. A pilot flew 800 mi. at 120 mph. In the same time (2.5)
a) how many miles would he have traveled at 150 mph? *Ans.* 1000 mi.
b) how fast must he go to fly 1200 mi.? *Ans.* 180 mph

8. A motorist finds in traveling 100 mi., that he is consuming gas at the rate of 15 mi. per gal. If he uses the same number of gallons, find (2.5)
a) how far he could travel if gas were consumed at the rate of 12 mi. per gal. *Ans.* 80 mi.
b) the rate of consumption if he covers 120 mi. *Ans.* 18 mi. per gal.

9. A salesman earned \$25 in commission when he sold \$500 worth of tools. If his rate of commission remains fixed, (2.6)
a) how much would his commission be if he sold \$700 worth of tools? *Ans.* \$35
b) how much must he sell to earn \$45 in commission? *Ans.* \$900

10. The annual dividends on \$6000 worth of stock is \$360. At the same dividend rate, what is the annual dividends on \$4500? *Ans.* \$270 (2.6)

11. (*1*) Express each equation in the form, $xy=k$. (3.1)
(*2*) State the variables that vary inversely as each other.

a) $3LW = 27$
b) $\dfrac{PV}{10} = 7$
c) $\dfrac{1}{y} = x$
d) $f = \frac{n}{d}$ when $n=25$
e) $A = \frac{1}{2}Dd$ when A is constant
f) $A = \pi ab$ when $A=30$

Ans. *a*) (*1*) $LW=9$, (*2*) L and W
b) (*1*) $PV=70$, (*2*) P and V
c) (*1*) $xy=1$, (*2*) x and y
d) (*1*) $fd=25$, (*2*) f and d
e) (*1*) $Dd=k$, (*2*) D and d
f) (*1*) $ab=\dfrac{30}{\pi}$, (*2*) a and b

12. Complete each: (3.2)
a) If $pv=20$ and p is doubled, then ( ). *Ans.* v is halved.
b) If $RT=176$ and R is tripled, then ( ). *Ans.* T is divided by 3 or multiplied by $\frac{1}{3}$,
c) If $240=RP$ and P is multiplied by $\frac{4}{3}$, then ( ). *Ans.* R is divided by $\frac{4}{3}$ or multiplied by $\frac{3}{4}$.
d) If $IR=E$, E is constant and I is divided by 15, then ( ). *Ans.* R is multiplied by 15.

13. (*1*) Complete each statement. **(3.3)**
(*2*) State the formula to which inverse variation applies.

a) For a fixed area, tripling the width of a rectangle ( ).
b) Taking twice as long to cover the same distance requires ( ).
c) If the pressure of an enclosed gas at constant temperature is reduced to one-half, then its volume ( ).
d) Quadrupling the length of a triangle with a fixed area ( ).
e) If the numerator remains the same, halving the denominator will ( ).

Ans. *a*) (*1*) divides the length by 3. (*2*) $LW=k$ since $A=LW$ and A is constant.
b) (*1*) half the rate. (*2*) $RT=k$ since $D=RT$ and D is constant.
c) (*1*) is doubled. (*2*) $PV=k$ (*Boyle's Law*).
d) (*1*) divides the height by 4. (*2*) $bh=k$ since $A=\frac{1}{2}bh$ and A is constant.
e) (*1*) double the fraction. (*2*) $fd=k$ since $f=\frac{n}{d}$ and n is constant.

14. Find each missing value: **(3.4)**
a) If y varies inversely as x, and $y=15$ when $x=2$, find y when $x=6$. *Ans.* 5
b) If t varies inversely as r, and $t=15$ when $r=8$, find t when $r=24$. *Ans.* 5
c) If b varies inversely as h, and $b=21$ when $h=3$, find b when $h=9$. *Ans.* 7
d) If R varies inversely as T, and $R=36$ when $T=2\frac{1}{2}$, find R when $T=2$. *Ans.* 45
e) If P varies inversely as V, and $P=12$ when $V=40$, find V when $P=10$. *Ans.* 48

15. Two weights of 40 lb. and 48 lb. are in balance on a lever bar. If the 40 lb. weight is 15 in. from the fulcrum, how far is the 48 lb. from the fulcrum? Consider the weight of the bar as negligible. *Ans.* $12\frac{1}{2}$ in. **(3.5)**

16. On a lever, a weight of 42 lb. is 20 in. from the fulcrum. Consider the weight of the bar as negligible. **(3.5)**
a) What weight could be balanced three times as far from the fulcrum? *Ans.* 14 lb.
b) What weight could be balanced 5 in. nearer the fulcrum? *Ans.* 56 lb.
c) At what distance from the fulcrum should a weight which is two-thirds as heavy be placed for balance? *Ans.* 30 in.

17. A gear having 60 teeth is meshed with one having 48 teeth. **(3.6)**
a) When the larger makes 28 revolutions, how many revolutions does the smaller make? *Ans.* 35
b) When the smaller makes 40 revolutions, how many revolutions does the larger make? *Ans.* 32

18. Two pulleys are connected by the same belt. One pulley has a diameter of 16 in. and a speed of 450 rpm. **(3.7)**
a) If the diameter of the other pulley is 10 in., what is its speed? *Ans.* 720 rpm
b) If the speed of the other pulley is 360 rpm, what is its diameter? *Ans.* 20 in.

19. (*1*) Express each equation in the form $z=kxy$. **(4.1)**
(*2*) State how the variables vary jointly.

a) $3PT=100I$ *Ans.* *a*) (*1*) $I=.03PT$, (*2*) I varies jointly as P and T.
b) $2A=dD$ *Ans.* *b*) (*1*) $A=\frac{1}{2}dD$, (*2*) A varies jointly as d and D.
c) $V=LWH$ and L is constant *Ans.* *c*) (*1*) $V=kWH$, (*2*) V varies jointly as W and H.

20. Complete each: **(4.2)**
a) If $A=\frac{1}{2}dd'$ and d and d' are each multiplied by 5, then ( ). *Ans.* A is multiplied by 25.
b) If $I=.03PR$, P is multiplied by 6 and R by $\frac{1}{2}$, then ( ). *Ans.* I is tripled.
c) If $V=10WH$ and W and H are each halved, then ( ). *Ans.* V is divided by 4.
d) If $A=\pi ab$, a is quadrupled and b is divided by 4, then A is ( ). *Ans.* A remains constant.

21. Applying the rule, distance equals the product of rate and time, complete each of the following: **(4.2)**

a) Take four times as long and double rate, then ( ). *Ans.* distance is eight times as far
b) Double time and triple rate, then ( ). *Ans.* distance is six times as far
c) Halve time and double rate, then ( ). *Ans.* distance remains constant
d) Double time and ( ), then distance remains constant. *Ans.* halve rate
e) ( ) and quadruple rate, then distance is twice as far. *Ans.* halve time

22. Find each missing value: **(4.3)**

a) If z varies jointly as x and y, and $z=48$ when $x=4$ and $y=3$, find y when $z=96$ and $x=2$.
b) If A varies jointly as b and h, and $A=24$ when $b=4$ and $h=12$, find h when $A=48$ and $b=6$.
c) If V varies jointly as B and h, and $V=75$ when $B=25$ and $h=9$, find V when $B=15$ and $h=18$.

Ans. *a*) 12, *b*) 16, *c*) 90

23. Complete each: **(5.1)**

a) If $A = 4\pi R^2$ and R is doubled, then ( ). *Ans.* A is quadrupled
b) If $V = 10\pi h^2$ and h is divided by 5, then ( ). *Ans.* V is divided by 25
c) If $S = 4e^3$ and e is quadrupled, then ( ). *Ans.* S is multiplied by 64
d) If $I = \frac{k}{d^2}$ and d is divided by 10, then ( ). *Ans.* I is multiplied by 100

24. After the brakes have been applied, the distance an automobile goes before stopping varies directly as the square of the speed of the car. For a car going 20 mph, the stopping distance is 15 ft. **(5.2)**

a) What is the stopping distance of the same car going 60 mph? *Ans.* 135 ft.
b) What is the speed of the car when its stopping distance is 240 ft.? *Ans.* 80 mph

25. The energy of a moving body varies directly as the square of its speed. If a body has an energy of 50 units at a speed of 5 ft. per sec., **(5.2)**

a) What is its energy at a speed of 10 ft. per sec.? *Ans.* 200 units
b) What is its speed when its energy is 5000 units? *Ans.* 50 ft. per sec.

26. The amount of heat received by a body varies inversely as the square of its distance from a source. How many times as much heat will be received if the body is moved to a point **(5.3)**
a) twice as far, *b*) one-third as far, *c*) four-thirds as far?

Ans. *a*) one-fourth as much, *b*) nine times as much, *c*) nine-sixteenths as much.

27. If y varies directly as the square of x, and $y=9$ when $x=6$, *a*) find y when $x=18$, *b*) find x when $y=36$. *Ans.* *a*) $y=81$, *b*) $x=12$ **(5.4)**

28. If y varies inversely as the square of x, and $y=36$ when $x=15$, *a*) find y when $x=30$, *b*) find x when $y=4$. *Ans.* *a*) $y=9$, *b*) $x=45$ **(5.5)**

29. For each of the following, indicate the type of variation that applies: **(2.1, 3.1, 4.1)**

a) $D=45T$	*Ans.* direct	*e*) $S=2\pi Rl$	*Ans.* joint	*i*) $y=\frac{2}{3}x$	*Ans.* direct
b) $15=RT$	*Ans.* inverse	*f*) $\frac{C}{D}=\pi$	*Ans.* direct	*j*) $\frac{x}{y}=10$	*Ans.* direct
c) $2\frac{1}{2}R=D$	*Ans.* direct	*g*) $15=2\pi rh$	*Ans.* inverse	*k*) $\frac{10}{y}=x$	*Ans.* inverse
d) $RD=7$	*Ans.* inverse	*h*) $A=\frac{1}{2}bh$	*Ans.* joint	*l*) $PV=kT$	*Ans.* joint

30. For each of the following, indicate the type of variation that applies: **(2.1, 3.1, 4.1)**

a) $A=LW$	*Ans.* joint	*e*) $V=LWH$ when L is constant	*Ans.* joint
b) $A=LW$ when L is constant	*Ans.* direct	*f*) $V=LWH$ when L and W are constant	*Ans.* direct
c) $A=LW$ when W is constant	*Ans.* direct	*g*) $V=LWH$ when V and L are constant	*Ans.* inverse
d) $A=LW$ when A is constant	*Ans.* inverse	*h*) $V=LWH$ when V and H are constant	*Ans.* inverse

Chapter 15

INDIRECT MEASUREMENT

1. INDIRECT MEASUREMENT: USING TRIANGLES DRAWN TO SCALE

By indirect measurement, the measure of a quantity is obtained by measuring another quantity instead.

Thus, the distance between two towns, A and B, is 80 mi. if the line AB which represents this distance on a map is 4 units long and each unit represents 20 mi.

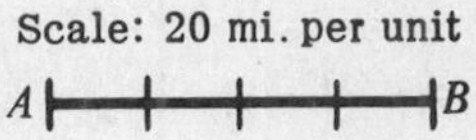

(A scale of 20 mi. per unit means that each map unit represents 20 mi.)

Using a Scale Triangle to Measure Lines and Angles Indirectly

1. The **actual distance**, D, represented by a line equals the product of the scale, S, and the number, N, of the units in the line: $D = SN$.

   (See section 2 for an extended treatment of $D = SN$.)

   Thus, in the scale triangle ABC which has a scale of 120 miles per unit:

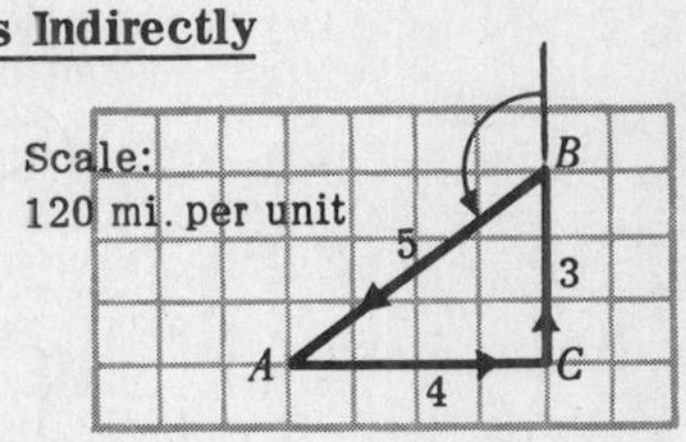

   $AC = (120)(4)$ or 480 mi., $BC = (120)(3)$ or 360 mi., $AB = (120)(5)$ or 600 mi.

   (The length of AB can be found by placing the ends of a compass on A and B and then laying off this opening along a line of the graph.)

2. The **actual angle** represented by an angle equals the number of degrees in the angle.

   Thus, since $\angle A$ in the scale triangle $= 37°$, the actual $\angle A = 37°$.

   (Use a protractor to measure $\angle A$ in the scale triangle.)

1.1. Measuring Indirectly, Using Triangles Drawn to Scale

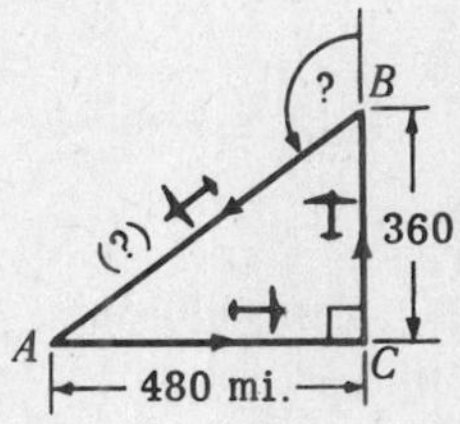

An aviator flew 480 miles due east from A to C. Then he flew 360 miles due north from C to B. To fly straight back to A,

(1) what distance must be flown and

(2) what angle of turn must be made, to the nearest degree?

Solution:

Construct scale triangle ABC as follows:

1. Choose a convenient scale, say 60 mi. per unit.
2. Lay off 8 units for $AC = 480$ mi.
3. Lay off 6 units for $BC = 360$ mi.
4. Draw AB.

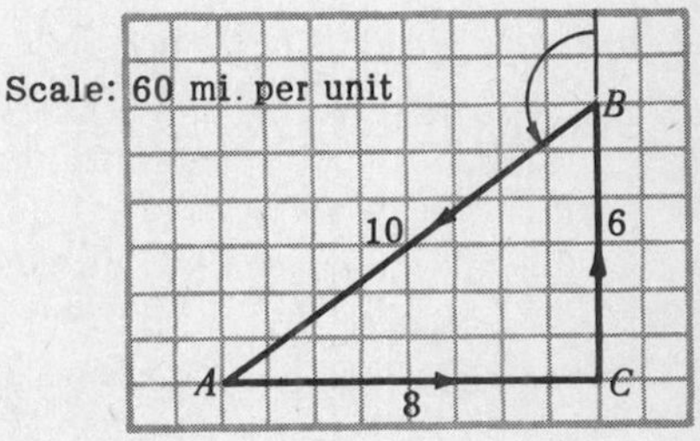

Find distance from A **to** B:

By measurement, $AB = 10$ units.

Since scale = 60 mi. per unit, distance AB = 600 mi.

Find angle of turn at B:

Using protractor, angle of turn at $B = 127°$.

Ans. (1) Distance from A to $B = 600$ mi.
(2) Angle of turn at $B = 127°$.

2. D = SN: A FORMULA OF INDIRECT MEASUREMENT FOR FIGURES DRAWN TO SCALE ON MAPS, GRAPHS, MODELS AND BLUEPRINTS

Rule. A line on a scaled figure such as a map represents a distance, D, equal to the product of the scale, S, and the number of units, N, in the line.

Formula: $D = SN$ **Other forms:** $N = \frac{D}{S}$, $S = \frac{D}{N}$

Note. To use this formula, $D = SN$, S must be in distance units per map unit or graph unit, such as miles per inch, or feet per graph unit.

Thus, for a distance of 100 miles represented by a 5 inch line, the scale is 20 **miles per inch.**

2.1. Finding Distance Being Represented: $D = SN$

Find the distance, D, represented by a line AB on a map or graph

a) if AB = 5 in. and the scale of the map is 20 mi. per in.
b) if AB = 8 cm. and the scale of the map is 12 mi. per cm.
c) if AB = 20 units and the scale of the graph is $2\frac{1}{2}$ ft. per unit.
d) if AB = $4\frac{1}{2}$ units and the scale of the graph is 6 yd. per unit.

Solutions: (Using $D = SN$)

a) $D = (20)(5) = 100$ *Ans.* 100 mi.
b) $D = (12)(8) = 96$ *Ans.* 96 mi.
c) $D = (2\frac{1}{2})(20) = 50$ *Ans.* 50 ft.
d) $D = (6)(4\frac{1}{2}) = 27$ *Ans.* 27 yd.

2.2. Finding Number of Units Needed: $N = \frac{D}{S}$

On a map or graph, a distance of 60 mi. is to be represented by a line CD. Find the length needed for CD, if the scale of the

a) map is 20 mi. per inch.
b) map is 30 mi. per cm.
c) graph is 4 mi. per graph unit.
d) graph is $\frac{1}{2}$ mi. per graph unit.

Solutions: (Using $N = \frac{D}{S}$)

a) $N = \frac{60}{20} = 3$ *Ans.* 3 in.
b) $N = \frac{60}{30} = 2$ *Ans.* 2 cm.
c) $N = \frac{60}{4} = 15$ *Ans.* 15 units
d) $N = 60 \div \frac{1}{2} = 120$ *Ans.* 120 units

2.3. Finding Scale: $S = \frac{D}{N}$

On a map or graph, a distance of 30 mi. is represented by EF. Find the scale being used if EF equals *a*) 3 in., *b*) 3 yd., *c*) 8 graph units, *d*) $\frac{2}{3}$ graph unit.

Solutions: (Using $S = \frac{D}{N}$)

a) $S = \frac{30}{3} = 10$ *Ans.* 10 mi. per in.
b) $S = \frac{30}{3} = 10$ *Ans.* 10 mi. per yd.
c) $S = \frac{30}{8} = 3\frac{3}{4}$ *Ans.* $3\frac{3}{4}$ mi. per unit
d) $S = 30 \div \frac{2}{3} = 45$ *Ans.* 45 mi. per unit

2.4. Finding Distance and Direction

The scale triangle shown was drawn by a navigator to find the distance between two cities, A and B, and the angle of turn at B. He knew that the actual distance from A to C and from C to B was 100 mi. and that BC was at right angles to AC. What is the distance AB and the angle of turn shown at B?

Solution:

The scale $S = \frac{D}{N} = \frac{100}{10} = 10$. The scale is **10 mi. per unit.**

By measurement, $AB = 14$ units. Hence the distance $AB = (14)(10)$ or 140 mi. Using protractor, angle of turn at $B = 135°$.

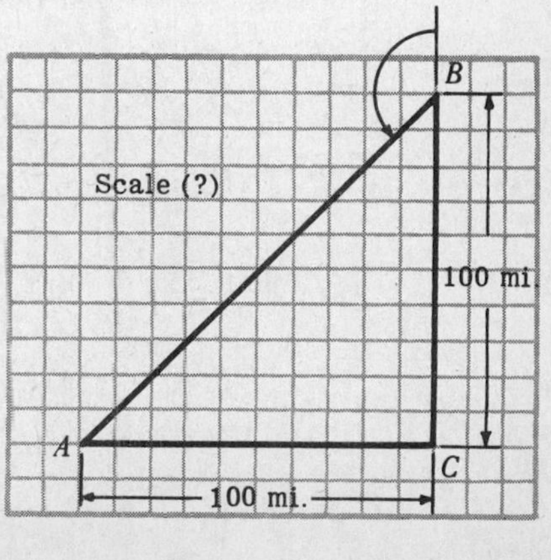

Ans. AB = 140 mi., angle of turn at B = 135°.

SUPPLEMENTARY PROBLEMS

1. Solve the following, using triangles drawn to scales of 15 and 30 mi. per unit: **(1.1)**

An aviator flew 240 mi. due south from G to H. Then he flew 450 mi. due west from H to J. To fly straight back to G,

(*1*) what distance must be flown and

(*2*) what angle of turn must be made, to the nearest degree?

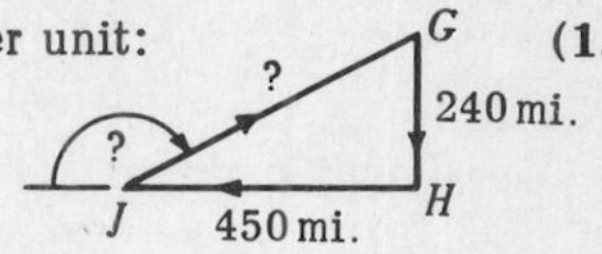

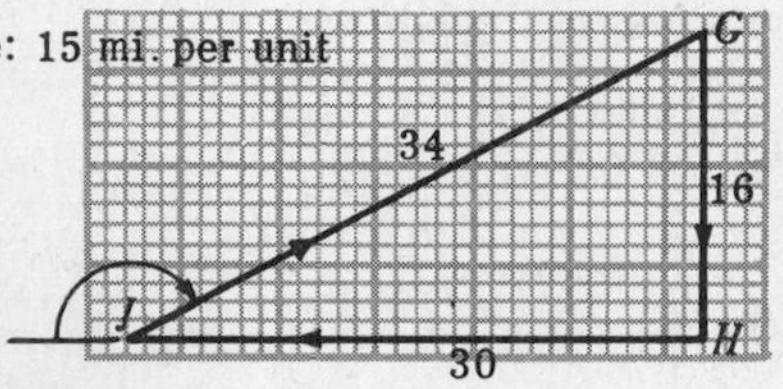

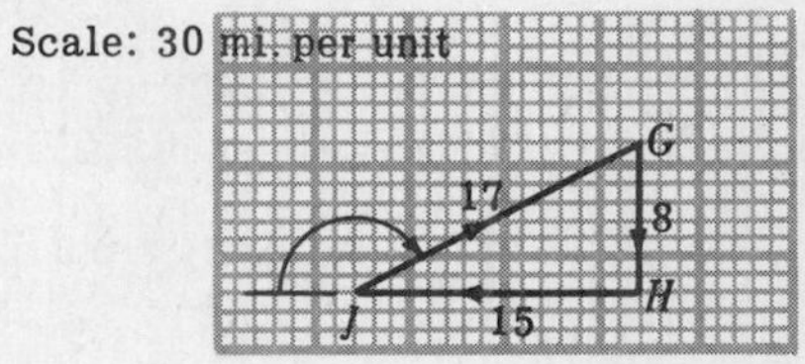

Ans. (*1*) 510 mi. (*2*) 152°

2. On a graph, a line AB is 8 units long. Find the distance AB represents, using a scale of **(2.1)**

a) 30 mi. per unit, *b*) 8 rd. per unit, *c*) $4\frac{1}{2}$ ft. per unit, *d*) 5.1 yd. per unit.

Ans. *a*) 240 mi. *b*) 64 rd. *c*) 36 ft. *d*) 40.8 yd.

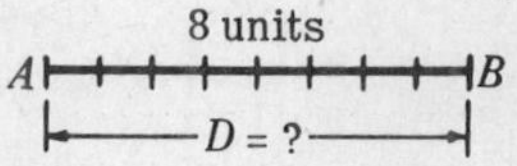

3. On a map, a distance of 450 mi. is to be represented by CD. Find the length of CD, using a scale of **(2.2)**

a) 50 mi. per in., *b*) 150 mi. per ft., *c*) 200 mi. per ft., *d*) 900 mi. per yd.

Ans. *a*) $\frac{450}{50}$ or 9 in. *b*) $\frac{450}{150}$ or 3 ft. *c*) $\frac{450}{200}$ or $2\frac{1}{4}$ ft. *d*) $\frac{450}{900}$ or $\frac{1}{2}$ yd.

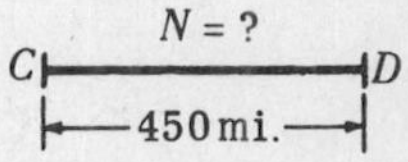

4. In a house plan, find the number of units needed to represent **(2.2)**

a) a length of 40 ft., using a scale of 5 ft. per unit. *Ans.* $\frac{40}{5} = 8$

b) a length of 40 ft., using a scale of 6 in. per unit. *Ans.* $40 \div \frac{1}{2} = 80$

c) a length of 10 ft., using a scale of $\frac{1}{4}$ ft. per unit. *Ans.* $10 \div \frac{1}{4} = 40$

d) a length of 24 ft., using a scale of $\frac{3}{4}$ ft. per unit. *Ans.* $24 \div \frac{3}{4} = 32$

5. On a graph, find the number of units required to represent a distance of 200 mi. if the scale used is

a) 400 mi. per unit, *b*) 150 mi. per unit, *c*) 80 mi. per unit, *d*) $33\frac{1}{3}$ mi. per unit. **(2.2)**

Ans. *a*) $\frac{1}{2}$ *b*) $1\frac{1}{3}$ *c*) $2\frac{1}{2}$ *d*) 6

6. Find the scale being used on a graph **(2.3)**

a) if a distance of 12 yd. is represented by a line of 8 units. *Ans.* $\frac{12}{8}$ or $1\frac{1}{2}$ yd. per unit

b) if a line of 10 units represents a distance of 75 ft. *Ans.* $\frac{75}{10}$ or $7\frac{1}{2}$ ft. per unit

c) if a line of $3\frac{1}{4}$ units represents a distance of 39 in. *Ans.* $39 \div \frac{13}{4}$ or 12 in. per unit

7. To determine the distance from G to M, a navigator draws the scale triangle shown and finds GM to be 13 units. Find the distance GM **(2.4)**

a) if the scale used is $3\frac{1}{2}$ mi. per unit. *Ans.* $45\frac{1}{2}$ mi.

b) if GK represents 36 mi. *Ans.* 39 mi.

c) if MK represents 55 mi.. *Ans.* 143 mi.

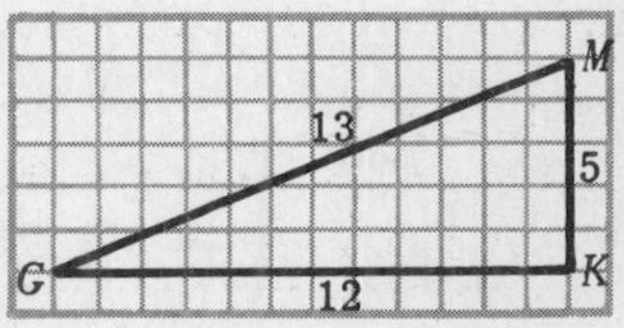

8. To find the distance from P to S, a navigator used the scale triangle shown. **(2.4)**

a) Find the scale used.

b) Find PS to the nearest number of units.

c) Using the answer in (*b*), find the distance PS.

Ans. *a*) 3 mi. per unit, *b*) 19 units, *c*) 57 mi.

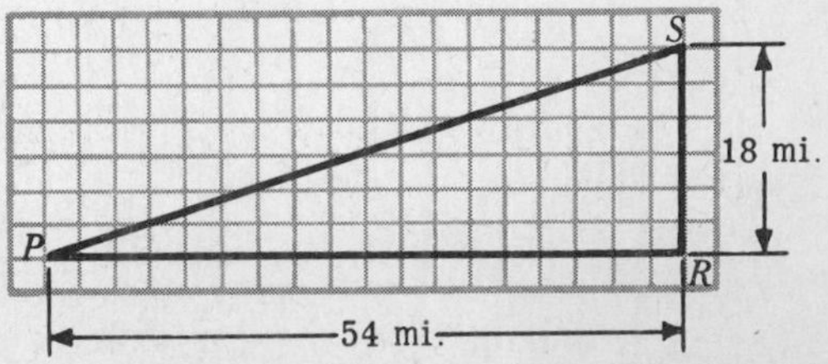

Chapter 16

LAW of PYTHAGORAS, PROPORTIONS and SIMILAR TRIANGLES

1. LAW OF PYTHAGORAS

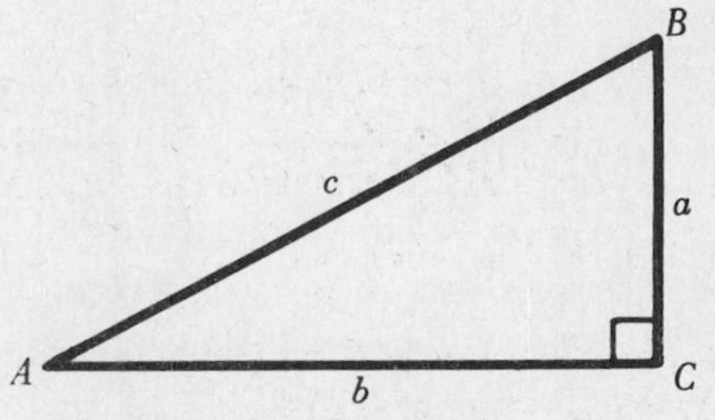

The square of the hypotenuse:

In right triangle ABC, if C is a right angle,

$$c^2 = a^2 + b^2$$

Note. In a triangle, the small letter for a side should agree with the capital letter for the vertex of the angle opposite that side. Thus, side a is opposite angle A, etc.

> **Law of Pythagoras:** In a right triangle, the **square** of the hypotenuse equals the **sum of the squares** of the two arms.

Pythagoras, a famous Greek mathematician and philosopher, lived about 500 B.C.

The square of either arm:

By transposition, $a^2 = c^2 - b^2$ and $b^2 = c^2 - a^2$

> **Transformed Law of Pythagoras:** In a right triangle the **square** of either arm equals the **difference of the squares** of the hypotenuse and the other arm.

To test for a right triangle, use the following rule:

> **Test Rule For a Right Triangle:** If $c^2 = a^2 + b^2$ applies to the three sides of a triangle, then the triangle is a right triangle; but if $c^2 \neq a^2 + b^2$, then the triangle is not a right triangle.

Distance Between Two Points on a Graph

If d is the distance between $P_1(x_1, y_1)$ and $P_2(x_2, y_2)$:

$$d^2 = (x_2 - x_1)^2 + (y_2 - y_1)^2$$

Thus, the distance equals 5 from the point (2,5) to the point (6,8), as follows:

	(x, y)	
P_2	$(6, 8)$	$\longrightarrow x_2 = 6, \quad y_2 = 8$
P_1	$(2, 5)$	$\longrightarrow x_1 = 2, \quad y_1 = 5$

$$d^2 = (x_2 - x_1)^2 + (y_2 - y_1)^2$$
$$d^2 = (6-2)^2 + (8-5)^2$$
$$d^2 = 4^2 + 3^2 = 25$$
$$d = 5$$

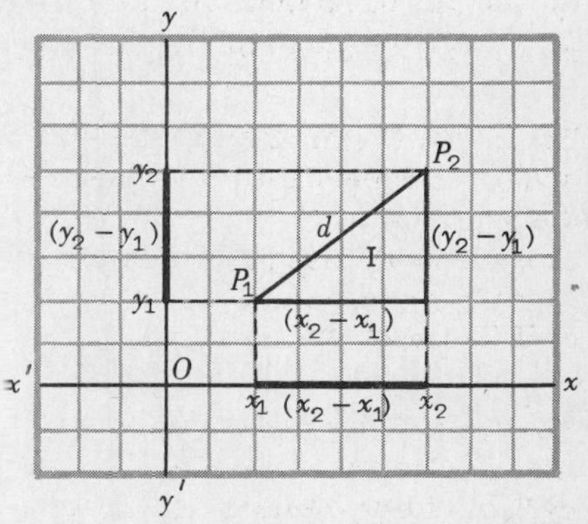

In the following exercises, express each irrational answer in simplest radical form, unless otherwise indicated.

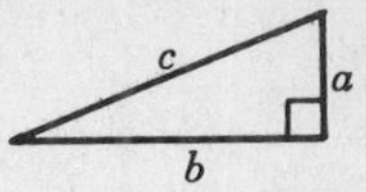

1.1. Finding the Hypotenuse of a Right Triangle: $c^2 = a^2 + b^2$

Find hypotenuse c in the right triangle shown when

a) $a=12$, $b=9$ b) $a=3$, $b=7$ c) $a=3$, $b=6$ d) $a=3$, $b=3\sqrt{3}$

Solutions:

a) $c^2 = a^2+b^2$, $c^2 = 12^2+9^2$, $c^2 = 144+81$, $c^2 = 225$, $c = \sqrt{225}$, $c = 15$ *Ans.*

b) $c^2 = a^2+b^2$, $c^2 = 3^2+7^2$, $c^2 = 9+49$, $c^2 = 58$, $c = \sqrt{58}$ *Ans.*

c) $c^2 = a^2+b^2$, $c^2 = 3^2+6^2$, $c^2 = 9+36$, $c^2 = 45$, $c = \sqrt{45}$, $c = 3\sqrt{5}$ *Ans.*

d) $c^2 = a^2+b^2$, $c^2 = 3^2+(3\sqrt{3})^2$, $c^2 = 9+27$, $c^2 = 36$, $c = \sqrt{36}$, $c = 6$ *Ans.*

Note. Since the hypotenuse is to be considered as positive only, reject the negative answer obtainable in each case. Thus, if $c^2 = 225$, c may equal 15 or -15. Reject the negative, -15.

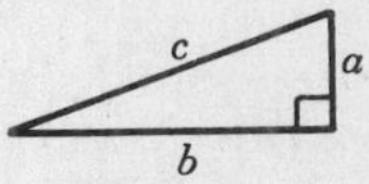

1.2. Finding an Arm of a Right Triangle: $a^2 = c^2 - b^2$ or $b^2 = c^2 - a^2$

In the right triangle shown, find each missing arm when

a) $b=5$, $c=13$ b) $a=24$, $c=25$ c) $b=6$, $c=8$ d) $a=4\sqrt{3}$, $c=8$

Solutions:

a) $a^2 = c^2 - b^2$, $a^2 = 13^2-5^2$, $a^2 = 169-25$, $a^2 = 144$, $a = \sqrt{144}$, $a = 12$ *Ans.*

b) $b^2 = c^2 - a^2$, $b^2 = 25^2-24^2$, $b^2 = 625-576$, $b^2 = 49$, $b = \sqrt{49}$, $b = 7$ *Ans.*

c) $a^2 = c^2 - b^2$, $a^2 = 8^2-6^2$, $a^2 = 64-36$, $a^2 = 28$, $a = \sqrt{28}$, $a = 2\sqrt{7}$ *Ans.*

d) $b^2 = c^2 - a^2$, $b^2 = 8^2-(4\sqrt{3})^2$, $b^2 = 64-48$, $b^2 = 16$, $b = \sqrt{16}$, $b = 4$ *Ans.*

1.3. Ratios in a Right Triangle

In a right triangle whose hypotenuse is 20, the ratio of the two arms is 3:4. Find each arm.

Solution:

Let $3x$ and $4x$ represent the two arms of the right triangle.

$$(3x)^2 + (4x)^2 = 20^2$$
$$9x^2 + 16x^2 = 400$$
$$25x^2 = 400$$
$$x^2 = 16$$
$$x = 4$$

If $x = 4$,
$3x = 12$
$4x = 16$

Ans. Arms are 12 and 16.

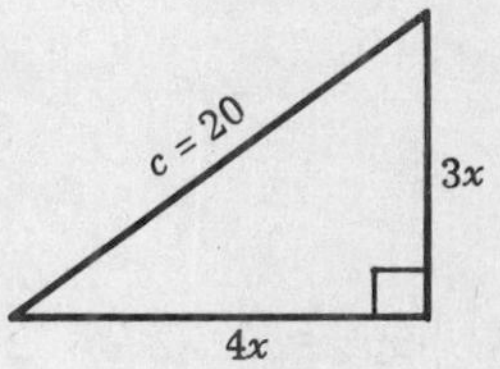

1.4. Applying Law of Pythagoras to a Rectangle

In a rectangle, find a) the diagonal if its sides are 9 and 40,
b) one side if the diagonal is 30 and the other side is 24.

Solution:

The diagonal of the rectangle is the hypotenuse of a right triangle.

a) $d^2 = 9^2 + 40^2$, $d^2 = 1681$, $d = 41$ *Ans.* 41

b) $h^2 = 30^2 - 24^2$, $h^2 = 324$, $h = 18$ *Ans.* 18

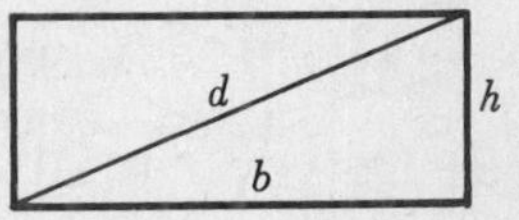

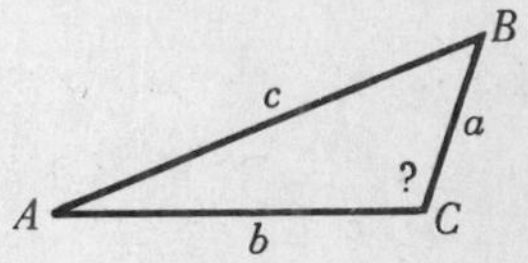

1.5. Testing for Right Triangles, Using the Test Rule

Using the three sides given, which triangles are right triangles?

ΔI: 8, 15, 17; ΔII: 6, 9, 11; ΔIII: $1\frac{1}{2}$, 2, $2\frac{1}{2}$.

Solutions: Rule. If $c^2 = a^2 + b^2$, ΔABC is a right triangle; but if $c^2 \neq a^2 + b^2$, ΔABC is not.

ΔI: $8^2 + 15^2 \stackrel{?}{=} 17^2$, $64 + 225 \stackrel{?}{=} 289$, $289 = 289$. ΔI is a rt.Δ

ΔII: $6^2 + 9^2 \stackrel{?}{=} 11^2$, $36 + 81 \stackrel{?}{=} 121$, $117 \neq 121$. ΔII is not a rt.Δ

ΔIII: $(1\frac{1}{2})^2 + 2^2 \stackrel{?}{=} (2\frac{1}{2})^2$, $2\frac{1}{4} + 4 \stackrel{?}{=} 6\frac{1}{4}$, $6\frac{1}{4} = 6\frac{1}{4}$. ΔIII is a rt.Δ

1.6. Finding Distance Between Two Points on a Graph

Find the distance between each of the following pairs of points:

a) from (3,4) to (6,8)

Solutions:

	(x,y)	
P_2	(6,8)	$\rightarrow x_2 = 6,\ y_2 = 8$
P_1	(3,4)	$\rightarrow x_1 = 3,\ y_1 = 4$

$$d^2 = (x_2-x_1)^2 + (y_2-y_1)^2$$
$$d^2 = (6-3)^2 + (8-4)^2$$
$$d^2 = 3^2 + 4^2 = 25$$
$$d = 5 \quad Ans.$$

b) from (3,4) to (6,10)

	(x,y)	
P_2	(6,10)	$\rightarrow x_2 = 6,\ y_2 = 10$
P_1	(3,4)	$\rightarrow x_1 = 3,\ y_1 = 4$

$$d^2 = (x_2-x_1)^2 + (y_2-y_1)^2$$
$$d^2 = (6-3)^2 + (10-4)^2$$
$$d^2 = 3^2 + 6^2 = 45$$
$$d = 3\sqrt{5} \quad Ans.$$

c) from (−3,2) to (9,−3)

	(x,y)	
P_2	(9,−3)	$\rightarrow x_2 = 9,\ y_2 = -3$
P_1	(−3,2)	$\rightarrow x_1 = -3,\ y_1 = 2$

$$d^2 = (x_2-x_1)^2 + (y_2-y_1)^2$$
$$d^2 = [9 - (-3)]^2 + (-3-2)^2$$
$$d^2 = 12^2 + (-5)^2 = 169$$
$$d = 13 \quad Ans.$$

Note. Since the distance is considered to be positive only, the negative answer obtainable in each case is to be rejected.

1.7. Using Law of Pythagoras to Derive Formulas

a) Derive a formula for the diagonal *d* of a square in terms of any side *s*.

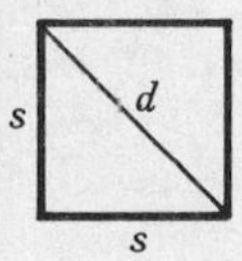

Solutions:

$$d^2 = s^2 + s^2$$
$$d^2 = 2s^2$$
$$d = s\sqrt{2} \quad Ans.$$

b) Derive a formula for the altitude *h* of any equilateral triangle in terms of any side *s*.

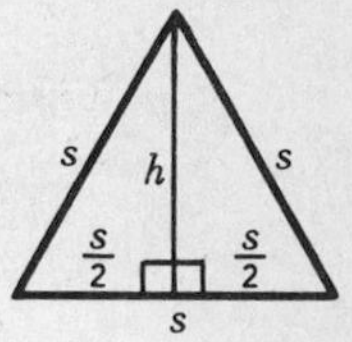

The altitude *h* of the equilateral triangle bisects the base *s*.

$$h^2 = s^2 - (\tfrac{s}{2})^2$$
$$h^2 = s^2 - \frac{s^2}{4} = \frac{3s^2}{4}$$
$$h = \tfrac{s}{2}\sqrt{3} \quad Ans.$$

1.8. Applying Law of Pythagoras to an Inscribed Square

The largest possible square is to be cut from a circular piece of cardboard having a diameter of 10 inches. Find the side of the square to the nearest inch.

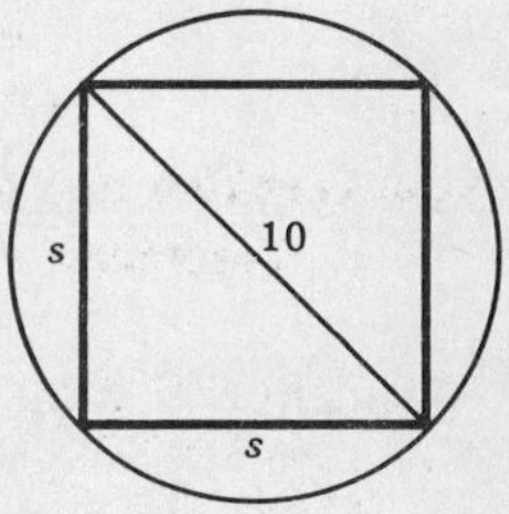

Solution:

The diameter of the circle will be the diagonal of the square.

Hence, $s^2 + s^2 = 100,\quad 2s^2 = 100$

$s^2 = 50,\quad s = 5\sqrt{2} = 7.07 \qquad Ans.\ 7$ in.

1.9. Applying Law of Pythagoras to an Isosceles Triangle

Find the altitude of the isosceles triangle shown if *a*) $a=25$ and $b=30$, *b*) $a=12$ and $b=8$.

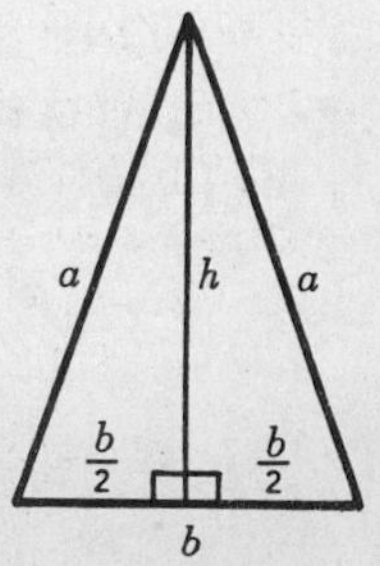

Solution:

The altitude *h* of the isosceles triangle bisects the base *b*.

Hence, $h^2 = a^2 - (\frac{b}{2})^2$.

a) $h^2 = 25^2 - 15^2,\quad h^2 = 400,\quad h = 20 \qquad Ans.\ 20$

b) $h^2 = 12^2 - 4^2,\quad h^2 = 128,\quad h = 8\sqrt{2} \qquad Ans.\ 8\sqrt{2}$

2. PROPORTIONS: EQUAL RATIOS

Understanding Proportions

A **proportion** is an equality of two ratios.

Thus, $2:5 = 4:10$ or $\frac{2}{5} = \frac{4}{10}$ is a proportion.

The fourth term of a proportion is the **fourth proportional** to the other three taken in order.

Thus, in $2:3 = 4:x$, x is the fourth proportional to 2, 3 and 4.

The **means** of a proportion are its middle terms; that is, its second and third terms.

The **extremes** of a proportion are its outside terms; that is, its first and fourth terms.

Thus, in $a:b = c:d$, the means are b and c, and the extremes are a and d.

Proportion Rule: If $a:b = c:d$, $ad = bc$.

Proof: If $\frac{a}{b} = \frac{c}{d}$, then $bd\left(\frac{a}{b}\right) = bd\left(\frac{c}{d}\right)$. Hence $ad = bc$.

Stating the Proportion Rule in Two Forms

Fraction Form

In any proportion, the cross-products are equal.

Thus, in $\frac{2}{5} = \frac{4}{10}$: $5 \times 4 = 20$, $2 \times 10 = 20$

Colon Form

In any proportion, the product of the means equals the product of the extremes.

Thus, in $2:5 = 4:10$: $5 \times 4 = 20$, $2 \times 10 = 20$

2.1. Finding Unknowns in Proportions Using Equal Cross-Products

Solve for x:

a) $\frac{x}{20} = \frac{3}{5}$	b) $\frac{3}{x} = \frac{2}{5}$	c) $\frac{x}{2x-3} = \frac{3}{5}$	d) $\frac{32}{x} = \frac{x}{2}$	e) $\frac{b}{a} = \frac{c}{x}$
Solutions:				
$5x = 60$	$2x = 15$	$5x = 6x - 9$	$x^2 = 64$	$bx = ac$
$x = 12$ *Ans.*	$x = 7\frac{1}{2}$ *Ans.*	$9 = x$ *Ans.*	$x = \pm 8$ *Ans.*	$x = \frac{ac}{b}$ *Ans.*

2.2. Finding Unknowns in Proportions Using Means and Extremes

Solve for x:

	Proportions	Product of Means	Product of Extremes	Product of Means = Product of Extremes	Answers
a)	$x:4 = 6:8$	$4(6) = 24$	$8x$	$8x = 24$	$x = 3$
b)	$3:5 = x:12$	$5x$	$3(12) = 36$	$5x = 36$	$x = 7\frac{1}{5}$
c)	$3:x = x:27$	$x \cdot x = x^2$	$3(27) = 81$	$x^2 = 81$	$x = \pm 9$
d)	$x:5 = 2x:x+3$	$5(2x) = 10x$	$x(x+3) = x^2 + 3x$	$x^2 + 3x = 10x$	$x = 0, 7$
e)	$x-2:4 = 7:x+2$	$4(7) = 28$	$(x-2)(x+2) = x^2 - 4$	$x^2 - 4 = 28$	$x = \pm 4\sqrt{2}$

2.3. Solving Fraction Problems Involving Proportions

The numerator of a fraction is 5 less than the denominator. If the numerator is doubled and the denominator is increased by 7, the value of the resulting fraction is $\frac{2}{3}$. Find the original fraction.

Solution:

Let x = denominator of the original fraction and $x - 5$ = numerator of the original fraction.

Then $\frac{2(x-5)}{x+7} = \frac{2}{3}$. Cross multiply: $6x - 30 = 2x + 14$, $4x = 44$, $x = 11$. Original fraction $= \frac{6}{11}$ *Ans.*

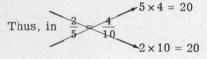

3. SIMILAR TRIANGLES

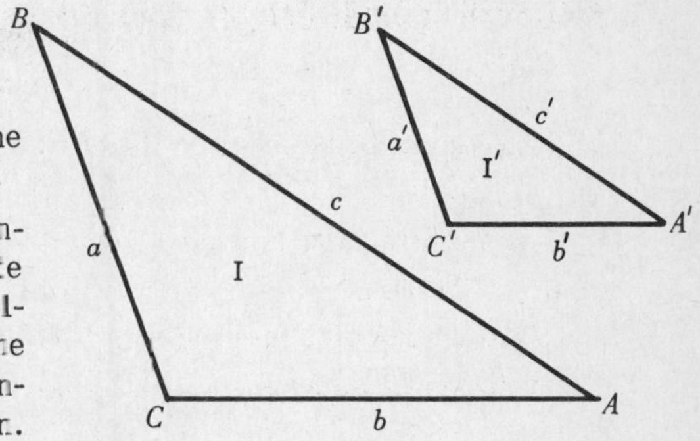

Similar Polygons Have the Same Shape

Thus, if ΔI and ΔI$'$ are similar, then they have the same shape although they need not have the same size.

Notation: "ΔI$\sim\Delta$I$'$" is to be read as "triangle I is similar to triangle I-prime". In the diagram, note how the sides and angles having the same relative position are designated by using the same letters and primes. **Corresponding sides** or angles are those having the same relative position.

Two Basic Properties of Similar Triangles

Rule 1: If two triangles are similar

a) **their corresponding angles are equal.**

Thus, if ΔI$\sim\Delta$I$'$, then $\angle C' = \angle C = 90°$, $\angle A' = \angle A = 40°$, $\angle B' = \angle B = 50°$.

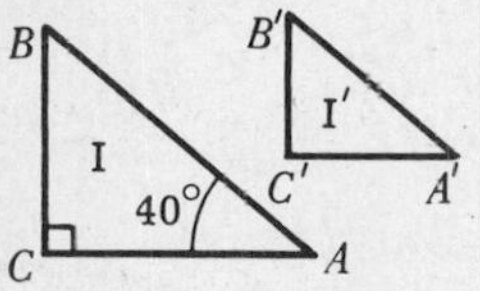

b) **the ratios of their corresponding sides are equal.**

Thus, if ΔI$\sim\Delta$I$'$, then $c = 15$ since $\frac{c}{5} = \frac{9}{3}$ and $b = 12$ since $\frac{b}{4} = \frac{9}{3}$.

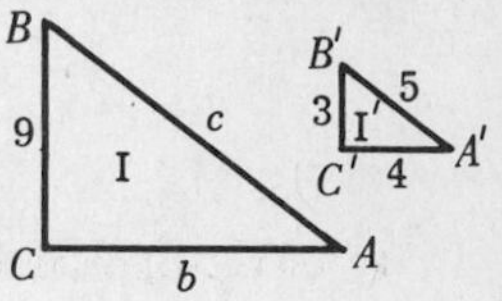

Three Basic Methods of Determining Similar Triangles

Rule 2: Two triangles are similar

a) **if two angles of one equal two angles of the other.**

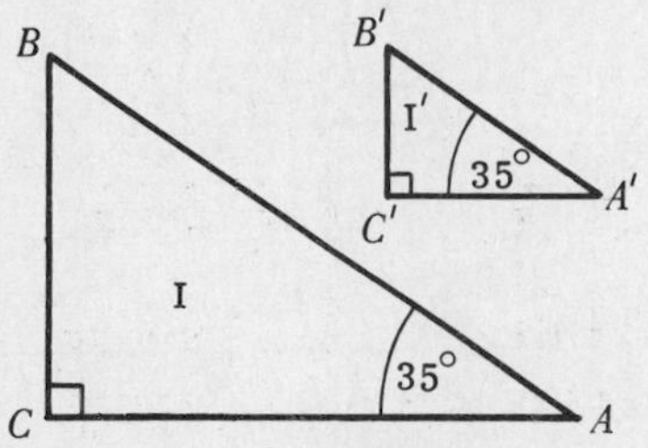

Thus, ΔI$\sim\Delta$I$'$ since $\angle C = \angle C'$ and $\angle A = \angle A'$

b) **if the three ratios of the corresponding sides are equal.**

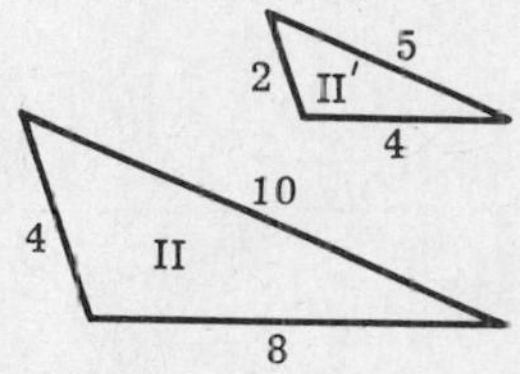

Thus, ΔII$\sim\Delta$II$'$ since $\frac{10}{5} = \frac{8}{4} = \frac{4}{2}$

c) **if two ratios of corresponding sides are equal and the angles between the sides are equal.**

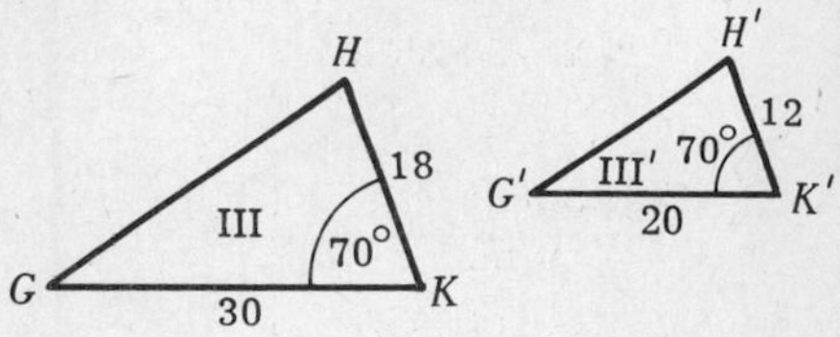

Thus, ΔIII$\sim\Delta$III$'$ since $\frac{30}{20} = \frac{18}{12}$ and $\angle K = \angle K'$

Rule 3: A triangle is similar to any one of its scale triangles.

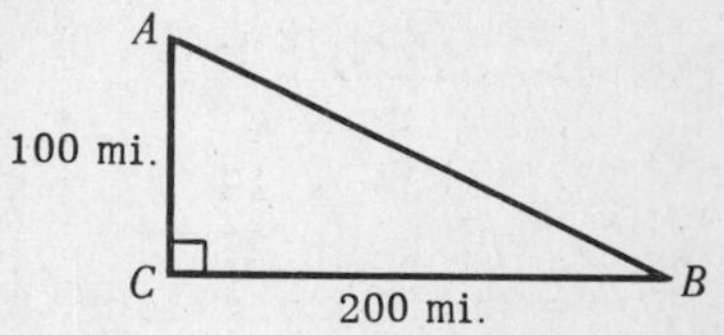

Thus, if ΔI and ΔI$'$ are drawn to scale to represent ΔABC, then they are similar to ΔABC and also to each other; that is,

$$\Delta ABC \sim \Delta \text{I} \sim \Delta \text{I}'$$

Note. Use **Rule 2*c*** to show that the triangles are similar. For example, ΔI$\sim\Delta ABC$ since

$$\frac{100}{2} = \frac{200}{4}$$

and the right angles which are between these sides are equal. For this reason a scale triangle may be constructed using only two sides and the included angle.

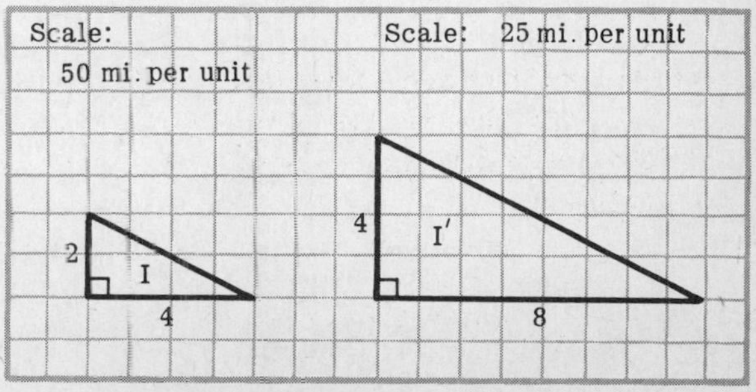

3.1. Rule 1*a*. Corresponding Angles of Similar Triangles are Equal

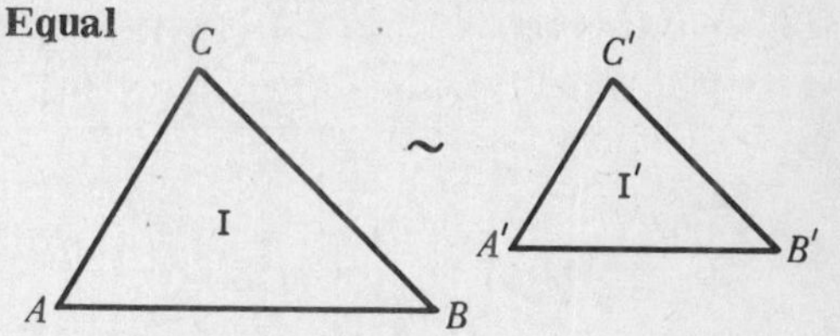

If $\triangle I' \sim \triangle I$ find $\angle C'$

a) if $\angle A = 60°$ and $\angle B = 45°$,

b) if $\angle A + \angle B = 110°$.

Solutions:

Using **Rule 1**, if $\triangle I' \sim \triangle I$ then $\angle C' = \angle C$.

a) Since the sum of the angles of a triangle equals 180°, $\angle C = 180° - 60° - 45° = 75°$. Hence, $\angle C' = 75°$.

b) Since the sum of the angles of a triangle equals 180°, $\angle C = 180° - 110° = 70°$. Hence, $\angle C' = 70°$.

3.2. Rule 1*b*. Ratios of Corresponding Sides of Similar Triangles are Equal

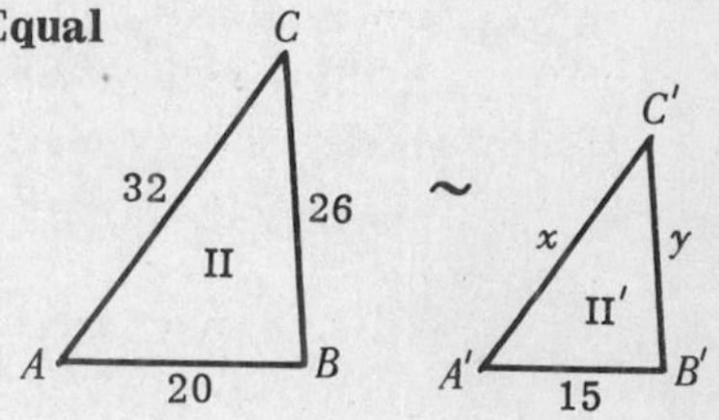

If $\triangle II' \sim \triangle II$, find x and y using the data indicated.

Solutions:

Since $\triangle II' \sim \triangle II$,

$\frac{x}{32} = \frac{15}{20}$	$\frac{y}{26} = \frac{15}{20}$
$x = \frac{15}{20}(32)$	$y = \frac{15}{20}(26)$
$x = 24$	$y = 19\frac{1}{2}$

Ans. 24, $19\frac{1}{2}$

3.3. Rule 2. Determining Similar Triangles

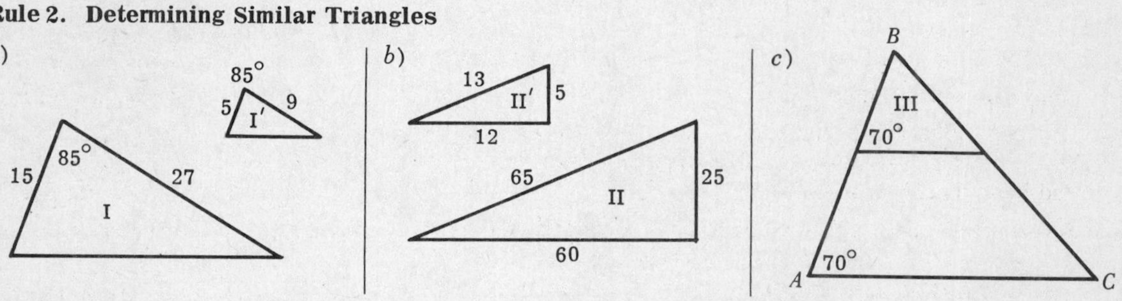

Which rule is needed in each case to show that both triangles are similar?

Solutions:

By **Rule 2*c***, $\triangle I' \sim \triangle I$ since each has an 85° angle and there are two equal ratios for the sides of these equal angles; i.e., $\frac{27}{9} = \frac{15}{5}$.

By **Rule 2*b***, $\triangle II' \sim \triangle II$ since $\frac{65}{13} = \frac{25}{5} = \frac{60}{12}$. Thus, there are three equal ratios of the corresponding sides.

By **Rule 2*a***, $\triangle III \sim \triangle ABC$ since there are two pairs of equal angles. Each triangle has $\angle B$ and a 70° angle.

3.4. Finding Heights Using Ground Shadows

A tree casts a 15 ft. shadow at a time when a nearby upright pole of 6 ft. casts a shadow of 2 ft. Find the height h of the tree if both tree and pole make right angles with the ground.

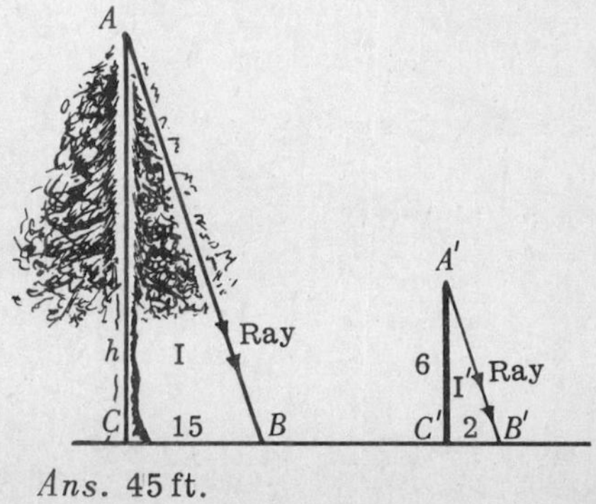

Solution:

At the same time in localities near each other, the rays of the sun strike the ground at equal angles; hence $\angle B = \angle B'$. Since the tree and pole make right angles with the ground, $\angle C = \angle C'$. Since there are two pairs of equal angles, $\triangle I' \sim \triangle I$. Hence, $\frac{h}{6} = \frac{15}{2}$, $h = \frac{15}{2}(6) = 45$.

Ans. 45 ft.

3.5. Using a Scale Triangle to Find Parts of a Triangle

If $\Delta I'$ is a scale triangle of ΔI:

a) Find a and c when $b=45$. *b*) Show that ΔABC is a right triangle.

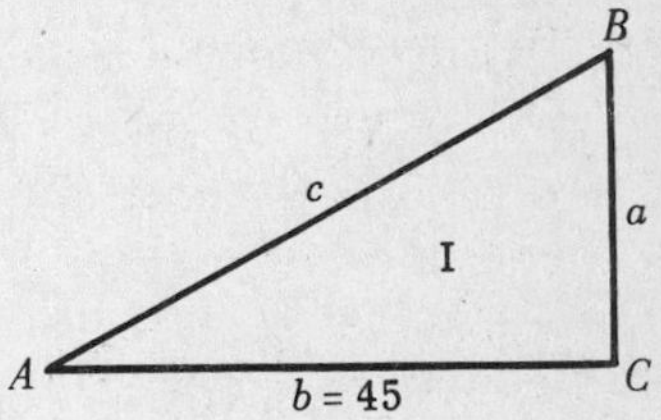

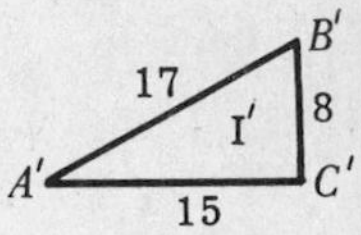

Solution:

a) Using **Rule 3**, $\Delta I \sim \Delta I'$; hence

$$\frac{a}{8} = \frac{45}{15} \quad \text{and} \quad \frac{c}{17} = \frac{45}{15}$$

$$a = 3(8) = 24 \qquad c = 3(17) = 51 \qquad \textit{Ans. } 24, 51$$

b) Since $8^2 + 15^2 = 17^2$, $\Delta I'$ is a right triangle.
Since $\Delta I \sim \Delta I'$, ΔI is a right triangle.

3.6. Using a Scale Triangle to Find a Distance

An aviator traveled east a distance of 150 mi. from A to C. He then traveled north for 50 mi. to B. Using a scale of 50 mi. per unit, find his distance from A, to the nearest mile.

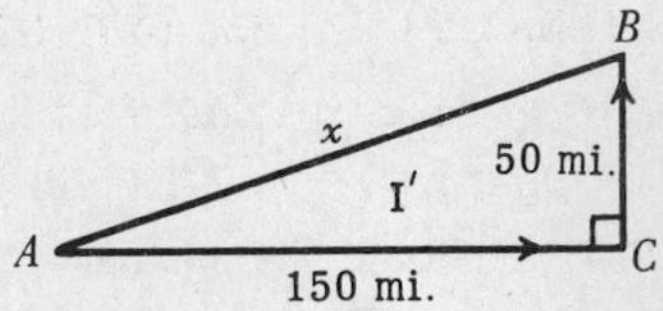

Solution:

By **Rule 3**, a triangle is similar to any one of its scale triangles.

Let x = length of AB in mi.

In ΔI, the scale triangle of $\Delta I'$, $c^2 = 3^2 + 1^2 = 10$. Hence, $c = \sqrt{10}$.

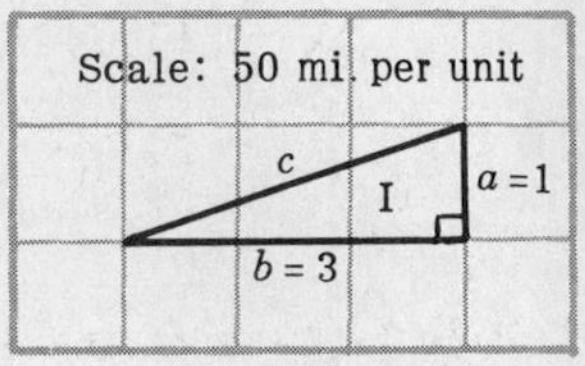

Since $\Delta I' \sim \Delta I$, $\frac{x}{c} = \frac{50}{1}$

$$x = 50c$$

$$x = 50\sqrt{10} = 50(3.162) = 158.1 \quad \textit{Ans. } 158 \text{ mi.}$$

3.7. Applying a Scale to a Square

A baseball diamond is a square 90 ft. on each side. Using a scale of 90 ft. per unit, find the distance from home plate to second base to the nearest foot.

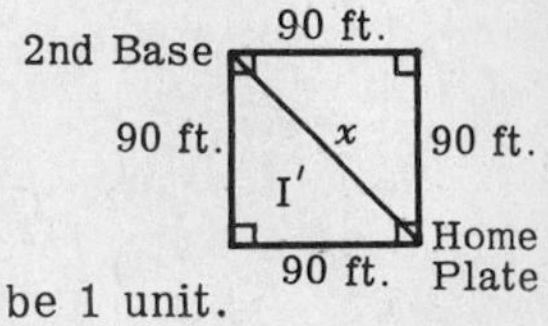

Solution: Using a scale of 90 ft. per unit, each side of the new square will be 1 unit.

Let x = distance from home plate to second base in ft.

In ΔI, the scale triangle of $\Delta I'$, $c^2 = 1^2 + 1^2 = 2$. Hence, $c = \sqrt{2}$.

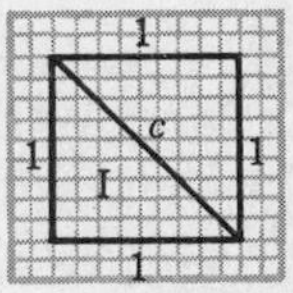

Since $\Delta I' \sim \Delta I$, $\frac{x}{c} = \frac{90}{1}$

$$x = 90c$$

$$x = 90\sqrt{2} = 90(1.414) = 127.26 \quad \textit{Ans. } 127 \text{ ft.}$$

SUPPLEMENTARY PROBLEMS

1. In a right triangle whose arms are a and b, find the hypotenuse c when: **(1.1)**

a) $a=15$, $b=20$ *Ans.* 25 *c*) $a=5$, $b=4$ *Ans.* $\sqrt{41}$ *e*) $a=7$, $b=7$ *Ans.* $7\sqrt{2}$

b) $a=15$, $b=36$ *Ans.* 39 *d*) $a=5$, $b=5\sqrt{3}$ *Ans.* 10 *f*) $a=8$, $b=4$ *Ans.* $4\sqrt{5}$

2. In the right triangle shown, find each missing arm when **(1.2)**

a) $a=12$, $c=20$ *c*) $b=15$, $c=17$ *e*) $a=5\sqrt{2}$, $c=10$

b) $b=6$, $c=8$ *d*) $a=2$, $c=4$ *f*) $a=\sqrt{5}$, $c=2\sqrt{2}$

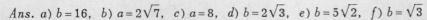

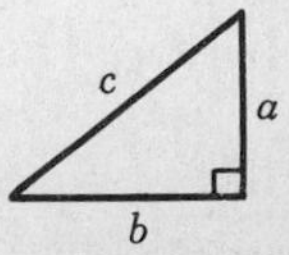

Ans. a) $b=16$, *b*) $a=2\sqrt{7}$, *c*) $a=8$, *d*) $b=2\sqrt{3}$, *e*) $b=5\sqrt{2}$, *f*) $b=\sqrt{3}$

3. Find the arms of a right triangle whose hypotenuse is c if these arms have a ratio of **(1.3)**
a) $3:4$ and $c=15$, *b*) $5:12$ and $c=26$, *c*) $8:15$ and $c=170$, *d*) $1:2$ and $c=10$.
Ans. *a*) 9, 12 *b*) 10, 24 *c*) 80, 150 *d*) $2\sqrt{5}$, $4\sqrt{5}$

4. In a rectangle, find the diagonal if its sides are **(1.4)**
a) 30 and 40, *b*) 9 and 40, *c*) 5 and 10, *d*) 2 and 6. *Ans.* *a*) 50, *b*) 41, *c*) $5\sqrt{5}$, *d*) $2\sqrt{10}$

5. In a rectangle, find one side if the diagonal is 15 and the other side is **(1.4)**
a) 9, *b*) 5, *c*) 10, *d*) 12. *Ans.* *a*) 12, *b*) $10\sqrt{2}$, *c*) $5\sqrt{5}$, *d*) 9

6. Using the three sides given, which triangles are right triangles? **(1.5)**
a) 33, 55, 44
b) 120, 130, 50
c) 4, $7\frac{1}{2}$, $8\frac{1}{2}$
d) 25, 7, 24
e) 5 in., 1 ft., 1 ft. 1 in.
f) 1 yd., 1 yd. 1 ft., 1 yd. 2 ft.
g) 11 mi., 60 mi., 61 mi.
h) 5 rd., 5 rd., 7 rd.
Ans. Only (*h*) is not a right triangle since $5^2+5^2 \neq 7^2$. In all the other cases, the square of the largest side equals the sum of the squares of the other two.

7. Find the distance between each of the following pairs of points: **(1.6)**
a) (4,1) and (7,5)
b) (3,3) and (3,5)
c) (1,7) and (10,7)
d) (−3,−6) and (3,2)
e) (2,3) and (−10,−2)
f) (2,2) and (5,5)
Ans. *a*) 5, *b*) 2, *c*) 9, *d*) 10, *e*) 13, *f*) $3\sqrt{2}$

8. Find the lengths of the sides of triangle DEF if its vertices are $D(2,5)$, $E(6,5)$ and $F(2,8)$. Show that $\triangle DEF$ is a right triangle. **(1.5, 1.6)**
Ans. $DE=4$, $DF=3$ and $EF=5$. Since $3^2+4^2=5^2$, $\triangle DEF$ is a right triangle.

9. Derive a formula for the side s of a square in terms of its diagonal d. **(1.7)**
Ans. $s = \dfrac{d}{\sqrt{2}}$ or $s = \dfrac{d}{2}\sqrt{2}$

10. Using formula $d=s\sqrt{2}$, express in radical form the diagonal of a square whose side is **(1.7)**
a) 5, *b*) 7.2, *c*) $\sqrt{3}$, *d*) 90 ft., *e*) 3.47 yd.
Ans. *a*) $5\sqrt{2}$, *b*) $7.2\sqrt{2}$, *c*) $\sqrt{6}$, *d*) $90\sqrt{2}$ ft., *e*) $3.47\sqrt{2}$ yd.

11. Using formula $h=\frac{s}{2}\sqrt{3}$, express in radical form the altitude of an equilateral triangle whose side is *a*) 6, *b*) 20, *c*) 11, *d*) 90 in., *e*) 4.6 yd. **(1.7)**
Ans. *a*) $3\sqrt{3}$, *b*) $10\sqrt{3}$, *c*) $\frac{11}{2}\sqrt{3}$, *d*) $45\sqrt{3}$ in., *e*) $2.3\sqrt{3}$ yd.

12. The largest possible square is to be cut from a circular piece of wood. Find the side of the square, to the nearest inch, if the diameter of the circle is *a*) 30 in., *b*) 14 in., *c*) 17 in. **(1.8)**
Ans. *a*) 21 in., *b*) 10 in., *c*) 12 in.

13. Find the altitude of an isosceles triangle if one of its two equal sides is 10 and its base is **(1.9)**
a) 12, *b*) 16, *c*) 18, *d*) 10. *Ans.* *a*) 8, *b*) 6, *c*) $\sqrt{19}$, *d*) $5\sqrt{3}$

14. Solve for x: **(2.1)**
a) $\frac{5}{7}=\frac{15}{x}$
b) $\frac{7}{x}=\frac{3}{2}$
c) $\frac{3}{x}=\frac{x}{12}$
d) $\frac{x}{5}=\frac{15}{x}$
e) $\frac{x+2}{5}=\frac{6}{3}$
f) $\frac{x-1}{3}=\frac{5}{x+1}$
g) $\frac{2x}{x+7}=\frac{3}{5}$
h) $\frac{a}{x}=\frac{x}{b}$
Ans. *a*) 21, *b*) $4\frac{2}{3}$, *c*) ± 6, *d*) $\pm 5\sqrt{3}$, *e*) 8, *f*) ± 4, *g*) 3, *h*) $\pm\sqrt{ab}$

15. Solve for x: **(2.2)**
a) $x:6=8:3$
b) $5:4=20:x$
c) $9:x=x:4$
d) $x:2=10:x$
e) $(x+4):3=3:(x-4)$
f) $(2x+8):(x+2)=(2x+5):(x+1)$
g) $a:b=c:x$
h) $x:2y=18y:x$
Ans. *a*) 16, *b*) 16, *c*) ± 6, *d*) $\pm 2\sqrt{5}$, *e*) ± 5, *f*) 2, *g*) $\frac{bc}{a}$, *h*) $\pm 6y$

16. *a*) The denominator of a fraction is 1 more than twice the numerator. If 2 is added to both the numerator and denominator, the value of the fraction is $\frac{3}{5}$. Find the original fraction. **(2.3)**

b) A certain fraction is equivalent to $\frac{2}{5}$. If the numerator of this fraction is decreased by 2 and its denominator is increased by 1, the resulting fraction is equivalent to $\frac{1}{3}$. Find the original fraction. **(2.3)**

Ans. *a*) $\frac{1}{3}$ $\left[\text{Solve: } \frac{x+2}{(2x+1)+2} = \frac{3}{5}\right]$, *b*) $\frac{14}{35}$ $\left[\text{Solve: } \frac{2x-2}{5x+1} = \frac{1}{3}\right]$

17. If $\triangle I \sim \triangle I'$, find $\angle B$ **(3.1)**

a) if $\angle A' = 120°$ and $\angle C' = 25°$,

b) if $\angle A' + \angle C' = 127°$.

Ans. *a*) 35°, *b*) 53°

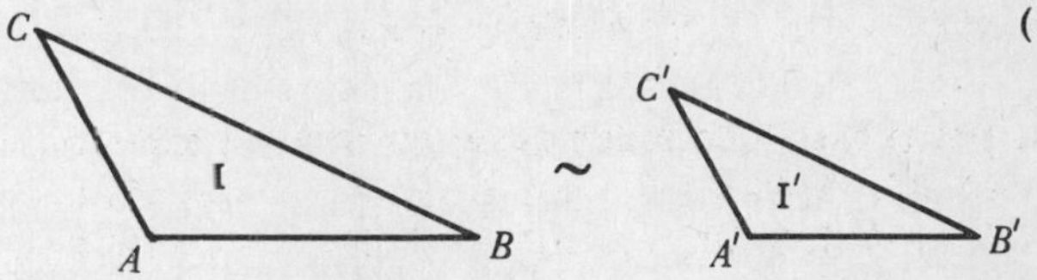

18. If $\triangle I \sim \triangle I'$, using the data shown **(3.2)**

a) find a if $c = 24$. *Ans.* $\frac{a}{4} = \frac{24}{6}$, $a = 16$

b) find b if $a = 20$. *Ans.* $\frac{b}{3} = \frac{20}{4}$, $b = 15$

c) find c if $b = 63$. *Ans.* $\frac{c}{6} = \frac{63}{3}$, $c = 126$

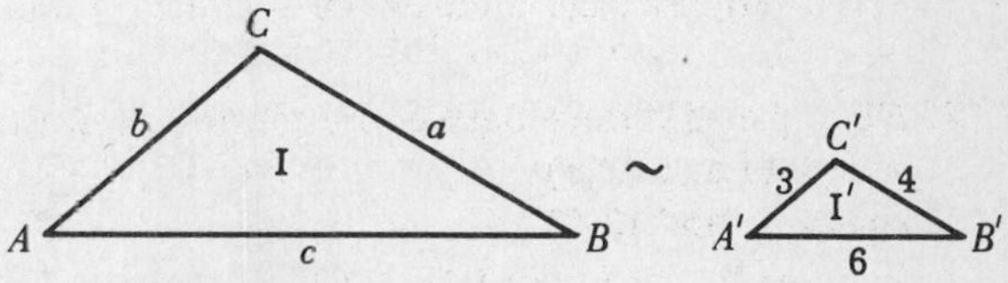

19. Which rule is needed in each case to show both triangles are similar? **(3.3)**

a) Show $\triangle I \sim \triangle ABC$.

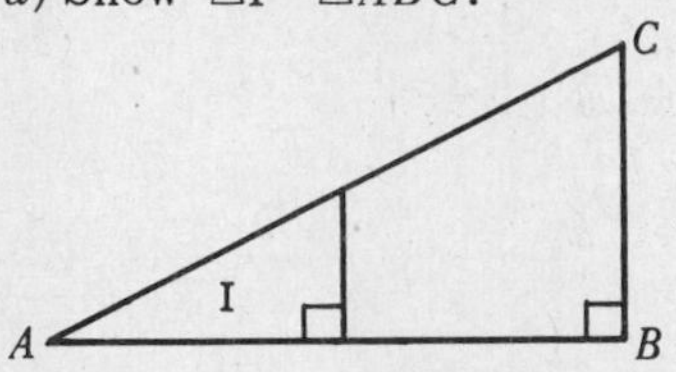

Ans. *a*) **Rule 2*a*.**

b) Show $\triangle II \sim \triangle PQR$.

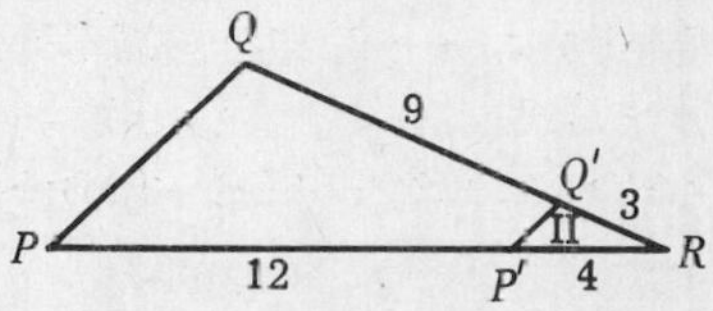

b) **Rule 2*c***, since $\frac{3}{12} = \frac{4}{16}$ and each triangle has $\angle R$.

c) Show $\triangle III \sim \triangle FGH$

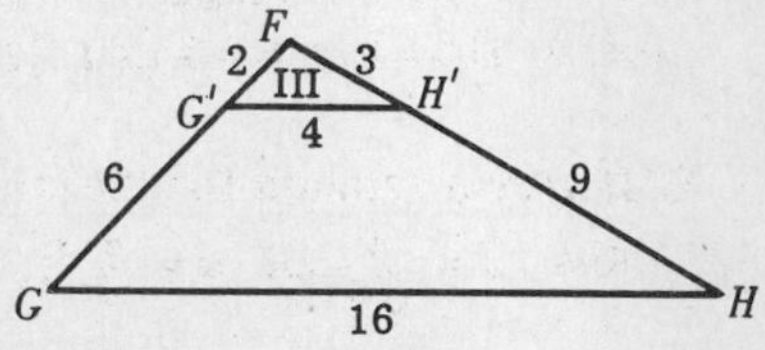

c) **Rule 2*b***, since $\frac{2}{8} = \frac{3}{12} = \frac{4}{16}$.

20. A 7 ft. upright pole near a vertical tree casts a 6 ft. shadow. At that time, **(3.4)**

a) find the height of the tree if its shadow is 36 ft. *Ans.* 42 ft.

b) find the shadow of the tree if its height is 77 ft. *Ans.* 66 ft.

21. If $\triangle I'$ is a scale triangle of $\triangle I$, **(3.5)**

a) find a and b when $c = 125$,

b) show that $\triangle I$ is a right triangle.

Ans. *a*) 35, 120

b) $\triangle I$ is a right triangle since $7^2 + 24^2 = 25^2$ and $\triangle I \sim \triangle I'$.

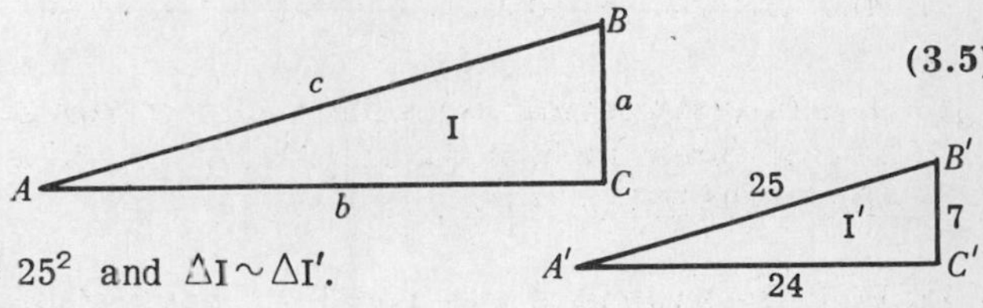

22. Two planes leave an airport at the same time, one going due east at 250 mph and the other due north at 150 mph. Using a scale of 200 mi. per unit, find the distance between them at the end of 4 hr., to the nearest mile. *Ans.* 1166 mi. **(3.6)**

23. A square lot, 50 ft. on each side, has a diagonal path. Using a square drawn to a scale of 50 ft. per unit, find the length of the path to the nearest foot. *Ans.* 71 ft. **(3.7)**

Chapter 17

TRIGONOMETRY

1. UNDERSTANDING TRIGONOMETRIC RATIOS

Trigonometry means "measurement of triangles". Consider its parts: "tri" means three, "gon" means "angle", and "metry" means "measure". Thus, in trigonometry we study the measurement of triangles.

In trigonometry the following ratios are used in a right triangle to relate the sides and either acute angle:

1. tangent ratio, abbreviated "tan".
2. sine ratio, abbreviated "sin".
3. cosine ratio, abbreviated "cos".

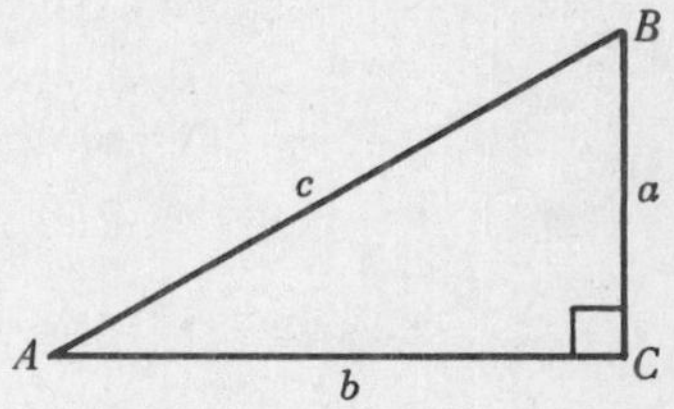

Trigonometry Rules and Formulas

Rules	Formulas
1. The tangent of an acute angle equals the leg opposite the angle divided by the leg adjacent to the angle.	$\tan A = \frac{\text{leg opp. } A}{\text{leg adj. } A} = \frac{a}{b}$ $\tan B = \frac{\text{leg opp. } B}{\text{leg adj. } B} = \frac{b}{a}$
2. The sine of an acute angle equals the leg opposite the angle divided by the hypotenuse.	$\sin A = \frac{\text{leg opp. } A}{\text{hyp.}} = \frac{a}{c}$ $\sin B = \frac{\text{leg opp. } B}{\text{hyp.}} = \frac{b}{c}$
3. The cosine of an acute angle equals the leg adjacent to the angle divided by the hypotenuse.	$\cos A = \frac{\text{leg adj. } A}{\text{hyp.}} = \frac{b}{c}$ $\cos B = \frac{\text{leg adj. } B}{\text{hyp.}} = \frac{a}{c}$

Rules: If A and B are the acute angles of a right triangle:

$\sin A = \cos B$ ↘↗ $\cos A = \sin B$ $\qquad \tan A = \frac{1}{\tan B}$

$\sin B = \cos A$ ↙↖ $\cos B = \sin A$ $\qquad \tan B = \frac{1}{\tan A}$

To Find an Angle to the Nearest Degree

Find x, to the nearest degree, if $\sin x = .6350$.

Procedure:	**Solution:**
1. Find nearest tabular values:	1. $\sin x = .6350$ is between $\sin 40^\circ = .6428$ and $\sin 39^\circ = .6293$
2. Find differences as shown:	2. $\sin 40^\circ = .6428$ $\sin x = .6350$ $\sin 39^\circ = .6293$ **Differences**: .0078 (between sin 40° and sin x); .0057 (between sin x and sin 39°)
3. Find angle to the nearest degree:	3. Since $\sin x$ is nearer to $\sin 39^\circ$, $x = 39^\circ$ to the nearest degree. *Ans.* 39°

1.1. Using Table of Sines, Cosines and Tangents

The following values are taken from the **Table of Sines, Cosines and Tangents** (page 292). State each indicated value in proper trigonometric form and complete the last line:

		Sine	Cosine	Tangent
(1)	1°	.0175	.9998	.0175
(2)	30°	.5000	.8660	.5774
(3)	60°	.8660	.5000	1.7321
(4)	(?)	(?)	.3420	(?)

Solution:

(1) Since .0175 is aligned with "Sine" and "1°", $\sin 1^\circ = .0175$.
Similarly, $\cos 1^\circ = .9998$ and $\tan 1^\circ = .0175$.

(2) From the 30° row, $\sin 30^\circ = .5000$, $\cos 30^\circ = .8660$ and $\tan 30^\circ = .5774$.

(3) From the 60° row, $\sin 60^\circ = .8660$, $\cos 60^\circ = .5000$ and $\tan 60^\circ = 1.7321$.

(4) Since $.3420 = \cos 70^\circ$, the angle is 70°. The other values in this row are $\sin 70^\circ = .9397$ and $\tan 70^\circ = 2.7475$.

1.2. Finding Angles to the Nearest Degree

Using the table of sines, cosines and tangents, find x to the nearest degree if

a) $\sin x = .9235$

b) $\cos x = \frac{21}{25}$

c) $\tan x = \frac{\sqrt{5}}{10}$

Solutions:

a)

Using table:

	Differences
$\sin 68^\circ = .9272$	
$\sin x = .9235$	.0037
$\sin 67^\circ = .9205$	.0030

Since $\sin x$ is nearer to $\sin 67^\circ$, $x = 67^\circ$ to the nearest degree. *Ans.* 67°

b) Since $\frac{21}{25} = .8400$, $\cos x = .8400$

Using table:

	Differences
$\cos 32^\circ = .8480$	
$\cos x = .8400$	.0080
$\cos 33^\circ = .8387$	.0013

Since $\cos x$ is nearer to $\cos 33^\circ$, $x = 33^\circ$ to the nearest degree. *Ans.* 33°

c) Since $\frac{\sqrt{5}}{10} = .2236$, $\tan x = .2236$

Using table:

	Differences
$\tan 13^\circ = .2309$	
$\tan x = .2236$	.0073
$\tan 12^\circ = .2126$	.0110

Since $\tan x$ is nearer to $\tan 13^\circ$, $x = 13^\circ$ to the nearest degree. *Ans.* 13°

1.3. Finding Trigonometric Ratios

In each right triangle, state trigonometric values of each acute angle:

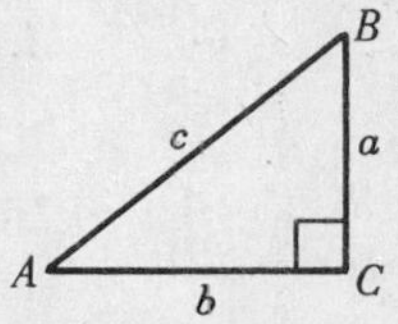

a)

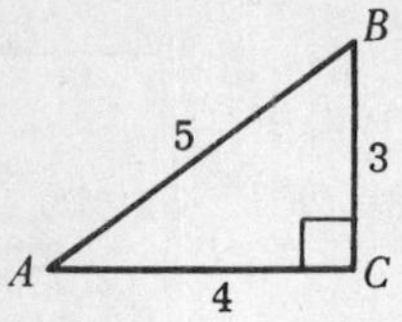

b)

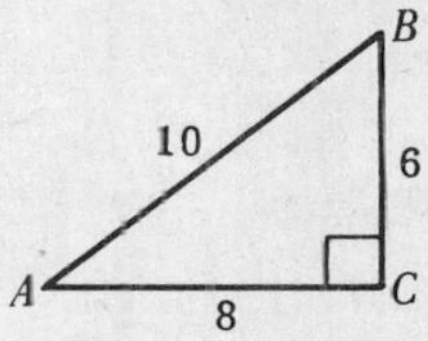

c)

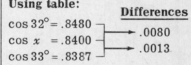

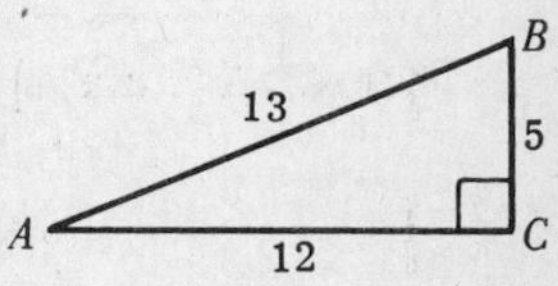

Formulas	$a=3,\ b=4,\ c=5$	$a=6,\ b=8,\ c=10$	$a=5,\ b=12,\ c=13$
$\tan A = \frac{a}{b}$	$\tan A = \frac{3}{4}$	$\tan A = \frac{6}{8} = \frac{3}{4}$	$\tan A = \frac{5}{12}$
$\tan B = \frac{b}{a}$	$\tan B = \frac{4}{3}$	$\tan B = \frac{8}{6} = \frac{4}{3}$	$\tan B = \frac{12}{5}$
$\sin A = \frac{a}{c}$	$\sin A = \frac{3}{5}$	$\sin A = \frac{6}{10} = \frac{3}{5}$	$\sin A = \frac{5}{13}$
$\sin B = \frac{b}{c}$	$\sin B = \frac{4}{5}$	$\sin B = \frac{8}{10} = \frac{4}{5}$	$\sin B = \frac{12}{13}$
$\cos A = \frac{b}{c}$	$\cos A = \frac{4}{5}$	$\cos A = \frac{8}{10} = \frac{4}{5}$	$\cos A = \frac{12}{13}$
$\cos B = \frac{a}{c}$	$\cos B = \frac{3}{5}$	$\cos B = \frac{6}{10} = \frac{3}{5}$	$\cos B = \frac{5}{13}$

1.4. Finding Angles by Trigonometric Ratios

Find A, to the nearest degree, in each:

a)

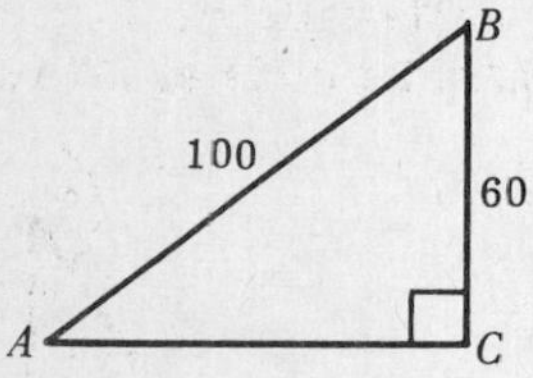

b)

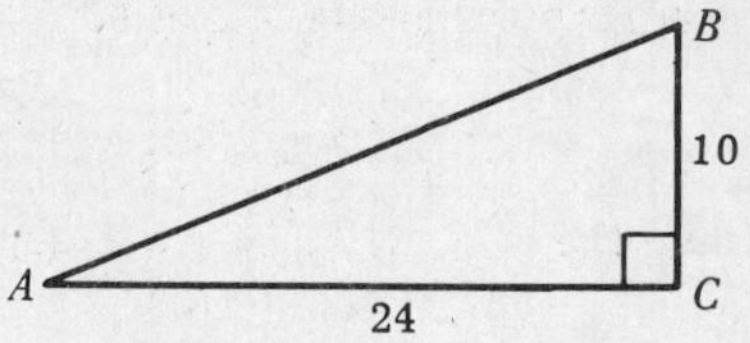

c)

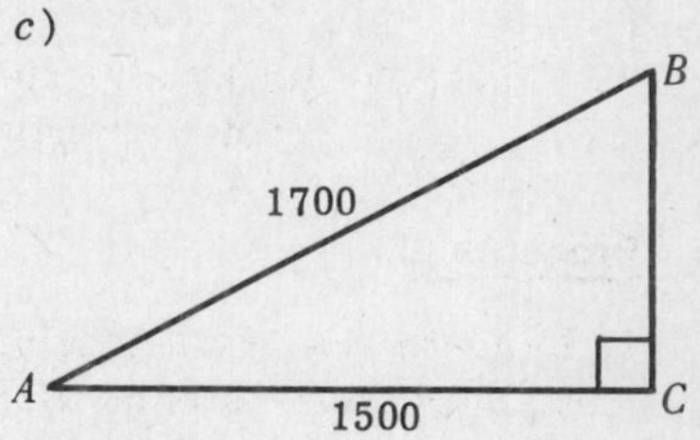

Solutions:

a) $\sin A = \frac{60}{100} = .6000$

Since $\sin 37° = .6018$ is nearest sine value,

$A = 37°$ *Ans.*

b) $\tan A = \frac{10}{24} = \frac{5}{12} = .4167$

Since $\tan 23° = .4245$ is nearest tangent value,

$A = 23°$ *Ans.*

c) $\cos A = \frac{1500}{1700} = \frac{15}{17} = .8824$

Since $\cos 28° = .8829$ is nearest cosine value,

$A = 28°$ *Ans.*

1.5. Deriving Trigonometric Values for 30° and 60° Ratios

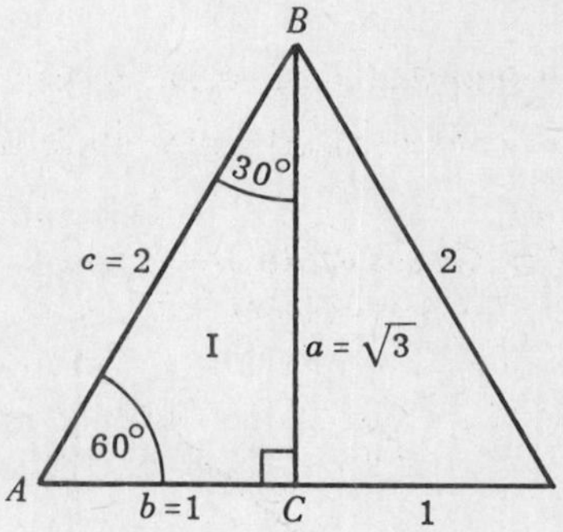

Show a) $\tan 30° = .577$ d) $\tan 60° = 1.732$

b) $\sin 30° = .5$ e) $\sin 60° = .866$

c) $\cos 30° = .866$ f) $\cos 60° = .5$

Solutions:

The trigonometric ratios for 30° and 60° may be obtained by using an equilateral triangle. Draw the altitude from the midpoint of the base to the opposite vertex. Consider each side equal to 2 units.

In right triangle I, $c = 2$ and $b = 1$.

By Law of Pythagoras, $a^2 = 2^2 - 1^2 = 3$. Hence, $a = \sqrt{3}$.

Using $\triangle$I:

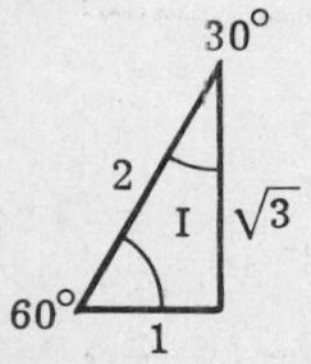

a) $\tan 30° = \frac{1}{\sqrt{3}} = \frac{1}{\sqrt{3}} \cdot \frac{\sqrt{3}}{\sqrt{3}} = \frac{\sqrt{3}}{3} = .577$

d) $\tan 60° = \frac{\sqrt{3}}{1} = 1.732$

b) $\sin 30° = \frac{1}{2} = .5$

e) $\sin 60° = \frac{\sqrt{3}}{2} = .866$

c) $\cos 30° = \frac{\sqrt{3}}{2} = \frac{1.732}{2} = .866$

f) $\cos 60° = \frac{1}{2} = .5$

1.6. Finding Acute Angles of a Right Triangle Whose Sides Have a Given Ratio

To the nearest degree, find each acute angle of any right triangle whose sides are in the ratio of 3 : 4 : 5.

Solution:

Let $3x$, $4x$ and $5x$ represent the three sides of $\triangle$I as shown in the adjoining figure.

Now, $\tan A = \frac{3x}{4x} = \frac{3}{4} = .7500$.

Since $\tan 37° = .7536$ is nearest tangent value, $A = 37°$.

$B = 90° - 37° = 53°$. *Ans.* 37° and 53°

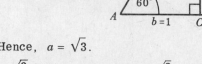

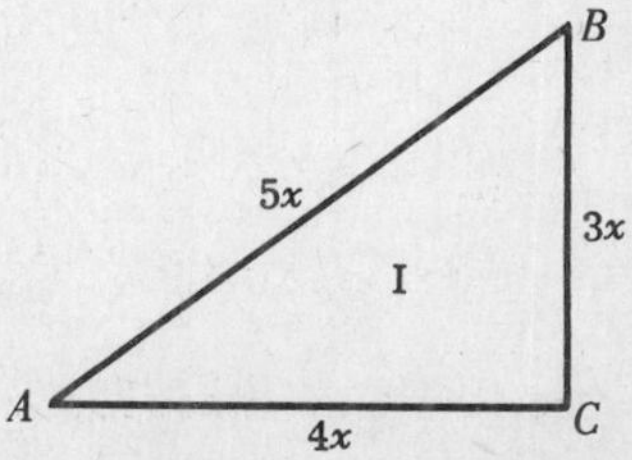

2. SOLVING TRIGONOMETRY PROBLEMS

To Solve Problems by Using Trigonometry

Problem: An aviator flew 70 mi. east from A to C. From C, he flew 100 mi. north to B. Find the angle of turn, to the nearest degree, that must be made at B to return to A.

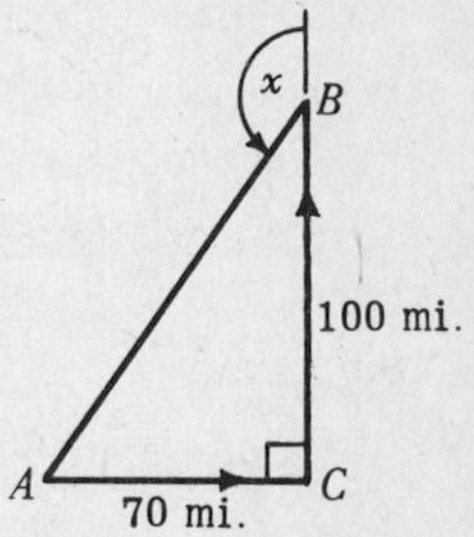

Procedure:	**Solution:**
1. Indicate the values given and the unknown on a labeled diagram: (*A diagram should be drawn to scale.*)	1. Let x = angle of turn at B.
2. Write a trigonometric equation involving the given values:	2. $\tan B = \frac{70}{100} = .7000$.
3. Find the unknown:	3. Since $B = 35°$ to nearest degree, angle of turn $= 180° - 35° = \underline{145°}$ *Ans.*

2.1. Obtaining Trigonometric Equations Given an Acute Angle and a Side

In each triangle, state the trigonometric equation needed to solve for x and y:

(*Avoid division where possible.*)

a) **Given acute angle and adjacent leg**

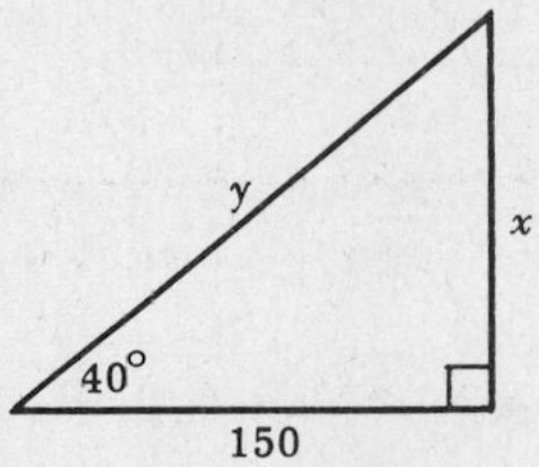

Solution:

(*1*) **To find** x:

Since $\tan 40° = \frac{x}{150}$,

$x = 150 \tan 40°$

(*2*) **To find** y:

Since $\cos 40° = \frac{150}{y}$,

$y \cos 40° = 150$

** Hence, $y = \frac{150}{\cos 40°}$

b) **Given acute angle and opposite leg**

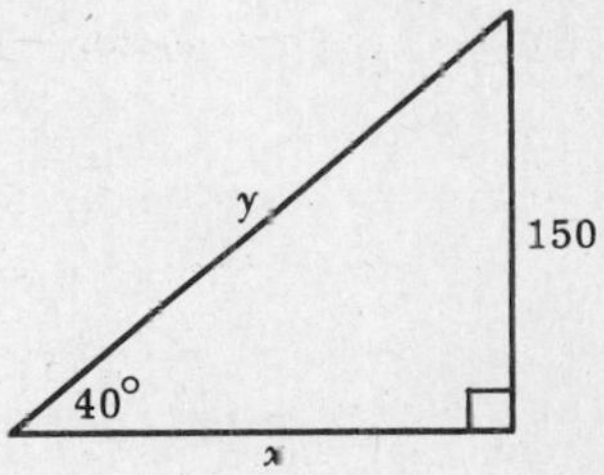

*(*1*) **To find** x, use 50° angle:

Since $\tan 50° = \frac{x}{150}$,

$x = 150 \tan 50°$

(*2*) **To find** y:

Since $\sin 40° = \frac{150}{y}$,

$y \sin 40° = 150$

** Hence, $y = \frac{150}{\sin 40°}$

c) **Given acute angle and hypotenuse**

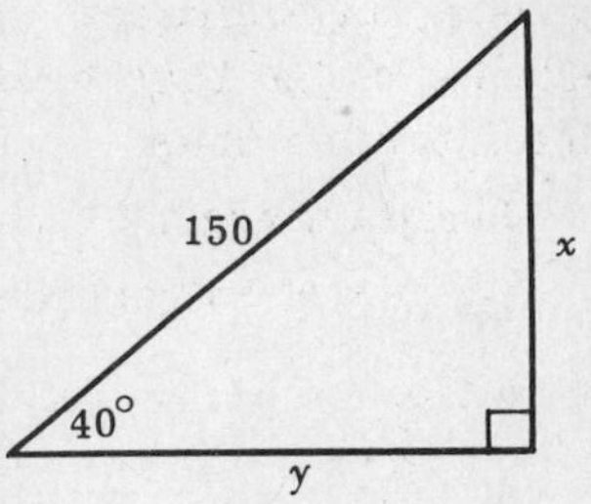

(*1*) **To find** x:

Since $\sin 40° = \frac{x}{150}$,

$x = 150 \sin 40°$

(*2*) **To find** y:

Since $\cos 40° = \frac{y}{150}$

$y = 150 \cos 40°$

***Note 1.** To find x in (*b*), if $\tan 40°$ were used, division would be needed:

$$\text{Since } \tan 40° = \frac{150}{x}, \quad x \tan 40° = 150, \quad x = \frac{150}{\tan 40°}.$$

****Note 2.** To find y in either (*a*) or (*b*), division cannot be avoided, using sine or cosine.

2.2. Solving Trigonometric Equations Derived in 2.1

Solve each trigonometric equation. Find unknown to nearest integer:

(*Equations were obtained in* **2.1.**)

a) (*1*) $x = 150 \tan 40^\circ$

(*2*) $y = \dfrac{150}{\cos 40^\circ}$

b) (*1*) $x = 150 \tan 50^\circ$

(*2*) $y = \dfrac{150}{\sin 40^\circ}$

c) (*1*) $x = 150 \sin 40^\circ$

(*2*) $y = 150 \cos 40^\circ$

Solutions:

a)

(*1*) $x = 150 \tan 40^\circ$
$x = 150(.8391)$
$x = 125.87$
$x = 126$ *Ans.*

(*2*) $y = \dfrac{150}{\cos 40^\circ}$
$y = \dfrac{150}{.7660}$
$y = 195.8$
$y = 196$ *Ans.*

b)

(*1*) $x = 150 \tan 50^\circ$
$x = 150(1.1918)$
$x = 178.77$
$x = 179$ *Ans.*

(*2*) $y = \dfrac{150}{\sin 40^\circ}$
$y = \dfrac{150}{.6428}$
$y = 233.4$
$y = 233$ *Ans.*

c)

(*1*) $x = 150 \sin 40^\circ$
$x = 150(.6428)$
$x = 96.42$
$x = 96$ *Ans.*

(*2*) $y = 150 \cos 40^\circ$
$y = 150(.7660)$
$y = 114.9$
$y = 115$ *Ans.*

2.3. Finding Legs of a Right Triangle

An aviator takes off at A and ascends at a fixed angle of 22° with level or horizontal ground. After flying 3000 yd., find, to the nearest 10 yd.,

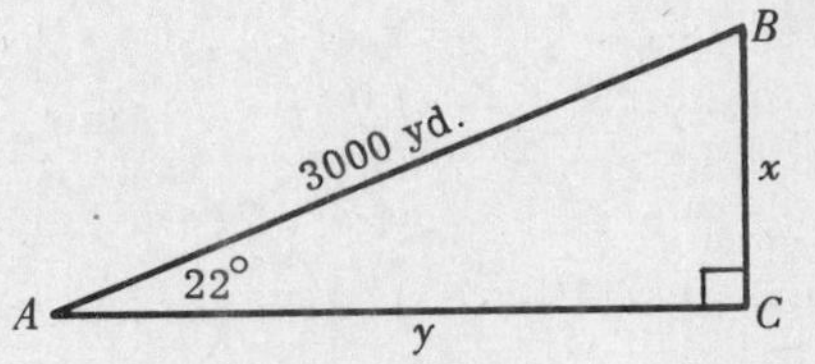

a) the altitude of the plane.

b) the distance from A to C which is on a level with A and directly under the plane.

Solutions:

a) Let x = altitude of plane in yd.

Since $\sin 22^\circ = \dfrac{x}{3000}$,

$x = 3000(.3746) = 1123.8$

Ans. 1120 yd., to nearest 10 yd.

b) Let y = required distance in yd.

Since $\cos 22^\circ = \dfrac{y}{3000}$,

$y = 3000(.9272) = 2781.6$

Ans. 2780 yd., to nearest 10 yd.

2.4. Finding an Angle and then the Hypotenuse of a Right Triangle

A road is to be constructed so that it will rise 105 ft. for each 1000 ft. of horizontal distance. Find

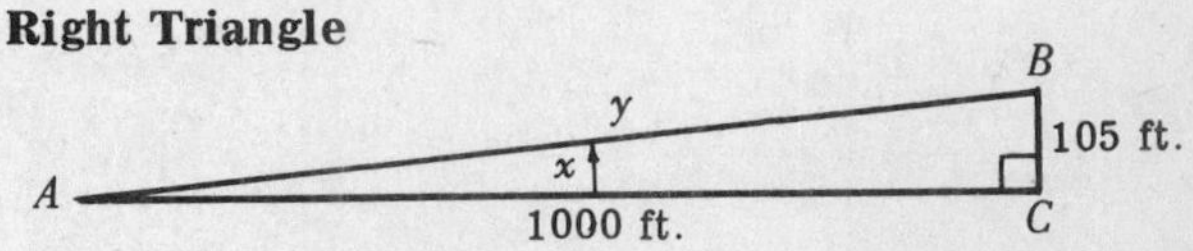

a) its angle of rise, x, to the nearest degree.

b) the length of road, y, to the nearest ft., for each 1000 ft. of horizontal distance, using the value of x found in (*a*).

Solutions:

a) Since $\tan x = \dfrac{105}{1000} = .1050$,

$x = 6^\circ$, to nearest degree. *Ans.*

b) Since $\cos x = \dfrac{1000}{y}$, $\cos 6^\circ = \dfrac{1000}{y}$.

Hence, $y = \dfrac{1000}{\cos 6^\circ} = \dfrac{1000}{.9945}$ or 1005.5.

Ans. 1006 ft., to nearest ft.

3. ANGLES OF ELEVATION AND DEPRESSION

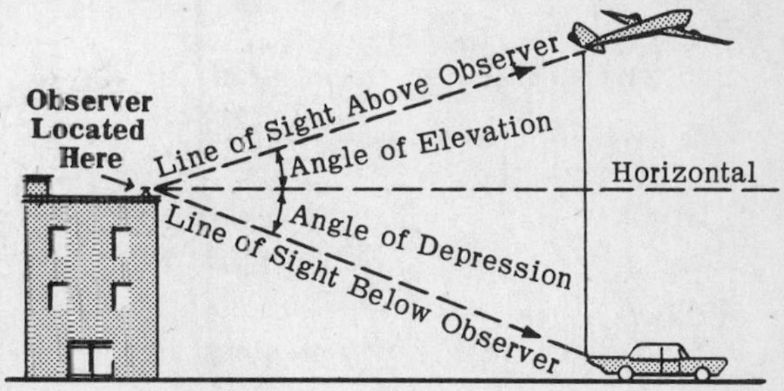

The line of sight is the line from the eye of the observer to the object sighted.

A horizontal line is a line level with the surface of water, or a line at right angles to a vertical line.

Thus, by centering a bubble of air in a water chamber, the table of a transit or sextant may be horizontally leveled.

An angle of elevation (or depression) is an angle formed by a horizontal line and a line of sight above (or below) the horizontal line and in the same vertical plane.

Thus, in the figure above, the observer is sighting an airplane above the horizontal. **The angle of elevation** is found by **elevating** or tilting upward the sighting tube of the transit or sextant above the horizontal position.

Also, in the diagram above, the observer is sighting an automobile below the horizontal. The **angle of depression** is found by **depressing** or tilting downward the sighting tube of the transit or sextant below the horizontal position.

3.1. Using a Transit to Find a Distance

Sighting the top of a building, Henry found the angle of elevation to be 21°. The ground is level. The transit is 5 ft. above the ground and 200 ft. from the house. Find the height of the building, to the nearest ft.

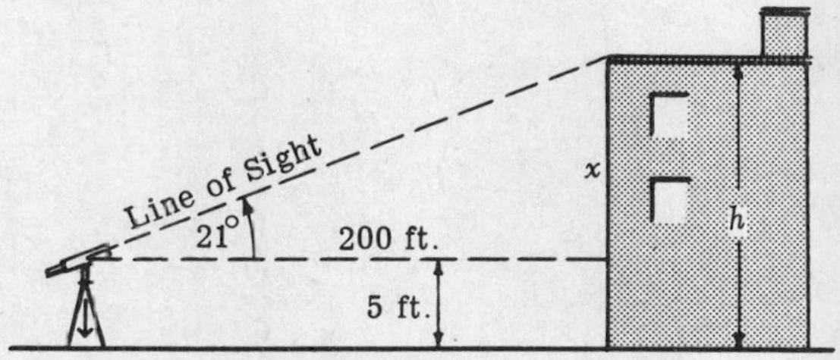

Solution:

Let x = the height, in ft., of the top of the building above the transit.

Then, $\tan 21° = \frac{x}{200}$

$.3839 = \frac{x}{200}$

$200(.3839) = x$

$x = 77$, to nearest ft.

To find the height h in ft. of the building, add 5 to x:

$h = x + 5$

$h = 77 + 5 = 82$

Ans. 82 ft., to the nearest ft.

3.2. Using Angle of Elevation of Sun

If the angle of elevation of the sun at a certain time is 42°, find, to the nearest foot,

a) the height h of a tree whose shadow s is 25 ft. long.

b) the shadow s of a tree along level ground if its height h is 35 ft.

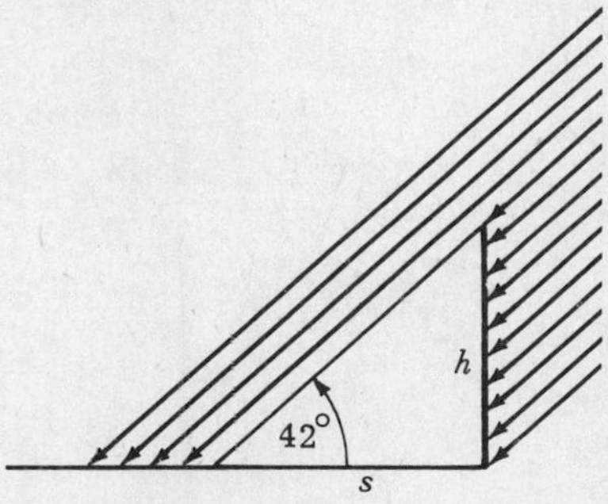

Solutions:

a) $\tan 42° = \frac{h}{s}$

$h = s \tan 42°$

$h = 25(.9004)$

$h = 23$, to nearest unit

Ans. 23 ft.

b) $\tan 48° = \frac{s}{h}$

$s = h \tan 48°$

$s = 35(1.1106)$

$s = 39$, to nearest unit

Ans. 39 ft.

3.3. Using Both an Angle of Elevation and an Angle of Depression

At the top of a lighthouse 200 ft. high, a lighthouse keeper sighted an airplane, and a ship directly beneath the plane. Sighting the plane, the angle of elevation was 25°. Sighting the ship, the angle of depression was 32°. Find

a) the distance d, to the nearest 10 ft., of the boat from the foot of the lighthouse.

b) the height h, to the nearest 10 ft., of the plane above the water.

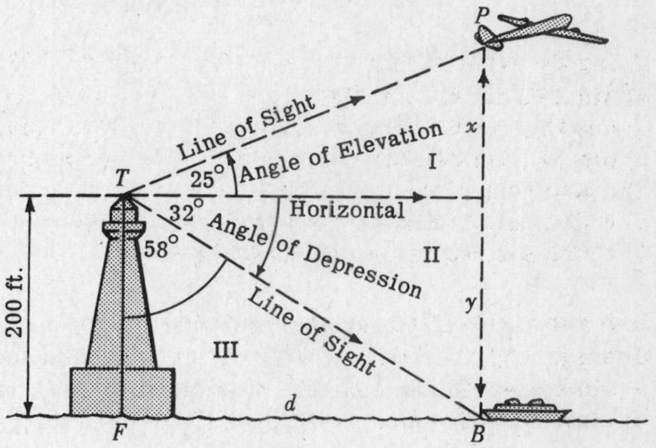

Solutions:

a)

$$\tan 58^\circ = \frac{d}{200}$$

$$1.6003 = \frac{d}{200}$$

$$200(1.6003) = d$$

$$320.06 = d$$

Ans. 320 ft., to nearest 10 ft.

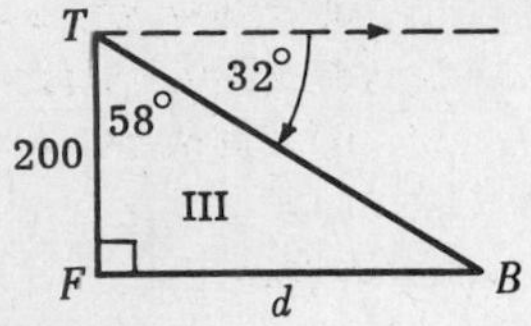

b) Using $d = 320$:

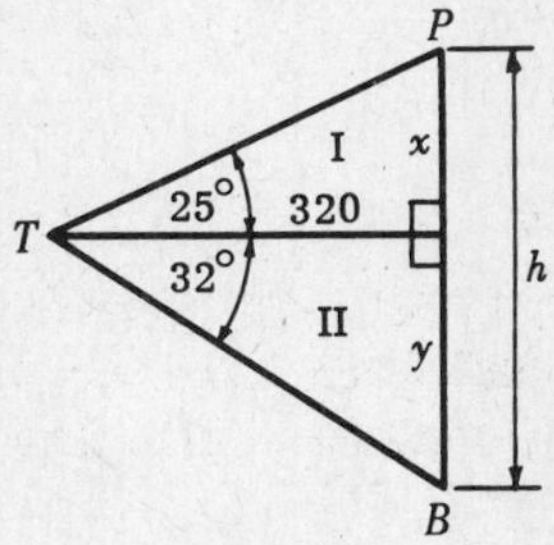

In ΔI, $\tan 25^\circ = \frac{x}{320}$

$$.4663 = \frac{x}{320}$$

$$320(.4663) = x$$

$$149.216 = x$$

In ΔII, $\tan 32^\circ = \frac{y}{320}$

$$.6249 = \frac{y}{320}$$

$$320(.6249) = y$$

$$199.968 = y$$

Since $h = x + y$

$$h = 149 + 200 = 349$$

Ans. 350 ft., to nearest 10 ft.

3.4. Using Two Angles of Depression

An observer on the top of a hill 250 ft. above the level of a lake sighted two boats directly in line. Find, to the nearest ft., the distance between the boats if the angles of depression noted by the observer were 11° and 16°.

Solution: (*Separate* ΔI and ΔII *as shown.*)

$$BB' = CB' - CB$$
$$= 250 \tan 79^\circ - 250 \tan 74^\circ$$
$$= 250(\tan 79^\circ - \tan 74^\circ)$$
$$= 250(5.1446 - 3.4874)$$
$$= 250(1.6572)$$
$$= 414.3$$

Ans. 414 ft., to nearest ft.

$CB' = 250 \tan 79^\circ$
$CB = 250 \tan 74^\circ$

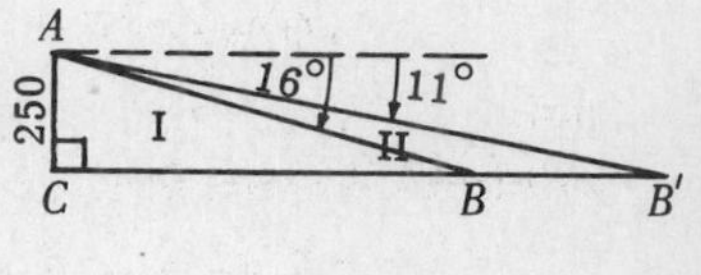

$CB = 250 \tan 74^\circ$

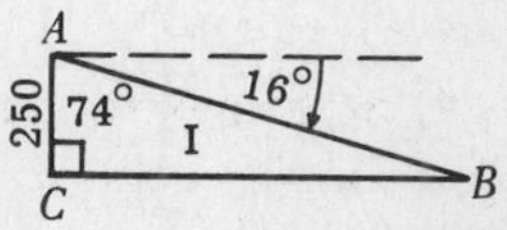

$CB' = 250 \tan 79^\circ$

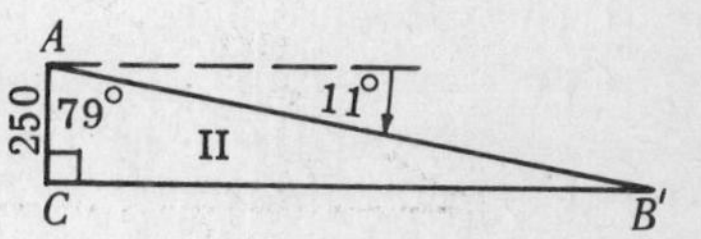

SUPPLEMENTARY PROBLEMS

1. Using the table of sines, cosines and tangents, find **(1.1)**

a) sin 25°, sin 48°, sin 59°, sin 89° *Ans.* .4226, .7431, .8572, .9998
b) cos 15°, cos 52°, cos 74°, cos 88° *Ans.* .9659, .6157, .2756, .0349
c) tan 4°, tan 34°, tan 55°, tan 87° *Ans.* .0699, .6745, 1.4281, 19.0811
d) which sets of values increase as the angle increases from 0° to 90°. *Ans.* sine and tangent
e) which set of values decreases as the angle increases from 0° to 90°. *Ans.* cosine
f) which set has values greater than 1. *Ans.* tangent

2. Using the table, find the angle in each: **(1.1)**

a) $\sin x = .3420$ *Ans.* $x = 20°$
b) $\sin A = .4848$ *Ans.* $A = 29°$
c) $\sin B = .9455$ *Ans.* $B = 71°$
d) $\cos A' = .9336$ *Ans.* $A' = 21°$
e) $\cos y = .7071$ *Ans.* $y = 45°$
f) $\cos Q = .3584$ *Ans.* $Q = 69°$
g) $\tan W = .3443$ *Ans.* $W = 19°$
h) $\tan B' = 2.3559$ *Ans.* $B' = 67°$
i) $\tan R = 28.6363$ *Ans.* $R = 88°$

3. Using the table, find x, to the nearest degree if **(1.2)**

a) $\sin x = .4400$
b) $\sin x = .7280$
c) $\sin x = .9365$
d) $\cos x = .9900$
e) $\cos x = .7650$
f) $\cos x = .2675$
g) $\tan x = .1245$
h) $\tan x = .5200$
i) $\tan x = 5.6745$
j) $\sin x = \frac{11}{50}$
k) $\sin x = \frac{\sqrt{2}}{2}$
l) $\cos x = \frac{3}{25}$
m) $\cos x = \frac{3}{8}$
n) $\cos x = \frac{\sqrt{3}}{2}$
o) $\tan x = \frac{2}{7}$
p) $\tan x = \frac{\sqrt{3}}{10}$

Ans. *a*) 26°, *b*) 47°, *c*) 69°, *d*) 8°, *e*) 40°, *f*) 74°, *g*) 7°, *h*) 27°, *i*) 80°, *j*) 13° since $\sin x = .2200$,
k) 45° since $\sin x = .707$, *l*) 59° since $\cos x = .5200$, *m*) 68° since $\cos x = .3750$, *n*) 30° since $\cos x = .866$, *o*) 16° since $\tan x = .2857$, *p*) 10° since $\tan x = .1732$.

4. In each right triangle, find $\sin A$, $\cos A$ and $\tan A$: (*Leave answer in radical form.*) **(1.3)**

a)

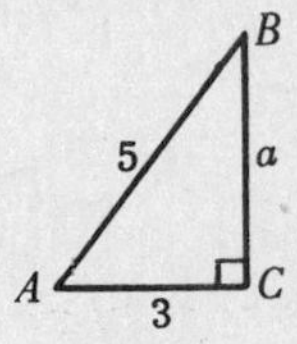

b)

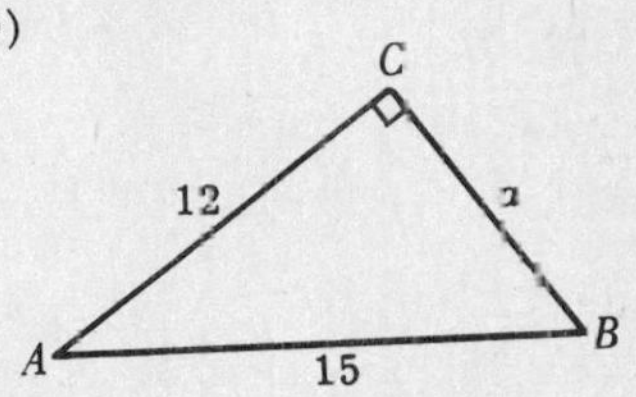

c)

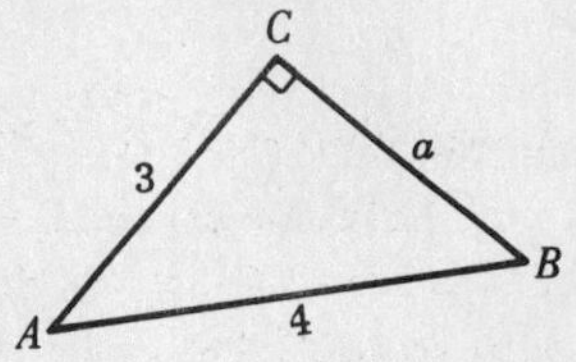

Ans.
a) $\sin A = \frac{4}{5}$, $\cos A = \frac{3}{5}$, $\tan A = \frac{4}{3}$
b) $\sin A = \frac{3}{5}$, $\cos A = \frac{4}{5}$, $\tan A = \frac{3}{4}$
c) $\sin A = \frac{\sqrt{7}}{4}$, $\cos A = \frac{3}{4}$, $\tan A = \frac{\sqrt{7}}{3}$

5. Find A, to the nearest degree, in each: **(1.4)**

a)

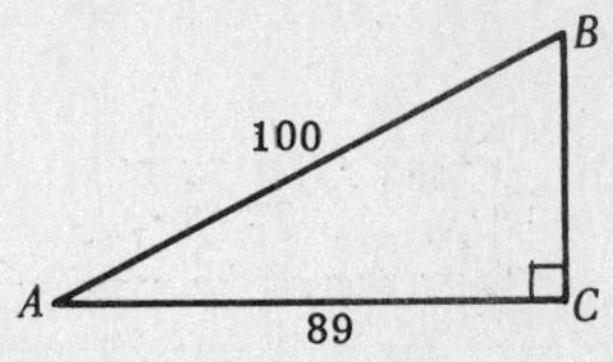

b)

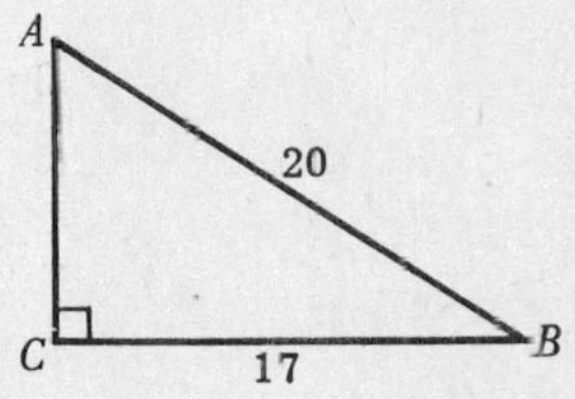

c)

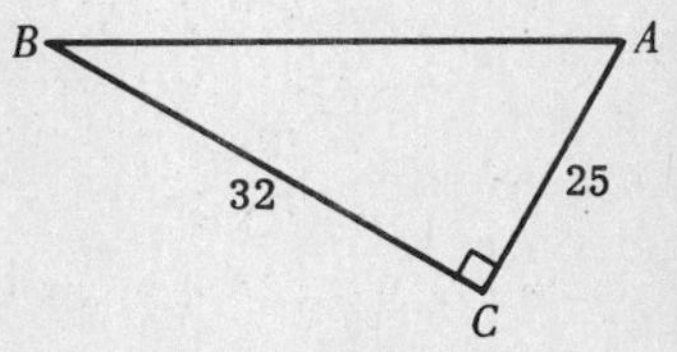

Ans.
a) $A = 27°$ since $\cos A = .8900$
b) $A = 58°$ since $\sin A = .8500$
c) $A = 52°$ since $\tan A = 1.2800$

6. Find B to the nearest degree, **(1.4)**

a) if $b = 67$ and $c = 100$. *Ans.* $B = 42°$ since $\sin B = .6700$
b) if $a = 14$ and $c = 50$. *Ans.* $B = 74°$ since $\cos B = .2800$
c) if $a = 22$ and $b = 55$. *Ans.* $B = 68°$ since $\tan B = 2.500$
d) if $a = 3$ and $b = \sqrt{3}$ *Ans.* $B = 30°$ since $\tan B = .577$

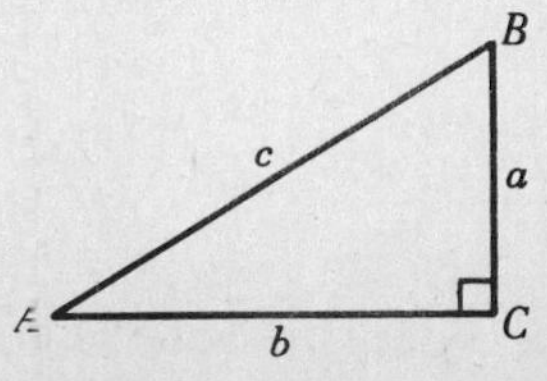

7. Using a square with a side of 1 unit, show that **(1.5)**

a) diagonal $c = \sqrt{2}$,

b) $\tan 45° = 1$,

c) $\sin 45° = \cos 45° = .707$.

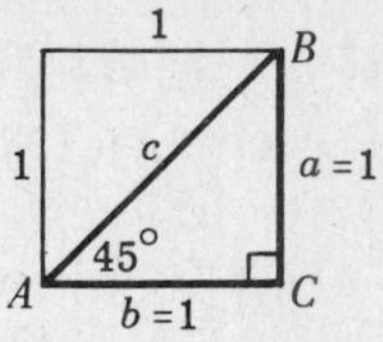

8. To the nearest degree, find each acute angle of any right triangle whose sides are in the ratio of

a) 5 : 12 : 13 *b*) 8 : 15 : 17 *c*) 7 : 24 : 25 *d*) 11 : 60 : 61 **(1.6)**

Ans. *a*) 23°, 67° *b*) 28°, 62° *c*) 16°, 74° *d*) 10°, 80°

9. In each triangle, state the trigonometric equation needed to solve for x and y: **(2.1)**

(*Avoid division where possible.*)

a)

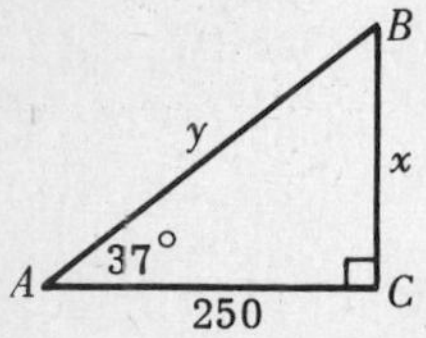

b)

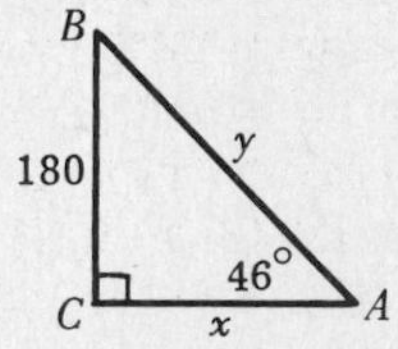

c)

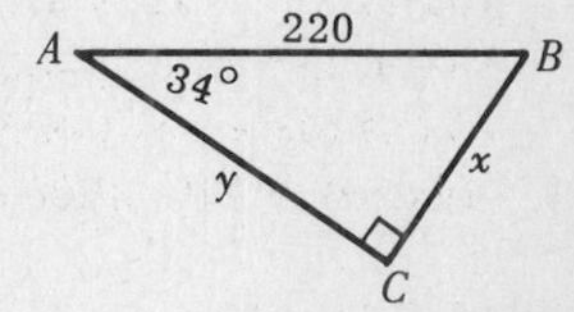

Ans. *a*) (*1*) $x = 250 \tan 37°$ (*2*) $y = \dfrac{250}{\cos 37°}$

b) (*1*) $x = 180 \tan 44°$, using $B = 44°$ (*2*) $y = \dfrac{180}{\sin 46°}$

c) (*1*) $x = 220 \sin 34°$ (*2*) $y = 220 \cos 34°$

10. Solve each equation. Find the unknown, to the nearest integer. **(2.2)**

a) (*1*) $x = 250 \tan 37°$ (*2*) $y = \dfrac{250}{\cos 37°}$

b) (*1*) $x = 180 \tan 44°$ (*2*) $y = \dfrac{180}{\sin 46°}$

c) (*1*) $x = 220 \sin 34°$ (*2*) $y = 220 \cos 34°$

Ans. *a*) $x = 188$, $y = 313$ *b*) $x = 174$, $y = 250$ *c*) $x = 123$, $y = 182$

(*In examples* **11-23**, *the trigonometric expression used to obtain the answer is shown in parentheses.*)

11. A ladder leans against the side of a building and makes an angle of 70° with the ground. **(2.1*a*, 2.2*a*)** The foot of the ladder is 30 feet from the building. Find, to the nearest foot,

a) how high up on the building the ladder reaches,

b) the length of the ladder.

Ans. *a*) 82 ft., ($30 \tan 70°$) *b*) 88 ft., ($\dfrac{30}{\cos 70°}$)

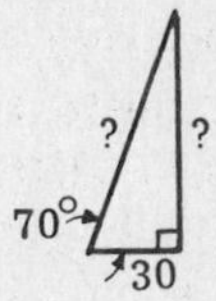

12. To find the distance across a swamp, a surveyor took measurements as shown. AC is at right angles to BC. **(2.1*a*, 2.2*a*)**

If $A = 24°$ and $AC = 350$ ft., find the distance BC across the swamp.

Ans. 156 ft., ($350 \tan 24°$)

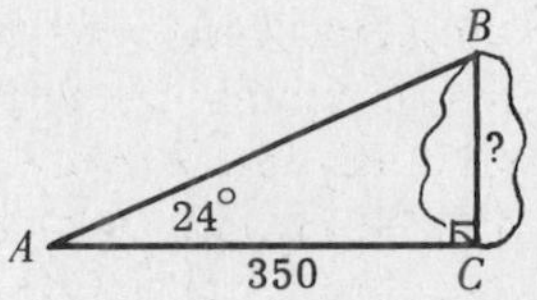

13. A plane rises from a take-off and flies at a fixed angle of 9° with the horizontal ground. When it has gained 400 ft. in altitude, find, to the nearest 10 ft., *a*) the horizontal distance flown, *b*) the distance the plane has actually flown. **(2.1*b*, 2.2*b*)**

Ans. *a*) 2530 ft., ($AC = 400 \tan 81°$)

b) 2560 ft., ($AB = \dfrac{400}{\sin 9°}$)

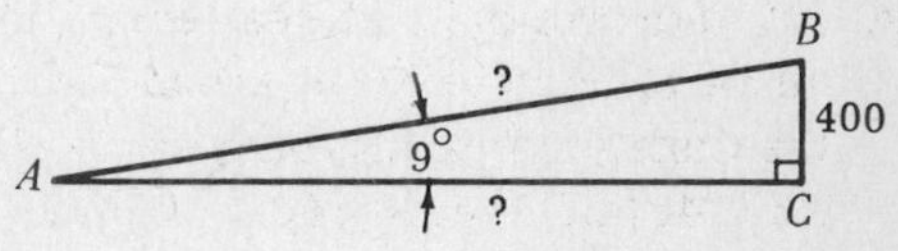

14. If the base angle of an isosceles triangle is 28° and each leg is 45 in., find, to the nearest inch, **(2.1*c*, 2.2*c*)**

a) the altitude drawn to the base,

b) the base.

Ans. *a*) 21 in., ($45 \sin 28°$)

b) 79 in., [$2(45 \cos 28°) = 90 \cos 28°$]

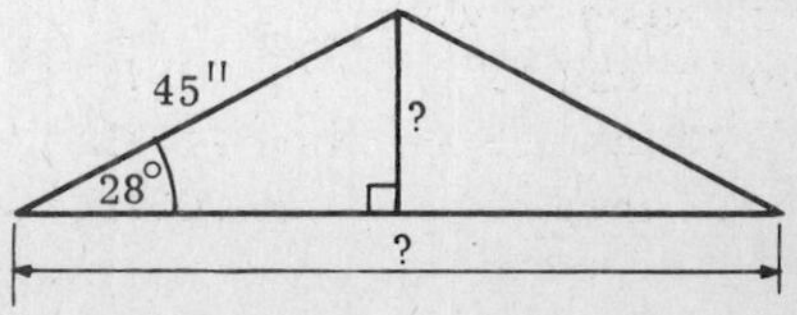

15. A road is inclined uniformly at an angle of 6° with the horizontal. After driving 10,000 ft. along this road, find, to the nearest 10 ft., the (2.3)
a) increase in the altitude of the driver,
b) horizontal distance that has been driven.
Ans. a) 1050 ft., (10,000 sin 6°) b) 9950 ft., (10,000 cos 6°)

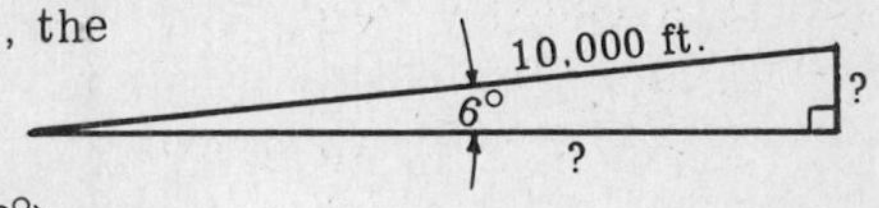

16. An airplane travels 15,000 ft. through the air at a uniform angle of climb, thereby gaining 1900 ft., in altitude. Find its angle of climb. (2.4)
Ans. 7° ($\sin x = \frac{1900}{15{,}000}$ where x is the angle of climb)

17. The angle of elevation of the top of a building from a point 500 ft. from the top is 21°. Disregarding the height of the transit, find to the nearest foot, (3.1)
a) the height of the building,
b) the distance along level ground from the observer to the foot of the building.
Ans. a) 179 ft., (500 sin 21°) b) 467 ft., (500 cos 21°)

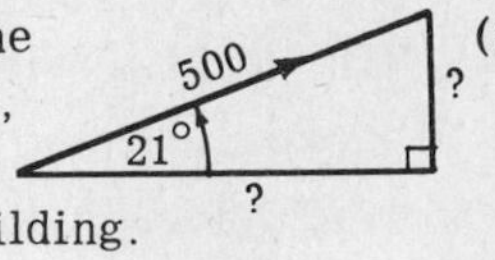

18. Sighting the top of a monument, William found the angle of elevation to be 16°. The ground is level and the transit is 5 ft. above the ground. If the monument is 86 ft. high, find to the nearest foot the distance from William to the foot of the monument. (3.1)
Ans. 282 ft., (81 tan 74°)

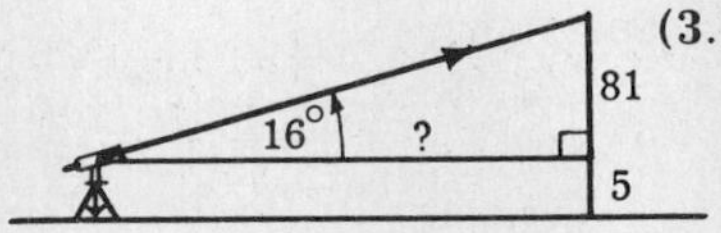

19. Find to the nearest degree the angle of elevation of the sun when a tree 60 ft. high casts a shadow of (3.2)
a) 10 ft., b) 20 ft., c) 40 ft., d) 60 ft.
Ans. a) 81°, ($\tan x = 6$) c) 56°, ($\tan x = 1.5$)
b) 72°, ($\tan x = 3$) d) 45°, ($\tan x = 1$)

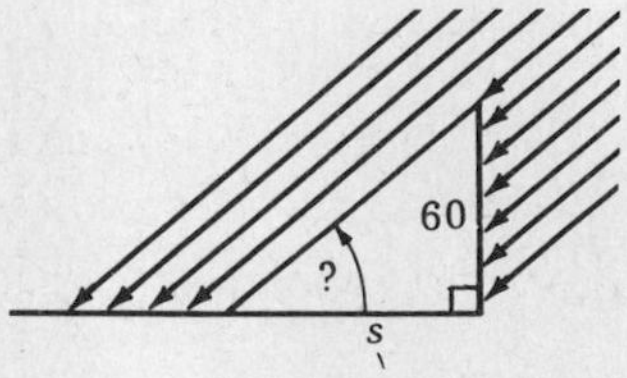

20. At a certain time of day, the angle of elevation of the sun is 34°. Find to the nearest foot the shadow cast by (3.2)
a) a 15 ft. vertical pole. *Ans.* 22 ft., (x = 15 tan 56°)
b) a building 70 ft. high. *Ans.* 104 ft., (x = 70 tan 56°)
c) a monument 450 ft. high. *Ans.* 667 ft., (x = 450 tan 56°)

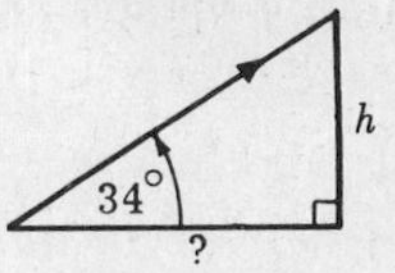

21. A light at C is projected vertically to a cloud at B. An observer at A, 1000 ft. from C, notes the angle of elevation of B. Find the height of the cloud, to the nearest foot, if (3.3)
a) $A = 20^\circ$. *Ans.* 364 ft., (1000 tan 20°)
b) $A = 37^\circ$. *Ans.* 754 ft., (1000 tan 37°)
c) $A = 49^\circ$. *Ans.* 1150 ft., (1000 tan 49°)

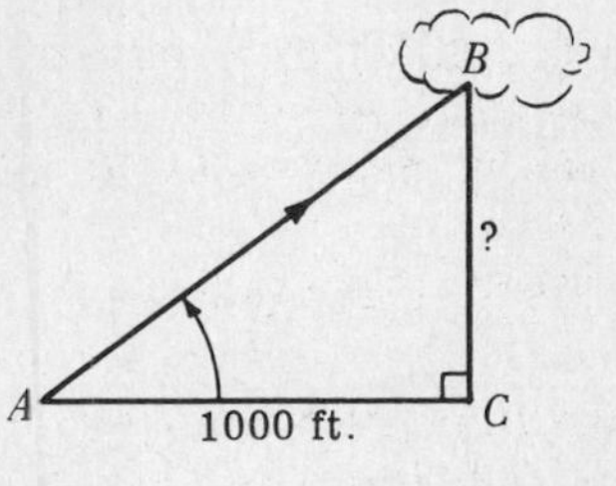

22. A lighthouse built at sea level is 180 ft. high. From its top, the angle of depression of a buoy is 24°. Find, to the nearest foot the distance from the buoy to the foot of the lighthouse. (3.3)
Ans. 404 ft., (180 tan 66°)

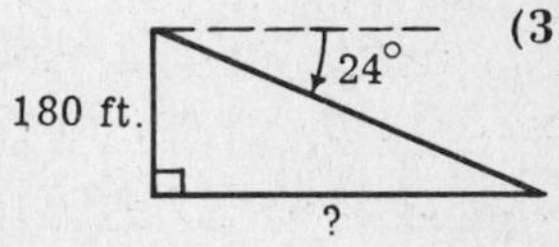

23. An observer on the top of a hill 300 ft. above the level of a lake, sighted two ships directly in line. Find, to the nearest ft., the distance between the boats if the angles of depression noted by the observer were (3.4)
a) 20° and 15°. *Ans.* 295 ft., [300 (tan 75° − tan 70°)]
b) 35° and 24°. *Ans.* 245 ft., [300 (tan 66° − tan 55°)]
c) 9° and 6°. *Ans.* 960 ft., [300 (tan 84° − tan 81°)]

Chapter 18

SUPPLEMENTARY TOPICS

1. SOLVING PROBLEMS GRAPHICALLY

Problems may be solved graphically. For this purpose, obtain a table of values directly from the problem relationships rather than from equations.

1.1. Solving a Work Problem Graphically

Solve Graphically: Abe can dig a certain ditch in 6 hr. while Zeke requires 3 hr. If both start together from opposite ends, how long would they take to complete the digging of the ditch?

Solution:

TABLE of VALUES for ABE

Hours Worked	Position Along Ditch
0	0
3	$\frac{1}{2}$
6	1

TABLE of VALUES for ZEKE

Hours Worked	Position Along Ditch
0	1
$1\frac{1}{2}$	$\frac{1}{2}$
3	0

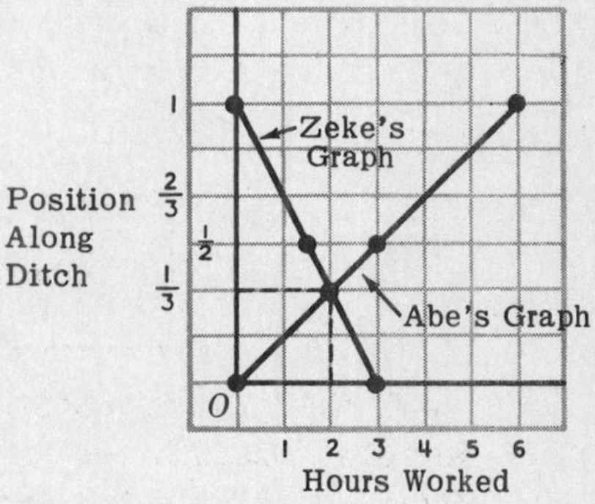

In the table and on the graph, "*O*" represents the beginning of the ditch where Abe started digging, while "1" represents the end of the ditch where Zeke began.

The common solution shows that they required 2 hr. At that time Abe had completed $\frac{1}{3}$ of the job while Zeke finished the remaining $\frac{2}{3}$ of the work. *Ans.* 2 hr.

1.2. Solving a Motion Problem Graphically

Solve Graphically: At 12 noon, Mr. Pabst left New York. After traveling at 30 mph for 2 hr., he rested for 1 hr. and then continued at 35 mph. Mr. Mayer wishes to overtake him. If Mr. Mayer starts at 3 P.M. from the same place and travels along the same road at 50 mph, when will they meet?

Solution:

TABLE of VALUES for Mr. Pabst

Time (P.M.)	Distance (mi.) from N.Y.	
(*start*)		
12 noon	0	30 mph
1 P.M.	30	
2 P.M.	60	
(1 hour of rest.)		
3 P.M.	60	35 mph
4 P.M.	95	
5 P.M.	130	
6 P.M.	165	
7 P.M.	200	

TABLE of VALUES for Mr. Mayer

Time (P.M.)	Distance (mi.) from N.Y.	
(*start*)		
3 P.M.	0	50 mph
4 P.M.	50	
5 P.M.	100	
6 P.M.	150	
7 P.M.	200	

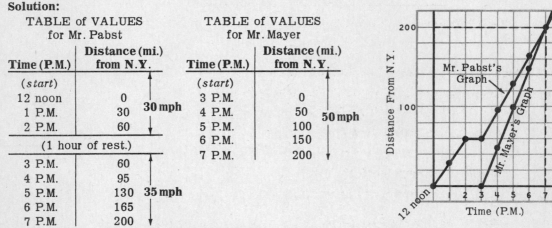

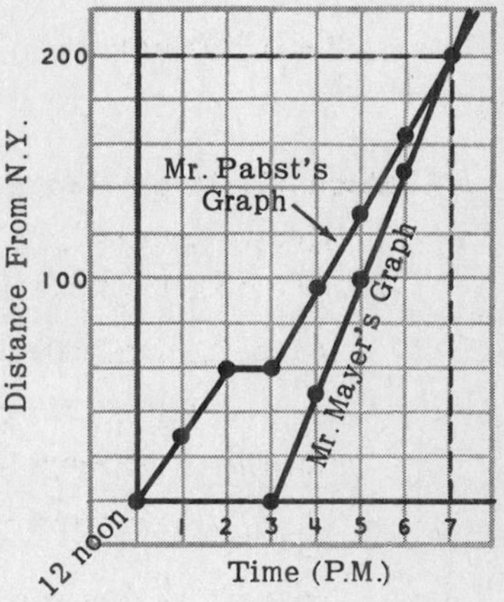

The common solution shows that Mr. Mayer will overtake Mr. Pabst at 7 P.M. when they are 200 mi. from New York. *Ans.* 7 P.M.

2. UNDERSTANDING SLOPE OF A LINE

THINK! **Slope of Line** $= \dfrac{y_2 - y_1}{x_2 - x_1} = \dfrac{\Delta y}{\Delta x} = m = \tan i$

The following are three slope rules for the **slope of a line through two points:**

(1) By Definition:

If line passes through $P_1(x_1, y_1)$ and $P_2(x_2, y_2)$:

Rule 1. **Slope of** $P_1P_2 = \dfrac{y_2 - y_1}{x_2 - x_1}$

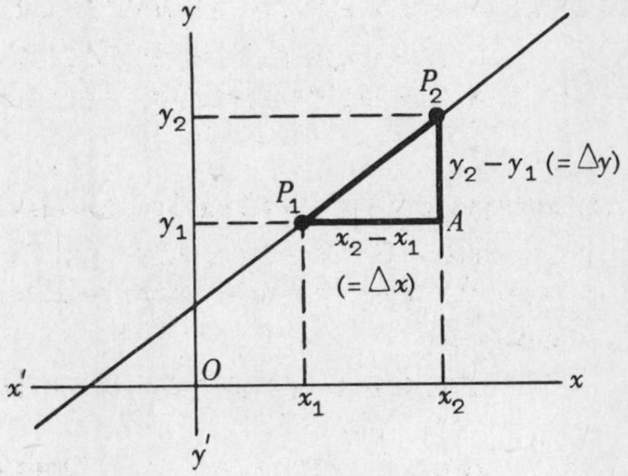

Note. $y_2 - y_1$ is the difference of the y values. In the diagram, the length of P_2A equals $y_2 - y_1$. Similarly, $x_2 - x_1$ is the corresponding difference of the x values. The length of P_1A equals $x_2 - x_1$.

Delta Form of Slope Definition

Slope of $P_1P_2 = \dfrac{\Delta y}{\Delta x}$

Δy means $y_2 - y_1$.
Δx means $x_2 - x_1$.

Δ, the fourth letter of the Greek alphabet, corresponds to d, the fourth letter of the English alphabet. Read Δy as "delta y" and Δx as "delta x". If you think of "delta y" as "y-difference", you will see why "Δy" is used to replace "$y_2 - y_1$".

(2) Slope of a line $= m$ if its equation is in form $y = mx + b$.

Rule 2. $m =$ **Slope of a line whose equation is** $y = mx + b$

Thus, the slope of $y = 3x + 2$ equals 3.

(3) Slope of a line = tangent of its inclination.

The inclination of a line is the angle it makes with the positive direction of the x-axis.

If i is the inclination of P_1P_2:

Rule 3. **Slope of** $P_1P_2 = \tan i$

Note in the diagram that $\tan i = \dfrac{\Delta y}{\Delta x}$.

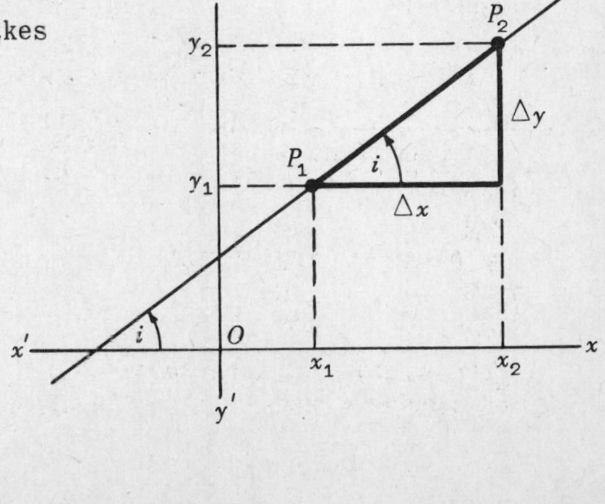

2.1. Applying Rule 1: Slope of line $= \dfrac{y_2 - y_1}{x_2 - x_1} = \dfrac{\Delta y}{\Delta x}$

Graph and find the slope of the line through
a) (0,0) and (2,4), b) (−2,−1) and (4,3).

Procedure:

1. **Plot** the points and draw the line which passes through them.
2. **Find** Δy **and** Δx, the corresponding differences of y and x:
3. **Find slope of line** $= \dfrac{\Delta y}{\Delta x}$:

Solutions:

a)

1.

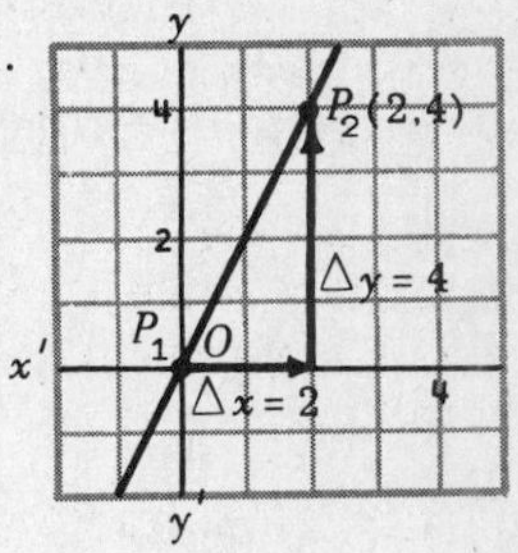

2. $P_2(2,4) \to x_2 = 2,\ y_2 = 4$
$P_1(0,0) \to x_1 = 0,\ y_1 = 0$
$\Delta x = 2,\ \Delta y = 4$

3. Slope of P_1P_2
$= \dfrac{\Delta y}{\Delta x} = \dfrac{4}{2} = 2$

Ans. 2

b)

1.

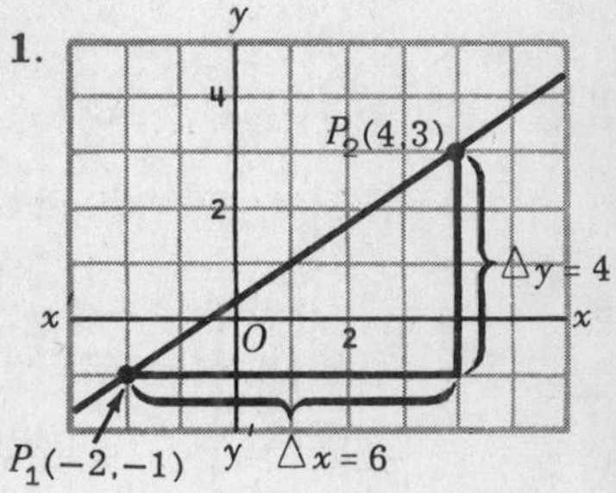

2. $P_2(4,3) \to x_2 = 4,\ y_2 = 3$
$P_1(-2,-1) \to x_1 = -2,\ y_1 = -1$
$\Delta x = 6,\ \Delta y = 4$

3. Slope of P_1P_2
$= \dfrac{\Delta y}{\Delta x} = \dfrac{4}{6} = \dfrac{2}{3}$

Ans. $\frac{2}{3}$

2.2. Applying Rule 2: m = slope of line, $y = mx + b$.

Find the slope of the line whose equation is
a) $2y = 6x - 8$, b) $3y - 4x = 15$.

Procedure:

1. **Transform** equation into the form, $y = mx + b$:
2. **Find slope of line** $= m$, the coefficient of x:

Solutions:

a)

1. $\mathbf{D_2}$ $2y = 6x - 8$
$y = 3x - 4$

2. Slope = 3.
Ans. 3

b)

1. **Tr** $3y - 4x = 15$
$\mathbf{D_3}$ $3y = 4x + 15$
$y = \frac{4}{3}x + 5$

2. Slope $= \frac{4}{3}$.
Ans. $\frac{4}{3}$

2.3. Applying Rule 3: Slope of line = tangent of its inclination.

Graph each line and find its inclination, to the nearest degree:
a) $y = 2x + 3$ b) $y = \frac{1}{2}x - 1$

Procedure:

1. **Graph line:**
2. **Find slope of line,** using m = slope of $y = mx + b$:
3. **Find inclination** i, using $\tan i$ = slope of line:

Solutions:

a)

1.

y	3	5
x	0	1

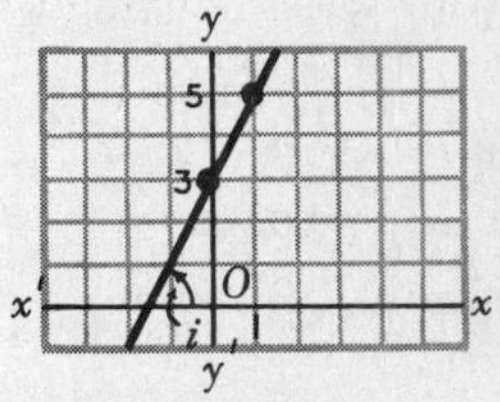

2. Slope = 2

3. $\tan i = 2.0000$
To nearest degree,
$i = 63°$. *Ans.* 63°

b)

1.

y	−1	0
x	0	2

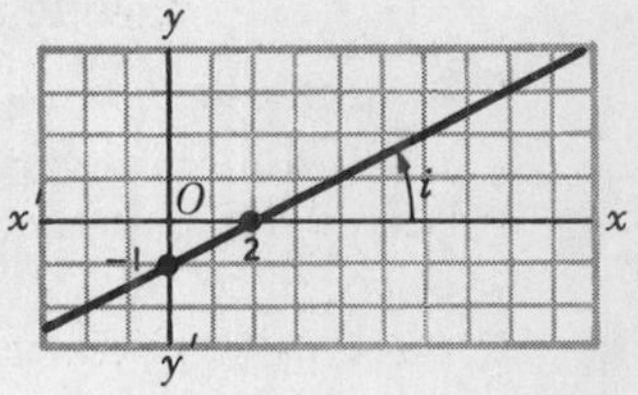

2. Slope $= \frac{1}{2}$

3. $\tan i = \frac{1}{2} = .5000$
To nearest degree,
$i = 27°$. *Ans.* 27°

3. UNDERSTANDING DESCRIPTIVE STATISTICS

Understanding Statistical Terminology

Frequency Distribution of Final Mathematics Marks

Marks	Tally	Frequency
100	III	3
95	~~IIII~~ III	8
90	~~IIII~~ ~~IIII~~ ~~IIII~~	15
85	~~IIII~~ I	6
80	III	3

1. **A rank order** is an arrangement of items into ordered sets. Thus, the final marks in the table have been distributed into separate sets beginning with 100's, then 95's, etc.
2. **A frequency distribution** is a distribution which indicates the tally or frequency of each set of items. Thus, the frequency or number of 100's is 3, the number of 95's is 8, etc.
3. **The range of a group of items** is the interval between the smallest and largest sets. Thus, the range of final marks is from 80 to 100.
4. **The mode** is the set of items having the largest frequency. Thus, 90 is the mode since it has the largest frequency, 15.
5. **The arithmetic mean** is the average mark of the group. Thus, $90\frac{2}{7}$ is the average of the 35 marks. This can be obtained by dividing the sum total of all the marks by 35, as follows:

$$\textbf{Average} = \frac{3(100) + 8(95) + 15(90) + 6(85) + 3(80)}{35} = 90\frac{2}{7}$$

6. **The median** is the middle item of an ordered group. Thus, 90 is the median of the 35 marks since it is the 18th or middle mark.
7. **The measures of central tendency** of a frequency distribution are the **mode, median** and **arithmetic mean.**

Frequency Distribution of Grouped Items

The table shown is a frequency distribution of the following 29 contest scores: 96, 87, 85, 76, 74, 71, 70, 68, 67, 67, 64, 64, 63, 60, 59, 57, 57, 57, 54, 53, 52, 52, 48, 46, 46, 42, 41, 38, 32.

These were tallied in seven groups as shown.

Thus, 52 was tallied twice in the 49.5 to 59.5 group.

Frequency Distribution of Grouped Contest Scores

Group Interval	Group Midpoint	Tally	Frequency
89.5 – 99.5	94.5	I	1
79.5 – 89.5	84.5	II	2
69.5 – 79.5	74.5	IIII	4
59.5 – 69.5	64.5	~~IIII~~ II	7
49.5 – 59.5	54.5	~~IIII~~ III	8
39.5 – 49.5	44.5	~~IIII~~	5
29.5 – 39.5	34.5	II	2
			Total: 29

Frequency Distribution Graphs: Histogram and Frequency Polygon

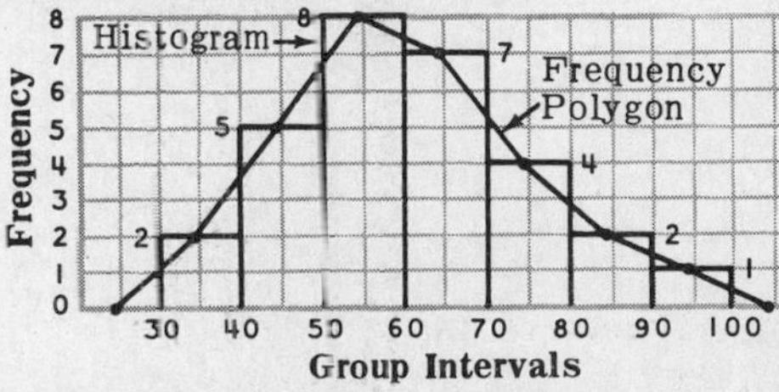

The bar graph shown is the **histogram** of the grouped contest scores above. The height of each vertical bar is the frequency of the group being represented. The group intervals are placed along the horizontal axis.

The broken-line graph of lines connecting the group midpoints is the **frequency polygon** of the grouped scores.

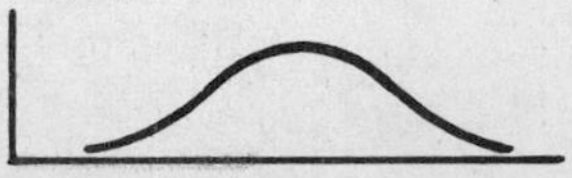

For a very large number of groups of scores, frequency polygons tend to be bell-shaped in the form of the **normal probability curve,** as shown in the diagram at the left.

3.1. Tallying Items

Make a frequency distribution of the following 51 readings, representing the maximum daily temperatures during a 51-day period. State its range, mode, median and arithmetic mean.

66, 69, 66, 66, 65, 68, 61, 62, 65, 63, 67, 62, 63, 63, 65, 65, 66
64, 65, 64, 68, 68, 67, 67, 64, 64, 63, 62, 64, 65, 65, 69, 66, 68
65, 64, 64, 65, 67, 66, 68, 67, 65, 66, 61, 62, 63, 60, 65, 67, 67

Solution:

Maximum Temperatures	69	68	67	66	65	64	63	62	61	60	
Tally	II	~~IIII~~	~~IIII~~ II	~~IIII~~ II	~~IIII~~ ~~IIII~~ I	~~IIII~~ II	~~IIII~~	IIII	II	I	
Frequency	2	5	7	7	11	7	5	4	2	1	**Total:** 51

The range is from 60 to 69.
The mode (most frequent) = 65.
The median (middle score is 26th in rank) = 65.
The average or arithmetic mean

$$= \frac{69(2)+68(5)+67(7)+66(7)+65(11)+64(7)+63(5)+62(4)+61(2)+60(1)}{51} = 65.04$$

3.2. Tallying Groups of Items

Make a frequency distribution of the 51 temperatures of the previous problem by tallying them in the following five groups:

59.5 – 61.5, 61.5 – 63.5, 63.5 – 65.5, 65.5 – 67.5, 67.5 – 69.5

Solution:

Grouped Frequency Distribution

Group Interval	Group Midpoint	Tally	Frequency
67.5 – 69.5	68.5	~~IIII~~ II	7
65.5 – 67.5	66.5	~~IIII~~ ~~IIII~~ IIII	14
63.5 – 65.5	64.5	~~IIII~~ ~~IIII~~ ~~IIII~~ III	18
61.5 – 63.5	62.5	~~IIII~~ IIII	9
59.5 – 61.5	60.5	III	3

3.3. Making Histogram and Frequency Polygon

Make a histogram and frequency polygon of the 5 groups of the previous problem:

Solution:

Histogram and Frequency Polygon

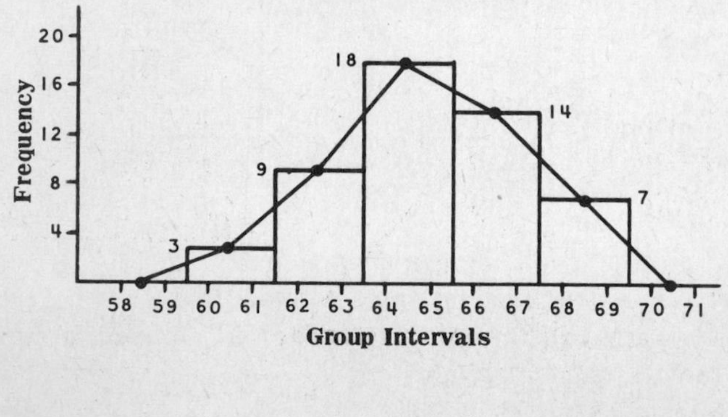

The histogram is the bar graph.
The frequency polygon is the broken line graph.

4. UNDERSTANDING CONGRUENT TRIANGLES

Congruent triangles have exactly the same size and shape. If triangles are congruent,

(1) their corresponding sides are equal and (2) their corresponding angles are equal.

Thus, in Fig. 1, congruent triangles I and I′ have equal corresponding sides and angles; that is, three pairs of equal sides and three pairs of equal angles.

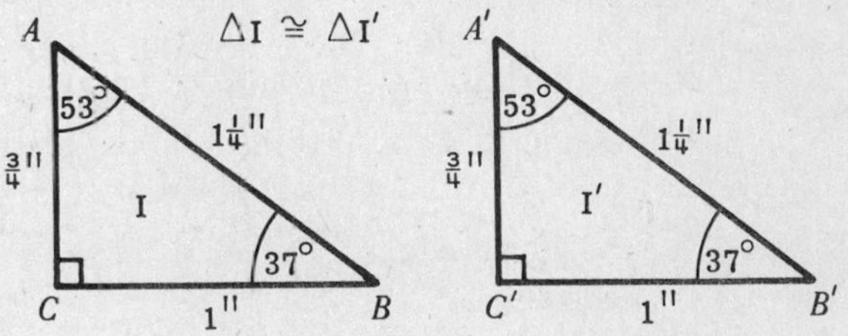

Fig. 1

Read "ΔI ≅ ΔI′ " as "triangle I is congruent to triangle I-prime".

Congruent triangles are exact duplicates of each other and may be made to fit together or coincide with one another. The aim of mass production is to produce exact duplicates or congruent objects.

Thus, if ΔI′ above is cut out and placed on ΔI, each side or angle of ΔI′ may be made to fit the corresponding part of ΔI.

The following three tests are used to determine when two triangles are congruent to each other.

Rule 1.: [s.s.s.= s.s.s.]

Two triangles are **congruent** if three sides of one triangle are equal respectively to three sides of the other.

Thus, ΔII ≅ ΔII′ (Fig. 2) since

(1) $DE = D'E'$ (= 4″)
(2) $DF = D'F'$ (= 5″)
and (3) $EF = E'F'$ (= 6″).

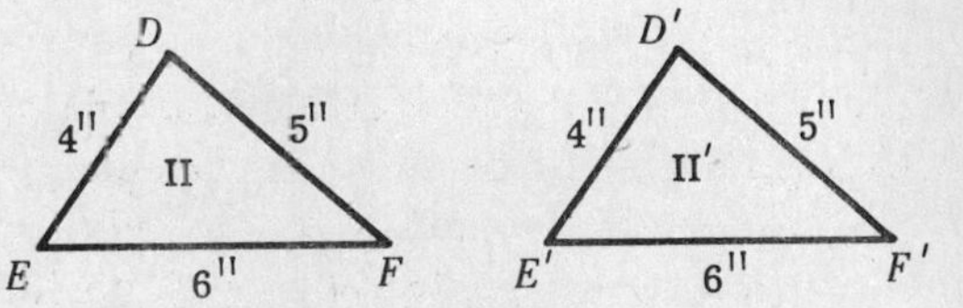

Fig. 2

Rule 2.: [a. s. a.= a. s. a.]

Two triangles are **congruent** if two angles and the included side of one triangle are equal respectively to two angles and the included side of the other.

Thus, ΔIII ≅ ΔIII′ (Fig. 3) since

(1) $\angle H = \angle H'$ (= 85°)
(2) $\angle J = \angle J'$ (= 25°)
and (3) $HJ = H'J'$ (= 10″)

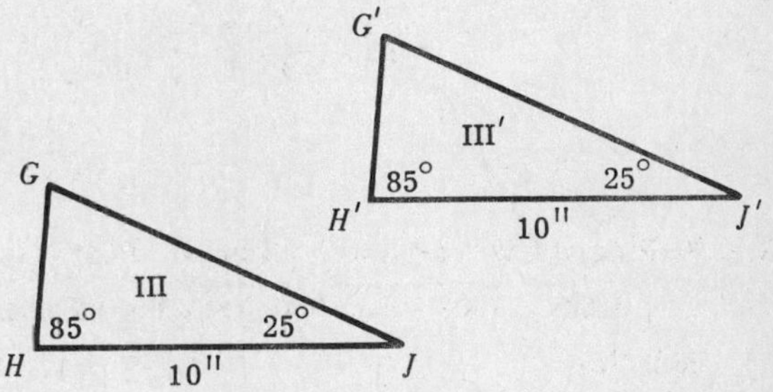

Fig. 3

Rule 3.: [s.a.s.= s.a.s.]

Two triangles are **congruent** if two sides and the included angle of one triangle are equal respectively to two sides and the included angle of the other.

Thus, ΔIV ≅ ΔIV′ (Fig. 4) since

(1) $KL = K'L'$ (= 9″)
(2) $\angle L = \angle L'$ (= 55°)
and (3) $LM = L'M'$ (= 12″)

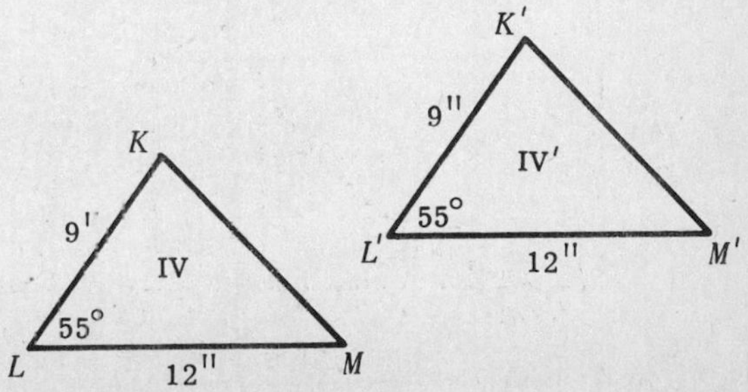

Fig. 4

Duplicating Lines, Angles and Triangles

1. To duplicate or copy a given line:

A line $A'B'$ may be constructed equal to a given line AB, as follows:

Using A' on a working line as a center, draw an arc of a circle whose radius equals AB.

Thus, in Fig. 5, $A'B' = AB$, by construction.

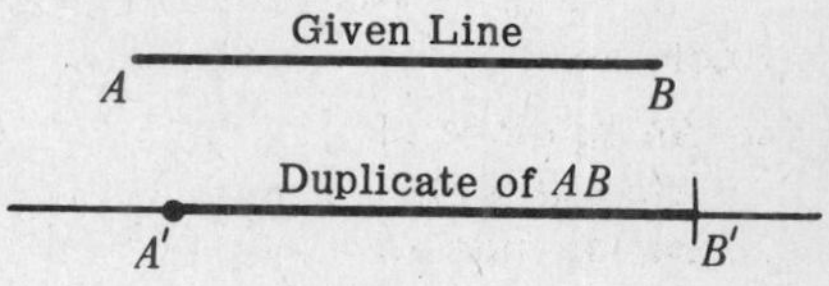

Fig. 5

2. To duplicate or copy a given angle:

An angle A' may be constructed equal to a given angle A, as follows:

(1) Using A as center and with a convenient radius, draw arc (a) intersecting AP in B and AQ in C.

(2) Using A' as center and with same radius, draw arc (a') intersecting $A'P'$ in B'.

(3) Using B' as center and with BC as a radius, draw arc (b') intersecting arc (a') in C'.

(4) Draw $A'C'$.

Thus, in Fig. 6, $\angle A' = \angle A$, by construction.

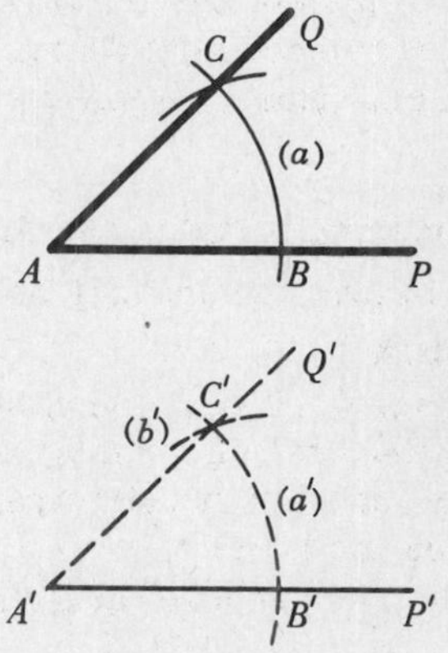

Fig. 6

To duplicate an angle of a given number of degrees, use your protractor.

Thus, to duplicate an angle of 30° at A on BC, set the protractor, as shown in Fig. 7.

Fig. 7

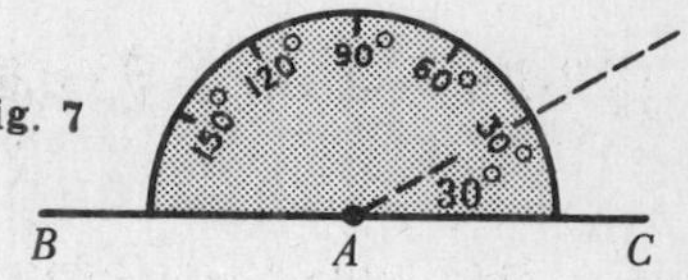

3. To duplicate or copy a given triangle:

Using each of the three rules for showing that triangles are congruent, a given $\triangle ABC$ may be duplicated.

Using **Rule 1**, $\triangle \text{I} \cong \triangle ABC$, by s.s.s. = s.s.s.

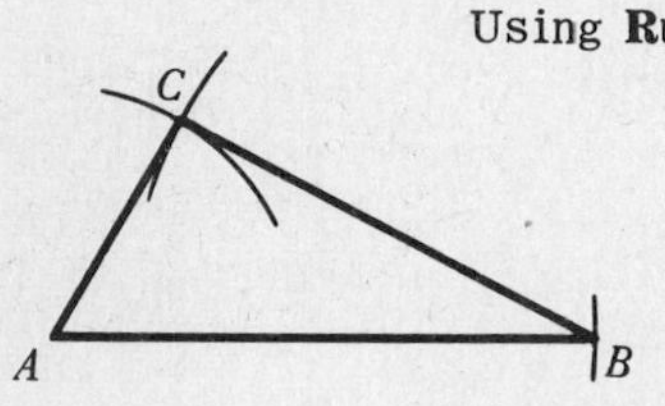

As shown, construct:

(1) $A'B' = AB$
(2) $B'C' = BC$
(3) $A'C' = AC$

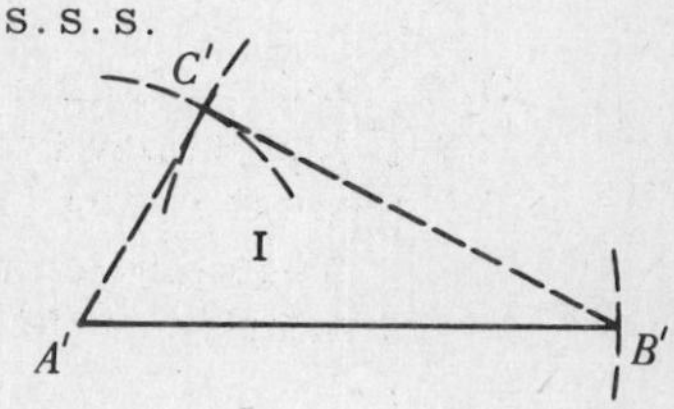

Using **Rule 2**, $\triangle \text{II} \cong \triangle ABC$, by a.s.a. = a.s.a.

As shown, construct:

(1) $A'B' = AB$
(2) $\angle B' = \angle B$
(3) $\angle A' = \angle A$

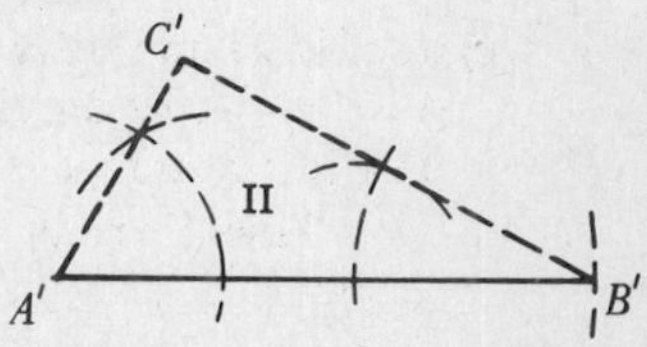

Using **Rule 3**, $\triangle \text{III} \cong \triangle ABC$, by s.a.s. = s.a.s.

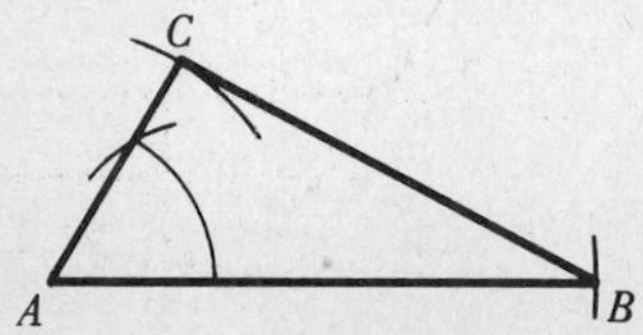

As shown, construct:

(1) $A'B' = AB$
(2) $\angle A' = \angle A$
(3) $A'C' = AC$
(Join B' and C')

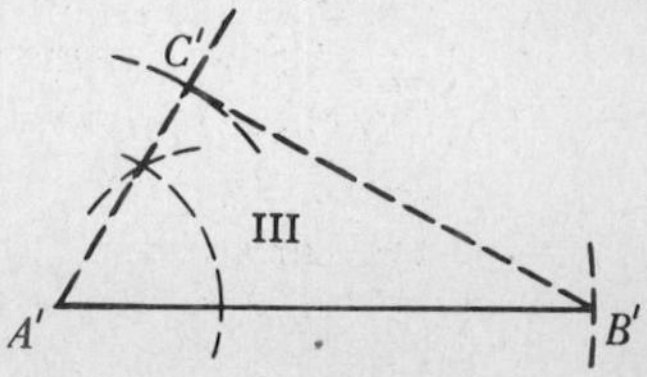

4.1. Constructing Congruent Triangles

If each of the following triangles are duplicated, why would new triangles be congruent in each case?

a) An equilateral triangle which has a side of 1''.

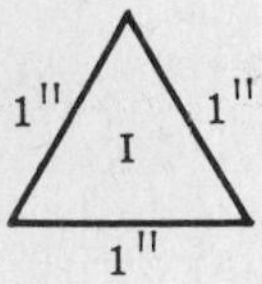

b) A right triangle having a leg of $1\frac{1}{2}$'' and a 36° angle adjacent to that leg.

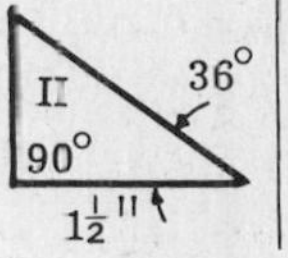

c) An isosceles triangle having two equal sides of 2'' and an included angle of 120°.

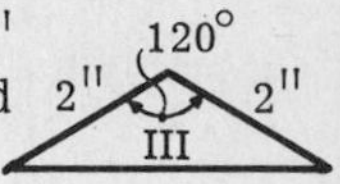

Solutions:

a) A duplicate equilateral triangle would be congruent to ΔI by s.s.s. = s.s.s.
b) A duplicate right triangle would be congruent to ΔII by a.s.a. = a.s.a.
c) A duplicate isosceles triangle would be congruent to ΔIII by s.a.s. = s.a.s.

4.2. Determining Lengths and Angles Indirectly

Using the data indicated, find AD and DB.

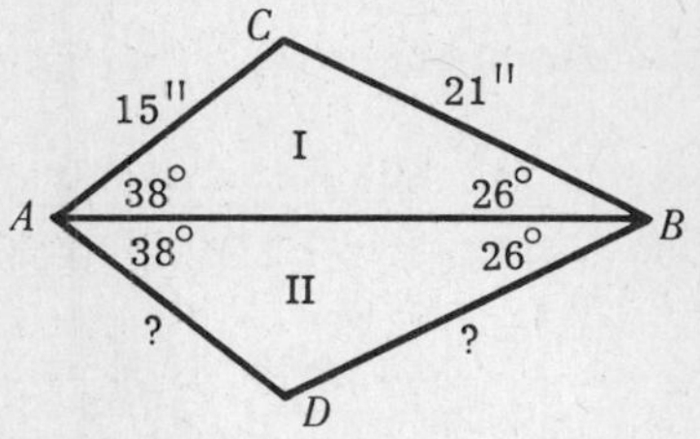

Solution:

Since AB is a side of both ΔI and ΔII, these triangles are congruent by a.s.a.= a.s.a. Hence, since the corresponding parts of congruent triangles are equal,

$AD = AC = 15''$ and $BD = BC = 21''$. *Ans.* 15'', 21''

5. UNDERSTANDING SYMMETRY

Point Symmetry: Center of Symmetry

1) A point is the **center of symmetry of two points** if it is midway between the two.

Thus, on a number scale, the origin is a center of symmetry with respect to the points representing opposites, such as +1 and −1, +2 and −2, etc.

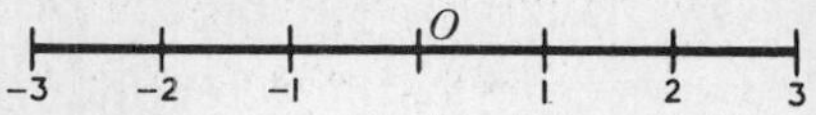

2) A point is the **center of symmetry of a figure** if any line joining two points of the figure and going through the point is bisected by the point.

Thus, the center of an ellipse is its center of symmetry. O is the center of AC, BD, etc.

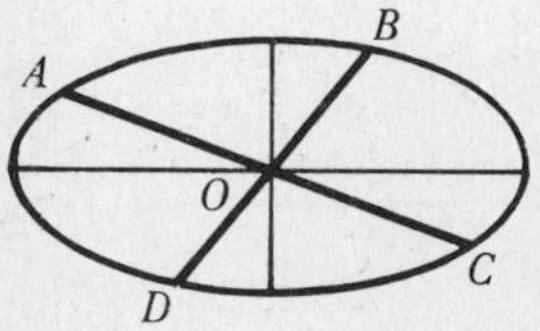

Line Symmetry: Axis of Symmetry

1) A line is the **axis of symmetry of two points** if it is the perpendicular bisector of the line joining the two points.

Thus, BC is the axis of symmetry of points A and A'.

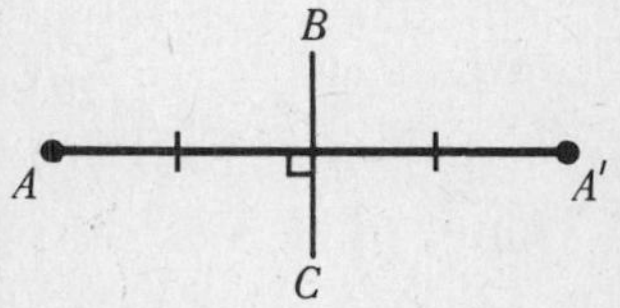

2) A line is an **axis of symmetry of a figure** if any line perpendicular to it is bisected by the figure.

Thus, FE is the axis of symmetry of the curve (parabola) shown. FE is perpendicular to and bisects AA', BB', etc. FE is the "folding line" of the parabola.

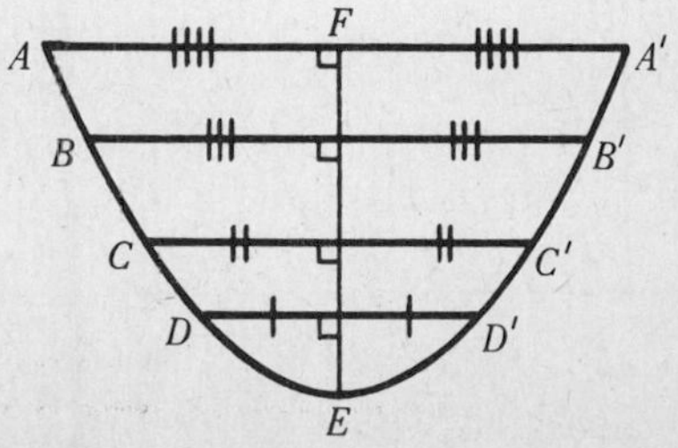

6. USING A SYMBOL TO SIMPLIFY VARIATION

The following is designed to further an understanding of the principles of Chap.14, The Variable. For this purpose a new symbol is introduced.

A New Ratio Symbol: $\nprec x)$

If x varies from a value x_1 to a new value x_2, the ratio of the second value to the first may be symbolized as $\nprec x)$; that is,

$$\nprec x) = \frac{x_2}{x_1} \qquad \text{Read } \nprec x) \text{ as "}x\text{-ratio".}$$

Similarly, the symbol may be used for any variable-ratio. Thus, $\nprec y) = \frac{y_2}{y_1}$; read $\nprec y)$ as "y-ratio".

Understanding the New Symbol

$\nprec x) = 2$ means that x is doubling in value. Compare this with $x = 2$ which means that x has a value of 2. Similarly, $\nprec y) = 3$ means that y is tripling in value. In a motion problem, $\nprec R) = 3$ and $\nprec T) = 4$ mean that a speed is being tripled and a time of travel quadrupled. In such a case, $\nprec D) = 12$; that is, the distance traveled is 12 times as far.

The Ratio of Any Two Values of a Constant Equals 1 : $\nprec k) = 1$

Since a constant remains fixed in value, the ratio of any two of its values must be 1.

Comparing the Meanings of $\nprec x)$ and Δx: $\nprec x) = \frac{x_2}{x_1}$ vs. $\Delta x = x_2 - x_1$

If x changes in value from 3 to 9, $\Delta x = 9 - 3 = 6$; that is, x becomes 6 more. In this case, $\nprec x) = \frac{9}{3} = 3$; that is, x becomes three times as much.

Using the Ratio Symbol to Replace Factors

> **In an equation or formula containing only factors,**
> **(1)** replace each variable factor by its ratio, and
> **(2)** replace each constant factor by 1.

Thus, if $z = kxy$ and k is constant, then $\nprec z) = \nprec x)\nprec y)$; that is, $\frac{z_2}{z_1} = \frac{x_2 y_2}{x_1 y_1}$.

If $V = \frac{4}{3}\pi R^3$, then $\nprec V) = \nprec R)^3$. Hence, if the radius of a sphere is made three times as large, its volume will be twenty-seven times as great.

SUMMARY OF THE VARIATION FORMULAS

Type of Variation	Formula of Variation	Formula Using Ratio Symbol	Formula Using Subscripts
Direct	$y = kx$	$\nprec y) = \nprec x)$	$\frac{y_2}{y_1} = \frac{x_2}{x_1}$
Inverse	$xy = k$	$\nprec x)\nprec y) = 1$	$\frac{x_2 y_2}{x_1 y_1} = 1$ (See note on p.279.)
Joint	$z = kxy$	$\nprec z) = \nprec x)\nprec y)$	$\frac{z_2}{z_1} = \frac{x_2 y_2}{x_1 y_1}$
Direct square	$y = kx^2$	$\nprec y) = \nprec x)^2$	$\frac{y_2}{y_1} = (\frac{x_2}{x_1})^2$
Inverse square	$y = \frac{k}{x^2}$	$\nprec y) = \frac{1}{\nprec x)^2}$	$\frac{y_2}{y_1} = \frac{1}{(\frac{x_2}{x_1})^2} = (\frac{x_1}{x_2})^2$
Direct cube	$y = kx^3$	$\nprec y) = \nprec x)^3$	$\frac{y_2}{y_1} = (\frac{x_2}{x_1})^3$

Note: When $\frac{x_2 y_2}{x_1 y_1} = 1$, we can readily obtain $\frac{x_2}{x_1} = \frac{y_1}{y_2}$ or $\frac{y_2}{y_1} = \frac{x_1}{x_2}$. From this it can be understood that in **inverse variation**, the ratios are **inversely equal**. On the other hand, in **direct variation**, the ratios are **directly equal**

6.1. Expressing the Same Formula in Ratio and Subscript Forms

Express each formula in ratio and subscript forms:

a) $D = RT$

b) $D = RT$ when R is constant

c) $D = RT$ when T is constant

d) $D = RT$ when D is constant

e) $V = \frac{1}{3}\pi R^2 H$

f) $V = \frac{1}{3}\pi R^2 H$ when H is constant

g) $V = \frac{1}{3}\pi R^2 H$ when R is constant

h) $V = \frac{1}{3}\pi R^2 H$ when V is constant

Solutions:

a) $\prec D) = \prec R)\prec T)$; $\quad \frac{D_2}{D_1} = \frac{R_2 T_2}{R_1 T_1}$

b) $\prec D) = \prec T)$; $\quad \frac{D_2}{D_1} = \frac{T_2}{T_1}$

c) $\prec D) = \prec R)$; $\quad \frac{D_2}{D_1} = \frac{R_2}{R_1}$

d) $\prec R)\prec T) = 1$; $\quad \frac{R_2 T_2}{R_1 T_1} = 1$

e) $\prec V) = \prec R)^2 \prec H)$; $\quad \frac{V_2}{V_1} = \frac{R_2^2 H_2}{R_1^2 H_1}$

f) $\prec V) = \prec R)^2$; $\quad \frac{V_2}{V_1} = \frac{R_2^2}{R_1^2}$

g) $\prec V) = \prec H)$; $\quad \frac{V_2}{V_1} = \frac{H_2}{H_1}$

h) $\prec R)^2 \prec H) = 1$; $\quad \frac{R_2^2 H_2}{R_1^2 H_1} = 1$

6.2. Expressing Formulas in Ratio Form. Express each formula in ratio form:

a) $A = \frac{1}{2}bh$

b) $A = 6e^2$

c) $V = LWH$

d) $V = \frac{4}{3}\pi R^3$

e) $E = \frac{1}{2}mv^2$ and m is constant

f) $F = \frac{kmm'}{d^2}$ and k, m and m' are constant

Solutions:

a) $\prec A) = \prec b)\prec h)$

b) $\prec A) = \prec e)^2$

c) $\prec V) = \prec L)\prec W)\prec H)$

d) $\prec V) = \prec R)^3$

e) $\prec E) = \prec v)^2$

f) $\prec F) = \frac{1}{\prec d)^2}$

6.3. Using the Ratio Symbol in Problem Solving

Complete each of the following:

a) If a man doubles his speed and triples his time, then he will go (   ).

b) Tripling the radius multiplies the circumference by ( ) and the area by ( ).

c) Applying Boyle's Law, $PV = k$, if the pressure of an enclosed gas becomes four-thirds as large, then the volume becomes (   ).

d) Applying Newton's Law of Gravitation, $F = \frac{kmm'}{d^2}$, if the distance between two constant masses is doubled, then the force of attraction is (   ).

e) If the stopping distance of a car varies directly as the square of its speed, then one who goes four times as fast will require a stopping distance (   ).

Solutions:

a) Since $D = RT$, $\prec D) = \prec R)\prec T)$. Hence, if $\prec R) = 2$ and $\prec T) = 3$, $\prec D) = (2)(3) = 6$.
Ans. Six times as far.

b) Since $C = 2\pi R$ and $A = \pi R^2$, then $\prec C) = \prec R)$ and $\prec A) = \prec R)^2$. Hence, if $\prec R) = 3$, then $\prec C) = 3$ and $\prec A) = 3^2 = 9$. *Ans.* Multiplies the circumference by 3 and the area by 9.

c) Since $PV = k$, $\prec P)\prec V) = 1$. Hence, if $\prec P) = \frac{4}{3}$, $\prec V) = \frac{3}{4}$. *Ans.* Three-fourths as large.

d) Here, $\prec F) = \frac{1}{\prec d)^2}$. Hence, if $\prec d) = 2$, then $\prec F) = \frac{1}{2^2} = \frac{1}{4}$. *Ans.* One-fourth as much.

e) Here, $\prec D) = \prec R)^2$ where D represents the stopping distance of the car and R its rate of speed. Hence, if $\prec R) = 4$, $\prec D) = 4^2 = 16$. *Ans.* Sixteen times as long.

SUPPLEMENTARY PROBLEMS

1. *Solve Graphically*: How long will it take Tom and George to mow a lawn together if Tom takes 12 hr. to do the job alone while George requires *a*) 6 hr., *b*) 4 hr., *c*) 3 hr.? **(1.1)**
Ans. *a*) 4 hr., *b*) 3 hr., *c*) $2\frac{2}{5}$ hr.

2. *Solve Graphically*: A motorist traveling at 60 mph seeks to overtake another who started 3 hr. earlier. Both are traveling along the same road. In how many hours will the slower motorist be overtaken if he is traveling at *a*) 40 mph, *b*) 30 mph, *c*) 20 mph? *Ans.* *a*) 6 hr., *b*) 3 hr., *c*) $1\frac{1}{2}$ hr. **(1.2)**

3. *Solve Graphically*: Towns *A* and *B* are 170 miles apart. Mr. Cahill left town *A* and is traveling to town *B* at an average rate of 40 mph. Mr. Fanning left town *B* one hour later and is traveling to town *A* along the same road at an average rate of 25 mph. When and where will they meet? **(1.2)**
Ans. They will meet, as shown graphically, 120 miles from *A*, three hours after Mr. Cahill started.

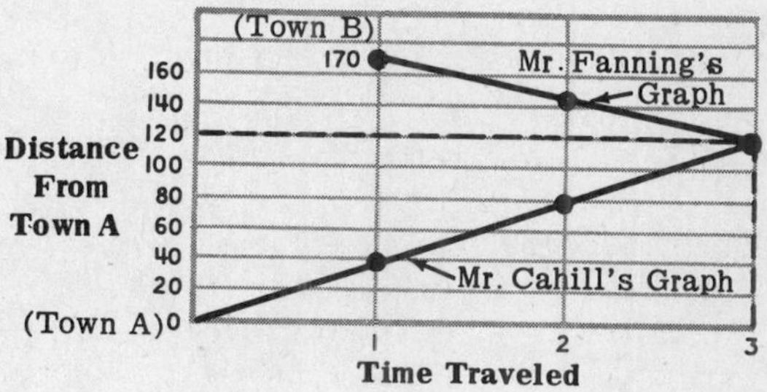

4. Find the slope of the line through **(2.1)**
a) (0,0) and (6,15) *b*) (2,3) and (6,15) *c*) (3,−4) and (5,6) *d*) (−2,−3) and (2,1) *e*) (8,10) and (0,−2) *f*) (−1,2) and (5,14)

a) $\frac{15-0}{6-0} = \frac{15}{6}$ *Ans.* $\frac{5}{2}$
b) $\frac{15-3}{6-2} = \frac{12}{4}$ *Ans.* 3
c) $\frac{6-(-4)}{5-3} = \frac{10}{2}$ *Ans.* 5
d) $\frac{1-(-3)}{2-(-2)} = \frac{4}{4}$ *Ans.* 1
e) $\frac{-2-10}{0-8} = \frac{-12}{-8}$ *Ans.* $\frac{3}{2}$
f) $\frac{14-2}{5-(-1)} = \frac{12}{6}$ *Ans.* 2

5. Find the slope of the line whose equation is **(2.2)**
a) $y = 3x-4$ *b*) $y = 5x$ *c*) $y = 5$ *d*) $2y = 6x-10$ *e*) $3y = 21-6x$ *f*) $4y-3x = 16$ *g*) $\frac{y}{2} = x-3$ *h*) $\frac{y}{3}+2x = 8$ *i*) $x-\frac{2}{3}y = 12$

a) $y = 3x-4$ *Ans.* 3
b) $y = 5x+0$ *Ans.* 5
c) $y = 0x+5$ *Ans.* 0
d) $y = 3x-5$ *Ans.* 3
e) $y = -2x+7$ *Ans.* −2
f) $y = \frac{3}{4}x+4$ *Ans.* $\frac{3}{4}$
g) $y = 2x-6$ *Ans.* 2
h) $y = -6x+24$ *Ans.* −6
i) $y = \frac{3}{2}x-18$ *Ans.* $\frac{3}{2}$

6. Find the inclination, to the nearest degree, of each line: **(2.3)**
a) $y = 3x-1$ *b*) $y = \frac{1}{3}x-1$ *c*) $y = \frac{5}{2}x+5$ *d*) $y = \frac{2}{5}x+5$ *e*) $5y = 5x-3$ *f*) $y = -3$

a) $\tan i = 3$ *Ans.* 72°
b) $\tan i = .3333$ *Ans.* 18°
c) $\tan i = 2.5$ *Ans.* 68°
d) $\tan i = .4$ *Ans.* 22°
e) $\tan i = 1$ *Ans.* 45°
f) $\tan i = 0$ *Ans.* 0°

7. Make a frequency distribution of the following 47 class registers in a school. State its range, mode, median and arithmetic mean. **(3.1)**

30, 32, 35, 31, 36, 28, 29, 34, 34, 32, 35, 31, 30, 29, 30, 28
33, 33, 34, 33, 32, 34, 29, 30, 33, 33, 34, 33, 30, 34, 35, 27
32, 31, 34, 31, 33, 32, 31, 29, 32, 33, 31, 32, 31, 32, 33

Ans.

Range = 27 to 36. Mode = 33.
Average or arithmetic mean = 31.87.
Median = 32 (24th score in rank).

Frequency Distribution

Registers	Frequency	Registers	Frequency
36	1	31	7
35	3	30	5
34	7	29	4
33	9	28	2
32	8	27	1

8. Make a frequency distribution of the 47 class registers of the previous problem by tallying them in the following five groups: 26.5 – 28.5, 28.5 – 30.5, 30.5 – 32.5, 32.5 – 34.5, 34.5 – 36.5.

Make a histogram and a frequency polygon of the five groups. **(3.2, 3.3)**

Ans.

Grouped Frequency Distribution

Group Intervals	Group Midpoints	Frequency
34.5 – 36.5	35.5	4
32.5 – 34.5	33.5	16
30.5 – 32.5	31.5	15
28.5 – 30.5	29.5	9
26.5 – 28.5	27.5	3

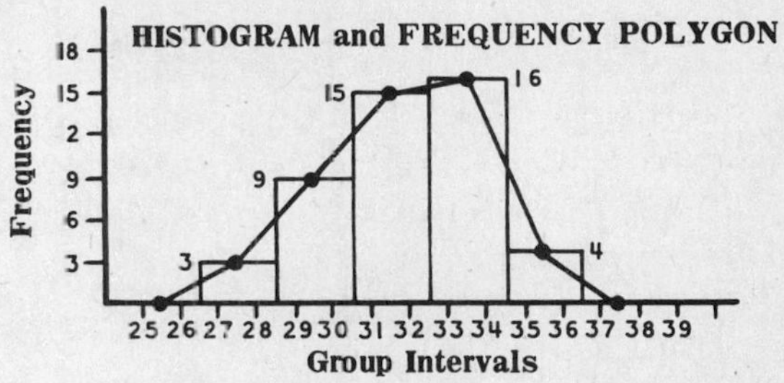

9. If each of the following triangles are duplicated, why would the new triangles be congruent in each case? **(4.1)**

a) A right triangle whose legs are 2″ and $1\frac{3}{4}$″.
b) An isosceles triangle having a base of 3″ and equal base angles of 35°.
c) An isosceles triangle having equal sides of 4″ and a vertical angle of 50°.
d) A right triangle having a leg of $3\frac{1}{2}$″ and a 55° angle opposite that leg.
e) A triangle having its sides in the ratio of 2 : 3 : 4 and 1″ for its smallest side.

Ans. *a*) s.a.s. = s.a.s., *b*) a.s.a. = a.s.a., *c*) s.a.s. = s.a.s.
d) a.s.a. = a.s.a., using 35° angle. *e*) s.s.s. = s.s.s., using sides of 1″, $1\frac{1}{2}$″, 2″.

10. In each, determine why the triangles are congruent and find the measurements indicated with a question mark. **(4.2)**

a)

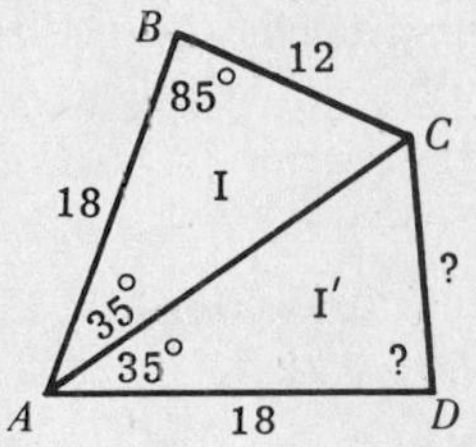

b)

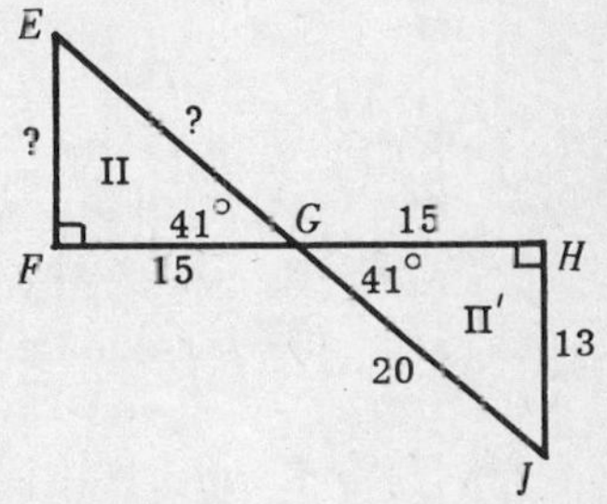

c)

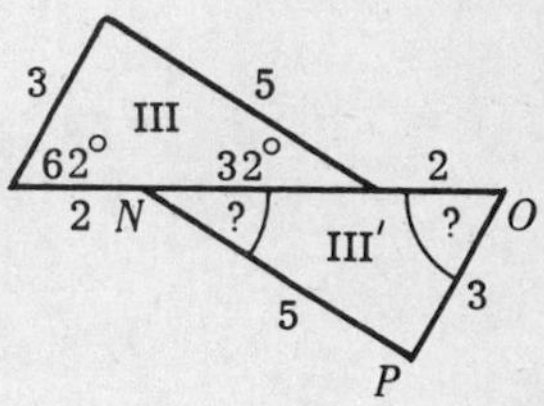

Ans.
a) $\Delta I \cong \Delta I'$ by s.a.s. = s.a.s. $\angle D = 85°$, $CD = 12$
b) $\Delta II \cong \Delta II'$ by a.s.a. = a.s.a. $EF = 13$, $EG = 20$
c) $\Delta III \cong \Delta III'$ by s.s.s. = s.s.s. $\angle O = 62°$, $\angle ONP = 32°$

11. State the point and line symmetry of each: **(5.)**
a) circle, *b*) parallelogram, *c*) rhombus, *d*) isosceles triangle.

Diagram		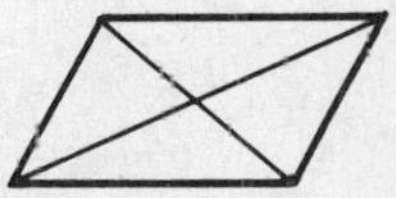	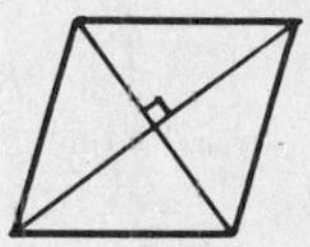	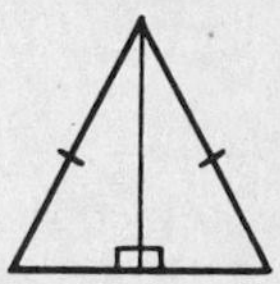
Figure	*a*) circle	*b*) parallelogram	*c*) rhombus	*d*) isosceles triangle
Center of Symmetry	center of circle	point of intersection of the diagonals	point of intersection of the diagonals	none
Axis of Symmetry	any diameter	none	each diagonal	altitude from vertex

Chapter 19

REVIEWING ARITHMETIC

1. REVIEWING WHOLE NUMBERS

A. Understanding Whole Numbers

(1) **Whole numbers are the numbers used in counting.**

Thus 3, 35 and 357 are whole numbers.

(2) **Place-value of each digit of a whole number:**

Depending on its **place in the whole number**, each digit of the number **represents** a number of units, tens, hundreds, thousands, and so on.

Thus in 35, 3 represents 3 tens; while in 357, 3 represents 3 hundreds.

(3) **Reading whole numbers using place-value.**

In reading a whole number, give place-value to its digits.

Thus, read 4444 as "four **thousand**, four **hundred**, forty (four **tens**) and four".

(4) **Rounding off a whole number: approximate value of a number.**

In rounding off a whole number to the nearest ten, replace the number by the multiple of 10 to which it is closest.

Thus, 37 to the nearest ten becomes 40. Here, 37 is between 30 and 40. However, 37 is closer to 40. Similarly round off a number to the nearest hundred, nearest thousand, etc.

Note. When a whole number is exactly halfway between two possible answers, round it off to the larger. Thus, round off 235 to the nearest ten as 240, not 230.

(5) **An approximate value of a whole number** is a value obtained by rounding off the whole number. Thus, 40 is an approximate value of 37.

(6) **Estimating answers by using approximate values of whole numbers.**

Procedure:	Estimate: *a*) 38×72	*b*) $\frac{4485}{503}$	*c*) $7495 - 2043$
1. **Round off** each number to a suitable multiple of 10, 100, etc.:	1. Round off 38 to 40, 72 to 70	1. Round off 4485 to 4500, 503 to 500	1. Round off 7495 to 7500, 2043 to 2000
2. **Perform** the indicated operation on the resulting **approximations:**	2. $40 \times 70 = 2800$	2. $\frac{4500}{500} = 9$	2. $7500 - 2000 = 5500$
	Ans. 2800	*Ans.* 9	*Ans.* 5500

(7) **Roman numbers.**

Roman numbers: M = 1000, D = 500, C = 100, L = 50, X = 10, V = 5, I = 1. In calculating a Roman number, **add** if a smaller digit follows a larger one but **subtract** if a smaller digit precedes a larger.

Thus, XI = 11 but IX = 9; MC = 1100 but CM = 900.

Hence, MCMLIX = M + CM + L + IX = 1000 + 900 + 50 + 9 = 1959.

B. Terms Used in the Fundamental Operations

Addition, subtraction, multiplication and division are the four fundamental operations performed on numbers. The names used in each operation must be memorized. Note them in each of the following.

(1) **Addition**: Add 77 and 20.

Terms Used in Addition

Addends are the numbers that are added.

Sum is the answer obtained in addition.

Addend	77
Addend	+20
Sum	97

(2) **Subtraction**: Subtract 20 from 77.

Terms Used in Subtraction

Minuend is the number from which we subtract.

Subtrahend is the number being subtracted.

Difference is the answer obtained in subtraction.
(*The answer in subtraction may be called* **remainder**.)

Minuend	77
Subtrahend	− 20
Difference	57

(3) **Multiplication**: Multiply 77 by 20.

Terms Used in Multiplication

Multiplicand is the number being multiplied.

Multiplier is the number by which we multiply.

Product is the answer obtained in multiplication.
(*Numbers being multiplied are also* **factors** *of their product.*)

Multiplicand	77
Multiplier	× 20
Product	1540

(4) **Division**: Divide 77 by 20.

Terms Used in Division

Dividend is the number being divided.

Divisor is the number by which we divide.

Quotient is the answer obtained in division.

Dividend $\frac{77}{20}$ **Divisor**

Quotient $= 3\frac{17}{20}$

(*Call the quotient a complete quotient to distinguish it from* 3, *the partial quotient.*)

Rule: Complete quotient = partial quotient + $\frac{\text{remainder}}{\text{divisor}}$

Thus, in $3\frac{17}{20}$ above, 3 is the partial quotient and 17 is the remainder.
Keep in mind, $3\frac{17}{20} = 3 + \frac{17}{20}$.

C. Checking the Fundamental Operations

(1) **Add** 25, 32 and 81. **Check** the addition.

Add either down the column or up the column. **To check addition, add** the col- in the reverse direction.

$$\text{Add down} \downarrow \quad \begin{array}{r} 25 \\ 32 \\ 81 \\ \hline 138 \end{array} \quad \uparrow \text{Check up}$$

(2) **Subtract** 32 from 85. **Check** the subtraction.

To check subtraction, add the difference and subtrahend. The answer thus obtained should be the minuend.

$$\begin{array}{r} 85 \\ -32 \\ \hline 53 \end{array} \qquad \text{Check: } \begin{array}{r} 53 \\ +32 \\ \hline 85 \end{array}$$

(3) **Multiply** 85 by 32. **Check** the multiplication.

To check multiplication, multiply the numbers after interchanging them.

$$\begin{array}{r} 85 \\ \times 32 \\ \hline 170 \\ 255 \\ \hline 2720 \end{array} \qquad \text{Check: } \begin{array}{r} 32 \\ \times 85 \\ \hline 160 \\ 256 \\ \hline 2720 \end{array}$$

(4) Divide 85 by 32. **Check** the division.

To check division, multiply the divisor by the partial quotient. To the result obtained, **add** the remainder. The final answer should be the dividend.

$$\begin{array}{r} 2 \\ 32\overline{)85} \\ \underline{64} \\ 21 \end{array}$$ 2 (Partial Quotient), 21 (Remainder)

Ans. $2\frac{21}{32}$

Check: $32 \times 2 = 64$; $64 + 21 = 85$

2. REVIEWING FRACTIONS

A. Understanding Fractions

Proper and Improper Fractions

(1) A **proper fraction** is a fraction whose value is less than 1. In a proper fraction, the numerator is less than the denominator.

Thus, $\frac{5}{8}$ and $\frac{69}{70}$ are proper fractions. Each has a value less than 1.

(2) An **improper fraction** is a fraction whose value is equal to or greater than 1. In an improper fraction, the numerator is equal to or greater than the denominator.

Thus, $\frac{8}{5}$ and $\frac{15}{15}$ are improper fractions; $\frac{8}{5}$ is greater than 1, and $\frac{15}{15}$ equals 1.

(3) The **terms of a fraction** are its numerator and denominator.

Thus, the terms of $\frac{25}{30}$ are 25 and 30.

(4) **Equivalent fractions** are fractions having the same value.

Thus, $\frac{1}{5}$, $\frac{2}{10}$, $\frac{5}{25}$ and $\frac{20}{100}$ are equivalent fractions; that is, $\frac{1}{5} = \frac{2}{10} = \frac{5}{25} = \frac{20}{100}$.

B. Changing Forms of Mixed Numbers and Improper Fractions

(1) A **mixed number** equals a whole number plus a fraction. Thus, $17\frac{2}{5} = 17 + \frac{2}{5}$.

(2) **Changing a mixed number to an improper fraction.**

Change to an improper fraction: *a*) $17\frac{2}{5}$ | *b*) $101\frac{2}{9}$

Procedure:	**Solutions:** *a*)	*b*)
1. **Multiply** whole number by denominator:	1. $17 \times 5 = 85$	1. $101 \times 9 = 909$
2. **Add** numerator to product:	2. $+2$ = 87	2. $+2$ = 911
3. **Form fraction** by placing result over denominator:	3. *Ans.* $\frac{87}{5}$	3. *Ans.* $\frac{911}{9}$

(3) **To change an improper fraction to a mixed number**, divide the numerator by the denominator.

Thus, $\frac{87}{5} = 87 \div 5 = 17\frac{2}{5}$.

C. Changing a Fraction to an Equivalent Fraction

(1) **Fundamental Law of Fractions:**

To change a fraction to an equivalent fraction, multiply or divide both the numerator and denominator by the same number.

Thus, $\frac{2}{5} = \frac{2 \times 10}{5 \times 10} = \frac{20}{50}$.   In turn, $\frac{20}{50} = \frac{20 \div 10}{50 \div 10} = \frac{2}{5}$.

(2) **To raise a fraction to higher terms**, multiply its terms by the same whole number. Thus, when $\frac{2}{5}$ is changed to $\frac{20}{50}$, it has been changed to higher terms.

(3) **To reduce a fraction to lower terms**, divide its terms by the same whole number. Thus, when $\frac{20}{50}$ is changed to $\frac{2}{5}$, it has been changed to lower terms.

(4) **To reduce a fraction to lowest terms**, divide both numerator and denominator by the highest common factor.

Thus, $\frac{20}{50}$ can be reduced to $\frac{10}{25}$, $\frac{4}{10}$ or $\frac{2}{5}$; but to reduce it to lowest terms, use $\frac{2}{5}$. To obtain $\frac{2}{5}$, divide terms by 10, the highest common factor of 20 and 50.

(5) To change a fraction to a new specified fraction.

Change to new fraction: $a)\ \frac{3}{5} = \frac{?}{45}$ | $b)\ \frac{3}{5} = \frac{45}{?}$

Procedure:	Solutions: a)	b)
1. **Divide** new denominator by old denominator, or divide new numerator by old numerator:	1. $45 \div 5 = 9$	1. $45 \div 3 = 15$
2. **Multiply** the result by old numerator or old denominator:	2. $\times 3$ = 27	2. $\times 5$ = 75
3. **Form fraction** needed:	3. *Ans.* $\frac{27}{45}$	3. *Ans.* $\frac{45}{75}$

(6) To compare the size of fractions.

Compare $\frac{5}{9}$ and $\frac{7}{12}$

Procedure:

1. **Change to equivalent fractions** having the same denominator. Use the lowest common denominator (**L.C.D.**). See page 192.
2. **Compare** new fractions:

Solution:

1. Change: $\frac{5}{9} = \frac{?}{36}$ and $\frac{7}{12} = \frac{?}{36}$.
   Here, $\frac{5}{9} = \frac{20}{36}$ and $\frac{7}{12} = \frac{21}{36}$.
2. Since $\frac{20}{36}$ is less than $\frac{21}{36}$,
   then $\frac{5}{9}$ is less than $\frac{7}{12}$.

D. Multiplying and Dividing Mixed Numbers and Fractions

Multiply and divide as indicated: $a)\ 1\frac{1}{6} \div 1\frac{5}{9}$ | $b)\ 1\frac{1}{7} \times 2\frac{11}{12} \div \frac{8}{9}$

Procedure:	Solutions: a)	b)
1. **Change** mixed numbers to fractions:	1. $\frac{7}{6} \div \frac{14}{9}$	1. $\frac{8}{7} \times \frac{35}{12} \div \frac{8}{9}$
2. **Multiply** by each inverted divisor:	2. $\frac{7}{6} \times \frac{9}{14}$	2. $\frac{8}{7} \times \frac{35}{12} \times \frac{9}{8}$
3. **Divide** a numerator and a denominator by a common factor of both:	3. $\frac{\overset{1}{\cancel{7}}}{\underset{2}{\cancel{6}}} \times \frac{\overset{3}{\cancel{9}}}{\underset{2}{\cancel{14}}}$	3. $\frac{\overset{1}{\cancel{8}}}{\underset{1}{\cancel{7}}} \times \frac{\overset{5}{\cancel{35}}}{\underset{4}{\cancel{12}}} \times \frac{\overset{3}{\cancel{9}}}{\underset{1}{\cancel{8}}}$
4. **Multiply** the remaing factors in the numerator and denominator separately:	4. $\frac{3}{2 \times 2} = \frac{3}{4}$ *Ans.*	4. $\frac{5 \times 3}{4} = \frac{15}{4}$ or $3\frac{3}{4}$ *Ans.*

E. Adding and Subtracting Fractions

(1) Combining fractions having same denominator.

Combine: $a)\ \frac{5}{12} + \frac{11}{12} - \frac{7}{12}$ | $b)\ \frac{15}{7} - \frac{3}{7} + \frac{2}{7}$

Procedure:	Solutions: a)	b)
1. **Keep** the common denominator and add or subtract the numerator as indicated:	1. $\frac{5+11-7}{12}$	1. $\frac{15-3+2}{7}$
2. **Reduce** result to the lowest terms:	2. $\frac{9}{12} = \frac{3}{4}$ *Ans.*	2. $\frac{14}{7} = 2$ *Ans.*

(2) Combining fractions having different denominators.

Combine: $\frac{5}{6} - \frac{7}{12} + \frac{23}{36}$

Procedure:	Solution:
1. **Change** the fractions to equivalent fractions having the same denominator, using the **L.C.D.**:	1. **L.C.D.** = 36 $\frac{30}{36} - \frac{21}{36} + \frac{23}{36}$
2. **Combine** the resulting fractions:	2. $\frac{30 - 21 + 23}{36}$
3. **Reduce** to lowest terms:	3. $\frac{32}{36} = \frac{8}{9}$ *Ans.*

F. Adding and Subtracting Mixed Numbers

Add or subtract as indicated: *a*) $15\frac{7}{8} + 9\frac{1}{4}$ | *b*) $15\frac{7}{8} - 9\frac{1}{4}$

Procedure:	Solutions: (a)	(b)
1. **Add** the fractions or **subtract** the fractions as indicated:	1. $15\frac{7}{8} \rightarrow \frac{7}{8}$, $+9\frac{1}{4} \rightarrow \frac{2}{8}$; $1\frac{1}{8}$	1. $15\frac{7}{8} \rightarrow \frac{7}{8}$, $-9\frac{1}{4} \rightarrow \frac{2}{8}$; $\frac{5}{8}$
2. **Add** the whole numbers or **subtract** the whole numbers as indicated:	2. $15 + 9$	2. $15 - 9$
3. **Add** both results:	3. $24 + 1\frac{1}{8}$; $25\frac{1}{8}$ *Ans.*	3. $6 + \frac{5}{8}$; $6\frac{5}{8}$ *Ans.*

Here are the complete solutions of (*a*) and (*b*) above!	*a*)	*b*)
	$15\frac{7}{8} \rightarrow \frac{7}{8}$	$15\frac{7}{8} \rightarrow \frac{7}{8}$
	$+9\frac{1}{4} \rightarrow \frac{2}{8}$	$-9\frac{1}{4} \rightarrow \frac{2}{8}$
	$24 + 1\frac{1}{8}$	$6 + \frac{5}{8}$
	$25\frac{1}{8}$ *Ans.*	$6\frac{5}{8}$ *Ans.*

Note: When the fraction of the minuend is smaller than the fraction of the fraction of the subtrahend, borrow one unit from the minuend to increase the smaller fraction.

Thus, $5\frac{2}{7} - 1\frac{5}{7} = 4\frac{9}{7} - 1\frac{5}{7} = 3\frac{4}{7}$. →

$$\begin{array}{rcr} 5\frac{2}{7} & = & 4\frac{9}{7} \\ -1\frac{5}{7} & = & -1\frac{5}{7} \\ \hline & & 3\frac{4}{7} \end{array}$$

3. REVIEWING DECIMALS

A. Understanding Decimals

(1) **A decimal or decimal fraction** is a fraction whose denominator is 10, 100, 1000 or some other power of 10.

Thus, $\frac{3}{100}$ may be written in decimal form as .03 .

(2) **Rounding off a decimal: approximate value of a decimal** .

In rounding off a decimal to the nearest tenth, replace the decimal by the tenth to which it is closest. Thus, .37 to the nearest tenth becomes .4. Here, .37 is between .30 and .40. However, .37 is closer to .40 or .4.

Similarly, round off a decimal to the nearest hundredth, nearest thousandth, and so on.

Note. When a decimal is exactly halfway between two possible answers, round it off to the larger. Thus, round off .25 to the nearest tenth as .3, not .2.

(3) **An approximate value of a decimal** is a value obtained by rounding off the decimal.

Thus, .4 is an approximate value of .37.

(4) **To estimate answers** of exercises or problems involving decimals, use rounded off values of these decimals.

Thus, to estimate .38×.72, round off to .4×.7 and use .28 as the estimated answer.

B. Changing Forms of Fractions and Decimals

(1) **To change a common fraction to a decimal fraction**, divide the numerator by the denominator. Carry the answer to the desired number of places if a remainder exists.

Thus, $\frac{3}{8} = \frac{3.000}{8} = .375$, $\frac{5}{16} = \frac{5.0000}{16} = .3125$.

Equivalent Fractions and Decimals

$\frac{1}{2} = .5$	$\frac{1}{3} = .33\frac{1}{3}$	$\frac{1}{8} = .125$	$\frac{1}{5} = .2$	$\frac{1}{7} = .14\frac{2}{7}$
$\frac{1}{4} = .25$	$\frac{2}{3} = .66\frac{2}{3}$	$\frac{3}{8} = .375$	$\frac{2}{5} = .4$	$\frac{1}{9} = .11\frac{1}{9}$
$\frac{3}{4} = .75$	$\frac{1}{6} = .16\frac{2}{3}$	$\frac{5}{8} = .625$	$\frac{3}{5} = .6$	$\frac{1}{12} = .08\frac{1}{3}$
$\frac{1}{10} = .1$	$\frac{5}{6} = .83\frac{1}{3}$	$\frac{7}{8} = .875$	$\frac{4}{5} = .8$	$\frac{1}{16} = .06\frac{1}{4}$

(2) **To change a decimal fraction to a common fraction**, change the decimal to a common fraction and reduce to lowest terms.

Thus, $.65 = \frac{65}{100} = \frac{13}{20}$.

C. Adding and Subtracting Decimals

Procedure:

1. Arrange the decimals vertically with decimal points directly under each other:

2. **Add or subtract** as in whole numbers, placing the decimal point in the result directly under the other decimal points:

a) Add 1.35 and .952

Solutions:

1. $\begin{array}{r} 1.35 \\ + .952 \\ \hline \end{array}$

2. 2.302

Ans. 2.302

b) Subtract .952 from 1.35

1. $\begin{array}{r} 1.350 \\ - .952 \\ \hline \end{array}$

2. .398

Ans. .398

D. Multiplying and Dividing Decimals

(1) Multiplying and dividing numbers by 10, 100, 1000 or some power of 10.

To multiply a decimal by a power of 10, move the decimal point as many places to the right as there are zeros in the power.

Thus, to multiply 5.75 by **1000**, move decimal point **three** places to the right. *Ans.* 5750.

To divide a decimal by a power of 10, move the decimal point as many places to the left as there are zeros in the power.

Thus, to divide 5.75 by **1000**, move decimal point **three** places to the left. *Ans.* .00575

(2) Multiplying Decimals

Multiply: *a*) 1.1 by .05

Procedure:

1. **Multiply** as in whole numbers:

2. **Mark off** decimal places in the product equal to the sum of the decimal places in the numbers multiplied:

Solutions:

1. $\begin{array}{r} 1.1 \ (\textit{one place}) \\ \times .05 \ (\textit{two places}) \\ \hline \end{array}$

2. .055 (*three places*)

Ans. .055

b) 3.71 × .014

1. $\begin{array}{r} 3.71 \ (\textit{2 places}) \\ \times .014 \ (\textit{3 places}) \\ \hline 1484 \\ 371 \\ \hline \end{array}$

2. .05194 (*5 places*)

Ans. .05194

(3) Dividing Decimals

Divide: *a*) .824 by .04   *b*) 5.194 by 1.4

Procedure:

1. **Move** the decimal point of the divisor to the right to make the divisor a whole number:

2. **Move** the decimal point of the dividend the same number of places to the right:

3. **Divide** as in whole numbers and mark off decimal places in the quotient equal to new number of places in the dividend:

Solutions:

1. $\dfrac{.824}{.04}$

2. $\dfrac{.824}{.04}$

3. $\dfrac{82.4}{4}$ = 20.6

b) **Vertically Arranged**

$$\begin{array}{r} 3.7\,1 \\ 1.4\,)\overline{5.1\,9\,4} \\ 4\,2 \\ \hline 9\,9 \\ 9\,8 \\ \hline 1\,4 \\ 1\,4 \\ \hline \end{array}$$

Note. If a remainder exists, the quotient may be carried to additional decimal places by adding zeros to the dividend.

Thus, $\frac{1}{7} = \dfrac{1.0000000000000}{7} = .142857142857\frac{1}{7}$

4. REVIEWING PER CENTS AND PERCENTAGE

A. Understanding Per Cents and Percentage

Per cent means hundredths. The per cent symbol, %, is a combination of a 1 and two zeros
Thus, 7% of a number means .07 or $\frac{7}{100}$ of the number.

Rule. Percentage = Rate × Base

The **percentage** is the answer obtained when a per cent is taken of a number.
The **rate** is the per cent taken of a number.
The **base** is the number of which a per cent is being taken.
Thus, since 2% of 400 = 8, 8 is the percentage, 2% is the rate and 400 is the base

B. Interchanging Forms of Per Cents, Decimals and Fractions

(1) Interchanging forms of per cents and fractions.

To Change a Per Cent to a Fraction

1. **Omit** the % sign.
2. **Divide** the number by 100.
3. **Reduce** to lowest terms.

Thus, $150\% = \frac{150}{100} = \frac{3}{2}$.

Also, $2\frac{1}{2}\% = \frac{5}{2} \times \frac{1}{100} = \frac{1}{40}$.

To Change a Fraction to Per Cent

1. **Add** the % sign.
2. **Multiply** the number by 100.

Thus, $\frac{3}{2} = (\frac{3}{2} \times 100)\% = 150\%$.

Also, $\frac{1}{40} = (\frac{1}{40} \times 100)\% = 2\frac{1}{2}\%$.

Equivalent Per Cents and Fractions

$\frac{1}{2} = 50\%$	$\frac{1}{3} = 33\frac{1}{3}\%$	$\frac{1}{8} = 12\frac{1}{2}\%$	$\frac{1}{5} = 20\%$	$\frac{1}{7} = 14\frac{2}{7}\%$
$\frac{1}{4} = 25\%$	$\frac{2}{3} = 66\frac{2}{3}\%$	$\frac{3}{8} = 37\frac{1}{2}\%$	$\frac{2}{5} = 40\%$	$\frac{1}{9} = 11\frac{1}{9}\%$
$\frac{3}{4} = 75\%$	$\frac{1}{6} = 16\frac{2}{3}\%$	$\frac{5}{8} = 62\frac{1}{2}\%$	$\frac{3}{5} = 60\%$	$\frac{1}{12} = 8\frac{1}{3}\%$
$\frac{1}{10} = 10\%$	$\frac{5}{6} = 83\frac{1}{3}\%$	$\frac{7}{8} = 87\frac{1}{2}\%$	$\frac{4}{5} = 80\%$	$\frac{1}{16} = 6\frac{1}{4}\%$

(2) Interchanging forms of per cents and decimals.

To Change a Per Cent to a Decimal

1. **Omit** the % sign.
2. **Move** the decimal point two places to the left.

Thus, 175% = 1.75.
Also, 12.5% = .125.

To Change a Decimal to a Per Cent

1. **Add** the % sign.
2. **Move** the decimal point two places to the right.

Thus, 1.75 = 175%.
Also, .125 = 12.5%.

C. Percentage Problems *(See algebraic solution of these types on pages 22 to 25)*

Three Types of Percentage Problems

Types and Their Rules	Problems and Their Solutions	Problems and Their Solutions
(1) Finding Percentage **Rule. Rate × Base = Percentage**	*a*) Find 5% of 400. $.05 \times 400 = 20$	*d*) Find 30% of 80. $.30 \times 80 = 24$
(2) Finding Base **Rule. Base = $\frac{\text{Percentage}}{\text{Rate}}$**	*b*) 5% of what no. is 20? $\frac{20}{.05} = \frac{2000}{5} = 400$	*e*) 30% of what no. is 24? $\frac{24}{.30} = \frac{240}{3} = 80$
(3) Finding Rate **Rule. Rate = $\frac{\text{Percentage}}{\text{Base}}$**	*c*) 20 is what % of 400? $\frac{20}{400} = \frac{1}{20} = .05 = 5\%$	*f*) 24 is what % of 80? $\frac{24}{80} = \frac{3}{10} = .3 = 30\%$

5. *REVIEWING STATISTICAL GRAPHS*

Statistical graphs, by picturing data, enable us to **compare the quantities involved**.

(1) In the **bar graph**, the heights of the bars are used to compare quantities. The bars used have the same width.

Thus, in the **bar graph** shown, quantity 1 is half of 2 and twice 3.

Bar Graph

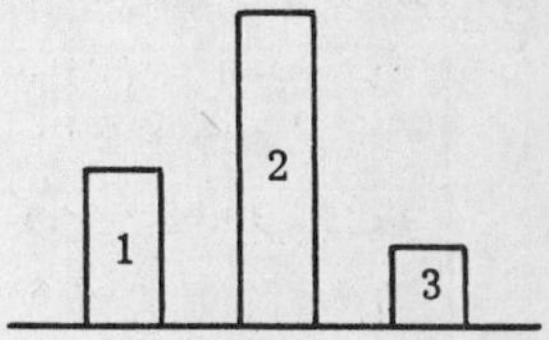

(2) In the **circle graph**, the central angles of sectors of the same circle are used to compare quantities. Since the sectors are parts of the same circle, each quantity may be compared with the whole.

Thus, in the **circle graph** shown, quantity 1 is half of 2, a quarter of 3, equal to 4 and one-eighth of the whole.

Circle Graph

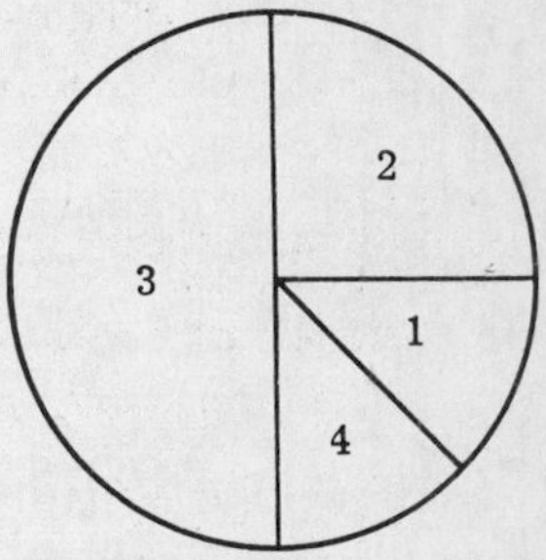

(3) In the **rectangle graph**, the bases of rectangles are used to compare quantities. Since the rectangles are parts of the entire rectangle, each quantity may also be compared with the whole.

Rectangle Graph

Thus, in the rectangle graph shown, quantity 1 is a third of 2, a third of 3, equal to 4 and one-eighth of the whole.

In each of the following graphs, comparisons should be made between the respective quantities, using ratios and differences.

The following bar graph pictures the average weights of American boys at ages indicated in the table:

Age (yr)	10	11	12	13	14	15
Weight (lb)	70	75	81	89	100	112

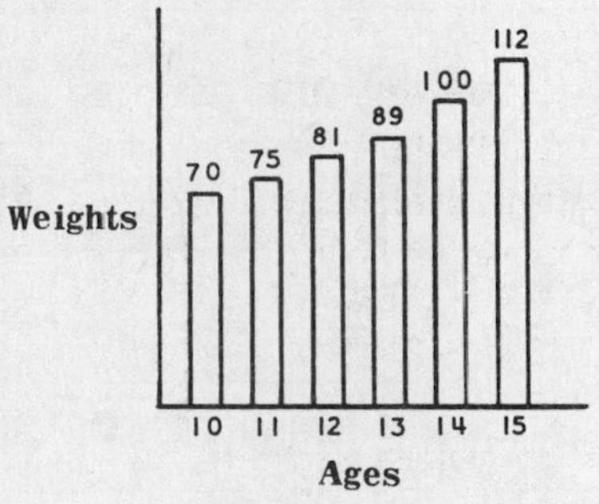

Weight Bar Graph

The following circle graph pictures the food dollar spent by an average American family in a certain month:

	Meat Eggs	Fruits Vegetables	Butter Fats	Cereals	Miscel-laneous	Whole Dollar
Costs	35¢	25¢	15¢	15¢	10¢	100¢
Angles	126°	90°	54°	54°	36°	360°

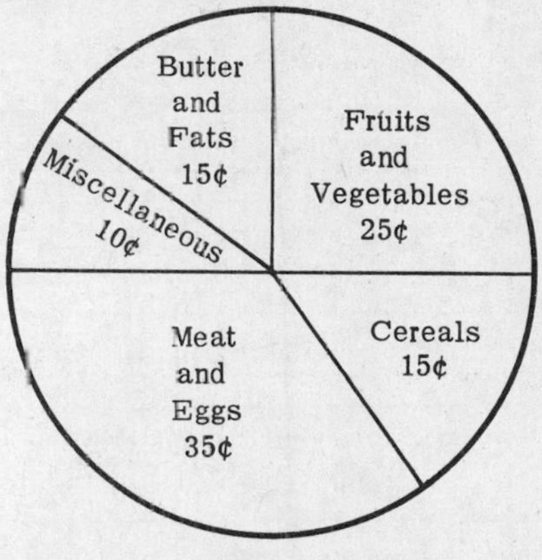

Food Dollar Circle Graph

The following rectangle graph pictures the rate at which various kinds of automobile accidents occurred in a certain month.

Kind of Accident	Rate
Collision with auto	40%
Collision with pedestrian	24%
Collision with train	4%
Collision with fixed object	10%
Miscellaneous	22%
All accidents	**100%**

Automobile Accidents Rectangle Graph

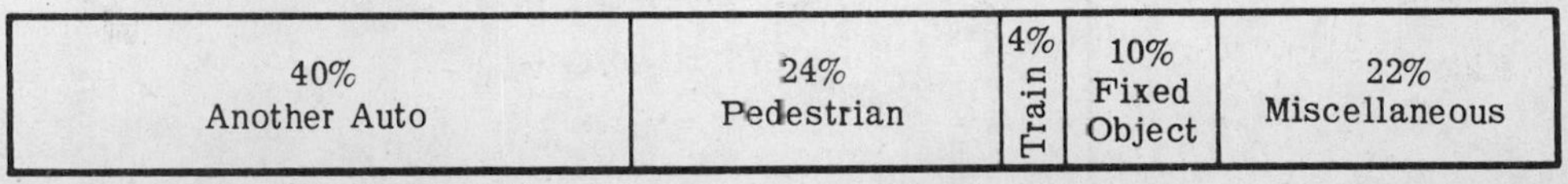

Table of Approximate Square Roots

N	√N	N	√N	N	√N	N	√N	N	√N
1	1.000	51	7.141	101	10.05	151	12.29	201	14.18
2	1.414	52	7.211	102	10.10	152	12.33	202	14.21
3	1.732	53	7.280	103	10.15	153	12.37	203	14.25
4	2.000	54	7.348	104	10.20	154	12.41	204	14.28
5	2.236	55	7.416	105	10.25	155	12.45	205	14.32
6	2.449	56	7.483	106	10.30	156	12.49	206	14.35
7	2.646	57	7.550	107	10.34	157	12.53	207	14.39
8	2.828	58	7.616	108	10.39	158	12.57	208	14.42
9	3.000	59	7.681	109	10.44	159	12.61	209	14.46
10	3.162	60	7.746	110	10.49	160	12.65	210	14.49
11	3.317	61	7.810	111	10.54	161	12.69	211	14.53
12	3.464	62	7.874	112	10.58	162	12.73	212	14.56
13	3.606	63	7.937	113	10.63	163	12.77	213	14.59
14	3.742	64	8.000	114	10.68	164	12.81	214	14.63
15	3.873	65	8.062	115	10.72	165	12.85	215	14.66
16	4.000	66	8.124	116	10.77	166	12.88	216	14.70
17	4.123	67	8.185	117	10.82	167	12.92	217	14.73
18	4.243	68	8.246	118	10.86	168	12.96	218	14.76
19	4.359	69	8.307	119	10.91	169	13.00	219	14.80
20	4.472	70	8.367	120	10.95	170	13.04	220	14.83
21	4.583	71	8.426	121	11.00	171	13.08	221	14.87
22	4.690	72	8.485	122	11.05	172	13.11	222	14.90
23	4.796	73	8.544	123	11.09	173	13.15	223	14.93
24	4.899	74	8.602	124	11.14	174	13.19	224	14.97
25	5.000	75	8.660	125	11.18	175	13.23	225	15.00
26	5.099	76	8.718	126	11.22	176	13.27	226	15.03
27	5.196	77	8.775	127	11.27	177	13.30	227	15.07
28	5.292	78	8.832	128	11.31	178	13.34	228	15.10
29	5.385	79	8.888	129	11.36	179	13.38	229	15.13
30	5.477	80	8.944	130	11.40	180	13.42	230	15.17
31	5.568	81	9.000	131	11.45	181	13.45	231	15.20
32	5.657	82	9.055	132	11.49	182	13.49	232	15.23
33	5.745	83	9.110	133	11.53	183	13.53	233	15.26
34	5.831	84	9.165	134	11.58	184	13.56	234	15.30
35	5.916	85	9.220	135	11.62	185	13.60	235	15.33
36	6.000	86	9.274	136	11.66	186	13.64	236	15.36
37	6.083	87	9.327	137	11.70	187	13.67	237	15.39
38	6.164	88	9.381	138	11.75	188	13.71	238	15.43
39	6.245	89	9.434	139	11.79	189	13.75	239	15.46
40	6.325	90	9.487	140	11.83	190	13.78	240	15.49
41	6.403	91	9.539	141	11.87	191	13.82	241	15.52
42	6.481	92	9.592	142	11.92	192	13.86	242	15.56
43	6.557	93	9.644	143	11.96	193	13.89	243	15.59
44	6.633	94	9.695	144	12.00	194	13.93	244	15.62
45	6.708	95	9.747	145	12.04	195	13.96	245	15.65
46	6.782	96	9.798	146	12.08	196	14.00	246	15.68
47	6.856	97	9.849	147	12.12	197	14.04	247	15.72
48	6.928	98	9.899	148	12.17	198	14.07	248	15.75
49	7.000	99	9.950	149	12.21	199	14.11	249	15.78
50	7.071	100	10.00	150	12.25	200	14.14	250	15.81

N	√N	N	√N	N	√N	N	√N	N	√N
251	15.84	301	17.35	351	18.73	401	20.02	451	21.24
252	15.87	302	17.38	352	18.76	402	20.05	452	21.26
253	15 91	303	17.41	353	18.79	403	20.07	453	21.28
254	15.94	304	17.44	354	18.81	404	20.10	454	21.31
255	15.97	305	17.46	355	18.84	405	20.12	455	21.33
256	16.00	306	17.49	356	18.87	406	20.15	456	21.35
257	16.03	307	17.52	357	18.89	407	20.17	457	21.38
258	16.06	308	17.55	358	18.92	408	20.20	458	21.40
259	16.09	309	17.58	359	18.95	409	20.22	459	21.42
260	16.12	310	17.61	360	18.97	410	20.25	460	21.45
261	16.16	311	17.64	361	19.00	411	20.27	461	21.47
262	16.19	312	17.66	362	19.03	412	20.30	462	21.49
263	16.22	313	17.69	363	19.05	413	20.32	463	21.52
264	16.25	314	17.72	364	19.08	414	20.35	464	21.54
265	16.28	315	17.75	365	19.10	415	20.37	465	21.56
266	16.31	316	17.78	366	19.13	416	20.40	466	21.59
267	16.34	317	17.80	367	19.16	417	20.42	467	21.61
268	16.37	318	17.83	368	19.18	418	20.45	468	21.63
269	16.40	319	17.86	369	19.21	419	20.47	469	21.66
270	16.43	320	17.89	370	19.24	420	20.49	470	21.68
271	16.46	321	17.92	371	19.26	421	20.52	471	21.70
272	16.49	322	17.94	372	19.29	422	20.54	472	21.73
273	16.52	323	17.97	373	19.31	423	20.57	473	21.75
274	16.55	324	18.00	374	19.34	424	20.59	474	21.77
275	16.58	325	18.03	375	19.36	425	20.62	475	21.79
276	16.61	326	18.06	376	19.39	426	20.64	476	21.82
277	16.64	327	18.08	377	19.42	427	20.66	477	21.84
278	16.67	328	18.11	378	19.44	428	20.69	478	21.86
279	16.70	329	18.14	379	19.47	429	20.71	479	21.89
280	16.73	330	18.17	380	19.49	430	20.74	480	21.91
281	16.76	331	18.19	381	19.52	431	20.76	481	21.93
282	16.79	332	18.22	382	19.54	432	20.78	482	21.95
283	16.82	333	18.25	383	19.57	433	20.81	483	21.98
284	16.85	334	18.28	384	19.60	434	20.83	484	22.00
285	16.88	335	18.30	385	19.62	435	20.86	485	22.02
286	16.91	336	18.33	386	19.65	436	20.88	486	22.05
287	16.94	337	18.36	387	19.67	437	20.90	487	22.07
288	16.97	338	18.38	388	19.70	438	20.93	488	22.09
289	17.00	339	18.41	389	19.72	439	20.95	489	22.11
290	17.03	340	18.44	390	19.75	440	20.98	490	22.14
291	17.06	341	18.47	391	19.77	441	21.00	491	22.16
292	17.09	342	18.49	392	19.80	442	21.02	492	22.18
293	17.12	343	18.52	393	19.82	443	21.05	493	22.20
294	17.15	344	18.55	394	19.85	444	21.07	494	22.23
295	17.18	345	18.57	395	19.87	445	21.10	495	22.25
296	17.20	346	18.60	396	19.90	446	21.12	496	22.27
297	17.23	347	18.63	397	19.92	447	21.14	497	22.29
298	17.26	348	18.65	398	19.95	448	21.17	498	22.32
299	17.29	349	18.68	399	19.97	449	21.19	499	22.34
300	17.32	350	18.71	400	20.00	450	21.21	500	22.36

Table of Approximate Square Roots

N	$\sqrt{N}$	N	$\sqrt{N}$	N	$\sqrt{N}$	N	$\sqrt{N}$	N	$\sqrt{N}$
501	22.38	551	23.47	601	24.52	651	25.51	701	26.48
502	22.41	552	23.49	602	24.54	652	25.53	702	26.50
503	22.43	553	23.52	603	24.56	653	25.55	703	26.51
504	22.45	554	23.54	604	24.58	654	25.57	704	26.53
505	22.47	555	23.56	605	24.60	655	25.59	705	26.55
506	22.49	556	23.58	606	24.62	656	25.61	706	26.57
507	22.52	557	23.60	607	24.64	657	25.63	707	26.59
508	22.54	558	23.62	608	24.66	658	25.65	708	26.61
509	22.56	559	23.64	609	24.68	659	25.67	709	26.63
510	22.58	560	23.66	610	24.70	660	25.69	710	26.65
511	22.61	561	23.69	611	24.72	661	25.71	711	26.66
512	22.63	562	23.71	612	24.74	662	25.73	712	26.68
513	22.65	563	23.73	613	24.76	663	25.75	713	26.70
514	22.67	564	23.75	614	24.78	664	25.77	714	26.72
515	22.69	565	23.77	615	24.80	665	25.79	715	26.74
516	22.72	566	23.79	616	24.82	666	25.81	716	26.76
517	22.74	567	23.81	617	24.84	667	25.83	717	26.78
518	22.76	568	23.83	618	24.86	668	25.85	718	26.80
519	22.78	569	23.85	619	24.88	669	25.87	719	26.81
520	22.80	570	23.87	620	24.90	670	25.88	720	26.83
521	22.83	571	23.90	621	24.92	671	25.90	721	26.85
522	22.85	572	23.92	622	24.94	672	25.92	722	26.87
523	22.87	573	23.94	623	24.96	673	25.94	723	26.89
524	22.89	574	23.96	624	24.98	674	25.96	724	26.91
525	22.91	575	23.98	625	25.00	675	25.98	725	26.93
526	22.93	576	24.00	626	25.02	676	26.00	726	26.94
527	22.96	577	24.02	627	25.04	677	26.02	727	26.96
528	22.98	578	24.04	628	25.06	678	26.04	728	26.98
529	23.00	579	24.06	629	25.08	679	26.06	729	27.00
530	23.02	580	24.08	630	25.10	680	26.08	730	27.02
531	23.04	581	24.10	631	25.12	681	26.10	731	27.04
532	23.07	582	24.12	632	25.14	682	26.12	732	27.06
533	23.09	583	24.15	633	25.16	683	26.13	733	27.07
534	23.11	584	24.17	634	25.18	684	26.15	734	27.09
535	23.13	585	24.19	635	25.20	685	26.17	735	27.11
536	23.15	586	24.21	636	25.22	686	26.19	736	27.13
537	23.17	587	24.23	637	25.24	687	26.21	737	27.15
538	23.19	588	24.25	638	25.26	688	26.23	738	27.17
539	23.22	589	24.27	639	25.28	689	26.25	739	27.18
540	23.24	590	24.29	640	25.30	690	26.27	740	27.20
541	23.26	591	24.31	641	25.32	691	26.29	741	27.22
542	23.28	592	24.33	642	25.34	692	26.31	742	27.24
543	23.30	593	24.35	643	25.36	693	26.32	743	27.26
544	23.32	594	24.37	644	25.38	694	26.34	744	27.28
545	23.35	595	24.39	645	25.40	695	26.36	745	27.29
546	23.37	596	24.41	646	25.42	696	26.38	746	27.31
547	23.39	597	24.43	647	25.44	697	26.40	747	27.33
548	23.41	598	24.45	648	25.46	698	26.42	748	27.35
549	23.43	599	24.47	649	25.48	699	26.44	749	27.37
550	23.45	600	24.49	650	25.50	700	26.46	750	27.39

N	$\sqrt{N}$	N	$\sqrt{N}$	N	$\sqrt{N}$	N	$\sqrt{N}$	N	$\sqrt{N}$
751	27.40	801	28.30	851	29.17	901	30.02	951	30.84
752	27.42	802	28.32	852	29.19	902	30.03	952	30.85
753	27.44	803	28.34	853	29.21	903	30.05	953	30.87
754	27.46	804	28.35	854	29.22	904	30.07	954	30.89
755	27.48	805	28.37	855	29.24	905	30.08	955	30.90
756	27.50	806	28.39	856	29.26	906	30.10	956	30.92
757	27.51	807	28.41	857	29.27	907	30.12	957	30.94
758	27.53	808	28.43	858	29.29	908	30.13	958	30.95
759	27.55	809	28.44	859	29.31	909	30.15	959	30.97
760	27.57	810	28.46	860	29.33	910	30.17	960	30.98
761	27.59	811	28.48	861	29.34	911	30.18	961	31.00
762	27.60	812	28.50	862	29.36	912	30.20	962	31.02
763	27.62	813	28.51	863	29.38	913	30.22	963	31.03
764	27.64	814	28.53	864	29.39	914	30.23	964	31.05
765	27.66	815	28.55	865	29.41	915	30.25	965	31.06
766	27.68	816	28.57	866	29.43	916	30.27	966	31.08
767	27.69	817	28.58	867	29.44	917	30.28	967	31.10
768	27.71	818	28.60	868	29.46	918	30.30	968	31.11
769	27.73	819	28.62	869	29.48	919	30.32	969	31.13
770	27.75	820	28.64	870	29.50	920	30.33	970	31.14
771	27.77	821	28.65	871	29.51	921	30.35	971	31.16
772	27.78	822	28.67	872	29.53	922	30.36	972	31.18
773	27.80	823	28.69	873	29.55	923	30.38	973	31.19
774	27.82	824	28.71	874	29.56	924	30.40	974	31.21
775	27.84	825	28.72	875	29.58	925	30.41	975	31.22
776	27.86	826	28.74	876	29.60	926	30.43	976	31.24
777	27.87	827	28.76	877	29.61	927	30.45	977	31.26
778	27.89	828	28.77	878	29.63	928	30.46	978	31.27
779	27.91	829	28.79	879	29.65	929	30.48	979	31.29
780	27.93	830	28.81	880	29.66	930	30.50	980	31.30
781	27.95	831	28.83	881	29.68	931	30.51	981	31.32
782	27.96	832	28.84	882	29.70	932	30.53	982	31.34
783	27.98	833	28.86	883	29.72	933	30.55	983	31.35
784	28.00	834	28.88	884	29.73	934	30.56	984	31.37
785	28.02	835	28.90	885	29.75	935	30.58	985	31.38
786	28.04	836	28.91	886	29.77	936	30.59	986	31.40
787	28.05	837	28.93	887	29.78	937	30.61	987	31.42
788	28.07	838	28.95	888	29.80	938	30.63	988	31.43
789	28.09	839	28.97	889	29.82	939	30.64	989	31.45
790	28.11	840	28.98	890	29.83	940	30.66	990	31.46
791	28.12	841	29.00	891	29.85	941	30.68	991	31.48
792	28.14	842	29.02	892	29.87	942	30.69	992	31.50
793	28.16	843	29.03	893	29.88	943	30.71	993	31.51
794	28.18	844	29.05	894	29.90	944	30.72	994	31.53
795	28.20	845	29.07	895	29.92	945	30.74	995	31.54
796	28.21	846	29.09	896	29.93	946	30.76	996	31.56
797	28.23	847	29.10	897	29.95	947	30.77	997	31.58
798	28.25	848	29.12	898	29.97	948	30.79	998	31.59
799	28.27	849	29.14	899	29.98	949	30.81	999	31.61
800	28.28	850	29.15	900	30.00	950	30.82	1000	31.62

Table of Natural Trigonometric Functions

Angle	Sine	Cosine	Tangent	Angle	Sine	Cosine	Tangent
1°	.0175	.9998	.0175	46°	.7193	.6947	1.0355
2°	.0349	.9994	.0349	47°	.7314	.6820	1.0724
3°	.0523	.9986	.0524	48°	.7431	.6691	1.1106
4°	.0698	.9976	.0699	49°	.7547	.6561	1.1504
5°	.0872	.9962	.0875	50°	.7660	.6428	1.1918
6°	.1045	.9945	.1051	51°	.7771	.6293	1.2349
7°	.1219	.9925	.1228	52°	.7880	.6157	1.2799
8°	.1392	.9903	.1405	53°	.7986	.6018	1.3270
9°	.1564	.9877	.1584	54°	.8090	.5878	1.3764
10°	.1736	.9848	.1763	55°	.8192	.5736	1.4281
11°	.1908	.9816	.1944	56°	.8290	.5592	1.4826
12°	.2097	.9781	.2126	57°	.8387	.5446	1.5399
13°	.2250	.9744	.2309	58°	.8480	.5299	1.6003
14°	.2419	.9703	.2493	59°	.8572	.5150	1.6643
15°	.2588	.9659	.2679	60°	.8660	.5000	1.7321
16°	.2756	.9613	.2867	61°	.8746	.4848	1.8040
17°	.2924	.9563	.3057	62°	.8829	.4695	1.8807
18°	.3090	.9511	.3249	63°	.8910	.4540	1.9626
19°	.3256	.9455	.3443	64°	.8988	.4384	2.0503
20°	.3420	.9397	.3640	65°	.9063	.4226	2.1445
21°	.3584	.9336	.3839	66°	.9135	.4067	2.2460
22°	.3746	.9272	.4040	67°	.9205	.3907	2.3559
23°	.3907	.9205	.4245	68°	.9272	.3746	2.4751
24°	.4067	.9135	.4452	69°	.9336	.3584	2.6051
25°	.4226	.9063	.4663	70°	.9397	.3420	2.7475
26°	.4384	.8988	.4877	71°	.9455	.3256	2.9042
27°	.4540	.8910	.5095	72°	.9511	.3090	3.0777
28°	.4695	.8829	.5317	73°	.9563	.2924	3.2709
29°	.4848	.8746	.5543	74°	.9613	.2756	3.4874
30°	.5000	.8660	.5774	75°	.9659	.2588	3.7321
31°	.5150	.8572	.6009	76°	.9703	.2419	4.0108
32°	.5299	.8480	.6249	77°	.9744	.2250	4.3315
33°	.5446	.8387	.6494	78°	.9781	.2079	4.7046
34°	.5592	.8290	.6745	79°	.9816	.1908	5.1446
35°	.5736	.8192	.7002	80°	.9848	.1736	5.6713
36°	.5878	.8090	.7265	81°	.9877	.1564	6.3138
37°	.6018	.7986	.7536	82°	.9903	.1392	7.1154
38°	.6157	.7880	.7813	83°	.9925	.1219	8.1443
39°	.6293	.7771	.8098	84°	.9945	.1045	9.5144
40°	.6428	.7660	.8391	85°	.9962	.0872	11.4301
41°	.6561	.7547	.8693	86°	.9976	.0698	14.3007
42°	.6691	.7431	.9004	87°	.9986	.0523	19.0811
43°	.6820	.7314	.9325	88°	.9994	.0349	28.6363
44°	.6947	.7193	.9657	89°	.9998	.0175	57.2900
45°	.7071	.7071	1.0000	90°	1.0000	.0000	

INDEX

Abscissa, 109
Absolute value, 36-8, 40-1
Addends, 2, 283
Addition, checking, 2, 54-5, 283; equality rule, 20, 26; of decimals, 287; of fractions, 285-6; of monomials, 53-4; of polynomials, 54-5; of radicals, 210-11; of signed numbers, 40-2; words denoting, 4
Adjacent angles, 139-40
Age problems, 135-6
Algebraic expressions, 1, 4-9, 11-12
Angles, acute, 87; adjacent, 139-40; central, in circle graphs, 289; complementary, 139-40; equal, 86; finding an, in trigonometry, 260-2, 264-5; of climb, 269; of depression, of elevation, 265-6; obtuse, 87; problems about, 139-41; right, 87, 139; straight, 139; sum of, 139-41; supplementary, 139-40
Approximate numbers, 203, 207-8, 282, 286
Arc, 86, 91
Areas, 92-5
Arithmetic, 282-89; checking operations, 283-4; decimals, 286-7; fractions, 184, 185, 195, 284-6; mean, 273-4; per cent, percentage, 288; whole numbers, 282-4
Average, 273; speed, 148
Axes of a graph, 109
Axis of symmetry, 277, 281; parabola, 229

Bar, 57, 59
Bar graph, 273-4, 281, 289
Base, 87-9, 91-7, 103-6; in percentage, 288; having exponents, 9, 46; literal, 10
Binomials, 53; difference of two squares, 171-2; factors of trinomial, 174-5; product of, 171-3; square of, 175-6
Brace, 57-9
Bracket, 57-9
Broken-line graph, 273-4

Center of symmetry, 277, 281
Changing forms, of fractions and decimals, 286-7; of mixed numbers and improper fractions, 284; of per cents, decimals and fractions, 288
Checking, addition, 2, 54-5, 283; division, 62-4, 284; equations, 17, 21, 23, 25-8; identities, 18; multiplication, 3, 61, 283; roots of an equation, 17, 21-8, 222, 224; subtraction, 56-7, 283; verbal problems, 23, 25-7, 30, 130-2
Circle, 86, 89, 91-2, 94
Circle graph, 289

Clearing equations, of decimals, 78-9; of fractions, 72-3, 76-8; of parentheses, 75-6
Coefficients, 8-9, 11-12; in quadratic equations, 226; in trinomial, 174
Coin problems, 143-4
Combining, areas, 94; like terms, 11-2, 54-6; monomials, 11-2, 54-6; polynomials, 54-5; signed numbers, 40-3
Common denominator, 191-5, 285; factor, 188-9, 192-3; solution, 115-7, 124-8
Commutative law, 59
Complementary angles, 139-40
Complete factoring, 172, 177-8
Completing the square, 225-7
Complex fractions, 194-5
Cone, 88-9, 96
Congruent triangles, 275-7, 281
Consecutive integers, 133-4, 223
Consistent equations, 116-7, 121-2
Constant, 234; in power variation, 242-4; product in inverse variation, 238-40; ratio in direct and joint variation, 235-7, 241-2
Constructions of lines, angles and triangles, 276-7
Cosine, 260
Cross-products, 254, 258
Cube, 10, 88, 93, 95-6; root, 201
Cubic measure, 95-7
Cylinder, 88, 93, 95-7

Decimals, 286-7; as rational numbers, 203
Degree of an equation, 71, 124
Delta form of slope rule, 271-2
Denominator, 184, 284; lowest common, 77-8, 192-5, 285; rationalizing a, 213-4
Dependent equations, 116-18
Deriving equations from tables, 118-9
Descriptive statistics, 273-4, 280-1
Diagonal, 252-3, 257
Differences, 283; of two squares, 171-2; used to derive equations, 118-9, 123; used to determine distances, 251, 253; used to find slope, 271-2
Digits, 282
Directly proportional quantities, 235
Direct variation, 234-7; square, cube, 242-3
Direction, 149-52, 248-9
Distance, 44; between two points, 251, 253; by trigonometry, 265-6; indirect measurement of, 248-50
Distributive law, 60
Division, by zero, 4, 13; checking, 63, 284; in fraction form, 184; in variation, 236, 238, 241; of fractions, 191, 285; of monomials, 62-4; of polynomials, 63-5, 213; of radicals, 212-4; of radicands, 205-6; of signed numbers, 47-8; rule of equality, 20-3; to reduce a fraction, 188-9, 284; words denoting, 5
Duplicating, angles, lines, triangles, 275-7, 281; in square root computation, 206-8

Eliminating an unknown, 124-7
English into algebra, changing verbal statements into algebraic expressions, 5-6, 13-4; changing verbal statements into equations, 1, 12, 18, 31; expressing addition and subtraction, 4, 13; expressing multiplication and division, 5; from verbal problems to equations, 18, 31; in problem solving, 130-166; representing quantities by signed numbers, 37, 49
Equations, 17; consistent, 116-7, 121-2; containing parentheses, 75-6, 125, 221; dependent, 116, 118; deriving, from tables, 118-9; deriving, from sets of values, 98-9; first degree in one unknown, 71-85; first degree in two unknowns, 124-9; graphs of linear, 112-22; graphs of quadratic, 228-30; inconsistent, 116-7, 122; linear, 112-22; literal, 79-80; members of, 17; pairs of linear, 116-8, 124-9; quadratic, 221-33; radical, 214-5, 220-1, 223, 228; roots of, 17; simple, 17-35; solving pairs by addition and subtraction, 124-6, 128-9; solving by equality rules, 20-30; solving pairs by graphing, 116-8; solving pairs by substitution, 126-7; solving trigonometric, 264-6
Equilateral polygon, 86, 90; triangle, 87
Equivalent fractions, 185-6, 284-5
Estimating answers, 282, 286
Evaluation, 6-8; of an unknown in a formula, 101-2; of powers, 10-1; with signed numbers, 48-9
Exponents, 9-11; in division of powers, 62-3; in multiplication of powers, 59-60; in power of a power, 59; in square roots of numbers, 170; in squaring numbers, 169-70
Expressions, algebraic, 53
Extraneous roots, 215
Extremes of a proportion, 254

Factors, 3, 283; common, 168, 177-8, 188-93, 284; monomial, 167, 177-8; prime, 167

Factoring, 167-83; complete, 172, 177-8; difference of two squares, 171-2; highest common monomial factor in, 168-9, 177-8; perfect trinomial square, 176-7; to solve quadratic equations, 222-3; trinomials, 174-5
Formulas, 79-80, 86-108; area, 92-5; circle, 86, 89, 94; coin, 98-9, 143-4; containing radicals, 215, 220; deriving, 98-9, 253, 258; evaluating an unknown in a, 101-2; for circumference, 89, 91; from rules, 98-9; general, 98-9; in variation, 234-44; length, 99; motion, 99, 148-52; perimeter, 89-92, 141-2; quadratic, 227-8; solving, 79-80, 100-1; subject of, 100; time, 99; transforming, 100-1; value, 143-5; volume, 95-7; $xy=z$ type, 234
Fractions, 184-200, 284-6; adding, 191-4, 285-6; changing to equivalent, 185-6, 284-5; common, 286-7; complex, 194-5, dividing, 191, 285; equivalent, 185-6, 209-10, 284-5; fundamental principle of, 185; improper, 284; meanings of, 184-5; multiplying, 190-1, 285; powers of, 11; proper, 284; rationalizing the denominator of, 213-4; reciprocals, 186-7; reducing, 188-90, 284; simplifying complex, 194-5; square roots of, 209-10; subtracting, 191-4, 285
Frequency distribution, 273-4, 280-1
Fundamental operations, 283-4

Geometry, adjacent angles, 139-41; areas, 92-5; circles, 86, 89, 94; complementary angles, 139-41; congruent triangles, 275-7; coordinate geometry, 251, 271-2; formulas, 87-97; indirect measurement, 248-50; in variation, 236, 238, 240-1, 243; law of Pythagoras, 251-3; perimeters, 89-92; problems, 139-41; similar triangles, 255-7; solids, 88-9, 95-7; sum of angles of a triangle, 139-41; supplementary angles, 139-41; triangles, 87, 255-7, 275-7; triangles drawn to scale, 248-50, 255, 257; volumes, 95-7
Graphs, bar, 273-4, 281, 289; broken-line, 273-4, 281; circle, 289; curved line, 228-30, 233; in problem solving, 270, 280; of consistent equations, 116-7, 121-2; of dependent equations, 116, 118, 123; of first degree equations, 109-23; of inconsistent equations, 116-17, 122; of linear equations, 109-23; of quadratic equations, 228-30, 233; of square roots, 204-5; rectangle, 289; statistical, 273-4, 281, 289; straight line, 112-23; using intercepts, 112, 114-5
Grouped items, 273-4, 280-1
Grouping, symbols of, 57-9

Histogram, 273-4, 281
Hypotenuse, 87, 251-2

Identity, 17
Imaginary numbers, 202
Inclination, 271-2
Incomplete quadratic equations, 223-5
Inconsistent equations, 116-7, 122
Index of a root, 201, 210
Indirect measurement, 248-50
Integers, 133-4, 140, 203
Intercepts, 112, 114-5
Interest problems, 146-8, 237
Inverse operations, 19-20, 27-30, 71, 100
Inverse variation, 234, 238-40; square, 242-4
Inversely proportional quantities, 238-40
Investment problems, 146-8, 237
Irrational numbers, 202-4
Isosceles, triangle, 87, 142; trapezoid, 87, 142

Joint variation, 234, 241-2

Law of Pythagoras, 251-3
Lever problems, 240
Like terms, 11, 53
Line of sight, 265-6
Line symmetry, 277, 281
Linear, equations, 112-22; measure, 89-92
Literal, addends, 2; equations, 79-80
Lowest common denominator (**L.C.D.**), 192-5, 285

Map, 110, 249-50
Means, arithmetic, 273; of a proportion, 254
Measures, cubic, 95-7; linear, 89-92; of central tendency, 273-4; square, 92-5
Median, 273
Members of an equation, 17
Minuend, 283
Mixed numbers, 284-6
Mixture problems, 144-6
Mode, 273
Monomials, 53-70, 167-70; addition of, 53-4; division of, 62-4; multiplication of, 59-60; simplifying addition of, 53-4; square roots of, 170; squaring, 169-70; subtraction of, 55-6
Motion problems, 148-52, 237, 270, 280
Multiplicand, 283
Multiplication, by zero, 44-5, 47-8; checking, 3, 61, 283; in solving equations, 23-5; in variation, 236, 238, 241; of binomials, 171-3; of decimals, 287; of fractions, 285; of monomials, 59-60; of polynomials, 60-1, 212; of powers, 59; of radicals, 211-2; of radicands, 205-6; of signed numbers, 44-5; rule of equality, 20, 23; to obtain equivalent fractions, 185-6, 193-5, 284; words denoting, 5
Multiplier, 283

Negative numbers, see Signed Numbers.
Number scale, 37-9, 42-4, 50, 109
Numbers, absolute value of, 36-8, 40-1; approximate, 203, 207-8, 282, 286; imaginary, 202; irrational, 202-4; mixed, 284-6; order of operations on, 6-8; prime, 167; rational, 202-4; reciprocal, 186-8; Roman, 282; rounding off, 282, 286; signed, 36-52; square roots of, 201-20; squares, 9-10, 169-70; whole, 282-4

Opposites, 36, 42-3, 74
Order of operations, 6-8
Ordinate, 109
Origin, 109

Pairs of equations, graphing, 109-23; solving, 124-9
Parabola, 228-30, 233
Parallelogram, 87-8, 90, 92-3
Parentheses, 1, 4-8, 57-9, 75-6
Per cent, percentage, 288
Perimeters, 89-92
Pi, π, 89, 92
Place-value, 282
Plotting points, 109-12
Plus and minus signs, 201
Polygon, 86-94, 255
Polyhedron, 88
Polynomials, 53-72, 167-8; addition of, 54-5, 57-9; arranging terms of, 54-6, 64-5; division of, 63-5; factoring, 172, 174-8; having a common monomial factor, 168-9; multiplication of, 60-1, 172-3; subtraction of, 56-7; using grouping symbols for, 57-9
Positive numbers, see Signed numbers.
Powers, 9-11; ascending order, 54; descending order, 54-5; dividing, 62; evaluating, 10-11; multiplying, 59; of monomials, 169-70; of a power, 59; of signed numbers, 46
Prime factor, number, 167
Principal, 146-8
Principal square root, 170, 201-2
Prism, 88, 95-7
Problems involving, adjacent angles, 139-40; age, 135-6; area, 92-5, 223; checking, 23, 25-7, 30; coins, 143-4; combination, 153-4; complementary angles, 139-40; consecutive integers, 133-4, 140; cost, 144-6; gears, 240; geometry, 89-97, 139-42; indirect measurement, 248-50; interest, 146-8, 237; investment, 146-8, 237; law of Pythagoras, 252-3; lever, 240; maps, 249-50; mixture, 144-6; motion, 148-52, 237, 270, 280; number, 130-3, 138, 223; perimeter, 141-2; percentage, 22-5, 288; proportion, 237, 239-40, 242-4; pulleys, 240; ratio, 136-8; scale, 248-50, 257; solving graphically, 270, 280; stamps, 143-4; steps of problem-solving, 130; supplementary angles, 139-40; travel, 148-52; trigonometry, 263-6; value, 143-6; variation in, 237, 240, 242-4
Products, 1, 3-4, 283; special, 167-83; square roots of a, 208-9. See Multiplication.
Proportions, 254-7; in similar triangles, 255-7; in variation, 235, 237-44
Protractor, 276
Pyramid, 88, 96
Pythagoras, law of, 251-3

Quadrants, 109-10
Quadratic equations, 221-33; complete, 225-8; incomplete, 223-5; solved by completing the square, 225-7; solved by factoring, 222-5; solved by formula, 227-8; solved graphically, 228-30, 233; standard form of, 221, 227
Quadrilaterals, 86-7, 90, 142
Quotient, 1, 5, 47-8, 283-4

Radicals, 201-20; addition of, 210-1; combining, 210-1; division of, 212-3; equations, 214-5; like and unlike, 210-1; multiplication of, 211-2; principles of, 201-4; rationalizing denominators, 213-4; sign, 201; simplifying, 208-10; squaring, 212; subtraction of, 210-1
Radicand, 201, 203, 205-6
Range, 273-4
Rank order, 273
Rate, in percentage, 288
Ratios, 136-8, 185, 202-4; continued, 136-8; in a right triangle, 252; in direct variation, 235-7; in direct square variation, 243; in inverse variation, 238-40; in inverse square variation, 244; in joint variation, 241-2; in scale triangles, 255-7; in similar triangles, 255-7; trigonometric, 260-2; ways of expressing, 136
Rational numbers, 202-4
Rationalize a denominator, 213-4
Reciprocals, 186-8
Rectangle, 87, 90-4, 289
Rectangular solid, 88, 93, 95-6
Reducing a fraction, 188-90, 284-5
Regular polygon, 86, 90; pyramid, 88
Remainder, 283-4
Representation, 130
Rhombus, 87
Right triangle, 87, 251-3
Roman numbers, 282
Roots, 201-20; cube, 201-2; index of 201; negative, 73-4; of an equation, 17; of a quadratic equation, 222-5, 227-30; principal, 170; square, 170, 201-15
Rounding off numbers, 282-6

Satisfy an equation, 113, 115-6, 118
Scale, 248-50, 257
Scale triangles, 248-50, 255, 257
Sector, 86, 94
Signed numbers, 36-52; addition of, 40-2; combining, 40-3; comparing, 38; division of, 47-8; evaluation with, 48-9; in number scales, 37-9, 42-4, 50; meanings of, 38; multiplication of, 44-5; powers of, 46; simplifying addition of, 41-2; subtraction of, 42-4
Similar polygons, 255; triangles, 255-7
Simple equations, 17-35
Simultaneous equations, see Pairs of equations.
Sine, 260
Slope, 271-2, 280
Solids, 87, 93-4
Sphere, 88, 95-7
Square, 9-10, 87; area of, 92-3, 170; completing the, 225-7; difference of two, 171-2; measure, 92; of a binomial, 175-6; of a monomial, 169-70; of the hypotenuse, 251-3; unit, 87, 93
Square roots, 170, 201-2; adding, 210-1; approximate, 203-8; as rational or irrational numbers, 203-4; computing, 206-8; dividing, 212-3; from graph, 204-5; from table, 205-6; multiplying, 211-2; of a fraction, 209-10; of a monomial, 170; of a number, 201-20; of a perfect square trinomial, 176-7, 225-7; of a power, 208-9; of a product, 208-9; of a product of powers, 208-9; opposite, 202; principal, 170, 201; rationalizing, 213-4; simplifying, 208-11; subtracting, 210-1; table of, 290-1
Standard form, 119, 121-2, 221-2, 227
Statistical graphs, 273-4, 281, 289
Statistics, 273-4, 280-81
Substitution, checking by, 17; method of solving equations, 126-7, 129
Subtraction, 283; by number scale, 43-4; checking, 283; equality rule, 20, 25; of decimals, 287; of fractions, 285; of monomials, 55-6; of polynomials, 56-7; of radicals, 210-1; of signed numbers, 42-4; symbol for, 3, 42; words denoting, 4
Subtrahend, 283
Sum, 1, 283
Supplementary angles, 139-40
Surd, 203
Symbol for variable ratio, 278-9
Symmetry, 277, 281
Systems of equations, see Pairs of equations.

Tables, formulas from, 98-9; in problem-solving, 270, 280; of coordinate values, 112-8; of equivalent fractions and decimals, 287; of equivalent per cents and fractions, 288; of sines, cosines and tangents, 292; of square roots, 290-1; of values, 98, 118-9, 229; statistical, 273-4; trigonometric, 292
Tangent, 260; of inclination, 271-2
Terms, 8-9; combining like, 11-2; constant, 223-4; like and unlike, 11, 53; of a fraction, 184, 284; of a polynomial, 53
Transforming equations, 79-80, 127; formulas, 100-2
Transit, 265, 269
Translation, see English into algebra.
Transposition, 74-5
Trapezoid, 87, 92, 142
Triangles, 86-7, 139; acute, 87; area of, 92-3; congruent, 275-7, 281; equilateral, 87, 90; isosceles, 87, 142, 253; obtuse, 87; perimeter, 141; right, 87, 251-2; scale, 248-50, 255, 257; scalene, 87; similar, 255-7
Trigonometric ratios, 260-2; finding angles by, 261-2; table of, 292
Trigonometry, 260-9; in slope, 271-2
Trinomial, 53; factoring a, 174-7; in form of ax^2+bx+c, 174-5; in form of x^2+bx+c, 174; perfect square, 176-7

Units, cubic, 95; square, 92
Unknowns, 17
Unlike terms, 11, 53-4

Value, absolute, 36-8, 40-1; approximate, 282, 286; of a fraction, 185-6, 209-10; total, 143-5
Variables, 234-47; a symbol for variable-ratio, 278-9; measuring change in a, 234-5; multiplying or dividing, 235-9, 241
Variation, 234-47; direct, 234-7; direct square, 242-3; direct cube, 242-3; inverse, 238-40; inverse square, 242-4; joint, 241-2
Vectors, 37-9, 42-4, 50
Verification, 130
Volume, 95-7

Whole numbers, 282-4

X-axis, 109

Y-axis, 109

Zero, division by, 4, 13, 47, 184-5; multiplication by, 44, 47; when a product equals, 222